Springer Collected Works in Mathematics

For further volumes:
http://www.springer.com/series/11104

At Marriage Ceremony, 1939

In Mathematical Sciences Research Institute, 1985

Shiing-Shen Chern

Selected Papers IV

Reprint of the 1989 Edition

Springer

Shiing-Shen Chern
(1911 Jiaxing, China – 2004 Tianjin, China)
The University of Chicago
Chicago, IL
USA

ISSN 2194-9875
ISBN 978-1-4614-9085-2 (Softcover)
978-0-387-96820-9 (Hardcover)
DOI 10.1007/978-1-4614-9451-5
Springer New York Heidelberg Dordrecht London

Library of Congress Control Number: 2012954381

Mathematics Subject Classification: 01A75 53-02

Printed on acid-free paper

Springer is part of Springer Science+Business Media (www.springer.com)

Contents

Bibliography of the Publications of S. S. Chern vii

[123]* Geometrical Interpretation of the sinh-Gordon Equation 1
[125] Remarks on the Riemannian Metric of a Minimal Submanifold (with Robert Osserman) 9
[126] A Simple Proof of Frobenius Theorem (with Jon G. Wolfson) 37
[127] Foliations on a Surface of Constant Curvature and the Modified Korteweg-de Vries Equations (with Keti Tenenblat) 41
[128] On the Bäcklund Transformations of KdV Equations and Modified KdV Equations (with Chia-kuei Peng) 45
[129] Web Geometry 51
[130] Projective Geometry, Contact Transformations, and *CR*-Structures 59
[131] Minimal Surfaces by Moving Frames (with Jon G. Wolfson) 65
[132] On Surfaces of Constant Mean Curvature in a Three-Dimensional Space of Constant Curvature 91
[133] Deformation of Surfaces Preserving Principal Curvatures 95
[134] On Riemannian Metrics Adapted to Three-Dimensional Contact Manifolds (with R. S. Hamilton) 105
[135] Harmonic Maps of S^2 into a Complex Grassmann Manifold (with Jon Wolfson) 135
[136] Moving Frames 139
[139] Pseudospherical Surfaces and Evolution Equations (with Keti Tenenblat) 151
[140] On a Conformal Invariant of Three-Dimensional Manifolds 181
[142] Harmonic Maps of the Two-Sphere into a Complex Grassmann Manifold II (with Jon G. Wolfson) 189
[143] Tautness and Lie Sphere Geometry (with Thomas E. Cecil) 225
[144] Vector Bundles with a Connection 245
[145] Dupin Submanifolds in Lie Sphere Geometry (with Thomas E. Cecil) 269

Topics in Differential Geometry, Institute for Advanced Study 331
Minimal Submanifolds in a Riemannian Manifold 399
Permissions 463

*Numbers in brackets refer to the Bibliography of the Publications of S.S. Chern (see pages vii–xiv).

Bibliography of the Publications of S.S. Chern

Note: **Boldface** numbers at the end of each entry denote the volume in which the entry appears.

I. *Books and Monographs*

1. *Topics in Differential Geometry* (mimeographed), Institute for Advanced Study, Princeton (1951), 106 pp. **(IV)**
2. *Differentiable Manifolds* (mimeographed), University of Chicago, Chicago (1953), 166 pp.
3. *Complex Manifolds*
   a. University of Chicago, Chicago (1956), 195 pp.
   b. University of Recife, Recife, Brazil (1959), 181 pp.
   c. Russian translation, Moscow (1961), 239 pp.
4. *Studies in Global Geometry and Analysis* (Editor), Mathematical Association of America (1967), 200 pp.
5. *Complex Manifolds without Potential Theory*, van Nostrand (1968), 92 pp. Second edition, revised, Springer-Verlag (1979) 152 pp.
6. *Minimal Submanifolds in a Riemannian Manifold* (mimeographed), University of Kansas, Lawrence (1968), 55 pp. **(IV)**
7. (with Wei-huan Chen) *Differential Geometry Notes*, in Chinese, Beijing University Press (1983), 321 pp.
8. *Studies in Global Differential Geometry* (Editor), Mathematical Association of America (1988), 350 pp.

II. *Papers*

1932

[1] Pairs of plane curves with points in one-to-one correspondence. *Science Reports Nat. Tsing Hua Univ.* **1** (1932) 145–153. **(II)**

1935

*[2] Triads of rectilinear congruences with generators in correspondence. *Tohoku Math. J.* **40** (1935) 179–188.

[3] Associate quadratic complexes of a rectilinear congruence. *Tohoku Math. J.* **40** (1935) 293–316. **(II)**

[4] Abzählungen für Gewebe. *Abh. Math. Sem. Univ. Hamburg* **11** (1935) 163–170. **(I)**

1936

[5] Eine Invariantentheorie der Dreigewebe aus r-dimensionalen Mannigfaltigkeiten im R_{2r}. *Abh. Math. Sem. Univ. Hamburg* **11** (1936) 333–358. **(I)**

* Does not appear in these volumes.

BIBLIOGRAPHY OF THE PUBLICATIONS OF S.S. CHERN

1937

[6] Sur la géométrie d'une équation différentielle du troisième ordre. *C. R. Acad. Sci. Paris* **204** (1937) 1227–1229. **(II)**

[7] Sur la possibilitè de plonger un espace à connexion projective donné dans un espace projectif. *Bull. Sci. Math.* **61** (1937) 234–243. **(I)**

1938

[8] On projective normal coördinates. *Ann. of Math.* **39** (1938) 165–171. **(II)**

[9] On two affine connections. *J. Univ. Yunnan* **1** (1938) 1–18. **(II)**

1939

[10] Sur la géométrie d'un système d'équations différentielles du second ordre. *Bull. Sci. Math* **63** (1939) 206–212. **(II)**

1940

*[11] The geometry of higher path-spaces. *J. Chin. Math. Soc.* **2** (1940) 247–276.

[12] Sur les invariants intégraux en géométrie. *Science Reports Nat. Tsing Hua Univ.* **4** (1940) 85–95. **(II)**

[13] The geometry of the differential equation $y''' = F(x, y', y'')$. *Science Reports Nat. Tsing Hua Univ.* **4** (1940) 97–111. **(I)**

[14] Sur une généralisation d'une formule de Crofton. *C.R. Acad. Sci. Paris* **210** (1940) 757–758. **(II)**

[15] (with C.T. Yen) Sulla formula principale cinematica dello spazio ad n dimensioni. *Boll. Un. Mat. Ital.* **2** (1940) 434–437. **(II)**

*[16] Generalization of a formula of Crofton. *Wuhan Univ. J. Sci.* **7** (1940) 1–16.

1941

[17] Sur les invariants de contact en géométrie projective différentielle. *Acta Pontif. Acad. Sci.* **5** (1941) 123–140. **(II)**

1942

[18] On integral geometry in Klein spaces. *Ann. of Math.* **43** (1942) 178–189. **(I)**

[19] On the invariants of contact of curves in a projective space of N dimensions and their geometrical interpretation. *Acad. Sinica Sci. Record* **1** (1942) 11–15. **(I)**

[20] The geometry of isotropic surfaces. *Ann. of Math.* **43** (1942) 545–559. **(II)**

[21] On a Weyl geometry defined from an $(n-1)$-parameter family of hypersurfaces in a space of n dimensions. *Acad. Sinica Sci. Record* **1** (1942) 7–10. **(I)**

1943

[22] On the Euclidean connections in a Finsler space. *Proc. Nat. Acad. Sci. USA*, **29** (1943) 33–37. **(II)**

[23] A generalization of the projective geometry of linear spaces. *Proc. Nat. Acad. Sci. USA*, **29** (1943) 38–43. **(I)**

1944

[24] Laplace transforms of a class of higher dimensional varieties in a projective space of n dimensions. *Proc. Nat. Acad. Sci. USA*, **30** (1944) 95–97. **(II)**

[25] A simple intrinsic proof of the Gauss–Bonnet formula for closed Riemannian manifolds. *Ann of Math.* **45** (1944) 747–752. **(I)**

[26] Integral formulas for the characteristic classes of sphere bundles. *Proc. Nat. Acad. Sci. USA* **30** (1944) 269–273. **(II)**

*[27] On a theorem of algebra and its geometrical application. *J. Indian Math. Soc.* **8** (1944) 29–36.

BIBLIOGRAPHY OF THE PUBLICATIONS OF S.S. CHERN

1945

*[28] On Grassmann and differential rings and their relations to the theory of multiple integrals. *Sankhya* **7** (1945) 2–8.

[29] Some new characterizations of the Euclidean sphere. *Duke Math. J.* **12** (1945) 279–290. **(II)**

[30] On the curvature integra in a Riemannian manifold. *Ann. of Math.* **46** (1945) 674–684. **(I)**

*[31] On Riemannian manifolds of four dimensions. *Bull. Amer. Math. Soc.* **51** (1945) 964–971.

1946

[32] Some new viewpoints in the differential geometry in the large. *Bull. Amer. Math. Soc.* **52** (1946) 1–30. **(II)**

[33] Characteristic classes of Hermitian manifolds. *Ann. of Math.* **47** (1946) 85–121. (I)

1947

[34] (with H.C. Wang). Differential geometry in symplectic space I. *Science Report Nat. Tsing Hua Univ.* **4** (1947) 453–477. **(II)**

[35] Sur une classe remarquable de variétés dans l'espace projectif à *N* dimensions. *Science Reports Naț. Tsing Hua Univ.* **4** (1947) 328–336. **(I)**

*[36] On the characteristic classes of Riemannian manifolds. *Proc. Nat. Acad. Sci USA*, **33** (1947) 78–82.

*[37] Note of affinely connected manifolds. *Bull. Amer. Math. Soc.* **53** (1947) 820–823; correction ibid **54** (1948) 985–986.

*[38] On the characteristic ring of a differentiable manifold. *Acad. Sinica. Sci. Record* **2** (1947) 1–5.

1948

[39] On the multiplication in the characteristic ring of a sphere bundle. *Ann. of Math.* **49** (1948) 362–372. **(I)**

[40] Note on projective differential line geometry. *Acad. Sinica Sci. Record* **2** (1948) 137–139. **(II)**

*[41] (with Y.L. Jou) On the orientability of differentiable manifolds. *Science Reports Nat. Tsing Hua Univ.* **5** (1948) 13–17.

[42] Local equivalence and Euclidean connections in Finsler spaces. *Science Reports Nat. Tsing Hua Univ.* **5** (1948) 95–121. **(II)**

1949

[43] (with Y.F. Sun). The imbedding theorem for fibre bundles. *Trans. Amer. Math. Soc* **67** (1949) 286–303. **(II)**

*[44] (with S.T. Hu) Parallelisability of principal fibre bundles. *Trans. Amer. Math. Soc.* **67** (1949) 304–309.

1950

[45] (with E. Spanier). The homology structure of sphere bundles. *Proc. Nat. Acad. Sci. USA*, **36** (1950) 248–255. **(II)**

[46] Differential geometry of fiber bundles. *Proc. Int. Congr. Math.* (1950) **II** 397–411. **(II)**

1951

[47] (with E. Spanier). A theorem on orientable surfaces in four-dimensional space. *Comm. Math. Helv.* **25** (1951) 205–209. **(I)**

1952

[48] On the kinematic formula in the Euclidean space of N dimensions. *Amer. J. Math* **74** (1952) 227–236. **(II)**

[49] (with C. Chevalley). Elie Cartan and his mathematical work. *Bull. Amer. Math. Soc.* **58** (1952) 217–250. **(II)**

[50] (with N.H. Kuiper) Some theorems on the isometric imbedding of compact Riemann manifolds in Euclidean space. *Ann. of Math.* **56** (1952) 422–430. **(II)**

1953

[51] On the characteristic classes of complex sphere bundles and algebraic varieties. *Amer. J. of Math.*, **75** (1953) 565–597. **(I)**

*[52] Some formulas in the theory of surfaces. *Boletin de la Sociedad Matematica Mexicana*, **10** (1953) 30–40.

[53] Relations between Riemannian and Hermitian geometries. *Duke Math. J.*, **20** (1953) 575–587. **(II)**

1954

[54] Pseudo-groupes continus infinis *Colloque de Geom. Diff.* Strasbourg (1954) 119–136. **(I)**

[55] (with P. Hartman and A. Wintner) On isothermic coordinates. *Comm. Math. Helv.* **28** (1954) 301–309. **(I)**

1955

[56] La géométrie des sous-variétés d'un espace euclidien à plusieurs dimensions. *l'Ens. Math.*, **40** (1955) 26–46. **(II)**

[57] An elementary proof of the existence of isothermal parameters on a surface. *Proc. Amer. Math. Soc.*, **6** (1955) 771–782. **(II)**

[58] On special *W*-surfaces. *Proc. Amer. Math. Soc.*, **6** (1955) 783–786. **(II)**

[59] On curvature and characteristic classes of a Riemann manifold. *Abh. Math. Sem. Univ. Hamburg* **20** (1955) 117–126. **(II)**

1956

*[60] Topology and differential geometry of complex manifolds. *Bull. Amer. Math. Soc.*, **62** (1956) 102–117.

1957

[61] On a generalization of Kähler geometry. *Lefschetz jubilee volume.* Princeton Univ. Press (1957) 103–121. **(I)**

[62] (with R. Lashof) On the total curvature of immersed manifolds. *Amer. J. of Math.* **79** (1957) 306–318. **(I)**

[63] (with F. Hirzebruch and J-P. Serre) On the index of a fibered manifold. *Proc. Amer. Math. Soc.*, **8** (1957) 587–596. **(I)**

[64] A proof of the uniqueness of Minkowski's problem for convex surfaces. *Amer. J. of Math.*, **79** (1957) 949–950. **(II)**

1958

*[65] Geometry of submanifolds in complex projective space. *Symposium International de Topologia Algebraica* (1958) 87–96.

[66] (with R.K. Lashof) On the total curvature of immersed manifolds, II. *Michigan Math. J.* **5** (1958) 5–12. **(II)**

[67] Differential geometry and integral geometry. *Proc. Int. Congr. Math. Edinburgh* (1958) 441–449. **(II)**

BIBLIOGRAPHY OF THE PUBLICATIONS OF S.S. CHERN

1959

[68] Integral formulas for hypersurfaces in Euclidean space and their applications to uniqueness theorems. *J. of Math. and Mech.* **8** (1959) 947–956. **(I)**

1960

[69] (with J. Hano and C.C. Hsiung) A uniqueness theorem on closed convex hypersurfaces in Euclidean space. *J. of Math. and Mech.* **9** (1960) 85–88. **(I)**

[70] Complex analytic mappings of Riemann surfaces I. *Amer. J. of Math.* **82** (1960) 323–337. **(I)**

[71] The integrated form of the first main theorem for complex analytic mappings in several complex variables. *Ann. of Math.* **71** (1960) 536–551. **(I)**

*[72] Geometrical structures on manifolds. *Amer. Math. Soc. Pub.* (1960) 1–31.

*[73] La géométrie des hypersurfaces dans l'espace euclidean. *Seminaire Bourbaki*, **193** (1959–1960).

*[74] Sur les métriques Riemanniens compatibles avec une reduction du groupe structural. *Séminaire Ehresmann*, January 1960.

1961

[75] Holomorphic mappings of complex manifolds. *L'Ens. Math.* **7** (1961) 179–187. **(I)**

1962

*[76] Geometry of quadratic differential form. *J. of SIAM* **10** (1962) 751–755.

1963

[77] (with C.C. Hsiung) On the isometry of compact submanifolds in Euclidean space. *Math. Annalen* **149** (1963) 278–285. **(II)**

[78] Pseudo-Riemannian geometry and Gauss–Bonnet formula. *Academia Brasileira de Ciencias* **35** (1963) 17–26. **(I)**

1965

[79] Minimal surfaces in an Euclidean space of N dimensions. *Differential and Combinatorial Topology*, Princeton Univ. Press (1965) 187–198. **(I)**

[80] (with R. Bott) Hermitian vector bundles and the equidistribution of the zeroes of their holomorphic sections. *Acta. Math.* **114** (1965) 71–112. **(II)**

[81] On the curvatures of a piece of hypersurface in Euclidean space. *Abh. Math. Sem. Univ. Hamburg* **29** (1965) 77–91. **(III)**

[82] On the differential geometry of a piece of submanifold in Euclidean space. *Proc. of U.S.–Japan Seminar in Diff. Geom.* (1965) 17–21. **(III)**

1966

[83] Geometry of G-structures. *Bull. Amer. Math. Soc.* **72** (1966) 167–219. **(III)**

[84] On the kinematic formula in integral geometry. *J. of Math. and Mech.* **16** (1966) 101–118. **(III)**

*[85] Geometrical structures on manifolds and submanifolds. *Some Recent Advances in Basic Sciences*, Yeshiva Univ. Press (1966) 127–135.

1967

[86] (with R. Osserman) Complete minimal surfaces in Euclidean n-space. *J. de l'Analyse Math.* **19** (1967) 15–34. **(III)**

[87] Einstein hypersurfaces in a Kählerian manifold of constant holomorphic curvature. *J. Diff. Geom.* **1** (1967) 21–31. **(III)**

1968

[88] On holomorphic mappings of Hermitian manifolds of the same dimension. *Proc. Symp. Pure Math.* **11**. Entire Functions and Related Parts of Analysis (1968) 157–170. **(I)**

1969

[89] Simple proofs of two theorems on minimal surfaces. *L'Ens. Math.* **15** (1969) 53–61. **(I)**

1970

[90] (with H. Levine and L. Nirenberg) Intrinsic norms on a complex manifold. *Global analysis*, Princeton Univ. Press (1970) 119–139. **(I)**

[91] (with M. do Carmo and S. Kobayashi) Minimal submanifolds of a sphere with second fundamental form of constant length. *Functional Analysis and Related Fields*, Springer-Verlag (1970) 59–75. **(I)**

[92] (with R. Bott) Some formulas related to complex transgression. *Essays on Topology and Related Topics*, Springer-Verlag, (1970) 48–57. **(I)**

*[93] Holomorphic curves and minimal surfaces. *Carolina Conference Proceedings* (1970) 28 pp.

[94] On minimal spheres in the four–sphere, Studies and Essays Presented to Y. W. Chen, Taiwan, (1970) 137–150. **(I)**

[95] Differential geometry: Its past and its future. *Actes Congrès Intern. Math.* (1970) **1**, 41–53. **(III)**

[96] On the minimal immersions of the two-sphere in a space of constant curvature. *Problems in Analysis*, Princeton Univ. Press, (1970) 27–40. **(III)**

1971

[97] Brief survey of minimal submanifolds. *Differentialgeometrie im Grossen.* W. Klingenberg (ed.), **4** (1971) 43–60. **(III)**

*[98] (with J. Simons) Some cohomology classes in principal fibre bundles and their application to Riemannian geometry. *Proc. Nat. Acad. Sci. USA*, **68** (1971) 791–794.

1972

[99] Holomorphic curves in the plane. *Diff. Geom., in honor of K. Yano*, (1972) 73–94. **(III)**

*[100] Geometry of characteristic classes. *Proc. 13th Biennial Sem. Canadian Math. Congress*, (1972) 1–40. Also pub. in Russian translation.

1973

[101] Meromorphic vector fields and characteristic numbers. *Scripta Math.* **29** (1973) 243–251. **(I)**

*[102] The mathematical works of Wilhelm Blaschke. *Abh. Math. Sem. Univ. Hamburg* **39** (1973) 1–9.

1974

[103] (with. J. Simons) Characteristic forms and geometrical invariants. *Ann. of Math.* **99** (1974) 48–69. **(I)**

[104] (with M. Cowen, A. Vitter III) Frenet frames along holomorphic curves. *Proc. of Conf. on Value Distribution Theory*, Tulane Univ. (1974) 191–203. **(III)**

[105] (with J. Moser) Real hypersurfaces in complex manifolds. *Acta. Math.* **133** (1974) 219–271. **(III)**

1975

[106] (with S.I. Goldberg) On the volume decreasing property of a class of real harmonic mappings. *Amer. J. of Math.* **97** (1975) 133–147. **(III)**

[107] On the projective structure of a real hypersurface in C_{n+1}. *Math. Scand.* **36** (1975) 74–82. **(III)**

1976

[108] (with J. White) Duality properties of characteristic forms. *Inv. Math.* **35** (1976) 285–297. **(III)**

1977

*[109] Circle bundles. *Geometry and topology, III*. Latin Amer. School of Math, Lecture Notes in Math. Springer-Verlag, **597** (1977) 114–131.

*[110] (with P.A. Griffiths) Linearization of webs of codimension one and maximum rank. *Proc. Int. Symp. on Algebraic Geometry, Kyoto* (1977) 85–91.

1978

[111] On projective connections and projective relativity. *Science of Matter*, dedicated to Ta-you Wu, (1978) 225–232. **(III)**

[112] (with P.A. Griffiths) Abel's theorem and webs. *Jber. d. Dt. Math. Verein.* **80** (1978) 13–110. **(III)**

[113] (with P.A. Griffiths) An inequality for the rank of a web and webs of maximum rank. *Annali Sc. Norm. Super.–Pisa, Serie IV*, **5** (1978) 539–557. (III)

[114] Affine minimal hypersurfaces. *Minimal Submanifolds and Geodesics*. Kaigai Publications, Ltd. (1978) 1–14. **(III)**

1979

*[115] Herglotz's work on geometry. *Ges. Schriften Gustav Herglotz*, Göttingen (1979) xx–xxi.

[116] (with C.L. Terng) An analogue of Bäcklund's theorem in affine geometry. *Rocky Mountain J. Math.* **10** (1979) 105–124. **(III)**

[117] From triangles to manifolds. *Amer. Math. Monthly* **86** (1979) 339–349. **(III)**

[118] (with C.K. Peng) Lie groups and KdV equations. *Manuscripta Math.* **28** (1979) 207–217. **(III)**

1980

[119] General relativity and differential geometry. *Some Strangeness in the Proportion: A Centennial Symp. to Celebrate the Achievements of Albert Einstein, Harry Woolf* (ed.), Addison-Wesley Publ. (1980) 271–287. **(III)**

[120] (with W.M. Boothby and S.P. Wang) The mathematical work of H.C. Wang. *Bull. Inst. of Math,* **8** (1980) xiii–xxiv. **(III)**

*[121] Geometry and physics. *Math. Medley*, Singapore, **8** (1980) 1–6.

*[122] (with R. Bryant and P.A. Griffiths) Exterior differential systems. *Proc. of 1980 Beijing DD-Symposium,* (1980) 219–338.

1981

[123] Geometrical interpretation of the sinh-Gordon equation. *Annales Polonici Mathematici* **39** (1981) 63–69. **(IV)**

[124] (with P.A. Griffiths) Corrections and addenda to our paper: "Abel's theorem and webs." *Jber. d. Dt. Math.–Verein.* **83** (1981) 78–83. **(III)**

[125] (with R. Osserman) Remarks on the Riemannian metric of a minimal submanifold. *Geometry Symposium Utrecht 1980*, Lecture Notes in Math. Springer-Verlag **894** (1981) 49–90. **(IV)**

[126] (with J. Wolfson) A simple proof of Frobenius theorem. *Manifolds and Lie Groups, Papers in Honor of Y. Matsushima.* Birkhäuser (1981) 67–69. **(IV)**

[127] (with K. Tenenblat) Foliations on a surface of constant curvature and modified Korteweg–de Vries equations. *J. Diff. Geom.* **16** (1981) 347–349. **(IV)**

[128] (with C.K. Peng) On the Bäcklund transformations of KdV equations and modified KdV equations. *J. of China Univ. of Sci. and Tech.*, **11** (1981) 1–6. **(IV)**

1982

[129] Web geometry. *Proc. Symp. in Pure Math.* **39** (1983) 3–10. **(IV)**
[130] Projective geometry, contact transformations, and *CR*-structures. *Archiv der Math.* **38** (1982) 1–5. **(IV)**

1983

[131] (with J. Wolfson) Minimal surfaces by moving frames. *Amer. J. Math.* **105** (1983) 59–83. **(IV)**
[132] On surfaces of constant mean curvature in a three-dimensional space of constant curvature. *Geometric Dynamics*, Springer Lecture Notes **1007** (1983) 104–108. **(IV)**

1984

[133] Deformation of surfaces preserving principal curvatures, *Differential Geometry and Complex Analysis*, Volume in Memory of H. Rauch, Springer-Verlag (1984) 155–163. **(IV)**

1985

[134] (with R. Hamilton) On Riemannian metrics adapted to three-dimensional contact manifolds. *Arbeitstagung Bonn 1984* Springer Lecture Notes **1111** (1985) 279–308. **(IV)**
[135] (with J. Wolfson) Harmonic maps of S^2 into a complex Grassmann manifold. *Proc. Nat. Acad. Sci.* USA **82** (1985) 2217–2219. **(IV)**
[136] Moving frames, *Soc. Math. de France*, Astérisque, (1985) 67–77. **(IV)**
*[137] Wilhelm Blaschke and web geometry, Wilhelm Blaschke—Gesammelte Werke. **5**, Thales Verlag, (1985) 25–27.
*[138] The mathematical works of Wilhelm Blaschke—an update. Thales Verlag, (1985), 21–23.

1986

[139] (with K. Tenenblat) Pseudospherical surfaces and evolution equations. *Studies in Applied Math.* MIT **74** (1986) 55–83. **(IV)**
[140] On a conformal invariant of three-dimensional manifolds. *Aspects of Mathematics and Its Applications* Elsevier Science Publishers B.V. (1986) 245–252. **(IV)**
*[141] (with P.A. Griffiths) Pfaffian systems in involution. *Proceedings of 1982 Changchun Symposium on Differential Geometry and Differential Equations*, Science Press, China, (1986) 233–256.

1987

[142] (with J. Wolfson) Harmonic maps of the two-sphere into a complex Grassmann manifold II. *Ann. of Math.* **125** (1987) 301–335. **(IV)**
[143] (with T. Cecil) Tautness and Lie Sphere geometry *Math. Annalen*, Volume Dedicated to F. Hirzebruch **278** (1987) 381–399. **(IV)**

1988

[144] Vector bundles with a connection. *Studies in Global Differential Geometry*, MAA, no. 27 (1988), 1–26.

1989

[145] (with T. Cecil) Dupin submanifolds in Lie sphere geometry, to appear in *Differential Geometry and Topology*, Springer Lecture Notes 1989.
[146] Historical remarks on Gauss–Bonnet, to appear in Moser Volume, Academic Press.
[147] An introduction to Dupin submanifolds, to appear in Do Carmo Volume.

陳省身数学
論文選集

ANNALES
POLONICI MATHEMATICI
XXXIX (1981)

Geometrical interpretation of the sinh-Gordon equation

by SHIING-SHEN CHERN * (Berkeley, Calif.)

Stefan Bergman in memoriam

Abstract. In a three-dimensional pseudo-Riemannian manifold of constant curvature consider a spacelike (resp. timelike) surface of constant negative (resp. positive) Gaussian curvature. Then the asymptotic curves are everywhere real and distinct, and the function 2ψ (= the angle between the asymptotic directions) satisfies, relative to the Tchebycheff coordinates, a sine-Gordon (resp. sinh-Gordon) equation. An example of such a manifold is SL (2; R) with the biinvariant metric.

1. Introduction. It is well known that the sine-Gordon equation (SGE)

(1) $$u_{xx}-u_{tt} = \sin u$$

has a geometrical interpretation in terms of the surfaces of constant negative curvature in the three-dimensional euclidean space. We will show in this paper that by studying surfaces of constant Gaussian curvature in a three-dimensional pseudo-Riemannian manifold of constant curvature one is led to geometrical interpretations of (1) and of the sinh-Gordon equation (SHGE)

(2) $$u_{xx}-u_{tt} = \sinh u.$$

2. Pseudo-Riemannian geometry. In this section we will give a review of local pseudo-Riemannian geometry, using moving frames. Let M be a smooth manifold of dimension m, with the local coordinates x^α. (In this section all small Greek indices run from 1 to m.) A pseudo-Riemannian metric in M is given by the non-degenerate quadratic differential form

(3) $$ds^2 = \sum_{\alpha,\beta} G_{\alpha\beta}(x^1, \ldots, x^m)\, dx^\alpha dx^\beta, \qquad G_{\alpha\beta} = G_{\beta\alpha}.$$

The metric is called *Riemannian* if the form is positive definite and Lorentzian if it is of signature $+ \ldots + -$.

Let $x \in M$ and let T_x, T_x^* be respectively the tangent and cotangent spaces

* Work done under partial support of NSF grant MCS 77-23579.

of M at x. A frame at x is an ordered set of linearly independent vectors $e_\alpha \in T_x$. The essence of the method of moving frames is to free the frames from local coordinates, a freedom which gives handsome returns. To e_α is associated a dual coframe $\omega^\beta \in T_x^*$. When they are defined over a neighbourhood, ω^β can be identified with a linear differential form. Relative to ω^β we can write

$$ds^2 = \sum g_{\alpha\beta}\,\omega^\alpha \omega^\beta, \quad g_{\alpha\beta} = g_{\beta\alpha}. \tag{3a}$$

The Levi-Civita connection is given by

$$De_\alpha = \sum \omega_\alpha^\beta e_\beta, \tag{4}$$

where the connection forms ω_α^β are determined, uniquely, by the conditions

$$d\omega^\alpha = \sum \omega^\beta \wedge \omega_\beta^\alpha, \tag{5}$$

$$\omega_{\alpha\beta} + \omega_{\beta\alpha} = dg_{\alpha\beta}. \tag{6}$$

Geometrically the first condition (5) means the "absence of torsion". In the second condition (6) the $\omega_{\alpha\beta}$ are defined by

$$\omega_{\alpha\beta} = \sum g_{\beta\gamma}\,\omega_\alpha^\gamma \tag{7}$$

and the condition means the preservation of the scalar product of vectors under parallelism. We use $g_{\alpha\beta}$ to lower indices, as in classical tensor analysis.

The curvature forms are defined by

$$\Omega_\alpha^\beta = d\omega_\alpha^\beta - \sum_\gamma \omega_\alpha^\gamma \wedge \omega_\gamma^\beta, \tag{8}$$

$$\Omega_{\alpha\beta} = \sum g_{\beta\gamma}\,\Omega_\alpha^\gamma. \tag{9}$$

It can be proved that

$$\Omega_{\alpha\beta} + \Omega_{\beta\alpha} = 0. \tag{10}$$

The pseudo-Riemannian metric (3a) is said to be of *constant curvature* c if

$$\Omega_{\alpha\beta} = -c\,\omega_\alpha \wedge \omega_\beta, \tag{11}$$

where

$$\omega_\alpha = \sum g_{\alpha\beta}\,\omega^\beta. \tag{12}$$

In applications it will be advantageous to use frames, where $g_{\alpha\beta} = \text{const}$, such as orthonormal frames in the Riemannian case. Then (6) becomes

$$\omega_{\alpha\beta} + \omega_{\beta\alpha} = 0. \tag{13}$$

3. Surfaces in three-dimensional manifolds. Let M be a three-dimensional pseudo-Riemannian manifold and

$$f\colon S \to M \tag{14}$$

be an immersed surface. In a neighbourhood on S we take frames $xe_1 e_2 e_3$, so that e_1, e_2 are tangent vectors to S at $x \in S$. Restricted to these frames we have

$$\omega^3 = 0, \tag{15}$$

and the induced pseudo-Riemannian metric on S is

$$I = g_{11}(\omega^1)^2 + 2g_{12}\,\omega^1\omega^2 + g_{22}(\omega^2)^2, \tag{16}$$

when I stands for first fundamental form. The surface S is said to be *spacelike* (resp. *timelike*) at x if I is definite (resp. indefinite).

By exteriorly differentiating (15) and using (5), we get

$$\omega^1 \wedge \omega_1^3 + \omega^2 \wedge \omega_2^3 = 0. \tag{17}$$

It follows that

$$\omega_i^3 = \sum h_{ik}\,\omega^k, \quad 1 \leqslant i,\, k \leqslant 2, \tag{18}$$

where

$$h_{12} = h_{21}. \tag{19}$$

From now on we use orthonormal frames, so that

$$\begin{aligned} g_{\alpha\beta} &= 0, \quad \alpha \neq \beta,\ 1 \leqslant \alpha,\, \beta \leqslant 3, \\ g_{\alpha\alpha} &= \pm 1. \end{aligned} \tag{20}$$

By (13) the matrix

$$(\omega_{\alpha\beta}) = \begin{pmatrix} g_{11}\,\omega_1^1 & g_{22}\,\omega_1^2 & g_{33}\,\omega_1^3 \\ g_{11}\,\omega_2^1 & g_{22}\,\omega_2^2 & g_{33}\,\omega_2^3 \\ g_{11}\,\omega_3^1 & g_{22}\,\omega_3^2 & g_{33}\,\omega_3^3 \end{pmatrix} \tag{21}$$

is anti-symmetric. This implies in particular

$$\omega_1^1 = \omega_2^2 = \omega_3^3 = 0. \tag{22}$$

The second fundamental form is defined by

$$\begin{aligned} II &= -(dx, De_3) = -(g_{11}\,\omega^1\omega_3^1 + g_{22}\,\omega^2\omega_3^2) \\ &= g_{33}(\omega^1\omega_1^3 + \omega^2\omega_2^3) \\ &= g_{33}\{h_{11}(\omega^1)^2 + 2h_{12}\,\omega^1\omega^2 + h_{22}(\omega^2)^2\}. \end{aligned} \tag{23}$$

The curves defined by

$$II = 0 \tag{24}$$

are the asymptotic curves.

The principal directions of S are determined by the equations

(25) $g_{33}(h_{11}\omega^1+h_{12}\omega^2) = \lambda g_{11}\omega^1, \quad g_{33}(h_{12}\omega^1+h_{22}\omega^2) = \lambda g_{22}\omega^2,$

where λ is the corresponding principal curvature. It follows that the principal curvatures are the roots of the equation

(26) $$\begin{vmatrix} g_{33}h_{11}-\lambda g_{11} & g_{33}h_{12} \\ g_{33}h_{12} & g_{33}h_{22}-\lambda g_{22} \end{vmatrix} = 0.$$

The Gaussian curvature of S is the product of the principal curvatures and is given by

(27) $$K = \frac{1}{g_{11}g_{22}}(h_{11}h_{22}-h_{12}^2).$$

Put

(28) $$\varepsilon = g_{11}g_{22} = \pm 1,$$

so that

(29) $$K = \varepsilon(h_{11}h_{22}-h_{12}^2).$$

The surface S is spacelike or timelike according as $\varepsilon = +1$ or -1.

4. Surfaces of constant Gaussian curvature. From now on we suppose M to be of constant curvature and the surface S to be of constant Gaussian curvature

(30) $$K = -\varepsilon b^2, \quad b = \text{const} \neq 0,$$

i.e., of constant negative or positive Gaussian curvature, according as S is spacelike or timelike. In both cases the asymptotic directions are real and distinct.

We choose frames so that e_1, e_2 are along the principal directions, i.e.,

(31) $$h_{12} = 0.$$

Then the asymptotic directions are defined by

(32) $$h_{11}(\omega^1)^2+h_{22}(\omega^2)^2 = 0.$$

The tangent plane T_x has a definite or indefinite metric according as S is spacelike or timelike. Let 2ψ be the angle between the asymptotic directions relative to that metric. Since $h_{11}h_{22} = -b^2$, we have the two cases:

Case 1. S is spacelike ($\varepsilon = 1$). Then we have

(33) $$h_{11} = b\cot\psi, \quad h_{22} = -b\tan\psi.$$

Case 2. S is timelike ($\varepsilon = -1$). We have

(34) $$h_{11} = b\coth\psi, \quad h_{22} = -b\tanh\psi.$$

Over S we have defined a field of orthonormal frames $xe_1e_2e_3$ such that e_3 is along the normal and e_1, e_2 along the principal directions at x. We wish to show that there are local coordinates, to be called *Tchebycheff coordinates*, relative to which the connection forms of this field of frames take a simple form. At this stage we relate the frames to local coordinates.

Restricted to the field of frames described above we have, since M is of constant curvature,

$$\Omega_{13} = \Omega_{23} = 0, \tag{35}$$

whence

$$\Omega_1^3 = \Omega_2^3 = 0. \tag{36}$$

Equations (5) and (8) give respectively

$$\begin{aligned} d\omega^1 &= \omega^2 \wedge \omega_2^1, & d\omega^2 &= \omega^1 \wedge \omega_1^2, \\ d\omega_1^3 &= \omega_1^2 \wedge \omega_2^3, & d\omega_2^3 &= \omega_2^1 \wedge \omega_1^3. \end{aligned} \tag{37}$$

On the other hand, the anti-symmetry of the matrix (21) gives

$$\omega_2^1 = -\varepsilon\omega_1^2. \tag{38}$$

Hence we can write

$$d\omega^1 = -\varepsilon\omega^2 \wedge \omega_1^2, \quad d\omega_2^3 = -\varepsilon\omega_1^2 \wedge \omega_1^3. \tag{39}$$

Equation (8) also gives

$$d\omega_1^2 = \omega_1^3 \wedge \omega_3^2 + \Omega_1^2 = -g_{22}g_{33}\,\omega_1^3 \wedge \omega_2^3 + \Omega_1^2, \tag{40}$$

where

$$\Omega_1^2 = g_{22}\,\Omega_{12} = -g_{22}\,c\omega_1 \wedge \omega_2 = -g_{11}\,c\omega^1 \wedge \omega^2. \tag{41}$$

Hence the above equation can be written

$$d\omega_1^2 = (g_{22}g_{33}\,b^2 - g_{11}\,c)\,\omega^1 \wedge \omega^2. \tag{42}$$

Let u, v be local coordinates such that

$$\omega^1 = A du, \quad \omega^2 = C dv. \tag{43}$$

Geometrically this means using the lines of curvature (= integral curves of principal directions) as parametric curves. Then

$$d\omega^1 = -A_v\,du \wedge dv = -\varepsilon C dv \wedge \omega_1^2, \quad d\omega^2 = C_u\,du \wedge dv = A du \wedge \omega_1^2.$$

It follows that

$$\omega_1^2 = -\varepsilon\frac{A_v}{C}\,du + \frac{C_u}{A}\,dv. \tag{44}$$

By (18) we have

(45) $$\omega_1^3 = h_{11}\, A du, \qquad \omega_2^3 = h_{22}\, C dv.$$

Substituting (44), (45) into (37), (39), we get

(46) $$(Ah_{11})_v = \varepsilon h_{22}\, A_v, \qquad (Ch_{22})_u = \varepsilon h_{11}\, C_u.$$

We now consider the two cases separately:

Case 1. $\varepsilon = +1$. Equation (46) gives

$$(A \cot\psi)_v = -\tan\psi A_v, \qquad (C \tan\psi)_u = -\cot\psi C_u.$$

The first equation can be written

$$\left(\log \frac{A}{\sin\psi}\right)_v = 0,$$

so that $A/\sin\psi$ is a function of u only. Absorbing this function into u, we can suppose

$$A = \sin\psi.$$

Similarly, we can choose v such that

$$C = \cos\psi.$$

These (u, v)-coordinates are called the *Tchebycheff coordinates*. By (44), we have

$$\omega_1^2 = -\psi_v\, du - \psi_u\, dv,$$

and (42) gives

(47) $$\psi_{uu} - \psi_{vv} = (-g_{22}\, g_{33}\, b^2 + g_{11}\, c) \sin\psi \cos\psi.$$

By choosing

$$c = 0, \quad b = 1, \quad g_{\alpha\alpha} = 1, \quad u = t, \quad v = x,$$

and considering 2ψ as the dependent variable, this reduces to (1).

Case 2. $\varepsilon = -1$. Equation (46) gives

$$(A \coth\psi)_v = +\tanh\psi A_v, \qquad (C \tanh\psi)_u = \coth\psi C_u.$$

Exactly the same manipulations as in Case 1 show that we can choose u, v, so that

$$A = \sinh\psi, \qquad C = \cosh\psi.$$

By (44) we have then

$$\omega_1^2 = \psi_v\, du + \psi_u\, dv.$$

It follows from (42) that

(48) $$\psi_{uu} - \psi_{vv} = (+g_{22}\, g_{33}\, b^2 - g_{11}\, c) \sinh\psi \cosh\psi.$$

This is essentially the SHGE, the expression in the parenthesis being a constant.

We summarize the results in the following theorem:

In a three-dimensional pseudo-Riemannian manifold of constant curvature consider a spacelike (resp. timelike) surface of constant negative (resp. positive) Gaussian curvature. Then the asymptotic directions are everywhere real and distinct, and the function 2ψ *(= the angle between the asymptotic directions) satisfies, relative to the Tchebycheff coordinates, a SGE (resp. SHGE).*

5. SL $(2; R)$. An important example of a three-dimensional pseudo-Riemannian manifold of constant curvature is given by the special linear group in two real variables:

$$\text{(49)} \qquad \mathrm{SL}\,(2; R) = \left\{ X = \begin{pmatrix} x & y \\ z & t \end{pmatrix} \,\middle|\, xt - yz = 1 \right\},$$

provided with the biinvariant metric. The latter is defined by

$$\text{(50)} \qquad ds^2 = \tfrac{1}{2}\,\mathrm{Tr}\,(dXX^{-1}\,dXX^{-1}),$$

and is Lorentzian. This metric has curvature -1, according to the definition of Section 2.

Reçu par la Rédaction le 6. 12. 1978

Added April, 1988. Tilla Milnor called to my attention that, for the results of the paper to be valid, I need the additional assumption $H^2 > K$. This will ensure that the principal directions are defined and distinct at every point. Cf. Tilla Klotz Milnor, Harmonic maps and classical surface theory in Minkowski space, Transactions of American Mathematical Society, 280 (1983), 161–185.

Remarks on the Riemannian Metric of a Minimal Submanifold

S.-S. CHERN
Department of Mathematics
University of California
Berkeley CA 94720 / USA

R. OSSERMAN
Department of Mathematics
Stanford University
Stanford CA 94305 / USA

The general question that served as the starting point for this paper was to characterize those Riemannian metrics that arise as the induced metrics on minimal submanifolds of some euclidean space. We are, however, far from having a complete answer to that question, and we content ourselves here with a number of related results and remarks that may be of interest in their own right.

We note at the outset that the original question has two quite different aspects, depending on whether or not one specifies the codimension. In particular, the codimension-one case plays as usual a prominent role. The background in that case is the following.

First, Ricci ([17], p. 411) made the surprising discovery that there are simple necessary and sufficient conditions on a two-dimensional metric for it to be realizable on a minimal surface in E^3. (See Theorem 1.2 below). For higher-dimensional minimal submanifolds, various *necessary* conditions on the metric have been given by Pinl-Ziller [16] and Barbosa-Do Carmo [3], but they are clearly far from sufficient. We give here (Proposition 1.3) a much stronger necessary condition, directly generalizing that of Ricci. However, we note the anomaly that by a theorem of Thomas [19], for metrics of dimension at least four, the Codazzi equations (at least in the generic case) are consequences of the Gauss curvature equations. Thus the problem reduces to the purely algebraic one of determining when there exists a second fundamental form of trace zero satisfying the Gauss curvature equations for the given metric. In Theorem 3.1, we present (again in the generic case) one way of answering that question. The only remaining case is therefore that of three-dimensional metrics. There, the algebraic conditions for solving the Gauss equations are particularly simple to state (Proposition 3.2), but the Codazzi equations are not consequences, and they impose some further differential conditions. Those are described in Theorem 3.3, where necessary and sufficient conditions are stated for the metric on a generic minimal hypersurface in E^4.

The case of higher codimension presents still greater difficulties, even in finding effective necessary conditions. However, we note that for $n \geq 4$ and codimension

$p \le n/4$, there is a generalization of Thomas's theorem due to Allendoerfer [1]. Under the hypotheses of that theorem, the Gauss equations again imply the remaining equations needed for an embedding, and the problem is again reduced to an algebraic one. We give here (Theorem 2.4) an alternative proof of Allendoerfer's theorem. Finally, we consider the question of the "genericity" of the hypotheses in this theorem. We look at the case of a holomorphic hypersurface in $\mathbb{C}^m$ for $m \ge 5$, which may be viewed as a codimension-two minimal submanifold of $\mathbb{R}^{2m}$. We show that wherever the complex second fundamental form is non-singular, the assumptions of Theorem 2.4 are indeed satisfied.

In the special case when $n = 2$, if one does not fix the codimension, then a complete answer to our original question was given by Calabi [5]. Furthermore he gives explicit bounds on the effective codimensions that can arise. We note also that in selected dimensions, a paper of Dajczer and Rodriguez [8] extends to higher codimension some results of Pinl and Ziller [16]. They also correct a misstatement in the latter paper.

We note finally that Do Carmo and Dajczer [9] have carried further some of the ideas in the present paper. In particular, they have studied the case where the ambient manifold has arbitrary constant curvature.

Our order of presentation is the following.

Section 1 develops those ideas related to the original Ricci condition, including its generalization to hypersurfaces and the results of Calabi mentioned above.

Section 2, although motivated by the discussion in Section 1, is independent of it, and is devoted to general questions of local existence and rigidity of immersions. Special attention is paid to Allendoerfer's notion of "type" and its implications.

Finally, Section 3 combines the results of Section 1 and 2 in order to formulate necessary and sufficient conditions for the existence of a minimal hypersurface realizing a given metric.

1. The Ricci Condition and its Generalizations

We start by discussing the classical result of Ricci for minimal surfaces in E^3. We show how Ricci's condition has a direct generalization giving a necessary condition for a metric to be realized on a minimal hypersurface in E^n. We then discuss some Ricci-like conditions for two-dimensional minimal surfaces in E^n.

Lemma 1.1. Let M be a two-dimensional Ricci manifold with metric ds^2. Then at any point of M where the Gauss curvature K is negative, the following three conditions are equivalent:

i) the metric $d\hat{s}^2 = -K ds^2$ has constant curvature $\hat{K} \equiv 1$;
ii) the metric $d\tilde{s}^2 = \sqrt{-K ds^2}$ has constant curvature $\tilde{K} \equiv 0$;
iii) the curvature K satisfies

$$\Delta \log(-K) = 4K, \tag{1}$$

where Δ is the Laplace-Beltrami operator on M.

Proof. We use the well-known and easily-derived formula for the dependence of Gauss curvature on a conformal change of metric:

$$\bar{K} = f^2(K + \Delta \log f) \tag{2}$$

where

$$d\bar{s}^2 = ds^2/f^2. \tag{3}$$

Choosing in turn $f = 1/\sqrt{-K}$ and $f = 1/\sqrt[4]{-K}$, we find

$$\hat{K} = \frac{\Delta \log(-K)}{2K} - 1 \tag{4}$$

and

$$\tilde{K} = [K - \tfrac{1}{4}\Delta \log(-K)]/\sqrt{-K}. \tag{5}$$

Thus, each of the conditions (i) and (ii) is clearly equivalent to (iii).

Theorem 1.2. Let M be a minimal surface in E^3, and ds^2 the induced metric. Then the Gauss curvature K of M satisfies

a) $K \leq 0$,

and

b) equation (1) above, wherever $K < 0$.

Conversely, let M be a 2-manifold with metric ds^2. Any simply-connected domain on M where $K < 0$ and where any of the equivalent conditions of Lemma 1.1 holds, can be immersed isometrically as a minimal surface in E^3.

Remarks. 1. Theorem 1.2 is due to Ricci ([17], p. 411), who formulated it in terms of condition ii) of Lemma 1.1.

2. There is a slight discrepancy between the necessary and the sufficient conditions in the theorem, in that the latter assumes the strict inequality $K < 0$ at every point. The stronger assumption is in fact necessary, as noted by Lawson ([13], p. 364) who gave a counterexample where K vanishes at a single point.

Proof. On a minimal surface the principal curvatures k_1, k_2 satisfy $k_1 + k_2 = 0$, so that the Gauss curvature K satisfies $K = k_1 k_2 = -k_1^2 \leq 0$, proving (a). In any neighborhood where $K < 0$, the Gauss map is anti-conformal and the expression $-K ds^2$ represents the metric of the image under the Gauss map. Since that image lies on the unit sphere, its metric $d\hat{s}^2$ has constant curvature $\hat{K} \equiv 1$. Thus condition (i) of Lemma 1.1 holds, and by Lemma 1.1, equation (1) must hold too. This proves (b).

For the converse, we recall that in order to realize a metric ds^2 on a surface in E^3, it is sufficient (as well as necessary) to have 1-forms $\omega_1, \omega_2, \omega_{12}, \omega_{13}, \omega_{23}$ satisfying the fundamental equations

$$ds^2 = \omega_1^2 + \omega_2^2 \tag{6}$$

$$d\omega_1 = \omega_{12} \wedge \omega_2, \qquad d\omega_2 = \omega_1 \wedge \omega_{12} \tag{7}$$

(8) $$d\omega_{13} = \omega_{12} \wedge \omega_{23}, \qquad d\omega_{23} = \omega_{13} \wedge \omega_{12}$$

and

(9) $$d\omega_{12} = -\omega_{13} \wedge \omega_{23}.$$

The integrability conditions (8) and (9) are called the Codazzi and Gauss equations, respectively.

Suppose we have a metric ds^2 satisfying $K < 0$ and condition (i) of Lemma 1.1. Let e_1, e_2 be an orthonormal frame field for ds^2, locally, and let ω_1, ω_2 be the dual forms. Then (6) is satisfied. Set

(10) $$k = \sqrt{-K}$$

(11) $$\hat{\omega}_1 = k\omega_1, \qquad \hat{\omega}_2 = -k\omega_2.$$

Then

(12) $$\hat{\omega}_1^2 + \hat{\omega}_2^2 = k^2(\omega_1^2 + \omega_2^2) = -K ds^2 = d\hat{s}^2.$$

Let ω_{12} be the connection form associated with ω_1, ω_2, ds^2. Then (7) is satisfied. If $\hat{\omega}_{12}$ is the connection form associated with $\hat{\omega}_1, \hat{\omega}_2, d\hat{s}^2$, then

(13) $$\hat{K} \equiv 1 \Leftrightarrow d\hat{\omega}_{12} = -\hat{\omega}_1 \wedge \hat{\omega}_2 = -K\omega_1 \wedge \omega_2 = d\omega_{12}$$

by virtue of (10) and (11) and the fact that

$$K = -d\omega_{12}(e_1, e_2).$$

We would like to assert now that not only

(14) $$d\hat{\omega}_{12} = d\omega_{12},$$

as we have just seen, but even more:

(15) $$\hat{\omega}_{12} = \omega_{12}.$$

If that were the case, then we could set

(16) $$\omega_{13} = \hat{\omega}_1, \qquad \omega_{23} = \hat{\omega}_2;$$

the structure equations for $\hat{\omega}_1, \hat{\omega}_2$ are

$$d\hat{\omega}_1 = \hat{\omega}_{12} \wedge \hat{\omega}_2, \qquad d\hat{\omega}_2 = \hat{\omega}_1 \wedge \hat{\omega}_{12},$$

and

$$d\hat{\omega}_{12} = -\hat{K}\hat{\omega}_1 \wedge \hat{\omega}_2 = -\hat{\omega}_1 \wedge \hat{\omega}_2,$$

which by virtue of (15), (16) are precisely the Codazzi and Gauss equations (8) and (9). It turns out that equations (15) will *not* hold in general for our original choice of frame field, but it will always hold after adjusting the frame field by suitable rotations. Namely, from (14) we deduce that there exists locally a function α such that

(17) $$\hat{\omega}_{12} = \omega_{12} + d\alpha.$$

Consider the effect of rotating the field through angle θ, where θ is a smooth function. We set

$$\omega'_1 = (\cos\theta)\omega_1 + (\sin\theta)\omega_2$$
$$\omega'_2 = -(\sin\theta)\omega_1 + (\cos\theta)\omega_2,$$
$$\hat{\omega}'_1 = k\omega'_1 = (\cos\theta)\hat{\omega}_1 - (\sin\theta)\hat{\omega}_2$$
$$\hat{\omega}'_2 = -k\omega'_2 = (\sin\theta)\hat{\omega}_1 + (\cos\theta)\hat{\omega}_2.$$

Then a direct calculation yields

$$\omega'_{12} = \omega_{12} + d\theta, \qquad \hat{\omega}'_{12} = \hat{\omega}_{12} - d\theta,$$

and

$$\hat{\omega}'_{12} - \omega'_{12} = \hat{\omega}_{12} - \omega_{12} - 2d\theta = d(\alpha - 2\theta).$$

If we choose $\theta = \frac{\alpha}{2}$, then $\hat{\omega}'_{12} = \omega'_{12}$. Also

$$ds^2 = (\omega'_1)^2 + (\omega'_2)^2, \qquad d\hat{s}^2 = (\hat{\omega}'_1)^2 + (\hat{\omega}'_2)^2.$$

Thus, in the new frame field, we have equations (6), (7), (11) and (12) all holding, as well as (15). By the argument given above, (8) and (9) must also hold, using (16). We therefore conclude from the fundamental existence theorem that the given metric can be realized locally on a surface in E^3. The resulting surface is unique up to a euclidean motion. By a standard monodromy argument, we obtain an immersion of any simply-connected domain.

It only remains to show that the surface is minimal. But equations (11) and (16) may be interpreted to mean that on the immersed surface the second fundamental form has been diagonalized, so that the principal curvatures are precisely the quantities k and $-k$. This completes the proof of the theorem.

There are several remarks to be made on the above proof.

First, a shorter proof can be given using isothermal parameters. However, such a proof does not generalize to higher dimensions.

Second, concerning the idea behind the proof, we note that the fundamental problem in applying the existence theorem for surfaces (or for hypersurfaces more generally) is to describe the Weingarten map: the differential of the Guess map. In our context that means assigning the appropriate forms ω_{13}, ω_{23} to a given pair ω_1, ω_2. If one could determine intrinsically the principal curvature directions at a point, then choosing them for the frame field e_1, e_2, one would have $\omega_{13} = k_1\omega_1$, $\omega_{23} = k_2\omega_2$, where k_1, k_2 are the principal curvatures. Since we want a minimal surface, we set $k_2 = -k_1$, and then, since the Gauss curvature $K = k_1 k_2 = -k_1^2$ is determined intrisically, the forms ω_{13}, ω_{23} are uniquely determined (up to sign). That is the reason for the choice represented by equations (10), (11), (16). Now the curious fact is that although, as we shall see later, the principal curvature directions *are* determined intrinsically at each point by purely algebraic operations in the generic case for higher dimensional minimal hypersurfaces, they are *not* for two-dimensional minimal surfaces. On the other hand, what can be determined in the two-dimensional case is the *variation* of the principal curvature directions. In other words, one can single out a frame field with the property that after the surface is immersed, that frame field will correspond to the principal frame field rotated at each point by a fixed angle (independent of the point). Namely, examining the above proof, one sees that the function

α is determined up to a constant by (17), and then the rotation angle θ is determined up to a constant by the fact that $\alpha - 2\theta$ must be constant in order to satisfy (15). We see from (13) that the hypothesis $\hat{K} \equiv 1$ (or equivalently, by Lemma 1, the equation (1) for K) is precisely the integrability condition (14) that is indeed to solve the equation (15) together with (10) and (11), and hence to obtain the desired surface.

We next give the generalization to minimal hypersurfaces of the first half of Theorem 1.2.: the necessary conditions for realization of a metric on a minimal hypersurface.

Proposition 1.3. Let M be a minimal hypersurface in E^{n+1}, and let ds^2 be the induced metric. Denote by Ric_M the Ricci form associated to the metric ds^2. Then

a) Ric_M is negative semi-definite, and
b) at every point where Ric_M is negative definite, the metric $d\hat{s}^2$ defined by

$$d\hat{s}^2 = -\mathrm{Ric}_M$$

has constant sectional curvature $\hat{K} \equiv 1$.

Remark. For the case $n = 2$, $\mathrm{Ric}_M = K\, ds^2$. Thus, properties a) and b) above reduce precisely (via Lemma 1.1) to properties a) and b) respectively of Theorem 1.2.

Proof. The basic observation is that if we denote by III the third fundamental form of a hypersurface, that is to say, the pull-back under the Gauss map of the metric on the unit sphere, then for a minimal hypersurface M one has

$$\mathrm{Ric}_M = -\mathrm{III}. \tag{18}$$

Since, by its definition, III is positive semi-definite, part a) follows immediately. Furthermore, it follows from (18) that Ric_M is negative definite wherever the Gauss map is regular, and at such points the quadratic form

$$d\hat{s}^2 = -\,\mathrm{Ric}_M = \mathrm{III}$$

represents the metric of the unit sphere, which has constant sectional curvature $\hat{K} \equiv 1$, proving part b).

As for equation (18), a derivation will be included as part of our discussion of hypersurfaces at the beginning of §3.

We next review briefly the situation for two-dimensional surfaces in E^n. If the codimension is not prescribed, then a complete answer is possible, and is quite easy to formulate. The results are due to Calabi [4, 5], and in a later version, to Lawson [14, 15].

The key observation is that a metric ds^2 can be realized on a minimal surface in E^n if and only if it can be realized locally on a holomorphic curve in some $\mathbb{C}^m$. Namely, a holomorphic curve in $\mathbb{C}^m$ *is* a minimal surface in E^{2m} so that all metrics on holomorphic curves are metrics on minimal surfaces. Conversely, given a simply-connected domain on a minimal surface in E^n, if we represent it in isothermal parameters by a map $x(w)$, then each of the coordinate functions x_k is harmonic and has a harmonic conjugate y_k. The map $(x(w) + iy(w))/\sqrt{2}$ is then a holomorphic curve in $\mathbb{C}^n$ and has the same metric as the original minimal surface.

Thus the problem reduces to describing intrisically those metrics that arise on holomorphic curves. There the answer has been given by Calabi [4]. We give the following formulation, based on Lawson [15].

Theorem 1.4. Give a metric ds^2 with Gauss curvature K, define inductively the quantities

(19) $$K_1 = -K$$

(20) $$K_2 = \tfrac{1}{2}\Delta \log K_1 + 3K_1$$

(21) $$K_3 = \tfrac{1}{2}\Delta \log K_2 + 2K_2$$

$$\vdots$$

(22) $$K_{j+1} = \tfrac{1}{2}\Delta \log K_j + 2K_j - K_{j-1} + K_1.$$

If the metric arises on a holomorphic curve that lies in $\mathbb{C}^n$ and in no proper subspace, then each of the K_j, $j = 1, \ldots, n-1$, is positive except at isolated points, and $K_n \equiv 0$. Conversely, if each of the K_j defined above is positive for $j = 1, \ldots, n-1$ (thereby allowing the succeeding K_{j+1} to be defined), with $K_n \equiv 0$, then ds^2 can be realized as the metric on a holomorphic curve in $\mathbb{C}^n$.

Note the close analogy between this theorem and Theorem 1.2. Condition a) of Theorem 1.2. is simply that $K_1 \geq 0$, which in Theorem 1.4. extends to the inequalities $K_1 \geq 0, \ldots, K_{n-1} \geq 0$. Condition b) of Theorem 1.2. is replaced by the condition $K_n \equiv 0$, which is a differential equation for K involving iterated Laplacians.

Examining more closely the relation between Theorem 1.2. and Theorem 1.4., we find that in view of (19), equation (1) is equivalent to

(23) $$K_2 = K_1.$$

Substituting this in (21) and using (1) once more, we find $K_3 \equiv 0$. Thus, the hypotheses of Theorem 1.2. apply to the case $n = 3$ of Theorem 1.4. and allow us to deduce that the metric is induced on a holomorphic curve in $\mathbb{C}^3$, which is a minimal surface in E^6. One can in fact argue further, using another application of equation (1), that there exists locally an isometric image of the surface in E^3, thereby obtaining a different proof of Ricci's Theorem. For details, we refer to Lawson [14], pp. 165–167.

Note that if we fix the codimension in advance, then the problem becomes much more difficult. In view of the above result, the question reduces to that of finding all minimal surfaces isometric to a fixed holomorphic curve. A characterization of the space of such minimal surfaces has been given by Calabi [5]. (See also Lawson [14], pp. 153–158). In particular, Calabi shows that the range of integers n for which there exist minimal surfaces lying fully in E^n isometric to a holomorphic curve lying fully in $\mathbb{C}^m$ satisfies

$$m \leq n \leq 2m.$$

Although Theorem 1.4. provides necessary and sufficient conditions for the realization of a metric on a minimal surface, it would obviously be impractical to apply in the case of high codimension. It is therefore useful to have other conditions that are only necessary ones, but are easily verified. An important example is the following analog of Theorem 1.2.

Proposition 1.5. (Barbosa—Do Carmo [2]).
Let ds^2 be the metric induced on a two-dimensional minimal surface in E^n for some $n \geq 3$. Let K be the Gauss curvature. Then

a) $K \leq 0$

and

b) wherever $K < 0$, the metric $d\hat{s}^2 = -K\,ds^2$ has Gauss curvature $\hat{K}$ satisfying $\hat{K} \leq 2$.

Note that the conditions are totally independent of the value of the codimension (expect for codimension 1, where the stronger condition $\hat{K} \equiv 1$ of Theorem 1.2. is valid). The upper bound 2 is sharp for all $n \geq 4$, although the extreme case $\hat{K} \equiv 2$ occurs only when the original surface lies in some E^4 and is the real form of a holomorphic curve lying in $\mathbb{C}^2$ (see Hoffman-Osserman [12], §5).

The proof of Proposition 1.5. is based on the use of the generalized Gauss map into the Grassmannian of oriented 2-planes in E^n. That Grassmannian may be identified with the quadric Q_{n-2} in $\mathbb{C}P^{n-1}$, with the metric induced by the Fubini-Study metric in $\mathbb{C}P^{n-1}$ with constant holomorphic sectional curvature $\bar{K} \equiv 2$. The basic facts are

i) for a minimal surface S in E^n, the Gauss map into $\mathbb{C}P^{n-1}$ is anti-holomorphic (Chern [6]);

ii) the pull-back to S of the metric on $\mathbb{C}P^{n-1}$ satisfies

$$d\hat{s}^2 = -K\,ds^2 \tag{24}$$

(see Chern-Osserman [7]);

iii) the Gauss curvature of a holomorphic (or antiholomorphic) curve in a Kähler manifold is bounded above at each point by the corresponding sectional curvature of the ambient manifold at the point.

Combining these three facts (and the normalization $\bar{K} \equiv 2$, needed for (24)) yields Proposition 1.5.

2. Local Existence and Uniqueness Theorems

An obvious remark is that in order for a metric to be realizable on a minimal submanifold of E^n, it must first of all be realizable on *some* submanifold of E^n. By Nash's Theorem, that is no restriction if one allows the codimension to be large. But for small codimension it is a severe restriction. For example, for codimension one and dimension at least three, one cannot have at any point all sectional curvatures negative (by equation (25a) in §3, since each pair of k_i, k_j would have to have opposite signs... In fact the same equation shows that there always exists a frame field for which at least one third of the coordinate planes have non-negative sectional curvature).

The basic existence theorem states that given an n-dimensional metric, it is locally realizable in codimension p if and only if the fundamental equations of an immersion (the integrability conditions (26), (28), (29) below) can be solved. In many cases one also has a *rigidity* theorem, such as the Beez theorem for hypersurfaces and its generalization by Allendoerfer [1], stating that if an immersion exists, then it is unique up to euclidean motions. Wherever rigidity holds (see Proposition 2.3 below), the general Ricci problem (for fixed codimension) is in a sense superseded by the pair of questions: does there exist *any* isometric immersion in the given codimension, and if so, is the resulting submanifold minimal?

Turning to the general question of existence of immersions, we first note a curious fact. For codimension one, the basic integrability conditions are the Gauss curvature equations (9) and the Codazzi equations (14) in §3. Given the metric, the former are purely algebraic equations for the second fundamental form, whereas the latter are differential equations. However, when the dimension n is at least 4, it turns out that generically any solution of the Gauss curvature equation will automatically satisfy the Codazzi equation (Thomas [19]. See also Eisenbart [10], Appendix 22). Thus the problem becomes a purely algebraic one. We shall use that fact in the following section to show how the Ricci problem may be decided in the generic case for dimension $n \geq 4$ and codimension one.

A generalization of Thomas's theorem to higher codimension was given by Allendoerfer [1]. Although this result appears in the same paper as the well-known, and often-quoted rigidity theorem cited above, it appears to be not well-known. Since the theorem seems to us a fairly basic one, and since Allendoerfer's proof may be somewhat inaccessible to modern readers, we include a proof here using standard frame notation and methods.

We start by fixing the notation.

Let $xe_1 \ldots e_{n+p}$ be an orthonormal frame in E^{n+p}. For a smooth family of orthonormal frames or for the space of all orthonormal frames we have

(1) $$dx = \sum \tilde{\omega}_A e_A,$$

(2) $$de_A = \sum \tilde{\omega}_{AB} e_B, \qquad 1 \leq A, B, C \leq n + p,$$

where

(3) $$\tilde{\omega}_{AB} + \tilde{\omega}_{BA} = 0.$$

Exterior differentiation gives

(4) $$d\tilde{\omega}_A = \sum \tilde{\omega}_B \wedge \tilde{\omega}_{BA},$$

(5) $$d\tilde{\omega}_{AB} = \sum \tilde{\omega}_{AC} \wedge \tilde{\omega}_{CB}.$$

These are the structure equations of the group of motions of E^{n+p}. Equations (5) are the structure equations of the group $O(n + p)$ of orthogonal transformations in $n + p$ variables, which is the group of motions of E^{n+p} keeping a point 0 fixed.

The manifold of all the n-dimensional linear subspaces L through O is called the Grassmann manifold, to be denoted by $G(n, p)$, n being the dimension and p the codimension of L. $G(n, p)$ is of dimension np. By sending $Oe_1 \ldots e_{n+p}$ to the linear space $L = \{e_1, \ldots, e_n\}$ spanned by the first n vectors, we define a mapping

(6) $$O(n + p) \to G(n, p).$$

Consider a submanifold M of dimension n in E^{n+p}. Let $Q(M)$ (respectively P(M)) be the family of all frames $xe_1 \ldots e_{n+p}$ (respectively $xe_1 \ldots e_p$) such that $e_1, \ldots, e_n$ are tangent vectors to M at x. The basic situation is described by the Gauss diagram

$$\begin{array}{ccc} Q(M) & \xrightarrow{\tilde{\tilde{g}}} & O(n+p) \\ \lambda \downarrow & & \downarrow \lambda_0 \\ P(M) & \xrightarrow{\tilde{g}} & St(n, p) \\ \pi \downarrow & & \downarrow \pi_0 \\ M & \xrightarrow{g} & G(n, p), \end{array} \tag{7}$$

where

$$\begin{aligned} \lambda(xe_1 \ldots e_{n+p}) &= xe_1 \ldots e_n, & \pi(xe_1 \ldots e_n) &= x \\ \lambda_0(Oe_1 \ldots e_{n+p}) &= Oe_1 \ldots e_n, & \pi_0(Oe_1 \ldots e_n) &= \{e_1, \ldots, e_n\}, \end{aligned} \tag{8}$$

and $\tilde{\tilde{g}}$ sends $xe_1 \ldots e_{n+p}$ to $Oe_1 \ldots e_{n+p}$, etc. The mapping g is usually called the Gauss mapping, which maps $x \in M$ to the n-dimensional linear space through 0 which is parallel to the tangent space to M at x. We will write

$$\omega_{AB} = \tilde{\tilde{g}}^* \tilde{\omega}_{AB}, \tag{9}$$

and we will omit the pull-backs λ^*, π^* in our notations.

Since $e_1, \ldots, e_n$ are tangent vectors to M at x, the pull-back of (1) becomes

$$dx = \sum \omega_i e_i, \tag{10}$$

where, and throughout this section, we will agree on the following ranges of indices

$$1 \le i, j, l \le n, \qquad n + 1 \le \alpha, \beta, \gamma \le n + p. \tag{11}$$

We write the pull-back of (2) explicitly as

$$\begin{aligned} de_i &= \sum \omega_{ij} e_j + \sum \omega_{i\beta} e_\beta, \\ de_\alpha &= \sum \omega_{\alpha j} e_j + \sum \omega_{\alpha\beta} e_\beta. \end{aligned} \tag{12}$$

From (4) we get

$$\sum_i \omega_i \wedge \omega_{i\alpha} = 0.$$

By Cartan's lemma it follows that

$$\omega_{i\alpha} = \sum h_{ij\alpha} \omega_j, \qquad h_{ij\alpha} = h_{ji\alpha}. \tag{13}$$

We note that

$$\omega_{i\alpha} = -\omega_{\alpha i} = (de_i, e_\alpha) = -(de_\alpha, e_i).$$

Under changes of frames in the tangent and normal spaces:

$$e_i' = \sum u_{ij} e_j, \qquad e_\alpha' = \sum u_{\alpha\beta} e_\beta, \tag{15}$$

where (u_{ij}) and $(u_{\alpha\beta})$ are orthogonal matrices, we have

$$\omega_{i\alpha}' = (de_i', e_\alpha') = \sum u_{ij} u_{\alpha\beta} \omega_{j\beta}. \tag{16}$$

We have also

$$\omega_i' = (dx, e_i') = \sum u_{ij} \omega_j. \tag{17}$$

It follows that

$$\sum \omega_i' \omega_{i\alpha}' = \sum u_{\alpha\beta} \omega_j \omega_{j\beta}. \tag{18}$$

The quadratic differential forms

$$\mathrm{II}_\alpha := \sum \omega_i \omega_{i\alpha} = \sum h_{ij\alpha} \omega_i \omega_j \tag{19}$$

are called the second fundamental forms. If

$$v = \sum v_\alpha e_\alpha = \sum v_\alpha' e_\alpha' \tag{20}$$

is a normal vector, then

$$\mathrm{II}(v) = \sum v_\alpha \mathrm{II}_\alpha \tag{21}$$

is the second fundamental form in the direction v; the second fundamental form is thus a normal-valued quadratic differential form. In fact, differentiating (10), we have

$$(d^2 x, v) = \mathrm{II}(v), \tag{22}$$

so that $\mathrm{II}(v)$ is the projection of the second differential in the normal vector v. M is called a *minimal* submanifold if

$$\operatorname{Tr} \mathrm{II}(v) = 0 \tag{23}$$

for all normal vectors v, i.e., if

$$\sum_i h_{ii\alpha} = 0. \tag{23a}$$

The pull-back of the other structure equations in (4) gives

$$d\omega_i = \sum \omega_j \wedge \omega_{ji}, \tag{24}$$

which shows that (ω_{ij}) defines the torsionless connection, and therefore the Levi-Civita connection, in the tangent bundle. We write the pull-backs of the equations (5) in three sets:

$$d\omega_{ij} = \sum \omega_{ik} \wedge \omega_{kj} + \Omega_{ij}, \tag{25}$$

$$d\omega_{i\alpha} = \sum \omega_{ij} \wedge \omega_{j\alpha} + \sum \omega_{i\beta} \wedge \omega_{\beta\alpha}, \tag{26}$$

$$d\omega_{\alpha\beta} = \sum \omega_{\alpha\gamma} \wedge \omega_{\gamma\beta} + \Omega_{\alpha\beta}, \tag{27}$$

where

$$\Omega_{ij} = -\sum_\alpha \omega_{i\alpha} \wedge \omega_{j\alpha}, \tag{28}$$

$$\Omega_{\alpha\beta} = -\sum_i \omega_{i\alpha} \wedge \omega_{i\beta}. \tag{29}$$

Equation (25) defines Ω_{ij} as the curvature form of the Levi-Civita connection on M, and (28), called the Gauss equation, expresses this curvature in terms of the second fundamental form. Similarly, $(\omega_{\alpha\beta})$ defines a connection in the normal bundle and (29), called the Ricci equation, expresses its curvature in terms of II. Equation (26) is the Codazzi equation.

Suppose that among the II_α there are q independent ones. We can choose the frames so that

$$\mathrm{II}_{n+q+1} = \cdots = \mathrm{II}_{n+p} = 0. \tag{30}$$

Then $e_1, \ldots, e_{n+q}$ span the osculating space of M at x, i.e., the space spanned by the osculating planes (at x) of all the curves on M through x. The normal space spanned by $e_{n+1}, \ldots, e_{n+q}$ is called the *first normal space* of M at x. The matrix

$$\omega = (\omega_{ir}) = \begin{pmatrix} \omega_{1,n+1} \cdots \omega_{1,n+q} \\ \cdots \\ \omega_{n,n+1} \cdots \omega_{n,n+q} \end{pmatrix}, \qquad n+1 \leq r \leq n+q \tag{31}$$

plays a fundamental role in the extrinsic geometry of M. By (16), under a change of frames in the tangent space and the first normal space, the matrix ω undergoes the transformation

$$\omega \to \omega' = U\omega V, \tag{32}$$

where U and V are orthogonal matrices of orders n and q respectively. Among the fundamental extrinsic properties of M are those of ω which remain invariant under the transformation (32).

A fundamental quantity associated with the matrix (31) is its *type* τ, defined to be the maximum number of rows of (ω_{ir}) such that the τq forms in those rows are linearly independent. This notion was introduced by Allendoerfer [1], generalizing the type of a hypersurface, which is simply the rank of the second fundamental form. Note that the type is defined algebraically at each point, but that its definition depends on the dimension q of the first normal space. In all applications one needs the $\omega_{i\alpha}$ defined smoothly in a neighborhood, and that requires the dimension q of the first normal space to be constant in that neighborhood. We will make that assumption throughout.

We shall see later (Proposition 2.2) that although the dimension of the first normal space is clearly an extrinsic quantity, as soon as τ is at least two, q is intrinsically determined. (For more information on the notion of type and related matters, we refer to the recent survey paper of Gardner [11].)

We start with a useful lemma.

Lemma 2.1. Let ω in (31) be of type $\tau \geq 1$. Suppose that Φ_r are forms of degree $\leq \tau - 1$, such that

$$\sum \Phi_r \wedge \omega_{ir} = 0, \qquad 1 \leq i \leq n. \tag{33}$$

Then $\Phi_r = 0$.

Proof. For a fixed r we multiply (33) by

$$\omega_{i,n+1} \wedge \cdots \wedge \omega_{i,r-1} \wedge \omega_{i,r+1} \wedge \cdots \wedge \omega_{i,n+q}.$$

The result is

$$\Phi_r \wedge \omega_{i,n+1} \wedge \cdots \wedge \omega_{i,n+q} = 0. \tag{34}$$

Suppose the τq forms in the first τ rows of ω to be linearly independent. Then those τq forms can be completed to a basis of one-forms. Expressing Φ_r in that basis, we deduce from (34) that each term in Φ_r has a factor $\omega_{i\alpha}$ for some α, $n+1 \leq \alpha \leq n+q$. Since that is true for $1 \leq i \leq \tau$, it follows that if Φ_r is not zero, then its degree is at least τ.

Remark. Note that one does not need (33) to hold for all i, but just for τ rows with independent elements.

Proposition 2.2. Let M and M^* be isometric submanifolds of types ≥ 2. Then at corresponding points their first normal spaces have the same dimension.

Proof. Let the quantities pertaining to M^* be denoted by the same notation with asterisks. We will identify M and M^* by the isometry. Then their orthonormal tangent frames are identified, and we have

$$\omega_i^* = \omega_i, \qquad \omega_{ij}^* = \omega_{ij}, \qquad \Omega_{ij}^* = \Omega_{ij}, \qquad 1 \leq i, j \leq n.$$

By (28) the last equation gives

$$\sum_{n+1 \leq \alpha^* \leq n+q^*} \omega_{i\alpha^*}^* \wedge \omega_{j\alpha^*}^* = \sum_{n+1 \leq \alpha \leq n+q} \omega_{i\alpha} \wedge \omega_{j\alpha}.$$

Since M is of type ≥ 2, there exist i, j, such that the two-form at the right-hand side is of rank $2q$. It follows that $q^* \geq q$. By symmetry we have $q^* = q$, proving the result.

We next note that when $\tau \geq 2$, we can reduce the codimension from p to q.

Suppose the type $\tau \geq 2$. The frames can be chosen so that (30) holds. These equations can also be written

$$\omega_{iu} = 0,$$

where, as also later, we will use the ranges of indices

$$n+1 \leq r, s \leq n+q, \qquad n+q+1 \leq u, v \leq n+p.$$

It then follows from the Codazzi equations (26), that

$$\sum \omega_{ir} \wedge \omega_{ru} = 0, \qquad i = 1, \ldots, n.$$

Applying Lemma 2.1, we conclude $\omega_{ru} = 0$. Then equations (12) become

$$de_i = \sum \omega_{ij} e_j + \sum \omega_{ir} e_r,$$
$$de_r = \sum \omega_{rj} e_j + \sum \omega_{rs} e_s.$$

This means that the osculating space $xe_1 \ldots e_{n+q}$ is fixed, i.e., M lies on a linear space E^{n+q} of dimension $n+q$ of E^{n+p}.

In the following we will study submanifolds of type ≥ 2. We can suppose $E^{n+q} = E^{n+p}$, so that the first normal space at every point is the whole normal space.

When the type number is ≥ 3, we have the following theorem (cf. Spivak [18], pp. 364–7):

Proposition 2.3. Let M and M^* be isometric submanifolds of types ≥ 3. Then at corresponding points the second fundamental forms differ by an orthogonal transformation.

From this we can derive Allendoerfer's rigidity theorem that the isometry between M and M^* is the restriction of a Euclidean motion.

We come now to the main result, Allendoerfer's "Gauss implies Codazzi" Theorem.

Theorem 2.4. Let M be a Riemannian manifold. Let $e_1, \ldots, e_n$ be a local frame field, with corresponding dual forms ω_i, connection forms ω_{ij}, and curvature forms Ω_{ij}. Suppose that for some q there exists a matrix

$$\omega = (\omega_{i\alpha}), \qquad i = 1, \ldots, n, \qquad \alpha = n+1, \ldots, n+q,$$

of one-forms satisfying the Gauss equations (28), and suppose that the type τ of $(\omega_{i\alpha})$ satisfies $\tau \geq 4$. Then there exists a unique skew-symmetric matrix of one-forms

$$(\omega_{\alpha\beta}), \qquad \alpha, \beta = n+1, \ldots, n+q,$$

satisfying the Codazzi equations (26). The forms $\omega_{\alpha\beta}$ must then also satisfy the equations (27) and (29).

Corollary. Under the hypotheses of the theorem there exists locally an isometric immersion of M into E^{n+q}. By Proposition 2.3, that immersion is unique up to a rigid motion in E^{n+q}.

Proof. Define covariant derivatives:

$$D\omega_{i\alpha} = d\omega_{i\alpha} - \sum_j \omega_{ij} \wedge \omega_{j\alpha},$$

$$D\Omega_{ij} = d\Omega'_{ij} - \sum_k \omega_{ik} \wedge \Omega_{kj} + \sum \Omega_{ik} \wedge \omega_{kj}.$$

Then the Codazzi equations (26) may be written as

$$D\omega_{i\alpha} = \sum \omega_{i\beta} \wedge \omega_{\beta\alpha}, \tag{35}$$

while the Bianchi identity (obtained by taking the exterior derivative of the structure equation

$$d\omega_{ij} = \sum \omega_{ik} \wedge \omega_{kj} + \Omega_{ij} \tag{35}$$

and substituting this equation back in the result) is just

$$D\Omega_{ij} = 0. \tag{37}$$

But using (28) we have from (37) that

$$\begin{aligned} 0 = D\Omega_{ij} &= -\sum d\omega_{i\alpha} \wedge \omega_{j\alpha} + \sum \omega_{i\alpha} \wedge d\omega_{j\alpha} \\ &\quad + \sum \omega_{ik} \wedge \omega_{k\alpha} \wedge \omega_{j\alpha} - \sum \omega_{i\alpha} \wedge \omega_{k\alpha} \wedge \omega_{kj} \\ &= -\sum D\omega_{i\alpha} \wedge \omega_{j\alpha} + \sum \omega_{i\alpha} \wedge D\omega_{j\alpha}. \end{aligned} \tag{35}$$

Fix i and α, and write the above equation as

$$D\omega_{i\alpha} \wedge \omega_{j\alpha} = \sum_\beta \omega_{i\beta} \wedge D\omega_{j\beta} - \sum_{\beta \neq \alpha} D\omega_{i\beta} \wedge \omega_{j\beta}.$$

Multiplying both sides by $\bigwedge_\beta \omega_{i\beta} \wedge \bigwedge_{\beta\neq\alpha} \omega_{j\beta}$ gives

$$D\omega_{i\alpha} \wedge \bigwedge_\beta \omega_{i\beta} \wedge \bigwedge_\beta \beta_{j\beta} = 0. \tag{39}$$

Suppose now that the i, j and k^{th} rows of the matrix $(\omega_{i\alpha})$ consist of independent 1-forms. They can be completed to a basis of 1-forms, so that the 2-form $D\omega_{i\alpha}$ is a linear combination of terms of the form $\omega_{i\beta} \wedge \omega_{i\gamma}$, $\omega_{i\beta} \wedge \omega_{j\gamma}$, etc. By (39) the coefficient of any term containing neither an $\omega_{i\beta}$ or an $\omega_{j\beta}$ must be zero. However, (39) also holds with j replaced by k. It follows that every term in $D\omega_{i\alpha}$ contains a factor $\omega_{i\beta}$, so that

$$D\omega_{i\alpha} = \sum_\beta \pi_{i\alpha\beta} \wedge \omega_{i\beta}, \tag{40}$$

where $\pi_{i\alpha\beta}$ are 1-forms. The argument applies to the fixed value of i and any α. The same argument works with j or k instead of i. Thus (38) becomes

$$\begin{aligned} O &= -\sum_{\alpha,\beta} \pi_{i\alpha\beta} \wedge \omega_{i\beta} \wedge \omega_{j\alpha} + \sum_{\alpha,\beta} \omega_{i\alpha} \wedge \pi_{j\alpha\beta} \wedge \omega_{j\beta} \\ &= -\sum_{\alpha,\beta} (\pi_{i\alpha\beta} + \pi_{j\beta\alpha}) \wedge \omega_{i\beta} \wedge \omega_{j\alpha} \end{aligned} \tag{41}$$

and this implies that

$$\pi_{i\alpha\beta} + \pi_{j\beta\alpha} = \sum_\gamma a_{ij\alpha\beta\gamma}\omega_{i\gamma} + \sum_\gamma b_{ij\alpha\beta\gamma}\omega_{j\gamma}. \tag{42}$$

Interchanging i, j and α, β leaves the left-hand side of (42) fixed and gives

$$\pi_{i\alpha\beta} + \pi_{j\beta\alpha} = \sum_\gamma a_{ji\beta\alpha\gamma}\omega_{j\gamma} + \sum b_{ji\beta\alpha\gamma}\omega_{i\gamma}.$$

Hence $b_{ij\alpha\beta\gamma} = a_{ji\beta\alpha\gamma}$, and (42) becomes

$$\pi_{i\alpha\beta} + \pi_{j\beta\alpha} = \sum a_{ij\alpha\beta\gamma}\omega_{i\gamma} + \sum a_{ji\beta\alpha\gamma}\omega_{j\gamma}. \tag{43}$$

Permute $i \to j \to k \to i$ and $\alpha \to \beta \to \alpha$;

$$\begin{aligned} \pi_{j\beta\alpha} + \pi_{k\alpha\beta} &= \sum a_{jk\beta\alpha\gamma}\omega_{j\gamma} + \sum a_{kj\alpha\beta\gamma}\omega_{k\gamma} \\ \pi_{k\alpha\beta} + \pi_{i\beta\alpha} &= \sum a_{ki\alpha\beta\gamma}\omega_{k\gamma} + \sum a_{ik\beta\alpha\gamma}\omega_{i\gamma}. \end{aligned}$$

Add the first and third and subtract the second equation:

$$\begin{aligned} \pi_{i\alpha\beta} + \pi_{i\beta\alpha} &= \sum (a_{ij\alpha\beta\gamma} + a_{ik\beta\alpha\gamma})\omega_{i\gamma} \\ &\quad + \sum (a_{ji\beta\alpha\gamma} - a_{jk\beta\alpha\gamma})\omega_{j\gamma} \\ &\quad + \sum (a_{ki\alpha\beta\gamma} - a_{kj\alpha\beta\gamma})\omega_{k\gamma}. \end{aligned} \tag{44}$$

Now assuming that the l'th row is independent of the i, j, and k^{th} rows, we get the same equation with l instead of k, (i and j fixed). But the coefficient of $\omega_{i\gamma}$ depends only on i, α, β, and hence $a_{ik\beta\alpha\gamma} = a_{il\beta\alpha\gamma}$. The same holds with any permutation of i, j, k, l, and also with α, β transposed. Hence we may write (43) as

$$\pi_{i\alpha\beta} + \pi_{j\beta\alpha} = \sum c_{i\alpha\beta\gamma}\omega_{i\gamma} + \sum c_{j\beta\alpha\gamma}\omega_{j\gamma}, \tag{45}$$

where $c_{i\alpha\beta\gamma} = a_{ij\alpha\beta\gamma}$, or setting

$$(46)\qquad \begin{aligned} \pi'_{i\alpha\beta} &= \pi_{i\alpha\beta} - \sum c_{i\alpha\beta\gamma}\omega_{i\gamma}, \\ \pi'_{i\alpha\beta} + \pi'_{j\beta\alpha} &= 0, \qquad i \neq j. \end{aligned}$$

Also, (44) becomes

$$\pi'_{i\alpha\beta} + \pi'_{i\beta\alpha} = 0.$$

Thus $\pi'_{i\alpha\beta}$ is independent of i, and setting

$$\pi'_{i\alpha\beta} = \lambda_{\alpha\beta},$$

we have

$$(47)\qquad \lambda_{\alpha\beta} + \lambda_{\beta\alpha} = 0.$$

But

$$\pi_{i\alpha\beta} = \lambda_{\alpha\beta} + \sum c_{i\alpha\beta\gamma}\omega_{i\gamma}$$

and (40) becomes

$$(48)\qquad D\omega_{i\alpha} = \sum_{\beta} \lambda_{\alpha\beta} \wedge \omega_{i\beta} + \sum_{\beta,\gamma} c_{i\alpha\beta\gamma}\omega_{i\gamma} \wedge \omega_{i\beta}.$$

If we now substitute (45) into (41), we find that for fixed γ, β with $\gamma < \beta$, the term involving $\omega_{i\gamma} \wedge \omega_{i\beta} \wedge \omega_{j\alpha}$ occurs twice, and the sum of two coefficients is $c_{i\alpha\beta\gamma} - c_{i\alpha\gamma\beta}$. It follows that the difference is zero, and hence the second sum on the right of (48) is also zero. We thus have

$$(49)\qquad D\omega_{i\alpha} = \sum_{\beta} \lambda_{\alpha\beta} \wedge \omega_{i\beta},$$

with the $\lambda_{\alpha\beta}$ skew-symmetric by (47). But then the independence of the $\omega_{i\beta}$ implies that the $\lambda_{\alpha\beta}$ are uniquely determined. Hence the $\lambda_{\alpha\beta}$ are the $\omega_{\alpha\beta}$ of the theorem and we have shown that the Codazzi equations (35) hold for certain values of the index i. Namely, since we are assuming that the type is at least 4, we may re-order the indices so that the forms in the first four rows of the matrix $\omega_{i\alpha}$ are independent. We have then proved that (35) holds for $i = 1, 2, 3, 4$. We show now that it must also hold for all values of i. But for any i between 1 and 4 and for any $j > 4$, we may substitute (35) in (38), and we find

$$\sum \omega_{i\beta} \wedge \omega_{\beta\alpha} \wedge \omega_{j\alpha} - \sum \omega_{i\alpha} \wedge D\omega_{j\alpha} = 0.$$

Relabelling summation indices, we may write this as

$$\sum \omega_{i\beta} \wedge \Phi_{j\beta} = 0.$$

where

$$\Phi_{j\beta} = \sum \omega_{j\alpha} \wedge \omega_{\alpha\beta} - D\omega_{j\beta}.$$

Applying Lemma 2.1 and the Remark following it, we conclude that $\Phi_{j\beta} = 0$, which is precisely the Codazzi equation for $j > 4$.

The final step of the proof is to show that the forms $\omega_{\alpha\beta}$ also satisfy equations (27) and (29); that is, we must show

$$(50)\qquad d\omega_{\alpha\beta} = \sum \omega_{\alpha\gamma} \wedge \omega_{\gamma\beta} - \sum \omega_{j\alpha} \wedge \omega_{j\beta}.$$

But taking the exterior derivative of (26), and substituting (25), (26), (28) in the resulting equation yields

$$\begin{aligned}
0 &= \sum d\omega_{ij} \wedge \omega_{j\alpha} - \sum \omega_{ij} \wedge d\omega_{j\alpha} + \sum d\omega_{i\beta} \wedge \omega_{\beta\alpha} - \sum \omega_{i\beta} \wedge d\omega_{\beta\alpha} \\
&= \sum \omega_{ik} \wedge \omega_{kj} \wedge \omega_{j\alpha} - \sum \omega_{i\beta} \wedge \omega_{j\beta} \wedge \omega_{j\alpha} \\
&\quad - \sum \omega_{ij} \wedge \omega_{jk} \wedge \omega_{k\alpha} - \sum \omega_{ij} \wedge \omega_{j\beta} \wedge \omega_{\beta\alpha} \\
&\quad + \sum \omega_{ij} \wedge \omega_{j\beta} \wedge \omega_{\beta\alpha} + \sum \omega_{i\gamma} \wedge \omega_{\gamma\beta} \wedge \omega_{\beta\gamma} - \sum \omega_{i\beta} \wedge d\omega_{\beta\alpha} \\
&= \sum \omega_{i\beta} \wedge \left(\sum \omega_{\beta\gamma} \wedge \omega_{\gamma\alpha} - \sum \omega_{j\beta} \wedge \omega_{j\alpha} - d\omega_{\beta\alpha}\right)
\end{aligned}$$

since in the middle expression the first and third terms cancel, as do the fourth and fifth. Applying Lemma 2.1 once again gives equation (50) and completes the proof of the theorem.

We conclude this section with some remarks and examples concerning the "genericity" of the assumptions involving type in the previous theorems.

As noted earlier, for hypersurfaces the type is simply rank of the second fundamental form. Since in the generic case the second fundamental form has maximum rank, the type of a hypersurface is generically equal to the dimension of the manifold.

In the other direction, if we consider manifolds of arbitrarily high codimension, then the type will generically be equal to zero. The reason for that is that the dimension of the first normal space is generically greater than the dimension of the manifold, and hence and hence even *one* row of $\omega_{i\alpha}$, $\alpha = 1, \ldots, q$, cannot have all independent forms. On the other hand, if the codimension is restricted to the range where a given type τ is possible, then one might expect the type τ to be realized generically. For example, in order to have type 4, one needs the codimension to be at most one fourth the dimension. The first examples with codimension greater than one are therefore 8-dimensional manifolds in a 10-dimensional space. Among minimal submanifolds we have the class of complex hypersurfaces in $\mathbb{C}^5$. There it turns out that the type-4 hypothesis is indeed generic. Specifically, we have the following:

Proposition 2.5. Let M be a complex hypersurface in $\mathbb{C}^5$ considered as a real, 8-dimensional submanifold of E^{10}. Then M is of type 4 at all points where the complex second fundamental form is non-singular.

Proof. Given any point p of M, we may assume after an isometry of $\mathbb{C}^5$, that the point p is at the origin and that M may be represented locally in the form

$$z_5 = \frac{1}{2} \sum_{j=1}^{4} a_j z_j^2 + \text{higher order terms.}$$

The complex second fundamental form is non-singular at p if and only if all a_j are non-zero.

Choose a frame field such that at the origin:

$$e_1 = \frac{\partial}{\partial x_1}, \qquad e_2 = \frac{\partial}{\partial y_1}, \ldots, e_9 = \frac{\partial}{\partial x_5}, \qquad e_{10} = \frac{\partial}{\partial y_5}; \qquad z_j = z_j + iy_j.$$

Then the real second fundamental form is given by

$$\omega_{i\alpha} = \sum_{j=1}^{8} h_{ij\alpha}\omega_j, \qquad i = 1, \ldots, 8; \qquad \alpha = 9, 10,$$

where

$$h_{ij9} = \frac{\partial^2 x_5}{\partial u_i \partial u_j}(0), \qquad h_{ij,10} = \frac{\partial^2 y_5}{\partial u_i \partial u_j}(0); \qquad u_{2k-1} = x_k, \qquad u_{2k} = y_k.$$

Setting $a_j = \alpha_j + i\beta_j$, we find

$$x_5 = \sum_{j=1}^{4} [\tfrac{1}{2}\alpha_j(x_j^2 - y_j^2) - \beta_j x_j y_j] + \cdots,$$

$$y_5 = \sum_{j=1}^{4} [\alpha_j x_j y_j + \tfrac{1}{2}\beta_j(x_j^2 - y_j^2)] + \cdots,$$

so that at 0:

$$\frac{\partial^2 x_5}{\partial x_j^2} = \alpha_j = -\frac{\partial^2 x_5}{\partial y_j^2}, \qquad \frac{\partial^2 x_5}{\partial x_j \partial y_j} = -\beta_j,$$

$$\frac{\partial^2 x_5}{\partial x_j^2} = \beta_j = -\frac{\partial^2 x_5}{\partial y_j^2}, \qquad \frac{\partial^2 y_5}{\partial x_j \partial y_j} = -\alpha_j,$$

while all the other mixed derivatives are zero. Hence the matrix $(\omega_{i\alpha})$ takes the form

$$\begin{pmatrix} a_1\omega_1 - \beta_1\omega_2 & \beta_1\omega_1 - \alpha_2\omega_2 \\ -\beta_1\omega_1 - \alpha_1\omega_2 & \alpha_1\omega_1 - \beta_1\omega_2 \\ \alpha_2\omega_3 - \beta_2\omega_4 & \beta_2\omega_3 + \alpha_2\omega_4 \\ -\beta_2\omega_3 - \alpha_2\omega_4 & \alpha_2\omega_1 - \beta_2\omega_4 \\ \vdots & \vdots \end{pmatrix}$$

from which we conclude immediately:

1. any set of 4 rows containing both an odd and an even row contains dependent forms,
2. the 4 even rows and the 4 odd rows separately consist of independent forms if and only if all a_j are non-zero,

and hence

3. the type of $(\omega_{i\alpha})$ is 4 if and only if the complex second fundamental form is non-singular.

3. Minimal Hypersurfaces

We now specialize to the case of codimension one. Many of the considerations of the previous section simplify considerably in that case, including the proof of Theorem 2.4., which the reduces to the original theorem of Thomas [19].

Let us review now the basic formulas, as they appear in the case of hypersurfaces.

First of all, starting with an aribtrary metric on an n-dimensional Riemannian manifold, and choosing a local orthonormal frame field $e_1, \ldots, e_n$, with dual forms $\omega_1, \ldots, \omega_n$, there exists a unique set of ω_{ij} satisfying $\omega_{ji} = -\omega_{ij}$, and the structure equations

$$d\omega_i = \sum_{j=1}^{n} \omega_{ij} \wedge \omega_j, \qquad i = 1, \ldots, n. \tag{1}$$

In terms of those ω_{ij} one defines the curvature forms

$$\Omega_{ij} = d\omega_{ij} - \sum_{k=1}^{n} \omega_{ik} \wedge \omega_{kj}, \tag{2}$$

the components of the Riemann curvature tensor

$$R_{ijkl} = -\Omega_{ij}(\omega_k, \omega_l) \tag{3}$$

and the Ricci tensor

$$R_{ik} = \sum_{j=1}^{n} R_{ijkj}. \tag{4}$$

Finally, the Ricci form is defined by

$$\mathrm{Ric}_M = \sum_{i,j=1}^{n} R_{ij}\omega_i\omega_j. \tag{5}$$

In the case that M is a hypersurface of E^{n+1}, we start with an adapted orthonormal frame field along M: $e_1, \ldots, e_{n+1}$, where e_{n+1} is a unit normal vector. The Weingarten equation

$$de_{n+1} = -\sum_{i=1}^{n} \omega_{i,n+1} e_i \tag{6}$$

defines the forms $\omega_{i,n+1}$, $i = 1, \ldots, n$, on M essentially as the differential of the Gauss map $x \mapsto e_{n+1}(x)$ of M into the unit sphere. The equation

$$\omega_{i,n+1} = \sum_{j=1}^{n} h_{ij}\omega_j \tag{7}$$

defines the coefficients h_{ij} of the second fundamental form relative to the given frame field. In fact, we may define the first, second, and third fundamental forms by

$$\mathrm{I} = ds^2 = \sum_{i=1}^{n} \omega_i^2.$$

$$\mathrm{II} = \sum_{i=1}^{n} \omega_i\omega_{i,n+1} = \sum_{i,j=1}^{n} h_{ij}\omega_i\omega_j, \tag{8}$$

$$\mathrm{III} = \sum_{i=1}^{n} \omega_{i,n+1}^2 = \sum_{i,j,k=1}^{n} h_{ij}h_{ik}\omega_j\omega_k.$$

The Gauss curvature equations relate the intrinsic quantities R_{ijkl} to the coefficients of the second fundamental form h_{ij} by

(9) $$R_{ijkl} = h_{ik}h_{jl} - h_{il}h_{jk}.$$

Hence from (4),

(10) $$R_{ik} = \left(\sum_{j=1}^{n} h_{jj}\right) h_{ik} - \sum_{j=1}^{n} h_{ij}h_{jk}.$$

Finally, the mean curvature H of M is given by

(11) $$H = \frac{1}{n}\sum_{i=1}^{n} h_{ii}.$$

Combining (8), (10), (11), and using the symmetry of h_{ij}, gives the basic equation

(12) $$\mathrm{Ric}_M = n\,H\mathrm{II} - \mathrm{III}.$$

In particular, when M is minimal, $H \equiv 0$ and we have

(13) $$\mathrm{Ric}_M = -\mathrm{III},$$

which was equation (18) in the proof of Proposition 1.3.

The Codazzi equations (equation (26) of the previous section) now take the simple form

(14) $$d\omega_{i,n+1} = \sum_{j=1}^{n} \omega_{ij} \wedge \omega_{j,n+1}, \qquad i = 1, \ldots, n.$$

Note that by (6) and (7), the following conditions are equivalent at any point $p \in M$:

a) $\omega_{1,n+1}, \ldots, \omega_{n,n+1}$ are independent,
b) $\det(h_{ij}) \neq 0$,
c) the Gauss map is regular.

At such a point, if we set

(15) $$\hat{\omega}_i = \omega_{i,n+1},$$

then the metric

(16) $$d\hat{s}^2 = \sum_{i=1}^{n} \hat{\omega}_i^2 = \sum_{i=1}^{n} \omega_{i,n+1}^2 = \mathrm{III} = -\mathrm{Ric}_M$$

is the pull-back under the Gauss map of the metric on the unit sphere. For this metric the $\hat{\omega}_i$ form an orthonormal co-frame fields, and again there is a unique set of skew-symmetric connection forms $\hat{\omega}_{ij}$ satisfying

(17) $$d\hat{\omega}_i = \sum_{j=1}^{n} \hat{\omega}_{ij} \wedge \hat{\omega}_j, \qquad i = 1, \ldots, n,$$

with corresponding curvature forms

(18) $$\hat{\Omega}_{ij} = d\hat{\omega}_{ij} - \sum_{k=1}^{n} \hat{\omega}_{ik} \wedge \hat{\omega}_{kj}.$$

In view of (9), (10), (11), we may make the following observation:
The Codazzi equations (8) *are precisely equivalent to the equality*

(19) $$\hat{\omega}_{ij} = \omega_{ij}$$

between the connection forms ω_{ij} associated with the original co-frame field $\{\omega_i\}$ on M and the connection forms ω_{ij} associated to the co-frame field $\hat{\omega}_i = \omega_{i,n+1}$ of the metric (16) *defined by the third fundamental form; that is, the pull-back under the Gauss map of the metruc in the unit sphere* (assuming again that the Gauss map is regular).

Of course, an immediate consequence of (19) is the equality of the curvature forms:

(20) $$\hat{\Omega}_{ij} = \Omega_{ij}.$$

We next note that in view of (3) and (7) the Gauss curvature equation (9) may be written in the form

(21) $$\Omega_{ij} = -\omega_{i,n+1} \cap \omega_{j,n+1},$$

or by (15):

(22) $$\Omega_{ij} = -\hat{\omega}_i \wedge \hat{\omega}_j.$$

On the other hand, the equations

(23) $$\hat{\Omega}_{ij} = -\hat{\omega}_i \wedge \hat{\omega}_j$$

are precisely equivalent to the condition that the metric $d\hat{s}^2$ has constant sectional curvature $+1$.

Comparing equations (19), (20), (22), (23), we may summarize the situation as follows:

i) the Codazzi equations (14) are equivalent to (19);
ii) the equations (19) imply (20);
iii) in the presence of (20), the Gauss curvature equations (22) are equivalent to (23), which assert that the metric defined by the third fundamental form has constant curvature $+1$. Conversely, (22) and (23) together imply (20).

All these remarks hold for an arbitrary hypersurface in E^{n+1}. In the case of a minimal hypersurface, we know from (13) that the metric defined by the third fundamental form may be defined intrinsically as the negative of the Ricci form. That led us to the necessary conditions of Proposition 1.3.

We now ask whether those necessary conditions are sufficient, and if not, whether we can find additional necessary conditions.

What we must do, given a metric ds^2, is to find a second fundamental form matrix h_{ij} associated to a given frame field, such that the Gauss and Codazzi equations, (9), and (14), are satisfied. By the remarks above, if the Gauss equations are satisfied, then (22) holds, whereas the necessary conditions of Proposition 1.3. imply (23). As a consequence, (20) holds too. The question is whether we can reverse our steps and deduce the Codazzi equations in the form (19) from (20).

In the case of dimension $n = 2$, the second term on the right of (2) vanishes, and equation (20) reduces to

(24) $$d\hat{\omega}_{ij} = d\omega_{ij}.$$

As we saw in the proof of Theorem 1.2., one *cannot* deduce (19) from (24). On the other hand, starting from (24), one can choose a new frame field in which (19) does hold.

When $n = 3$, it seems unlikely that one could prove (19) without imposing further conditions. On the other hand, for $n \geq 4$ we can apply Thomas's theorem of the previous section to deduce the Codazzi equations from the Gauss equations. We therefore turn next to the question of solving the Gauss equations.

The basic observation is that at each point $p \in M$ we may diagonalize the second fundamental form matrix h_{ij} to obtain a principal curvature frame $e_1, \ldots, e_n$ with corresponding eigenvalues equal to the principal curvatures, which we may choose in order of decreasing magnitude:

(25) $$k_1 \geq k_2 \geq \cdots \geq k_n$$

The Gauss equations (9) then take the form

(26a) $$R_{ijij} = k_i k_j, \qquad i \neq j,$$

(26b) $$R_{ijkl} = 0, \qquad \text{unless } (k, l) = (i, j) \text{ or } (k, l) = (j, i).$$

It follows from (4) that

(27a) $$R_{ii} = \sum_{j \neq i} k_i k_j$$

(27b) $$R_{ij} = 0 \qquad \text{if } i \neq j.$$

In the case that M is minimal we have

$$\sum_{j=1}^{n} k_j = 0$$

or

(28) $$\sum_{j \neq i} k_j = -k_i,$$

whence

(29) $$R_{ii} = -k_i^2.$$

Thus from (5), the Ricci form becomes

(30) $$\mathrm{Ric}_M = \sum_{i=1}^{n} k_i^2 \omega_i^2.$$

In other words, *on a minimal hypersurface a principal curvature frame diagonalizes the Ricci form, and the diagonal elements are the squares of the principal curvatures.*

We arrive at the following result.

Theorem 3.1. Let M be a Riemannian manifold of dimension $n \geq 4$. If M is locally isometric to a minimal hypersurface in E^{n+1}, then the Ricci form of M is negative semi-definite, and at each point of M there exists an orthonormal frame $e_1, \ldots, e_n$ satisfying (26b) and

(31) $$K(e_i, e_l)K(e_j, e_l) = -K(e_i, e_j)\mathrm{Ric}(e_l, e_l), \qquad i, j, l \text{ distinct},$$

where

$$K(e_i, e_j) = R_{ijij} \tag{32}$$

is the sectional curvature of M for the plane spanned by e_i, e_j. Conversely, if the Ricci form of M has n distinct negative eigenvalues, let $e_1, \ldots, e_n$ be the (uniquely defined) smooth frame field diagonalizing it. If equations (26b) and (31) hold in a simply-connected neighborhood, then that neighborhood may be immersed isometrically as a minimal hypersurface in E^{n+1}.

Proof. The necessity is an immediate consequence of our previous discussion, since in a principal curvature frame, we have equations (26a) and (29) which imply (31).

For the converse, we have by assumption a unique smooth frame field $e_1, \ldots, e_n$ such that

$$-\operatorname{Ric}(e_i, e_j) = \lambda_i \delta_{ij}, \qquad \lambda_1 > \lambda_2 > \cdots > \lambda_n > 0. \tag{33}$$

Define

$$k_1 = \sqrt{\lambda_1}, \tag{34}$$

$$k_j = K(e_1, e_j)/k_1, \qquad j = 2, \ldots, n. \tag{35}$$

$$\hat{\omega}_i = k_i \omega_i, \qquad i = 1, \ldots, n. \tag{36}$$

Then (33) and (34) imply

$$\operatorname{Ric}(e_1, e_1) = -\lambda_1 = -k_1^2, \tag{37}$$

while (31) and (35) give

$$K(e_i, e_j) = -K(e_i, e_1)K(e_j, e_1)/\operatorname{Ric}(e_1, e_1) = k_i k_j. \tag{38}$$

It follows that the Gauss equations (26a), hold, and (26b) also hold by our hypothesis. Since the matrix

$$h_{ij} = k_i \delta_{ij} \tag{39}$$

has rank $n \geq 4$, we may apply Thomas's theorem to deduce that the Codazzi equations are also satisfied. There exists therefore an isometric immersion into E^{n+1}. Finally, from (37) and (35) we have

$$-k_1^2 = \operatorname{Ric}(e_1, e_1) = R_{11} = \sum_{j=2}^{n} R_{ijij} = \sum_{j=2}^{n} k_1 k_j$$

or

$$k_1 \left(\sum_{j=1}^{n} k_j \right) = 0.$$

Since $k_1 \neq 0$, it follows that

$$\sum_{j=1}^{n} k_j = \operatorname{tr}(h_{ij}) = 0.$$

Thus the immersed manifold is minimal, and the theorem is proved.

We conclude with some remarks for the case $n = 3$. There we may replace the equations (26b) and (31) by a simple necessary and sufficient condition in order that the Gauss equations may be satisfied by a second fundamental form matrix with trace zero.

Proposition 3.2. Let M be a minimal hypersurface in E^4. That at any point where the second fundamental form of M is nonsingular, the following holds:

a) the Ricci form of M is negative definite,

b) there is a unique unit vector e_1 satisfying

$$-\operatorname{Ric}(e_1, e_1) = \max_{|x|=1} \{-\operatorname{Ric}(x, x)\}, \tag{40}$$

c) the eigenvalues $\lambda_1 \geq \lambda_2 \geq \lambda_3 > 0$ of $-\operatorname{Ric}_M$ satisfy

$$\lambda_1 = \lambda_2 + \lambda_3 + 2\sqrt{\lambda_2 \lambda_3}. \tag{41}$$

Conversely, if M is a 3-dimensional Riemannian manifold for which a) and c) hold, then at each point there is a trace-zero matrix (h_{ij}) satisfying the Gauss curvature equations. Namely, set

$$h_{ij} = k_i \delta_{ij} \tag{42}$$

where

$$k_1 = \sqrt{\lambda_1}, \qquad k_2 = -\sqrt{\lambda_2}, \qquad k_3 = -\sqrt{\lambda_3}.$$

Proof. Starting with a minimal hypersurface, the principal curvatures must all be different from zero if the second fundamental form is nonsingular. Furthermore, since their sum is zero, there must be one positive and two negative, or the reverse. By reversing orientation if necessary we may assume that

$$k_1 > 0 > k_3 \geq k_2. \tag{44}$$

The minimality condition may be written

$$k_1 = -k_2 - k_3, \tag{45}$$

so that

$$k_1^2 = k_2^2 + k_3^2 + 2k_2 k_3. \tag{46}$$

But by (44), $k_2 \leq k_3 < 0$. Hence

$$k_1^2 > k_2^2 \geq k_3^2.$$

Since by (30) the eigenvalues of the negative of the Ricci form are the quantities k_i^2, it follows that

$$\lambda_1 = k_1^2 > \lambda_2 = k_2^2 \geq \lambda_3 = k_3^2 > 0. \tag{47}$$

This proves parts a) and b), while part c) follows immediately from (46) and (47).

For the converse, if we diagonalize Ric_M and define the quantities k_i by (43), then

$$-\operatorname{Ric}(e_1, e_j) = \delta_{ij} \lambda_j = \delta_{ij} k_j^2, \tag{48}$$

so that (41) is equivalent to (46) and hence to (45). Solving the equations

$$-k_1^2 = \mathrm{Ric}(e_1, e_1) = R_{1212} + R_{1313}$$

$$-k_2^2 = \mathrm{Ric}(e_2, e_2) = R_{2121} + R_{2323}$$

$$-k_3^2 = \mathrm{Ric}(e_3, e_3) = R_{3131} + R_{3232}$$

gives

(49) $$2R_{1212} = -k_1^2 - k_2^2 + k_3^2.$$

But rewriting (45) as

$$k_3 = -k_1 - k_2$$

yields

$$k_3^2 = k_1^2 + k_2^2 + 2k_1k_2.$$

Substituting in (49), we find

$$R_{1212} = k_1k_2.$$

In a similar manner we obtain all the Gauss curvature equations (26a). Also, since the e_i diagonalize. Ric_M, we have

$$0 = \mathrm{Ric}(e_1, e_2) = R_{12} = R_{1121} + R_{1222} + R_{1323} = R_{1323}$$

and in a similar manner, $R_{ikjk} = 0$ when i, j, k are all different. Since there are only three values possible for the indices, it follows that equations (26b) are all valid. We have thus solved the Gauss curvature equations, and by (45) the solution has trace zero. This proves Proposition 3.2.

Theorem 3.3. Let M be a 3-dimensional Riemannian manifold. Suppose that at some point $p \in M$ the eigenvalues of the Ricci form are all distinct and nonzero. Then necessary and sufficient that a neighborhood of p should have an isometric immersion as a minimal hypersurface in E^4 in that

a) the eigenvalues $\lambda_1 > \lambda_2 > \lambda_3$ of $-\mathrm{Ric}_M$ should be positive and satisfy equation (41).

b) the metric defined by $d\hat{s}^2 = -\mathrm{Ric}_M$ should have constant sectional curvature $\hat{K} \equiv 1$,

c) the connection forms ω_{ij} associated with the (uniquely defined) frame field e_1, e_2, e_3 of eigenvectors of the Ricci form must satisfy the algebraic conditions

(50) $$(k_1 - k_2)\omega_{12}(e_3) = (k_2 - k_3)\omega_{23}(e_1) = (k_3 - k_1)\omega_{31}(e_2),$$

where the k_i are given by (43), and

d) either one of the two following sets of differential conditions must be satisfied:

(51a) $$dk_i(e_j) = (k_i - k_j)\omega_{ij}(e_i), \qquad i = 1, 2, 3;\ j = i + 1 \pmod 3,$$

(51b) $$dk_j(e_i) = (k_i - k_j)\omega_{ij}(e_j), \qquad i = 1, 2, 3;\ j = i + 1 \pmod 3.$$

Proof. We know from Proposition 3.2. that condition a) implies that the Gauss curvature equations can be solved using (42), (43). Thus it only remains to show that the second fundamental form so defined also satisfies the Codazzi equations. But a calculation shows that when the second fundamental form is diagonalized, the Codazzi equations reduce to (50), (51a), (51b). Thus, conditions a) and c) together with *both* sets of equations (51a) and (51b) would guarantee an isometric immersion of the metric in E^4. It only remains to show that the basic necessary condition b) implies that the two sets of equations (51a) and (51b) are equivalent. Without going through the details of the computation, we note that as pointed out earlier, condition b) together with the Gauss curvature equations imply equation (20):

$$\hat{\Omega}_{ij} = \Omega_{ij},$$

where $\hat{\Omega}_{ij}$ are the curvature forms of the metric $d\hat{s}^2$ relative to the co-frame field

$$\hat{\omega}_i = \omega_{i4} = k_i \omega_i, \qquad i = 1, 2, 3.$$

Applying the Bianchi identity to both sets of curvature forms, one arrives after some computation at the three equations

$$k_l(dk_i(e_j) - (k_i - k_j)\omega_{ij}(e_i)] = -k_i[dk(e_j) - (k_j - k_l)\omega_{jl}(e_l)],$$

where $i = 1, j = 2, l = 3$, and their cyclic permutations. From these equations, in view of the fact that each $k_i \neq 0$, it follows immediately that either set of equations (51a) or (51b) implies the other.

This work was supported in part by NSF grant MCS77-23579 at the University of California, Berkeley, and NSF grant MCS78-04872 at Stanford Univerwsity.

REFERENCES

1. C. B. Allendoerfer, Rigidity for spaces of class greater than one, Amer. J. Math. 61 (1939), 633–644.
2. L. Barbosa and M. do Carmo, Stability of Minimal surfaces and eigenvalues of the Laplacian, Math. Z. 173 (1980), 13–28.
3. L. Barbosa and M. do Carmo, A necessary condition for a metric in M^n to be minimally immersed in $\mathbb{R}^{n+1}$, An. Acad. Bras. Cienc. 50 (1978), 445–454.
4. E. Calabi, Metric Riemann surfaces, in *Contributions to the Theory of Riemann Surfaces*, Annals of Math. Studies 30, Princeton University Press 1953, 77–85.
5. E. Calabi, Quelques applications de l'analyse complexe aux surfaces d'aire minima, *Topics in Complex Manifolds*, Presses de l'Université de Montréal 1968, 58–81.
6. S.-S. Chern, Minimal surfaces in an Euclidean space of N dimensions, *Differential and Combinatorial Topology*, Princeton University Press 1965, 187–198.
7. S.-S. Chern and R. Osserman, Complete minimal surfaces in Euclidean n-space, J. Analyse Math. 19 (1967), 15–34.
8. M. Dajczer and L. Rodriguez, On asymptotic directions of minimal immersions (preprint).
9. M. do Carmo and M. Dajczer, Necessary and sufficient conditions for existence of minimal hypersurfaces in spaces of constant curvature (preprint).
10. L. P. Eisenhart, *Riemannian Geometry*, Princeton University Press, Princeton, N. J. 1966.
11. R. B. Gardner, New viewpoints in the geometry of submanifolds of $\mathbb{R}^N$, Bull. Amer. Math. Soc. 83 (1977), 1–35.
12. D. A. Hoffman and R. Osserman, The geometry of the generalized Gauss map, Amer. Math. Soc. Memoir No. 236, 1980.

13. H. B. Lawson, Jr., Complete minimal surfaces in S^3, Ann. of Math. 92 (1970), 335–374.
14. H. B. Lawson, Jr., *Lectures on Minimal Surfaces*, IMPA, Rio de Janeiro 1970 (reprinted by Publish or Perish Press, Berkeley 1980).
15. H. B. Lawson, Jr., The Riemannian geometry of holomorphic curves, Proc. of Conference on Holomorphic Mappings and Minimal Surfaces, Bol. Soc. Bras. Mat. 2 (1971), 45–62.
16. M. Pinl and W. Ziller, Minimal hypersurfaces in spaces of constant curvature, J. Differential Geometry 11 (1976), 335–343.
17. G. Ricci-Curbastro, *Opere*, Vol. 1.
18. M. Spivak, *A Comprehensive Introduction to Differential Geometry*, Vol. 5, 2nd edition, Publish or Perish Press, Berkeley 1979.
19. T. Y. Thomas, Riemann spaces of class one and their characterization, Acta Math. 67 (1936), 169–211.

Reprinted from
Geometry Symposium Utrecht 1980, Springer Lecture Notes **894** Springer–Verlag (1981) 49–90.

A SIMPLE PROOF OF FROBENIUS THEOREM

Shiing-shen Chern[†] and Jon G. Wolfson[‡]

Frobenius Theorem, as stated in Y. Matsushima, *Differential Manifolds*, Marcel Dekker, N.Y., 1972, p. 167, is the following:

Let D *be an* r*-dimensional differential system on an* n*-dimensional manifold* M. *Then* D *is completely integrable if and only if for every local basis* $\{X_1,\ldots,X_r\}$ *of* D *on any open set* V *of* M, *there are* C^∞*-functions* c^k_{ij} *on* V *such that we have*

$$[X_i,X_j] = \sum_k c^k_{ij}X_k \quad , \qquad 1 \leqq i,j,k \leqq r \quad . \tag{1}$$

We recall that D is called completely integrable if, at each point $p \in M$, there is a local coordinate system $(x^1,\ldots,x^n)$ such that $\partial/\partial x^1,\ldots,\partial/\partial x^r$ form a local basis of D.

The theorem is of course a fundamental one in differential geometry and every mathematician has his favorite proof. We wish to record the following proof, because it is surprisingly simple and we have not found it in the literature.

The "only if" part of the theorem being trivial, we will prove the "if" part.

For $r=1$ condition (1) is automatically satisfied, and a stronger version of the theorem holds: *If a vector field* $X \neq 0$, *then there is at every point* $p \in M$ *a local coordinate system* $(x^1,\ldots,x^n)$ *such that* $X = \partial/\partial x^1$.

The proof of this statement is based on an existence theorem on ordinary differential equations. We will assume it. It turns out that this is the hardest part of the proof.

We suppose $r \geqq 2$. To apply induction we suppose the theorem be true for $r-1$. By the above statement there is a coordinate system $(y^1,\ldots,y^n)$ at p such that

$$X_r = \frac{\partial}{\partial y^r} \quad . \tag{2}$$

Let

$$X'_\lambda = X_\lambda - (X_\lambda y^r)X_r \quad , \qquad 1 \leq \lambda,\mu,\nu \leq r-1 \quad . \tag{3}$$

Then

$$X'_\lambda y^r = 0 \quad , \qquad X_r y^r = 1 \quad . \tag{4}$$

Let

$$[X'_\lambda, X'_\mu] \equiv a_{\lambda\mu} X_r \quad , \qquad \text{mod } X'_\nu \quad .$$

Applying both sides of the operator to the function y^r, we get $a_{\lambda\mu} = 0$. Hence the differential system $D' = \{X'_1,\ldots,X'_{r-1}\}$ satisfies the condition (1). By induction hypothesis there is a local coordinate system $(z^1,\ldots,z^n)$ at p such that

$$D' = \{X''_1,\ldots,X''_{r-1}\} \tag{5}$$

and

$$X''_\lambda = \frac{\partial}{\partial z^\lambda} \quad . \tag{6}$$

The X''_λ differ from X'_μ by a non-singular linear transformation. This has the consequence

$$X''_\lambda y^r = 0 \quad . \tag{7}$$

We have $D = \{X''_1,\ldots,X''_{r-1},X_r\}$ and condition (1) remains satisfied. Put

$$[X''_\lambda, X_r] \equiv b_\lambda X_r \quad , \qquad \text{mod } X''_\mu \quad .$$

Applying the operator on y^r, we get $b_\lambda = 0$. It follows that

$$[X''_\lambda, X_r] = \sum_\mu d^\mu_\lambda X''_\mu \quad . \tag{8}$$

In the z-coordinates let

$$X_r = \sum_A \xi^A \frac{\partial}{\partial z^A} \quad , \qquad 1 \leq A \leq n \quad . \tag{9}$$

Then

$$[X''_\lambda, X_r] = \sum_A \frac{\partial \xi^A}{\partial z^\lambda} \frac{\partial}{\partial z^A}$$

and condition (8) becomes

$$\frac{\partial \xi^\rho}{\partial z^\lambda} = 0 \qquad , \qquad 1 \leqq \lambda \leqq r-1, \quad r \leqq \rho \leqq n \quad , \tag{10}$$

which means that ξ^ρ are functions of $z^r,\ldots,z^n$ only. But D is also spanned by $X''_1,\ldots,X''_{r-1},X'_r$, where

$$X'_r = \sum_\rho \xi^\rho \frac{\partial}{\partial z^\rho} \qquad r \leqq \rho \leqq n \quad . \tag{11}$$

The last operator involves only the coordinates $z^r,\ldots,z^n$. By a change of coordinates $(z^r,\ldots,z^n) \to (w^r,\ldots,w^n)$, which will not affect $z^1,\ldots,z^{r-1}$, and hence not the equations (6), we can get

$$X'_r = \frac{\partial}{\partial w^r} \quad . \tag{12}$$

This completes the proof of the theorem. ■

University of California
Berkeley, California 94720, USA

†Work done under partial support of NSF Grant MC577-23579.

‡Supported by a postgraduate scholarship of NSERC of Canada.

(Received December 19, 1980)

Reprinted from
Manifolds and Lie Groups, Papers in Honor of Y. Matsushima, Birkhäuser (1981) 67–69.

J. DIFFERENTIAL GEOMETRY
16 (1981) 347–349

FOLIATIONS ON A SURFACE OF CONSTANT CURVATURE AND THE MODIFIED KORTEWEG–DE VRIES EQUATIONS

SHIING-SHEN CHERN & KETI TENENBLAT

Dedicated to Professor Buchin Su on his 80th birthday

ABSTRACT. The modified *KdV* equations are characterized as relations between local invariants of certain foliations on a surface of constant Gaussian curvature.

Consider a surface M, endowed with a C^∞-Riemannian metric of constant Gaussian curvature K. Locally let e_1, e_2 be an orthonormal frame field and ω_1, ω_2 be its dual coframe field. Then the latter satisfy the structure equations

$$(1)\qquad d\omega_1 = \omega_{12} \wedge \omega_2, \quad d\omega_2 = \omega_1 \wedge \omega_{12}, \quad d\omega_{12} = -K\omega_1 \wedge \omega_2,$$

where ω_{12} is the connection form (relative to the frame field). We write

$$(2)\qquad \omega_{12} = p\omega_1 + q\omega_2,$$

p, q being functions on M.

Given on M a foliation by curves. Suppose that both M and the foliation are oriented. At a point $x \in M$ we take e_1 to be tangent to the curve (or leaf) of the foliation through x. Since M is oriented, this determines e_2. The local invariants of the foliation are functions of p, q and their successive covariant derivatives. If the foliation is unoriented, then the local invariants are those which remain invariant under the change $e_1 \to -e_1$.

Under this choice of the frame field the foliation is defined by

$$(3)\qquad \omega_2 = 0,$$

and ω_1 is the element of arc on the leaves. It follows that p is the geodesic curvature of the leaves.

We coordinatize M by the coordinates x, t, such that

$$(4)\qquad \omega_2 = Bdt, \quad \omega_1 = \eta dx + Adt, \quad \omega_{12} = udx + Cdt,$$

Received September 22, 1981. The first author is supported partially by NSF Grant MCS-8023356.

where A, B, C, u are functions of x, t, and η ($\neq 0$) is a constant. Thus the leaves are given by $t = \text{const}$, and ηx and u/η are respectiviely the arc length and the geodesic curvature of the leaves. Substituting (4) into (1), we get

(5) $$A_x = uB, \quad B_x = \eta C - uA, \quad C_x - u_t = -K\eta B.$$

Elimination of B and C gives

(6) $$u_t = \left(\frac{A'_x}{u}\right)_{xx} + (uA')_x + \eta^2 K \frac{A'_x}{u},$$

where

(7) $$A' = A/\eta.$$

By choosing

(8) $$A' = -K\eta^2 + \frac{1}{2}u^2,$$

we get

(9) $$u_t = u_{xxx} + \frac{3}{2}u^2 u_x,$$

which is the modified Korteweg–de Vries (= $MKdV$) equation.

Condition (8) on the foliation can be expressed in terms of the invariants p, q as follows: By (2) and (4) we have

(10) $$u = \eta p, \quad C = Ap + Bq.$$

If we eliminate B, C in the second equation by using (5), it can be written

(11) $$\eta q = \left(\log \frac{A'_x}{u}\right)_x = (\log p_x)_x.$$

Introducing the covariant derivatives of p by

(12) $$dp = p_1\omega_1 + p_2\omega_2, \quad dp_1 = p_{11}\omega_1 + p_{12}\omega_2,$$

we have

(13) $$p_x = p_1\eta, \quad p_{xx} = p_{11}\eta^2.$$

Hence condition (11) can be written

(14) $$q = (\log p_1)_1.$$

A foliation will be called a K-foliation, if (14) is satisfied. We state our result in

Theorem. *The geodesic curvature of the leaves of a K-foliation satisfies, relative to the coordinates x, t described above, an $MKdV$ equation.*

The above argument can be generalized to $MKdV$ equations of higher order. The corresponding foliations are characterized by expressing q as a function of $p, p_1, p_{11}, p_{111}, \cdots$.

Is there a similar geometrical interpretation of the *KdV*-equation itself, which is

(15) $$u_t = u_{xxx} + uu_x?$$

We do not have a simple answer to this question. Unlike the *MKdV*-equation, the sign of the last term is immaterial, because it reverses when u is replaced by $-u$. It is therefore of interest to know that by a different foliation and a different coordinate system one can be led to a *MKdV*-equation (9) where the last term has a negative sign.

For this purpose we put

(16) $$\omega_2 = B dt, \quad \omega_1 = v dx + E dt, \quad \omega_{12} = \lambda dx + F dt,$$

where λ is a parameter. Substitution into (1) gives

(17) $$F_x = -KvB, \quad B_x = -\lambda E + vF, \quad E_x - v_t = \lambda B.$$

Suppose $K \neq 0$, we get, by eliminating B, E,

(18) $$v_t = \left(\frac{F'_x}{Kv}\right)_{xx} + (vF')_x + \frac{\lambda^2}{Kv} F'_x,$$

where

(19) $$F = F'\lambda.$$

The choice

(20) $$F' = \frac{K}{2} v^2 - \lambda^2$$

reduces (18) into

(21) $$v_t = v_{xxx} + \frac{3}{2} K v^2 v_x.$$

Here the sign of the second term depends on the sign of K.

It can be proved that the choice (20) corresponds to a foliation which is characterized by

(22) $$q = \frac{p_{11}}{p_1} - 3\frac{p_1}{p} = \left(\log \frac{p_1}{p^3}\right)_1.$$

References

[1] S. S. Chern & C. K. Peng, *Lie groups and KdV equations*, Manuscripta Math. **28** (1979) 207–217.

University of California, Berkeley
Universidade de Brasilia, Brasil

第11卷 第3期
1981年9月

中国科学技术大学学报
JOURNAL OF CHINA UNIVERSITY OF SCIENCE AND TECHNOLOGY

Vol.11. No.3
Sept., 1981

On the Bäcklund Transformations of KDV Equations and Modified KDV Equations

Chern Shiing-shen*
(University of California, Berkeley, U. S. A.)

Peng Chia-kuei**
(Department of Mathematics, University of Science and Technology of China)

1. Introduction

This is the continuation of our previous paper[1] which showed that both of Korteweg-de Vries equation and Modified KDV equation are special cases of the structrue equation of the Lie Group SL(2,R) (the special linear group of all (2×2)-real unimodular matrices). It leads naturally to the KDV and MKDV equations of higher order. Futhermore, there exists the Miura transformation which leads to the Bäcklund transformations of the KDV equations.

In the present paper, we will explain the Miura transformation from another viewpoint. Therefore we will derive the generalized Miura transformations between the family of MKDV equations and the family of other new equations. This leads to the Bäcklund transformation of the MKDV equations.

2. Miura Transformation and Bäcklund Transformation of KDV equation

According to the results of our previous paper, the family of the KDV equations satisfy the following recursion formula

received Feb. 19, 1981

* Work done under partial support of NSF grant MCS 77-23579.

** Work done when the author visited Berkeley, UC.

$$K_{n+1}(u)=TK_n(u),\ K_0=u_x,\tag{1}$$

where

$$T=\frac{1}{4}D^2-u-\frac{1}{2}u_xD^{-1},\qquad D=\frac{d}{dx},\qquad u_x=\frac{\partial u}{\partial x}.\tag{2}$$

Similarly, the family of the MKDV equations satisfy

$$M_{n+1}(v)=SM_n(v),\qquad M_0(v)=v_x,\tag{3}$$

where

$$S=\frac{1}{4}D^2-v^2-v_xD^{-1}v.\tag{4}$$

To obtain the Miura transformation between the KDV equation K_n and the MKDV equation M_n, we assume that

$$u=v_x+f(v),\tag{5}$$

where the function $f(v)$ will be determined by the commutative relation

$$T(v_x+f)\left(D+\frac{df}{dv}\right)=\left(D+\frac{df}{dv}\right)S(v).\tag{6}$$

The reason is as follows. Once the above equation (6) is satisfied one can apply it to $M_{n-1}(v)$ and it follows that

$$K_n(v_x+f)\left(D+\frac{df}{dv}\right)=\left(D+\frac{df}{dv}\right)M_n(v).\tag{7}$$

By a direct computation one can verify that if the function $v(x,t)$ is the solution of MKDV equation

$$\frac{\partial v}{\partial t}=M_n(v).\tag{8}$$

then the function $u(x,t)=v_x+f(v)$ will be the solution of the KDV equation

$$\frac{\partial u}{\partial t}=K_n(v).\tag{9}$$

So the question is reduced to finding the solution $f(v)$ of the equation(6). By expanding the equation(6) it gives

$$T(v_x+f)\left(D+\frac{df}{dv}\right)=a_3D^3+a_2D^2+a_1D+a_0+a_{-1}D^{-1}\left(\frac{df}{dv}\right),\tag{10}$$

where

$$\left.\begin{aligned}
a_3&=\frac{1}{4},\\
a_2&=\frac{1}{4}\,\frac{df}{dv},\\
a&=\frac{1}{2}\,\frac{d^2f}{dv^2}\cdot v_x-f-v_x,\\
a_0&=\frac{1}{4}\left(\frac{d^2f}{dv^2}v_x\right)_x-\frac{3}{2}\,\frac{df}{dv}\cdot v_x-f\cdot\frac{df}{dv}-\frac{1}{2}v_{xx},\\
a_{-1}&=-\frac{1}{2}\left(\frac{df}{dv}v_x+v_{xx}\right)\left(D^{-1}\left(\frac{df}{dv}\right)\right),
\end{aligned}\right\}\tag{11}$$

and similarly

$$\left(D+\frac{df}{dv}\right)S(v)=b_3D^3+b_2D^2+b_1D+b_0+b_{-1}D^{-1} \tag{12}$$

where

$$\left.\begin{aligned}
b_3&=\frac{1}{4},\\
b_2&=\frac{1}{4}\,\frac{df}{dv},\\
b_1&=-v^2,\\
b_0&=-3vv_x-v^2\frac{df}{dv},\\
b_{-1}&=-\left(v_{xx}+\frac{df}{dv}v_x\right).
\end{aligned}\right\} \tag{13}$$

By equating the coefficients a_i and b_i it is not difficult to obtain the solution

$$f(v)=v^2, \tag{14}$$

This implies that the transformation

$$u=v_x+v^2, \tag{15}$$

is the unique transformation satisfying the commutative relation(6),

By the same way, instead of the assumption(5), we set

$$u=-v_x+f(v). \tag{16}$$

It gives the same solution

$$f(v)=v^2. \tag{17}$$

Hence we obtain two commutative relations between KDV operator and MKDV operator

$$K_n(v_x+v^2)(D+2v)=(D+2v)M_n(v), \tag{18}$$

$$K_n(-v_x+v^2)(-D+2v)=(-D+2v)M_n(v). \tag{19}$$

here are just the Miura transformation. From this it is easy to derive to Bäcklund transfonmation of the KDV equation[1].

3. Generalized Miura Transformation and Bäcklund Transformation of MKDV Equation

In this section, we will find the generalized Miura traansformation between MKDV operators and the new operators which leads to the Bäcklund transformation of MKDV equations,

To do this we assume that

$$v=\varphi_x+f(\varphi) \tag{20}$$

which satisfies the following commutative relation

$$S(\varphi_x+f)\left(D+\frac{df}{d\varphi}\right)=\left(D+\frac{df}{d\varphi}\right)P(\varphi) \tag{21}$$

where f is the function of φ and the operator P will be determined. By expanding the left hand-side of the equation(21), it gives

$$S(\varphi_x+f)\left(D+\frac{df}{d\varphi}\right)=a_3D^3+a_2D^2+a_1D+a_0+a_{-1}D^{-1}\left(\varphi_{xx}-f\frac{df}{d\varphi}\right), \tag{22}$$

where

$$\left.\begin{aligned}
&a_3=\frac{1}{4},\\
&a_2=\frac{1}{4}\frac{df}{d\varphi},\\
&a_1=\frac{1}{2}\left(\frac{df}{d\varphi}\right)_x-(\varphi_x+f)^2,\\
&a_0=\frac{1}{4}\left(\frac{df}{d\varphi}\right)_{xx}-(\varphi_x+f)^2\frac{df}{d\varphi}-(\varphi_x+f)(\varphi_x+f)_x,\\
&a_{-1}=(\varphi_x+f)_x,
\end{aligned}\right\} \tag{23}$$

It is not difficult to find the operator P has the following form

$$P=\frac{1}{4}D^2+\frac{1}{2}\left(\frac{df}{d\varphi}\right)_x-(\varphi_x+f)^2+HD^{-1}\left(\varphi_{xx}-f\frac{df}{d\varphi}\right), \tag{24}$$

where the coefficient $H=H(\varphi,\varphi_x,\cdots)$ will be determined.

Substituting (24) into the equation(21) it gives

$$\left(D+\frac{df}{d\varphi}\right)P(\varphi)=b_3D^3+b_2D^2+b_1D+b_0+b_{-1}D^{-1}\left(\varphi_{xx}-f\frac{df}{d\varphi}\right), \tag{25}$$

where

$$\left.\begin{aligned}
&b_3=\frac{1}{4},\\
&b_2=\frac{1}{4}\frac{df}{d\varphi},\\
&b_1=\frac{1}{2}\left(\frac{df}{d\varphi}\right)_x-(\varphi_x+f)^2,\\
&b_0=\frac{1}{2}\left(\frac{df}{d\varphi}\right)_{xx}-(\varphi_x+f)^2\frac{df}{d\varphi}-2(\varphi_x+f)(\varphi_x+f)_x\\
&\qquad+\frac{1}{2}\frac{df}{d\varphi}\left(\frac{df}{d\varphi}\right)_x+H\left(\varphi_{xx}-f\cdot\frac{df}{d\varphi}\right),\\
&b_{-1}=H_x+H\frac{df}{d\varphi}.
\end{aligned}\right\} \tag{26}$$

The both sides of the equation (21) can be looked upon as the formal polynomial of the derivative D with the coefficients b_i and a_i.

By equating all of the coefficients it yields

$$b_i=a_i \qquad i=-1,0,1,2,3. \tag{27}$$

from $b_{-1}=a_{-1}$ we obtain

• 4 •

$$H=\Phi_x. \tag{28}$$

Substituting it into b_0 and from $b_0=a_0$ we obtain the equation

$$\frac{1}{4}\left(\frac{df}{d\varphi}\right)_{xx}-(\varphi_x+f)(\varphi_x+f)_x+\frac{1}{2}\frac{df}{d\varphi}\left(\frac{df}{d\varphi}\right)_x+\varphi_x\left(\varphi_{xx}-f\frac{df}{d\varphi}\right)=0, \tag{29}$$

namely,

$$\left(\frac{1}{4}\frac{d^3f}{d\varphi^3}-\frac{df}{d\varphi}\right)\varphi_{xx}+\left(\frac{1}{4}\frac{d^2f}{d\varphi^2}-f\right)\varphi_x^2+\frac{1}{2}\frac{df}{d\varphi}\left(\frac{d^2f}{d\varphi^2}-4f\right)\varphi_x=0. \tag{30}$$

Since the function f depends only on φ it gives an ordinary differential equation

$$\frac{d^2f}{d\varphi^2}+4f=0. \tag{31}$$

It is obvious that the above eqnation has two linear independent solutions

$$f_1=\cosh 2\varphi=\frac{1}{2}(e^{2\varphi}+e^{-2\varphi}), \tag{32}$$

$$f_2=\sinh 2\varphi=\frac{1}{2}(e^{2\varphi}-e^{-2\varphi}). \tag{33}$$

By substituting (32) and (33) into the expresion (24) of the operator P, we obtain

$$P=\frac{1}{4}D^2-(\varphi_x^2+\cosh^2 2\varphi)+\varphi_x D^{-1}(\varphi_{xx}-\sinh 4\varphi) \tag{34}$$

and

$$P=\frac{1}{4}D^2-(\varphi_x^2+\sinh^2 2\varphi)+\varphi_x D^{-1}(\varphi_{xx}-\sinh 4\varphi). \tag{35}$$

Now we can define a new family of the operator N which will be called Twice Modified KDV operators by satisfying the following recursion relation

$$N_{n+1}(\varphi)=PN_n(\varphi). \tag{36}$$

By the same argument as in the section 2, we obtain the commutative relation between MKDV operator M_n and Twice Modified KDV operator N_n

$$S(\varphi_x+\sinh 2\varphi)(D+2\cosh 2\varphi)=(D+2\cosh 2\varphi)P(\varphi). \tag{37}$$

Similarly we have

$$S(-\varphi_x+\sinh 2\varphi)(-D+2\cosh 2\varphi)=(-D+2\cosh 2\varphi)P(\varphi). \tag{38}$$

By applying (37) and (38) to N_{n-1}, it follows that

$$M_n(\varphi_x+\sinh 2\varphi)=(D+2\cosh 2\varphi)N_n(\varphi), \tag{39}$$

$$M_n(-\varphi_x+\sinh 2\varphi)=(-D+2\cosh 2\varphi)N_n(\varphi). \tag{40}$$

Let φ be the solution of the Twice MKDV equation

$$\frac{\partial\varphi}{\partial t}=N_n(\varphi). \tag{41}$$

Then from (39) and (40) it is easy to see that the functions

$$v(x,t)=\varphi_x+\sinh 2\varphi, \tag{42}$$

$$\tilde{v}(x,t)=-\varphi_x+\sinh 2\varphi, \tag{43}$$

are the two solutions of MKDV equation

$$\frac{\partial v}{\partial t}=M_n(v). \tag{44}$$

To pass from v to $\tilde{v}$, we set

$$v=w_x, \qquad \tilde{v}=\tilde{w}_x \tag{45}$$

then

$$v-\tilde{v}=(w-\tilde{w})_x=2\varphi_x, \tag{46}$$

and we can suppose

$$2\varphi=w-\tilde{w} \tag{47}$$

it follows that

$$\left.\begin{aligned}(w-\tilde{w})_t&=2N_n\left(\frac{1}{2}(w-\tilde{w})\right),\\(w+\tilde{w})_x&=2\sinh(w-\tilde{w}).\end{aligned}\right\} \tag{48}$$

Remark: If we set that f is another solution of the equation (32), we will obtain another set of Bäcklund transformation of MKDV equation.

Reference

[1] Chern Shiing-shen and Peng Chia-kuei, Lie Groups and KDV Equations, *Manuscripta Math.* 28(1979), 207—217.

论KDV方程及MKDV方程的Bäcklund变换

陈省身　　　　彭家貴

(美国加利福尼亚伯克利大学)　　(中国科学技术大学数学系)

摘　　要

从另一角度解释 Miura 变换并由此得到一类新的非綫性发展方程．給出变形 KDV 方程的 Bäcklund 变换．

Proceedings of Symposia in Pure Mathematics
Volume 39 (1983), Part 1

Web Geometry

SHIING-SHEN CHERN[1]

Introduction. Poincaré published two papers on surfaces of translation [10, 11].[2] They were among his lesser known papers. In the following pages I wish to show that the subject he touched is an exciting one and deserves further investigation.

1. Lie's theorem on surfaces of double translation and its developments. A surface M of translation in R^3 is defined by the parametric equations

$$x^\lambda = f^\lambda(u) + g^\lambda(v), \qquad 1 \leqslant \lambda \leqslant 3, \tag{1}$$

where x^λ are the coordinates in R^3 and f^λ, g^λ are arbitrary smooth functions. It is immediately seen that the tangent lines to the u-curves (respectively the v curves) are independent of v (resp. u) and define a curve C_u (resp. C_v) in the plane at infinity.

M is called a *surface of double translation* if it is a surface of translation in a second way, i.e., given also by the equations

$$x^\lambda = h^\lambda(s) + k^\lambda(t), \qquad 1 \leqslant \lambda \leqslant 3, \tag{2}$$

such that exactly two of the equations

$$f^\lambda(u) + g^\lambda(v) - h^\lambda(s) - k^\lambda(t) = 0, \qquad 1 \leqslant \lambda \leqslant 3, \tag{3}$$

are independent. In 1882 Sophus Lie proved the remarkable theorem [7]:

If *M is a surface of double translation* in R^3, *the four curves C_u, C_v, C_s, C_t in the plane at infinity defined by the tangent lines to the four families of parametric curves belong to the same algebraic curve of degree four.*

The theorem means that the solutions of the functional equations (3) on a surface arise from an algebraic structure. Lie's proof makes use of the integrability conditions of over-determined systems of partial differential equations. In fact, from (1) we have

$$\partial^2 x^\lambda / \partial u \partial v = 0, \tag{4}$$

which means that the parametric curves form a conjugate net, i.e., their tangent directions separate harmonically the asymptotic directions at each point. If the surface M is given in the nonparametric form

$$z = z(x, y), \tag{5}$$

Reprinted from Bulletin Amer. Math. Soc. (N.S.) 6 (1982), 1–8.
1980 *Mathematics Subject Classification.* Primary 53A60, 14D25.
[1]Work done under partial support of NSF grant MC577-23579.
[2]Numbers in brackets refer to the Bibliography at the end of the paper.

this condition is expressed by an equation

$$R(p, q)r + Q(p, q)s + T(p, q)t = 0 \tag{6}$$

where

$$p = z_x, \quad q = z_y, \quad r = z_{xx}, \quad s = z_{xy}, \quad t = z_{yy}. \tag{7}$$

A surface M of double translation satisfies, besides (6), another equation

$$R'(p, q)r + Q'(p, q)s + T'(p, q)t = 0. \tag{6'}$$

An investigation of the integrability conditions of the system of equations (6), (6′) involves long and tedious calculations. In particular, the fourth-order partial derivatives of z come into play. The work was a true tour de force, but Lie reached his goal.

Poincaré was quick to recognize the importance of Lie's work, and to observe its relation with the theory of abelian functions. In [**10**, **11**] he gave two proofs of Lie's theorem based on abelian functions and algebraic geometry rather than on partial differential equations.[3] Although the proofs are perhaps not complete, Poincaré introduced fresh ideas and new viewpoints. As a consequence of Poincaré's work it follows that a surface of double translation can be defined by equating a theta function to zero. As a result the surface

$$x_1x_2x_3 = a_1x_1 + a_2x_2 + a_3x_3, \tag{8}$$

where the a's are constants, is a surface of double translation. The best proof of Lie's theorem was given by Darboux, using the theory of residues [**5**].

It should be remarked that Lie started his program on surfaces of translation through his work on minimal surfaces. It was known to Monge that an analytic minimal surface is a surface of translation (1) with parametric curves which are minimal or isotropic curves.

Lie proceeded to study the high-dimensional case. A translation manifold in R^{n+1} is the hypersurface defined by the parametric equations

$$x^\lambda = \sum_{1 \leq \alpha \leq n} f^\lambda(u_\alpha), \qquad 1 \leq \lambda \leq n + 1, \tag{9}$$

where x^λ are the coordinates in R^{n+1} and the f's are smooth functions in the respective variables. Lie tried to determine all hypersurfaces of double translation and settled the case $n = 3$ in a long paper [**8**]. In the same paper he promised to return to the general case. He had several posthumous papers on the subject, without bringing the problem to a satisfactory conclusion [**9**]. The high-dimensional case was also considered by Poincaré. It was W. Wirtinger who in 1938 completely solved the problem, using Chow coordinates for projective varieties [**14**]. We will show below that web geometry offers a broader setting where the subject can be integrated.

---

[3]Lie was unhappy with Poincaré's intrusion: He said: "Unfortunately the distinguished author (i.e., Poincaré), whose achievements in other fields were recognized by nobody more than myself, failed to understand my investigations. I can only say that his works on translation surfaces and translation manifolds deal with results which are entirely special cases of my general theorems." (Ges. Abh, Bd II, Teil II, 527)

The conclusion that the functional equations imply an algebraic structure is a powerful one. This was utilized by B. St. Donat in 1975 to give a new proof of Torelli's theorem that a compact Riemann surface is determined up to isomorphism by its period matrix (or more exactly, by its polarized Jacobian variety) [12].

2. Web geometry of Blaschke-Bol [2]. Web geometry had its debut in 1926–27 on the beaches of Italy when W. Blaschke and G. Thomsen realized that the configuration of three foliations of the plane by curves, has local invariants. The distinguished geometric figure in this case is the hexagon (Figure 1). When all such hexagons are closed, for any point O and any neighboring point P on the first curve through O, the web is called *hexagonal*. Thomsen proved that a hexagonal web is locally homeomorphic to three families of parallel lines.

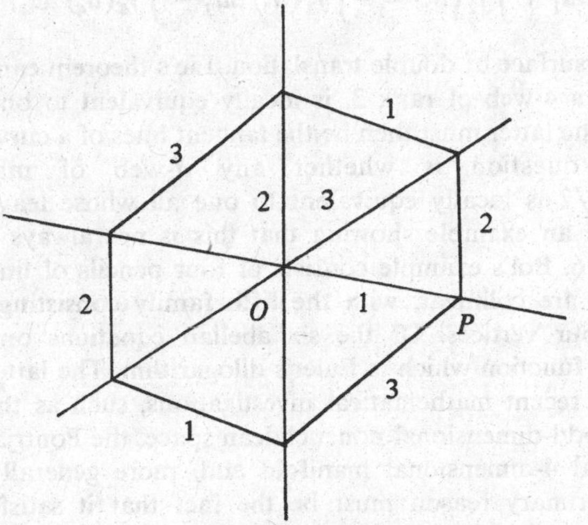

FIGURE 1

The relation of this subject to algebraic geometry is immediate. Before Blaschke and Thomsen started their work on web geometry, Graf and Sauer proved in 1924 a theorem which, in web geometry language, can be stated as follows: *If the curves of a hexagonal web are straight lines, they are the tangent lines of an algebraic curve of class three.*

Generally, a d-web in the plane is defined by d foliations in a neighborhood of the plane by curves such that through every point the d leaves have mutually distinct tangents. If x, y are the coordinates in the plane, a d-web can be defined by

$$u_i(x, y) = \text{const}, \qquad 1 \leqslant i \leqslant d. \tag{10}$$

We shall assume that the functions $u_i(x, y)$ are smooth and satisfying the condition grad $u_i \neq 0$.

An equation of the form

$$\sum_{1 \leqslant i \leqslant d} f_i(u_i)\, du_i = 0 \tag{11}$$

is called an *abelian equation*. The maximum number of linearly independent abelian equations, is called the rank of the web, to be denoted by π. It can be proved that

$$\pi \leq \tfrac{1}{2}(d-1)(d-2). \tag{12}$$

We see easily that for $d = 3$, the rank one webs are exactly the hexagonal webs. For $d = 4$, we have $\pi \leq 3$, and the 4-webs of rank 3 satisfy 3 linearly independent abelian equations

$$\sum f_i^\lambda(u_i)\, du_i = 0, \qquad 1 \leq \lambda \leq 3. \tag{13}$$

By setting

(14)

$$x^\lambda = \int f_1^\lambda(u_1)\, du_1 + \int f_2^\lambda(u_2) = -\int f_3^\lambda(u_3)\, du_3 - \int f_4^\lambda(u_4)\, du_4, \qquad 1 \leq \lambda \leq 3,$$

we get in R^3 a surface of double translation. Lie's theorem can be interpreted as saying that a 4-web of rank 3, is locally equivalent to one consisting of straight lines; the latter must then be the tangent lines of a curve of class 4.

The deep question is whether any d-web of maximum rank $(d-1)(d-2)/2$ is locally equivalent to one all whose leaves are straight lines. Bol gave an example showing that this is not always the case for a 5-web of rank 6. Bol's example consists of four pencils of lines, no three of whose vertices are collinear, with the fifth family consisting of the conics through the four vertices. Of the six abelian equations one involves the transcendental function which is Euler's dilogarithm. The latter has played a role in several recent mathematical investigations, such as the volume of a simplex in an odd-dimensional noneuclidean space, the Pontrjagin number of a combinatorial 4-dimensional manifold and, more generally, in algebraic K-theory. A primary reason must be the fact that it satisfies an abelian equation. For a recent work cf. [**15**].

3. High-dimensional webs. A d-web of dimension $N - k$ in R^N consists of d foliations of a neighborhood U of R^N by submanifolds of dimension $N - k$; k is called the codimension of the web. As an example consider an algebraic variety V of dimension k and degree d in a projective space P^m of dimension m. A linear subspace P^{m-k} of dimension $m - k$ meets V in d points through each of which pass $\infty^{k(m-k)}P^{m-k}$'s. This gives in the Grassmann manifold $G(m-k, m)$ of all P^{m-k}'s in P^m d foliations of dimension $k(m-k)$. Since

$$\dim G(m-k, m) = k(m-k+1),$$

the variety V gives rise to d foliations of codimension k in $G(m-k, m)$, which is locally R^{kn}, $n = m - k + 1$. In order to keep algebraic geometry in sight, we will consider on the basis of this example, only webs of codimension k in R^N, with $N = kn$. Even so the subject is a wide generalization of the geometry of projective algebraic varieties. Just as intrinsic algebraic varieties are generalized to Kähler manifolds and complex manifolds, such a generalization to web geometry seems justifiable.

The tangent spaces to the d leaves through a point $x \in R^{kn}$ give d linear subspaces of codimension k in the tangent space T_x to R^{kn} at x or, what is the same, d linear subspaces Ω_i of dimension k in the cotangent space T_x^*. We will suppose that they are in general position. For $k = 1$ the meaning of this is clear: no $kn = n$ of the d lines in T_x^* lie in a hyperplane of T_x^*. For $k > 1$ the right notion has to be introduced. We gave one in [4], again based on the example of projective varieties.

Analytically suppose the ith foliation of the web be defined by the equations

(15)
$$u_{i1}(x) = \text{const}, \quad \ldots, \quad u_{ik}(x) = \text{const}, \quad 1 \leqslant i \leqslant d, \quad x \in R^{kn}.$$

We will suppose the u's to be smooth functions satisfying

$$\text{(16)} \qquad \Omega_i(x) = du_{i1} \wedge \cdots \wedge du_{ik} \neq 0.$$

For a fixed i the functions $u_{i1}, \ldots, u_{ik}$ are defined up to a diffeomorphism and $\Omega_i(x)$ is defined up to a factor. The notation is so chosen that it defines the linear space $\Omega_i \subset T_x^*$ introduced above; we will call it the ith web normal.

An *abelian equation* is an equation of the form

$$\text{(17)} \qquad \sum_{1 \leqslant i \leqslant d} f_i(u_{i1}, \ldots, u_{ik})\Omega_i = 0.$$

The maximum number of the linearly independent abelian equations is called the rank of the web. In [4] we proved that the rank has an upper bound $\pi(d, n, k)$, which depends only on d, n, k. This bound is sharp. In particular, for $k = 1$ we have

(18)
$$\pi(d, n) \underset{\text{def}}{=} \pi(d, n, 1) = (1/2(n-1))\{(d-1)(d-n) + s(n-s-1)\}$$

where s is defined by

$$\text{(19)} \qquad s \equiv -d + 1 \mod n - 1, \qquad 0 \leqslant s \leqslant n - 2.$$

This number $\pi(d, n)$ has an important meaning in algebraic geometry. In fact, Castelnuovo proved that $\pi(d, n)$ is the maximum genus of an algebraic curve of degree d in P^n, which does not belong to a hyperplane P^{n-1}. By taking such a curve C^* in the dual space P^{*n}, we have through each point of P^n d hyperplanes belonging to C^* and it follows from Abel's theorem that the d-web so constructed is of rank $\pi(d, n)$. In relating web geometry to algebraic geometry it should be remarked that our web geometry is over real numbers while the corresponding notions of algebraic geometry refer to the complex field. The transition is not immediate, but it is possible owing mainly to the fact that we are dealing with local properties in web geometry.

It is clear that an important problem is to determine the d-webs of codimension 1 and maximum rank $\pi(d, n)$ in R^n. If the leaves are all hyperplanes, the answer is given by the following converse to Abel's [6] theorem (generalization of theorems of Graf-Sauer and Lie):

Consider a d-web of codimension 1 *in a neighborhood of* R^n *whose leaves are hyperplanes such that an abelian equation*

$$\sum_{1 \le i \le d} f_i(u_i)\, du_i = 0 \tag{20}$$

holds, with $f_i(u_i) \neq 0$. *Then the leaves belong to an algebraic curve of degree d in the dual projective space.*

For this theorem it is sufficient to have one abelian equation. The crucial question is thus the linearization problem: Is a d-web of codimension one and rank $\pi(d, n)$ in R^n linearizable, i.e., is it locally equivalent to one having hyperplanes as leaves? Bol's example in §2 shows that the answer is no for $n = 2, d = 5$.

Consider next the case $n \geqslant 3$. For $n + 1 \leqslant d \leqslant 2n - 1$ we have $\pi(d, n) = d - n$, and there are simple examples of nonlinearizable d-webs of rank $d - n$, which depend on arbitrary functions. Lie's case of hypersurfaces of double translation corresponds to $d = 2n$. In this case we have $\pi(2n, n) = n + 1$. For a $2n$-web of rank $n + 1$ the abelian equations

$$\sum f_i^\lambda(u_i)\, du_i = 0, \qquad 1 \leqslant \lambda \leqslant \pi = n + 1, \tag{21}$$

can be written

$$\sum \int_{1 \le i \le n} f_i^\lambda(u_i)\, du_i = -\sum \int_{n+1 \le i \le 2n} f_i^\lambda(u_i)\, du_i, \qquad 1 \leqslant \lambda \leqslant n + 1. \tag{22}$$

As a generalization of (14) these common expressions can be regarded as the coordinates in R^{n+1} of a hypersurface of double translation.

An essential step in the proof of the Lie-Wirtinger theorem on hypersurfaces of double translation is to show that a $2n$-web in R^n of codimension one and rank $n + 1$ is linearizable, from which the theorem follows from the converse to Abel's theorem. In this particular case the linearization follows by a simple argument, using an idea of Poincaré. It goes as follows: For $x \in U \subset R^n$ let $Z_i(x)$ be the point in a projective space of dimension $\pi - 1$, whose homogeneous coordinates are $[f_i^1(u_i), \dots, f_i^\pi(u_i)]$. The mapping which sends $x \in U \subset R^n$ to the space $\{Z_1, \dots, Z_d\} \subset P^{\pi-1}$ spanned by the Z_i's is called *Poincaré's mapping*. If $u_1, \dots, u_n$ are regarded as a local coordinate system in U, we have, from (21),

$$f_1^\lambda(u_1) + \sum_{n+1 \le i \le d} f_i^\lambda(u_i) \frac{\partial u_i}{\partial u_1} = 0, \qquad 1 \leqslant \lambda \leqslant \pi.$$

It follows that Z_1 is a linear combination of $Z_{n+1}, \dots, Z_d$. Since a similar equation holds for any Z_i (instead of Z_1), we have

$$\{Z_1, \dots, Z_d\} = d - n - 1. \tag{23}$$

In the case $d = 2n$, $\pi = n + 1$, the Poincaré mapping maps the points of U into hyperplanes of P^n, such that the ith leaf goes to the point Z_i. The Poincaré mapping followed by duality in P^n maps the points of U into the points of the dual space P^{*n} such that all the leaves of the web go to hyperplanes. This proves the linearization, and the Lie-Wirtinger theorem follows.

For $n = 3$ Bol proved the remarkable theorem: *In* R^3 *a d-web of codimension* 1 *and rank* $\pi(d, 3)$, $d \geqslant 6$, *is linearizable.*

Griffiths and I tried to generalize Bol's theorem to R^n [3]. So far we have only succeeded to prove the theorem under the additional condition of normality. The exact statement is: *In* R^n *a normal d-web of codimension* 1 *and rank* $\pi(d, n)$, $d \geqslant 2n$, *is linearizable.* For the definition of normality we refer to [3].

In recent years works on web geometry have also been done by M. A. Akivis and V. Goldberg in the Soviet Union. It is not clear to me how much these works are related to those of Poincaré. The reader is referred to [1] for further information.

Tschebotarow generalized Lie's idea to the study of surfaces which admit imprimitive systems relative to an arbitrary Lie group (instead of the group of translations) [13]. To my knowledge this generalization has not been further pursued.

4. Unsolved problems. In the following I wish to list a few most immediate unsolved problems:

1. Determine all d-webs of curves in the plane having maximum rank $\frac{1}{2}(d-1)(d-2)$, $d \geqslant 5$.

2. Is the above linearization theorem in R^n, $n \geqslant 4$, true without normality?

3 (Griffiths'). The hexagonal condition is meaningful for a 3-web of dimension k in R^{2k}. The construction described in the beginning of §3, with $m = k + 1$, so that $n = 2$, defines from a cubic hypersurface in P^{k+1} a 3-web of dimension k in R^{2k}. It can be shown that such a web is hexagonal. Is the converse true, i.e., does every hexagonal 3-web of dimension k in R^{2k} arise from such a construction (up to a local diffeomorphism)?

ADDED DURING PROOF. I am indebted to V. Goldberg that the answer to this question is "no".

4. Lie's treatment makes heavy use of overdetermined systems of partial differential equations. Give a proof of the Lie-Wirtinger theorem by PDE.

REFERENCES

1. M. A. Akivis, *Webs and almost Grassmann structures*, Soviet Math. Dokl. **21** (1980), no. 3. Further references to works by V. Goldberg and other Soviet mathematicians on the subject can be found in this paper.

2. W. Blaschke and G. Bol, *Geometrie der Gewebe*, Springer, Berlin, 1938.

3. S. S. Chern and P. A. Griffiths, *Abel's theorem and webs*, Jahresberichte der deut. Math. Ver. **80** (1978), 13–110; also, Corrections and Addenda, same Journal, **83** (1981), 78–83.

4. ______, *An inequality for the rank of a web and webs of maximum rank*, Ann. Scuola Norm. Pisa, Serie IV, **5** (1978), 539–557.

5. G. Darboux, *Théorie des surfaces*, t.1, (1914), 151–161.

6. P. A. Griffiths, *Variations on a theorem of Abel*, Invent. Math. **35** (1976), 321–390.

7. S. Lie, *Bestimmung aller Flächen, die in mehrfacher Weise durch Translationsbewegung einer Kurve erzeugt werden*, Arch. für Math. Bd. 7, Heft 2 (1882), 155–176; also Ges. Abhandlungen, Bd. 1, Abt 1, 450–467.

8. ______, *Das Abelsche Theorem und die Translationsmannigfaltigkeiten*, Leipziger Berichte 1897, pp. 181–248; also Ges. Abhandlungen, Bd. II, Teil II, paper XIV, pp. 580–639.

9. ______, Ges. Abhandlungen, Bd. 7.

10. H. Poincaré, *Fonctions abeliennes*, J. Math. Pures Appl., Serie 5, t.1, (1895), 219–314; also Oeuvres t.IV, pp. 384–472, in particular p. 430.

11. ______, *Sur les surfaces de translation et les fonctions abeliennes*, Bull. Soc. Math. France **29** (1901), 61–86; also Oeuvres t.VI, pp. 13–37.

12. B. Saint-Donat, *Variétés de translation et théorème de Torelli*, Comptes Rendus Paris **280** (1975), 1611–1612.

13. N. Tschebotarow, *Über Flächen welche Imprimitivitätssysteme in Bezug auf eine gegebene Kontinuirliche Transformationsgruppe enthalten*, Recueil Math. (Sbornik) **34** (1927), 149–204.

14. W. Wirtinger, *Lies Translationsmannigfaltigkeiten und Abelsche Integrale*, Monatshefte Math. Phys. **46** (1938), 384–431.

15. David B.. Damiano, *Webs, abelian equations, and characteristic classes*, Thesis, Brown University, 1980.

16. John B. Little, *Translation manifolds and the converse of Abel's theorem*, Thesis, Yale University, 1980.

DEPARTMENT OF MATHEMATICS, UNIVERSITY OF CALIFORNIA, BERKELEY, CALIFORNIA 94720

Sonderabdruck aus
ARCHIV DER MATHEMATIK
Vol. 38, 1982 BIRKHÄUSER VERLAG, BASEL UND STUTTGART Fasc. 1

Projective geometry, contact transformations, and CR-structures

By
SHIING-SHEN CHERN *)

Dedicated to MARTIN BARNER

1. Introduction. In a recent paper ([5], cf. references at the end) M. Kuranishi gave a new proof of a theorem of Kashiwara-Kawai-Sato and Boutet de Monvel that any non-degenerate CR-system on a conic neighborhood of its characteristic is conjugate under a canonical transformation to the CR-system derived from the Heisenberg structure. The theorem is clearly related to the classical theorem of Sophus Lie that a second-order differential equation

$$z'' = F(x, z, z') \tag{1}$$

in the (x, z)-plane has no local invariant under contact transformations. In this paper we shall give a generalization of Lie's theorem to high dimensions and apply it to CR-structures. Various notions of equivalence of CR-structures will be discussed.

2. A theorem on generalized projective geometry

In R^{n+1} or C^{n+1} consider ∞^{n+1} hypersurfaces

$$F(x^1, \ldots, x^{n+1}, a_1, \ldots, a_{n+1}) = 0, \tag{2}$$

where the x's are the coordinates and the a's are the parameters, and F is a C^∞- (resp analytic) function in all its arguments, with

$$\begin{aligned} &(\partial F/\partial x^1, \ldots, \partial F/\partial x^{n+1}) \neq 0, \\ &(\partial F/\partial a_1, \ldots, \partial F/\partial a_{n+1}) \neq 0. \end{aligned} \tag{3}$$

An example is given by

$$F = a_1 x^1 + \cdots + a_{n+1} x^{n+1} - 1, \tag{4}$$

when the hypersurfaces are hyperplanes.

We write $z = x^{n+1}$ and solve (2) for z:

$$z = z(x^\alpha, a_1, \ldots, a_{n+1}). \tag{2'}$$

Here, as later, we will agree that small Greek indices have the following range:

$$1 \leqq \alpha, \beta, \gamma, \cdots \leqq n. \tag{5}$$

---

*) Work done under partial support of NSF grant MC577-23579.

Differentiating (2′) and eliminating the a's, we get, under some generic hypothesis on the function F,

$$\frac{\partial^2 z}{\partial x^\alpha \partial x^\beta} = p_{\alpha\beta}(x, z, p), \tag{6}$$

where the arguments

$$(x, z, p) = (x^\alpha, z, p_\beta), \quad p_\beta = \frac{\partial z}{\partial x^\beta}. \tag{7}$$

It follows that the hypersurfaces (2) are the integral manifolds of the completely integrable differential system

$$\begin{aligned} \theta &\underset{\text{def}}{=} dz - \sum p_\alpha dx^\alpha = 0, \\ \theta_\alpha &\underset{\text{def}}{=} dp_\alpha - \sum p_{\alpha\beta}(x, z, p)\, dx^\beta = 0, \quad p_{\alpha\beta} = p_{\beta\alpha} \end{aligned} \tag{8}$$

in the $(2n + 1)$-dimensional space (x, z, p).

The space (x, z, p) has a geometrical meaning. In fact, we call a *hyperplane element* the figure consisting of a point and a hyperplane through it. The latter can be defined by

$$\theta = dz - \sum p_\alpha dx^\alpha = 0. \tag{9}$$

Thus the space (x, z, p) is the space of all hyperplane elements of R^{n+1} (or C^{n+1}). We generalize the notion of a hypersurface in R^{n+1} (or C^{n+1}) to be a family of ∞^n hyperplane elements satisfying (9). Then a point (x, z) is a hypersurface, namely the one consisting of all the hyperplane elements with this point as the origin. A *contact transformation* is a local diffeomorphism on (x^α, z, p_β) leaving invariant the equation (9).

We will prove the following theorem:

Theorem 1. *There is a contact transformation which carries* (2) *into the family of points.*

Proof. Let u^α, v be $n + 1$ independent first integrals of (8), so that

$$du^\alpha = a^\alpha \theta + \sum a^\alpha_\beta \theta^\beta, \quad dv = b\theta + \sum b_\beta \theta^\beta. \tag{10}$$

We can suppose that they are so chosen that

$$\det(a^\alpha_\beta) \neq 0. \tag{11}$$

We introduce q_α by the condition

$$dv - \sum q_\alpha du^\alpha \equiv 0, \quad \operatorname{mod} \theta,$$

i.e., the q_α satisfy

$$b_\beta = \sum q_\alpha a^\alpha_\beta,$$

and are thus uniquely determined. The transformation

$$(x^\alpha, z, p_\beta) \to (u^\alpha, v, q_\beta) \tag{12}$$

is a contact transformation carrying the system (8) to the system

(13) $du^\alpha = 0, \quad dv = 0.$ *QED*

In the special case $n = 1$ the system (8) becomes

$$dz - p\,dx = 0, \quad dp - r(x, z, p)\,dx = 0,$$

which is equivalent to (1). Thus we get Lie's theorem that the equation (1) has no local invariant under contact transformations.

The study of the invariants of (2) under point transformations, together with their geometrical interpretation, was first made by M. Hachtroudi, [4].

3. *CR*-structures and their equivalences.

In the study of functions of several complex variables we consider in C^{n+1} a real analytic hypersurface M defined by the equation

(14) $F(x^1, \ldots, x^{n+1}, \bar{x}^1, \ldots, \bar{x}^{n+1}) = 0,$

where F is a real-valued function. To M is associated, intrinsically, the family of Segre hypersurfaces

(15) $F(x^1, \ldots, x^{n+1}, a_1, \ldots, a_{n+1}) = 0.$

The aim is to study the local invariants of M under biholomorphic transformations of C^{n+1}, [1], [2]. B. Segre observed that among these are the invariants of the projective geometry defined by (15). E. Cartan stated, and J. Faran confirmed [3], that the Segre projective invariants are not complete. Nevertheless they have the advantage to be defined in the holomorphic category. Our theorem 1 says that the family (15) has no invariants under contact transformations.

The problem is best approached by the method of G-structures. We recall that on a manifold M of dimension m, a G-structure, corresponding to a given subgroup $G \subset GL(m)$, is given by a field of m linearly independent one-forms, defined up to a transformation of G. On the above real hypersurface M there is the real one-form

(16) $\omega = i\,\partial F = -i\,\bar{\partial} F,$

defined up to a real factor. Its exterior derivative can be written

(17) $d\omega = -i\,\partial\bar{\partial} F \equiv i \sum g_{\alpha\bar{\beta}}\,\omega^\alpha \wedge \omega^{\bar{\beta}}, \quad \bmod \omega,$

where ω^α are linear combinations of $dx^1, \ldots, dx^{n+1}$, and

(18) $\omega^{\bar{\beta}} = \overline{\omega^\beta}.$

Since ω is real, $g_{\alpha\bar{\beta}}$ is hermitian:

(19) $\bar{g}_{\alpha\bar{\beta}} = g_{\beta\bar{\alpha}}.$

The corresponding hermitian form is called the *Levi form* of M. We will suppose that it is non-degenerate, i.e.,

(20) $\det(g_{\alpha\bar{\beta}}) \neq 0$

By introducing the forms

(21) $\omega_\alpha = i \sum g_{\alpha\bar{\beta}}\,\omega^{\bar{\beta}},$

1*

equation (17) can be written

$$(22)\qquad d\omega \equiv \sum \omega^\alpha \wedge \omega_\alpha, \quad \bmod \omega.$$

The abstraction of the notion of a real hypersurface is that of a CR-manifold. We call a *contact manifold* a manifold M of dimension $2n+1$ on which there is a one-form ω, defined up to a factor, such that

$$(23)\qquad \omega \wedge (d\omega)^n \neq 0.$$

ω is called the contact form. A *strong CR-manifold* is a contact manifold on which there are given the contact form ω and the one-forms ω^α, ω_α, such that the following conditions are satisfied:

1) ω is a real one-form, while ω^α, ω_α are complex-valued one-forms such that ω^α differ from $\overline{\omega}_\beta = \omega_{\bar\beta}$ by a non-singular linear transformation.

2) The equation (22) is valid, so that ω, ω^α, ω_β are linearly independent.

3) The forms ω, ω_α are defined up to the transformation

$$(24)\qquad \begin{aligned} &\omega \to \omega^* = u\,\omega, \quad u \text{ real and } \neq 0,\\ &\omega_\alpha \to \omega_\alpha^* = \sum u_\alpha^\beta \omega_\beta + v_\alpha \omega, \quad \det(u_\alpha^\beta) \neq 0. \end{aligned}$$

4) The differential system

$$(25)\qquad \omega = \omega_\alpha = 0$$

is completely integrable.

By dropping the condition 1) we are led to a weak CR-manifold. Precisely, a *weak CR-manifold* is a (real or complex) contact manifold of dimension $2n+1$ on which there are given the contact form ω and the one-forms ω^α, ω_β, satisfying the following conditions:

1) The equation (22) is valid, so that ω, ω^α, ω_β are linearly independent.

2) The forms ω, ω^α, ω_β are defined up to the transformation

$$(26)\qquad \begin{aligned} &\omega \to \omega^* = u\,\omega, \quad u \neq 0,\\ &\omega^\alpha \to \omega^{*\alpha} = \sum t_\beta^\alpha \omega^\beta + s^\alpha \omega, \quad \det(t_\beta^\alpha) \neq 0,\\ &\omega_\alpha \to \omega_\alpha^* := \sum u_\alpha^\beta \omega_\beta + v_\alpha \omega, \quad \det(u_\alpha^\beta) \neq 0, \end{aligned}$$

subject to the condition that the equation (22) is preserved.

3) The differential system (25) is completely integrable.

Clearly an invariant of the weak CR-structure is an invariant of the strong CR-structure. The weak structure is defined without complex conjugation and is thus within the holomorphic category. Its invariants are an abstract form of the Segre invariants. On a weak CR-manifold a very weak CR-structure is defined, when we replace the middle equation of (26) by

$$(27)\qquad \omega^\alpha \to \omega^{*\alpha} = \sum t_\beta^\alpha \omega^\beta + s^\alpha \omega + \sum r^{\alpha\beta} \omega_\beta, \quad \det(t_\beta^\alpha) \neq 0.$$

A counter-part of Theorem 1 can be stated as follows:

Theorem 2. *A very weak CR-structure has no local invariants.*

Proof. Let ζ, ξ_α be the first integrals of the completely integrable system (25). Since ω is defined up to a factor, we can suppose

$$\omega = d\zeta - \sum \eta^\alpha d\xi_\alpha .$$

By (23) the functions $\xi_\alpha, \eta^\beta, \zeta$ are independent and can be regarded as local coordinates on M. We can also take

$$\omega_\alpha = d\xi_\alpha .$$

Then

$$d\omega = \sum d\xi_\alpha \wedge d\eta^\alpha = \sum \omega_\alpha \wedge d\eta^\alpha .$$

Comparing with (22), we get

$$\omega^\alpha + d\eta^\alpha \equiv 0 , \qquad \operatorname{mod} \omega , \omega_\beta .$$

By the transformation allowed in (27), we can choose $\omega^\alpha = - d\eta^\alpha$. *QED*

References

[1] S. S. Chern and J. K. Moser, Real hypersurfaces in complex manifolds. Acta Math. **133**, 219–271 (1974).
[2] S. S. Chern, On the projective structure of a real hypersurface in C_{n+1}. Math. Scandinavica **36**, 74–82 (1975).
[3] James J. Faran, Segre families and real hypersurfaces. Invent. Math. **60**, 135–172 (1980).
[4] M. Hachtroudi, Les espaces d'éléments à connexion projective normale. Paris **1937**.
[5] M. Kuranishi, On the equivalence of symbols of non-degenerate CR-structures. Comm. Pure Appl. Math. **33**, 1–21 (1980).

Eingegangen am 7. 4. 1981

Anschrift des Autors:

Shiing-shen Chern
Department of Mathematics
University of California
Berkeley, California 94720
USA

MINIMAL SURFACES BY MOVING FRAMES

SHIING-SHEN CHERN[1] and JON GORDON WOLFSON[2]

Dedicated to André Weil

In this paper we will develop, by moving frames, the basic formulas and some results on minimal surfaces, both in a Riemannian and in a Kählerian manifold. When the ambient Riemannian manifold is of constant curvature, we will show that a quartic form defined in a natural way from the second fundamental form is holomorphic, relative to the underlying complex structure of the surface. This will be applied to the study of minimal surfaces on the unit 4-sphere $S^4(1)$. We will show how to derive by this approach a theorem of R. Bryant that a compact Riemann surface can be conformally immersed as a minimal surface in $S^4(1)$.

When the ambient space is a Kählerian manifold, the first-order jet of the surface defines a scalar invariant, which is the Kähler form divided by the element of area of the surface. This will be used to normalize the frames attached to the surface. When the surface is minimal, an invariant cubic form appears naturally. We will show that it is holomorphic, if the ambient Kählerian manifold is of constant holomorphic sectional curvature. This will be applied to the complex projective plane $P_2(\mathbf{C})$ (with the Fubini-Study metric), to give a family of minimal surfaces in $P_2(\mathbf{C})$, and all its minimal two-spheres. A determination of all the minimal 2-sphere in $P_n(\mathbf{C})$ by this method has been carried out by the second author [6]. This latter problem was first studied by Din-Zakrewski [4] and a mathematician's treatment was given by Eells-Wood [5].

Throughout this paper we will adopt the following ranges of indices:

$$1 \leqq A, B, C \cdots \leqq N,$$

$$1 \leqq i, j, \ldots \leqq 2,$$

Manuscript received August 17, 1982

[1] Work done under partial support of NSF grant MCS80-23356.
[2] Partially supported by a postgraduate scholarship of NSERC of Canada.

$$1 \leqq \alpha, \beta, \gamma, \ldots \leqq n$$

$$3 \leqq r, s, t, \ldots \leqq N, \tag{0.1}$$

$$3 \leqq \lambda, \mu, \ldots \leqq n,$$

$$0 \leqq \rho, \sigma, \tau \leqq 4$$

$$0 \leqq a, b, c \leqq 2.$$

Before concluding this introduction we wish to thank Jost Eschenburg and Therese Langer for their critical comments and discussions.

1. Minimal surfaces in a Riemannian manifold. Consider a Riemannian manifold X of dimension N. Let e_A and θ_B be smooth fields of dual orthonormal frames and coframes (i.e., one-forms), so that we have the inner products

$$(e_A, e_B) = \delta_{AB} \tag{1.1}$$

and

$$ds^2 = \theta_1^2 + \cdots + \theta_N^2. \tag{1.2}$$

The Levi-Civita connection is given by an anti-symmetric matrix (θ_{AB}) of one-forms, the connection matrix, satisfying the conditions

$$d\theta_A = \Sigma \theta_B \wedge \theta_{BA}, \tag{1.3}$$

$$\theta_{AB} + \theta_{BA} = 0. \tag{1.4}$$

These θ_{AB} are the coefficients of the covariant differential

$$De_A = \Sigma \theta_{AB} e_B. \tag{1.5}$$

Its curvature Θ_{AB} is defined by the structure equations

$$d\theta_{AB} = \Sigma \theta_{AC} \wedge \theta_{CB} + \Theta_{AB}, \tag{1.6}$$

where

$$\Theta_{AB} + \Theta_{BA} = 0 \tag{1.7}$$

and we set

$$\Theta_{AB} = \tfrac{1}{2}\Sigma R_{ABCD}\theta_C \wedge \theta_D. \tag{1.8}$$

The R_{ABCD} satisfy the usual symmetry relations. X is of constant curvature c, if

$$\Theta_{AB} = -c\theta_A \wedge \theta_B. \tag{1.9}$$

Let $x: M \to X$ be an immersed (two-dimensional) surface. We choose the orthonormal frame field $e_A(p)$, $p \in M$, such that $e_i(p)$ are the tangent vectors and $e_r(p)$ are the normal vectors. (Notice the convention on indices in (0.1).) We call these *Darboux frames*. Restricted to M we have then

$$\theta_r = 0. \tag{1.10}$$

The first relative invariants of M are given by the second fundamental form, which is by definition

$$\mathrm{II} = \Sigma\theta_i(De_i)^{\perp} = \Sigma\theta_i\theta_{ir}e_r, \tag{1.11}$$

where $(De_i)^{\perp}$ is the normal component of De_i. II is thus a normal-valued symmetric quadratic differential form, and is clearly independent of the choice of the frame field. We will write

$$\mathrm{II} = \Sigma \mathrm{II}_r e_r, \tag{1.12}$$

where

$$\mathrm{II}_r = \Sigma\theta_i\theta_{ir}. \tag{1.13}$$

Under a change of the frame field the II_r's undergo an orthogonal transformation.

Taking the exterior derivative of (1.10) and using (1.3), we get

$$d\theta_r = \Sigma\theta_i \wedge \theta_{ir} = 0. \tag{1.14}$$

It follows that we can set

$$-\theta_{ir} = \theta_{ri} = \Sigma h_{rik}\theta_k, \tag{1.15}$$

where

$$h_{rik} = h_{rki}. \tag{1.16}$$

Then we have

$$-\mathrm{II}_r = \Sigma h_{rik}\theta_i\theta_k. \tag{1.17}$$

M is a minimal surface, if

$$-\mathrm{Tr}\mathrm{II}_r = \Sigma h_{rii} = 0. \tag{1.18}$$

In this case the second fundamental forms are intimately related to the complex structure of M induced by its Riemannian metric. In fact, the complex structure of M can be defined by the complex-valued one-form

$$\phi = \theta_1 + i\theta_2, \tag{1.19}$$

and we have, for a minimal surface,

$$-\mathrm{II}_r = h_{r11}(\theta_1^2 - \theta_2^2) + 2h_{r12}\theta_1\theta_2 = \mathrm{Re}\,\mathrm{II}_r', \tag{1.20}$$

where

$$\mathrm{II}_r' = \bar{H}_r\phi^2, \tag{1.21}$$

$$H_r = h_{r11} + ih_{r12}. \tag{1.22}$$

II_r' is a complex-valued ordinary form of type (2, 0), and is completely determined from II_r by (1.20), (1.21). Like II_r they are determined up to an orthogonal transformation. It follows that the 4-form

$$Q = (\Sigma \bar{H}_r^2)\phi^4, \tag{1.23}$$

of type (4, 0), is globally defined on M, independent of choice of frames.

We suppose

$$\Theta_{ir} = 0. \tag{1.24}$$

Taking the exterior derivative of (1.15) and using the structure equations (1.3), (1.6) we get

$$Dh_{rik} \wedge \theta_k = 0, \tag{1.25}$$

where

$$Dh_{rik} = dh_{rik} - \Sigma h_{rij}\theta_{kj} - \Sigma h_{rkj}\theta_{ij} - \Sigma h_{sik}\theta_{rs}, \tag{1.26}$$

By (1.25) we can set

$$Dh_{rik} = \Sigma h_{rikj}\theta_j \tag{1.27}$$

with

$$h_{rikj} = h_{rijk}. \tag{1.28}$$

Combined with (1.16), this means that h_{rikj} is symmetric in any two of the indices i, k, j. Utilizing this fact, and (1.18), we find, for M minimal,

$$d(\Sigma \bar{H}_r^2) - 4i(\Sigma \bar{H}_r^2)\theta_{12} = (h_{r111} - ih_{r112})\phi. \tag{1.29}$$

THEOREM 1. *Let $x: M \to X$ be a minimal surface, where X is a Riemannian manifold of constant curvature. Then the form Q defined in* (1.23) *is a holomorphic form of degree* 4.

Proof. The conclusion means that relative to a local coordinate ζ of the complex structure ϕ on M, $Q = f(\zeta)d\zeta^4$, where $f(\zeta)$ is a holomorphic function of ζ.

Since X is of constant curvature, the condition (1.24) is satisfied relative to Darboux frames, so that the above formulas are valid. We can write

$$\phi = \lambda d\zeta. \tag{1.30}$$

Taking its exterior derivative and using (1.3), we get

$$(d\lambda + i\theta_{12}\lambda) \wedge d\zeta = 0,$$

i.e.,

$$d\lambda + i\theta_{12}\lambda \equiv 0, \text{ mod } d\zeta. \tag{1.31}$$

By the definition (1.23)

$$Q = (\Sigma \bar{H}_r^2)\lambda^4 \cdot d\zeta^4.$$

From (1.29) and (1.31), we get

$$d(\lambda^4 \, \Sigma \bar{H}_r^2) \equiv 0, \text{ mod } d\zeta,$$

which means that the coefficient of $d\zeta^4$ in Q is a holomorphic function of ζ. This proves the theorem.

We cannot help feeling the mysterious relationship between minimal surface theory and complex function theory. It seems remarkable that the holomorphy of Q follows directly from the structure equations, assuming of course that X is of constant curvature.

Following R. Bryant we call a minimal surface M superminimal if $Q = 0$. Then we have the

COROLLARY. *A minimal two-dimensional sphere in a Riemannian manifold of constant curvature is superminimal.*

2. Minimal surfaces in a Kählerian manifold. Let Y be a Kählerian manifold of complex dimension n, whose metric we write as

$$ds^2 = \Sigma \omega_\alpha \bar{\omega}_\alpha = \Sigma \omega_\alpha \omega_{\bar{\alpha}}. \tag{2.1}$$

The forms ω_α are of type (1, 0) and are defined up to a unitary transformation. They constitute a unitary coframe. Relative to a coframe field ω_α, a unitary connection $\omega_{\alpha\bar{\beta}}$ is uniquely determined by the conditions

$$d\omega_\alpha = \Sigma \omega_{\alpha\bar{\beta}} \wedge \omega_\beta, \tag{2.2}$$

(2.3) $$\omega_{\alpha\bar{\beta}} + \omega_{\bar{\beta}\alpha} = 0 \; (\omega_{\bar{\beta}\alpha} = \bar{\omega}_{\beta\bar{\alpha}}).$$

Its curvature is given by

(2.4) $$d\omega_{\alpha\bar{\beta}} = \Sigma \omega_{\alpha\bar{\gamma}} \wedge \omega_{\gamma\bar{\beta}} + \Omega_{\alpha\bar{\beta}}.$$

Y is said to be of constant holomorphic sectional curvature 4ρ, if

(2.5) $$\Omega_{\alpha\bar{\beta}} = -\rho(\omega_\alpha \wedge \omega_{\bar{\beta}} + \delta_{\alpha\bar{\beta}} \Sigma \omega_\gamma \wedge \omega_{\bar{\gamma}}).$$

To get hold of the underlying Riemannian structure we set

(2.6) $$\omega_\alpha = \theta_\alpha + i\theta_{n+\alpha}.$$

Then (2.1) takes the form (1.2) with $N = 2n$, and the θ_A constitute an orthonormal coframe. The notion of a minimal surface in Y is defined in terms of its underlying Riemannian structure.

Consider now an immersed surface

(2.7) $$y: M \to Y.$$

We can choose a field of coframes over M satisfying

(2.8) $$\omega_\lambda = 0.$$

The induced metric on M is then

(2.9) $$ds_M^2 = \omega_1\bar{\omega}_1 + \omega_2\bar{\omega}_2.$$

Let the complex valued 1-form ϕ define the complex structure on M. We can modify ϕ by a real factor so that

(2.10) $$ds_M^2 = \phi\bar{\phi};$$

ϕ is then defined up to a complex factor of norm 1. We have, restricted to M,

(2.11) $$\omega_i = s_i\phi + t_i\bar{\phi}.$$

Substituting (2.11) into (2.9) and comparing with (2.10) we get

$$\begin{aligned} &|s_1|^2 + |s_2|^2 + |t_1|^2 + |t_2|^2 = 1, \\ &s_1\bar{t}_1 + s_2\bar{t}_2 = 0. \end{aligned} \tag{2.12}$$

The ω_i are defined up to a unitary transformation. In $\mathbf{C}^2$ we introduce the vectors

$$\vec{s} = (s_1, s_2), \qquad \vec{t} = (t_1, t_2).$$

Equations (2.12) show that they are orthogonal, with the sum of the squares of their norms equal to 1. By a unitary transformation at each point of M we can suppose

$$\vec{s} = \cos\frac{\alpha}{2}(1, 0), \quad \vec{t} = \sin\frac{\alpha}{2}(0, 1).$$

giving the "normalization"

$$\omega_1 = \cos\frac{\alpha}{2}\phi, \qquad \omega_2 = \sin\frac{\alpha}{2}\bar{\phi}. \tag{2.13}$$

If at a point $p \in M$ both $\vec{s}(p) \neq 0$ and $\vec{t}(p) \neq 0$ then in a neighborhood of p this normalization is smooth, that is, α is a smooth function. If either $\vec{s}(p) = 0$ or $\vec{t}(p) = 0$ then α can be defined only continuously.

However in all cases we have the normalization

$$\omega_1 = s\phi, \qquad \omega_2 = t\bar{\phi}, \tag{2.14}$$

where s and t are complex valued smooth functions which satisfy $|s|^2 + |t|^2 = 1$. In fact, if $\vec{s}(p) \neq 0$ then in a neighborhood of p, s can be taken to be real valued and similarly for t. Thus the normalization (2.13) is a special case of (2.14) for $st \neq 0$.

The quantity α has a simple geometric meaning. In fact, the Kähler form of the Hermitian metric is

$$\Omega = \frac{i}{2}(\omega_1 \wedge \bar{\omega}_1 + \omega_2 \wedge \bar{\omega}_2) = \frac{i}{2}\cos\alpha\phi \wedge \bar{\phi}, \tag{2.15}$$

where $(i/2)\phi \wedge \bar{\phi}$ is the element of area of M. This gives a geometric interpretation of $\cos\alpha$, hence of α. For a holomorphic (resp. anti-holomorphic) curve, we have $\sin(\alpha/2) = 0$. (resp. $\cos(\alpha/2) = 0$), and in both cases we have $\cos\alpha = \pm 1$, and Ω gives the element of area, up to sign. Without doubt α is the most fundamental invariant of an immersed real surface in an Hermitian manifold, as it is defined by the jet of the first order.

We will find the condition that M is a minimal surface. We set

$$\begin{aligned} \bar{s}\omega_1 + t\bar{\omega}_2 &= \theta_1 + i\theta_2, \\ \bar{t}\omega_1 - s\bar{\omega}_2 &= \theta_3 + i\theta_4, \\ \omega_\lambda &= \theta_{2\lambda-1} + i\theta_{2\lambda}. \end{aligned} \tag{2.16}$$

Then θ_A is an orthonormal coframe of the underlying Riemannian structure of Y. It is also a Darboux coframe of M, because along M we have

$$\bar{t}\omega_1 - s\bar{\omega}_2 = 0, \tag{2.17}$$

$$\omega_\lambda = 0. \tag{2.18}$$

By taking the exterior derivative of (2.17) and making use of (2.2), we get

$$[(sd\bar{t} - \bar{t}ds) + s\bar{t}(\omega_{1\bar{1}} + \omega_{2\bar{2}})] \wedge \phi + \omega_{1\bar{2}} \wedge \bar{\phi} = 0$$

which allows us to set

$$(sd\bar{t} - \bar{t}ds) + s\bar{t}(\omega_{1\bar{1}} + \omega_{2\bar{2}}) = a\phi + b\bar{\phi}, \tag{2.19}$$

$$\omega_{1\bar{2}} = b\phi + c\bar{\phi}. \tag{2.20}$$

Similarly, exterior differentiation of (2.18) gives

$$s\omega_{\lambda\bar{1}} \wedge \phi + t\omega_{\lambda\bar{2}} \wedge \bar{\phi} = 0$$

and we can write

$$\begin{aligned} s\omega_{\lambda\bar{1}} &= a_\lambda\phi + b_\lambda\bar{\phi}, \\ t\omega_{\lambda\bar{2}} &= b_\lambda\phi + c_\lambda\bar{\phi}. \end{aligned} \tag{2.21}$$

Instead of the $2n-2$ real second fundamental forms we can clearly consider the $n-1$ complex valued ones:

$$\mathrm{II}^{\mathbf{C}} = a\phi^2 + 2b\phi\bar{\phi} + c\bar{\phi}^2,$$

$$\mathrm{II}_\lambda^{\mathbf{C}} = a_\lambda\phi^2 + 2b_\lambda\phi\bar{\phi} + c_\lambda\bar{\phi}^2.$$

The condition for M to be minimal is the vanishing of the traces of $\mathrm{II}^{\mathbf{C}}$, $\mathrm{II}_\lambda^{\mathbf{C}}$, which is therefore

$$b = b_\lambda = 0. \tag{2.22}$$

In the case that M is minimal (2.19) yields interesting information about the zeroes of s and t. Let $p \in M$, then as $|s|^2 + |t|^2 = 1$ in a neighborhood of p either $s \neq 0$ or $t \neq 0$, say $s \neq 0$. (2.19) can be written

$$d(s\bar{t}) + s\bar{t}\left(\omega_{1\bar{1}} + \omega_{2\bar{2}} - 2\frac{ds}{s}\right) = a\phi. \tag{2.23}$$

Suppose ζ is a complex coordinate centered at p. Then (2.23) implies

$$\frac{\partial(s\bar{t})}{\partial\bar{\zeta}} + s\bar{t}\cdot h = 0, \tag{2.24}$$

where h is a C^∞ complex valued function. In fact $h\,d\bar{\zeta}$ is the (0, 1) part of $\omega_{1\bar{1}} + \omega_{2\bar{2}} - 2(ds/s)$. By the Theorem in Section 4 of [3] (2.24) implies that either $s\bar{t}$ vanishes identically or

$$s\bar{t} = \zeta^r \tilde{h}(\zeta),$$

where r is an integer ≥ 0 and $\tilde{h}$ is a C^∞ complex valued function such that $\tilde{h}(0) \neq 0$. In particular if M is minimal then s and t either vanish identically (in which case M is a holomorphic or antiholomorphic curve) or they have only isolated zeroes.

When M is minimal and Y has constant holomorphic sectional curvature (2.20) leads to an amazing theorem. Our coframe is defined up to the transformation

$$\omega_1 \to e^{ik_1}\omega_1,\ \omega_2 \to e^{ik_2}\omega_2;\quad k_1, k_2 \text{ real}$$

under which $\omega_{1\bar{2}}$ transforms as

$$\omega_{1\bar{2}} \to e^{ik_1}\omega_{1\bar{2}}e^{-ik_2}.$$

Hence the form

$$P = \omega_1\omega_{\bar{1}2}\omega_{\bar{2}} = s\bar{t}\bar{c}\phi^3 \tag{2.25}$$

which is a complex valued symmetric differential form of type (3, 0), is an invariant of the minimal surface M. It depends on the second-order jet.

THEOREM 2. *Let Y be a Kählerian manifold of constant holomorphic sectional curvature. Let $M \to Y$ be a minimal surface. Then the cubic form P is holomorphic.*

Proof. We use the local complex coordinate ζ as defined by (1.30). Let

$$\omega_1 = p_1 d\zeta, \qquad \omega_2 = p_2 d\bar{\zeta}, \qquad \omega_{1\bar{2}} = q d\bar{\zeta},$$

so that

$$P = (p_1\bar{p}_2\bar{q})d\zeta^3.$$

It remains to show that the product in the parentheses is a holomorphic function of ζ.

The structure equations give

$$d\omega_{1\bar{2}} = (\omega_{1\bar{1}} - \omega_{2\bar{2}}) \wedge \omega_{1\bar{2}} + \Sigma\omega_{1\bar{\lambda}} \wedge \omega_{\lambda\bar{2}} + \Omega_{1\bar{2}}.$$

By (2.21) $\omega_{1\bar{\lambda}} = -\bar{\omega}_{\lambda\bar{1}}$ and $\omega_{\lambda\bar{2}}$ are, for a minimal surface, both of type (0, 1), so that the middle sum is zero. On the other hand, $\Omega_{1\bar{2}} = 0$ since Y is of constant holomorphic sectional curvature. Substituting the expression for $\omega_{1\bar{2}}$ into this equation, we get

$$d\bar{q} \equiv \bar{q}(-\omega_{1\bar{1}} + \omega_{2\bar{2}}), \qquad \text{mod } d\zeta.$$

Similarly, by using the formulas for $d\omega_1$, $d\omega_2$, we derive

$$dp_1 \equiv p_1\omega_{1\bar{1}}, \qquad d\bar{p}_2 \equiv -\bar{p}_2\omega_{2\bar{2}}, \qquad \text{mod } d\zeta.$$

Hence we get

$$d(p_1\bar{p}_2\bar{q}) \equiv 0 \bmod d\zeta,$$

and the theorem is proved.

We know that if M is a minimal surface which is neither holomorphic nor antiholomorphic then s and t vanish only at isolated points. It follows that these points are the only zeroes of $\sin\alpha$ and that α is smooth everywhere except at these isolated points. For possible use in future investigations we wish to derive formulas for the Gaussian curvature K of M and for $\Delta\alpha$, Δ being the Laplacian, at points $p \in M$ where $\sin\alpha(p) \neq 0$. We restrict ourselves to the case that Y is of constant holomorphic sectional curvature.

In order to do these computations we utilize the normalization (2.13) to derive the second fundamental forms of M. We get, by the same computations as above:

$$\frac{1}{2}[d\alpha + \sin\alpha(\omega_{1\bar{1}} + \omega_{2\bar{2}})] = a\phi + b\bar{\phi}, \tag{2.26}$$

$$\omega_{1\bar{2}} = b\phi + c\bar{\phi}, \tag{2.27}$$

and

$$\begin{aligned} \cos\frac{\alpha}{2}\,\omega_{\lambda\bar{1}} &= a_\lambda\phi + b_\lambda\bar{\phi}, \\ \sin\frac{\alpha}{2}\,\omega_{\lambda\bar{2}} &= b_\lambda\phi + c_\lambda\bar{\phi}. \end{aligned} \tag{2.28}$$

As above the minimality of M is equivalent to $b = b_\lambda = 0$. Of course these formulas are only valid where $\sin\alpha \neq 0$.

To compute K notice that it is uniquely determined by the equations

$$d\phi = -i\theta_{12} \wedge \phi,$$

$$d\theta_{12} = -K\frac{i}{2}\phi \wedge \bar{\phi},$$

where θ_{12} is the connection form of the Riemannian metric on M and is a real one-form. By (2.13) we get

$$\phi = \cos\frac{\alpha}{2}\,\omega_1 + \sin\frac{\alpha}{2}\,\bar{\omega}_2,$$

from which we find by differentiation,

$$d\phi = \left(\cos^2\frac{\alpha}{2}\,\omega_{1\bar{1}} - \sin^2\frac{\alpha}{2}\,\omega_{2\bar{2}}\right)\wedge\phi.$$

It follows that

$$\theta_{12} = i\left(\cos^2\frac{\alpha}{2}\,\omega_{1\bar{1}} - \sin^2\frac{\alpha}{2}\,\omega_{2\bar{2}}\right). \tag{2.29}$$

Since α is real and $\omega_{i\bar{i}}$ are purely imaginary, the conjugate complex of (2.26) gives

$$d\alpha - \sin\alpha(\omega_{1\bar{1}} + \omega_{2\bar{2}}) = 2\overline{a\phi}.$$

It follows that

$$d\alpha = a\phi + \overline{a\phi},$$

and

$$\begin{aligned} \partial\alpha &= a\phi = \frac{1}{2}\{d\alpha + \sin\alpha(\omega_{1\bar{1}} + \omega_{2\bar{2}})\}, \\ \bar{\partial}\alpha &= \overline{a\phi} = \frac{1}{2}\{d\alpha - \sin\alpha(\omega_{1\bar{1}} + \omega_{2\bar{2}})\}, \end{aligned} \tag{2.30}$$

We also have

$$\Omega_{1\bar{1}} = -\rho\left(2\cos^2\frac{\alpha}{2} - \sin^2\frac{\alpha}{2}\right)\phi\wedge\bar{\phi},$$

$$\Omega_{2\bar{2}} = -\rho\left(\cos^2\frac{\alpha}{2} - 2\sin^2\frac{\alpha}{2}\right)\phi\wedge\bar{\phi}.$$

Differentiating (2.29), we find

$$\begin{aligned} -\frac{1}{2}K &= |a|^2 + |c|^2 + \Sigma|a_\lambda|^2 + |c_\lambda|^2 \\ &\quad + \left(-2 + \frac{3}{2}\sin^2\alpha\right)\rho. \end{aligned} \tag{2.31}$$

Similarly, the Laplacian of α is obtained by differentiating (2.30), and we get

$$\sin\alpha\,\partial\bar{\partial}\alpha = \cos\alpha\,\partial\alpha\wedge\bar{\partial}\alpha + 2\Big(-\sin^2\frac{\alpha}{2}\Sigma|a_\lambda|^2 + \cos^2\frac{\alpha}{2}\Sigma|c_\lambda|^2 + \frac{3}{2}\rho\sin^2\alpha\cos\alpha\Big)\phi\wedge\bar{\phi}, \tag{2.32}$$

from which $\Delta\alpha$ can be obtained through the formula

$$\partial\bar{\partial}\alpha = \frac{1}{4}\Delta\alpha\phi\wedge\bar{\phi}. \tag{2.33}$$

3. Minimal surfaces in the unit four-sphere. We wish to apply the results of Section 1 to the case that X is the sphere $S^4(1)$ of radius 1 in the euclidean space E^5 of dimension 5. We consider orthonormal frames $e_0, e_1, \ldots, e_4$ in E^5 and let $S^4(1)$ be described by its first vector e_0. We have

$$de_\rho = \Sigma\theta_{\rho\sigma}e_\sigma, \qquad 0 \leqq \rho, \sigma, \tau \leqq 4, \tag{3.1}$$

$$\theta_{\rho\sigma} + \theta_{\sigma\rho} = 0, \tag{3.2}$$

and, through exterior differentiation,

$$d\theta_{\rho\sigma} = \Sigma\theta_{\rho\tau}\wedge\theta_{\tau\sigma}. \tag{3.3}$$

The latter are the structure equations of the group $SO(5)$.

To recover the structure equations of $S^4(1)$, considered as a Riemannian manifold in the sense of Section 1, it suffices to set

$$\theta_{0A} = \theta_A, \qquad 1 \leqq A, B, C, \ldots \leqq 4. \tag{3.4}$$

Then equations (3.3) reduce to (1.6), with

$$\Theta_{AB} = -\theta_A\wedge\theta_B,$$

verifying the fact that $S^4(1)$ has constant curvature 1.

In this case our results in Section 1 have further geometrical interpretation. We complexify E^5 to $\mathbf{C}^5$, and extend the inner product $(\cdot,\cdot)$, so that it is complex bilinear in $\mathbf{C}^5$. Let

$$\pi:\mathbf{C}^5 - \{0\} \to P_4(\mathbf{C}) \tag{3.5}$$

be the map which sends $Z \in \mathbf{C}^5 - \{0\}$ to the point of the complex projective space with Z as its homogeneous coordinate vector. Then the equation

$$(Z, Z) = 0 \tag{3.6}$$

defines a non-degenerate hyperquadric Q_3 in $P_4(\mathbf{C})$. We will identify Z with the corresponding point in $P_4(\mathbf{C})$. Let

$$E_1 = e_1 + ie_2, \qquad E_2 = e_3 + ie_4. \tag{3.7}$$

Then

$$(E_1, E_1) = (E_1, E_2) = (E_2, E_2) = 0, \tag{3.8}$$

and the line E_1E_2 lies completely on Q_3.

Before studying minimal surfaces in $S^4(1)$ we will give a discussion of the projective geometry in question. By using the Plücker line coordinates in $P_3(\mathbf{C})$ we can identify the lines of $P_3(\mathbf{C})$ with the points of a non-degenerate hyperquadric Q_4 in $P_5(\mathbf{C})$. Our Q_3 is the intersection of Q_4 by a hyperplane in $P_5(\mathbf{C})$ and can hence be identified with thc lines of a linear line complex, L, i.e., all lines in $P_3(\mathbf{C})$ whose Plücker coordinates satisfy a linear equation. Moreover, L is a non-special linear complex, i.e., not consisting of lines meeting a given line. To a line λ of Q_3 corresponds a pencil of lines of L, i.e., all lines through a given point and lying in a given plane. By associating this point to λ we see that the manifold of all lines on Q_3 is three-dimensional and is biholomorphic to $P_3(\mathbf{C})$.

Let $Z = (z_0, z_1, \ldots, z_4)$. Then equation (3.6) can be written

$$z_0^2 + z_1^2 + \cdots + z_4^2 = 0. \tag{3.9}$$

Our Q_3, defined by this equation, has therefore the properties:

a) Q_3 contains no real point;
b) if $W \in Q_3$, its conjugate complex point $\bar{W} \in Q_3$.

It follows that if $\lambda \subset Q_3$, its conjugate complex line $\bar{\lambda} \subset Q_3$. λ and $\bar{\lambda}$ will not intersect, for their intersection would be a real point on Q_3. Hence they span a three-dimensional real projective space which, when joined to the origin of E^5, gives a real hyperplane $\{0, \lambda, \bar{\lambda}\}$ through the origin. Let x be the unit vector in E^5 perpendicular to it. Let E_1, E_2 be two points spanning λ. Then $x, E_1, E_2, \bar{E}_1, \bar{E}_2$ are linearly independent and the condition

$$(x, E_1, E_2, \bar{E}_1, \bar{E}_2) > 0 \tag{3.10}$$

uniquely determines x. It is also independent of the choice of E_1, E_2 on λ, for another choice E_1^*, E_2^* will differ from E_1, E_2 by a linear transformation and under such a change the determinant at the left-hand side will be multiplied by a positive factor. By assigning x to λ we have defined a mapping

$$p: P_3(\mathbf{C}) \to S^4(1). \tag{3.11}$$

To understand this mapping, let $x \in S^4(1)$. It defines an orthogonal hyperplane through the origin, which projects under π to a real hyperplane in P_4. The latter intersects Q_3 in a non-degenerate quadric Q_2. Q_2 has two families of lines such that lines of the same family do not intersect and two lines of different families always have a common point. As a result of this property a line and its conjugate complex line belong to the same family. All the lines of Q_2 are mapped by p into x or $-x$, and by continuity the lines of one family are mapped into the same point. Let λ be a line on Q_2 and take $E_1 \in \lambda$. Let E_2 be the point of intersection of the line of the first family through E_1 and the line of the second family through $\bar{E}_1$.

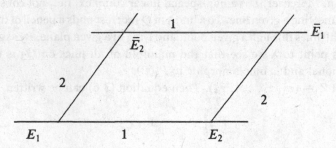

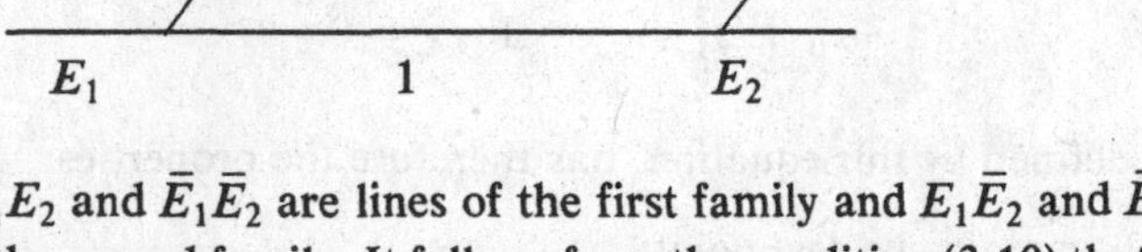

Then E_1E_2 and $\bar{E}_1\bar{E}_2$ are lines of the first family and $E_1\bar{E}_2$ and $\bar{E}_1E_2$ are lines of the second family. It follows from the condition (3.10) that the two families are mapped by p into x and $-x$ respectively, and $p^{-1}(x)$ is biholomorphically $P_1(\mathbf{C})$. Hence (3.11) defines a fibering of $P_3(\mathbf{C})$ with $P_1(\mathbf{C})$ as fiber. (3.11) is called the Penrose twistor map and our discussion gives a line-geometric interpretation of it.

We now return to our minimal surface in $S^4(1)$. We write $x = e_0$ and suppose that $xe_1e_2e_3e_4$ is a Darboux frame. With E_1, E_2 defined by (3.7), equations (3.1) can be written

$$
\begin{aligned}
dx &= \frac{1}{2}(\bar{\phi}E_1 + \phi\bar{E}_1) \\
dE_1 &= -\phi x - i\theta_{12}E_1 - \frac{1}{2}\bar{\phi}\{(H_3 - iH_4)E_2 \\
&\qquad + (H_3 + iH_4)\bar{E}_2\}, \\
dE_2 &= -i\theta_{34}E_2 + \frac{1}{2}(\bar{H}_3 + i\bar{H}_4)\phi E_1 \\
&\qquad + \frac{1}{2}(H_3 + iH_4)\bar{\phi}\bar{E}_1 .
\end{aligned} \tag{3.12}
$$

Suppose $M \to S^4(1)$ be superminimal. Then

$$H_3^2 + H_4^2 = 0, \tag{3.13}$$

which implies

$$H_3 \pm iH_4 = 0.$$

Following a discussion in [2], we can suppose that

$$H_3 + iH_4 = 0. \tag{3.14}$$

Then E_2 describes a holomorphic curve on Q_3, whose tangent lines E_2E_1 belong to Q_3. We call this curve the *directrix curve* and denote it by Δ.

Conversely, let a holomorphic (or anti-holomorphic) curve Δ on Q_3 be given, such that its tangent lines belong to Q_3. Identifying the left-hand side of (3.11) with the space of lines on Q_3, the map p defines a surface M on $S^4(1)$. We will show that this surface is minimal.

Suppose Δ be described by the point E_2. We normalize the coordinate vector, so that

(3.15) $$(E_2, \bar{E}_2) = 2,$$

in conformity with the above notation. Let E_2E_1 be the tangent to Δ at E_2. The condition that it belongs to Q_3 is

(3.16) $$(E_1, E_1) = (E_1, E_2) = (E_2, E_2) = 0.$$

The vector E_1 is defined up to the change

$$E_1 \to E_1 + sE_2.$$

After such a change if necessary, we can suppose

(3.17) $$(E_1, \bar{E}_2) = 0.$$

We further normalize E_1 so that

(3.18) $$(E_1, \bar{E}_1) = 2.$$

The image point x satisfies, by definition, the conditions

(3.19) $$(x, x) = 1, \qquad (x, E_1) = (x, E_2) = 0.$$

We can use as a base of $\mathbf{C}^5$ the vectors $x, E_1, E_2, \bar{E}_1, \bar{E}_2$. Then the inner products of any two of them are zero, except the following:

(3.20) $$(x, x) = 1, \qquad (E_1, \bar{E}_1) = (E_2, \bar{E}_2) = 2.$$

We wish to express dx, dE_1, dE_2 as linear combinations of $x, E_1, E_2, \bar{E}_1, \bar{E}_2$; $d\bar{E}_1, d\bar{E}_2$ are then obtained by complex conjugation. Since E_2E_1 is the tangent line to the holomorphic curve described by E_2 and E_2 is normalized by (3.15), dE_2 is a linear combination of E_1 and E_2, with the coefficient of E_2 a purely imaginary form. In conformity with (3.10), we write this relation as

(3.21) $$dE_2 = -i\theta_{34}E_2 + \psi E_1,$$

when θ_{34} is a real one-form and ψ is of type (1, 0).

Let

$$dE_1 = -\pi x + \pi_1 E_1 + \pi_2 E_2 + \pi_3 \bar{E}_1 + \pi_4 \bar{E}_2.$$

Taking its inner products with E_1, E_2 respectively, we get

$$\pi_3 = \pi_4 = 0.$$

Similarly, taking its inner products with $\bar{E}_1$, $\bar{E}_2$, we find that π_1 is purely imaginary and

$$\pi_2 = -\bar{\psi}.$$

We rewrite this relation as

$$dE_1 = -\pi x - i\theta_{12} E_1 - \bar{\psi} E_2, \tag{3.22}$$

where θ_{12} is real.

Finally, similar use of the inner products gives the equation

$$dx = \frac{1}{2}(\bar{\pi} E_1 + \pi \bar{E}_1). \tag{3.23}$$

From (3.21) and (3.22) we find

$$d(E_2 \wedge E_1) = -\pi(E_2 \wedge x) - i(\theta_{12} + \theta_{34})(E_2 \wedge E_1). \tag{3.24}$$

Since the lines E_2E_1 on Q_3 form a holomorphic curve Γ in the space $P_3(\mathbf{C})$ of lines on Q_3, π is of type $(1, 0)$. This shows that the surface M described by x has the same conformal structure as Γ or Δ. Comparing our equations with (3.12), we see that M is a superminimal surface on $S^4(1)$. The curve Γ, from which Δ can be constructed, can be interpreted as a curve belonging to a linear line complex in $P_3(\mathbf{C})$. This curve was defined in [2] for the case that M is a 2-sphere. For any M it was introduced by R. Bryant [1] and N. Hitchin (personal communication) by using Penrose's twistor theory. In [1] Bryant, by applying the Riemann-Roch Theorem, proved that any compact Riemann surface can be immersed as such as curve. As a result we have the following theorem first proved by R. Bryant.

THEOREM 3. *Any compact Riemann surface can be conformally immersed as a superminimal surface in* $S^4(1)$.

4. Minimal Surfaces in $P_2(\mathbf{C})$. We wish to apply the results of Section 2 to the case that Y is the complex projective plane $P_2(\mathbf{C})$ with the Fubini-Study metric. We begin by giving a description of the geometry of $P_2(\mathbf{C})$.

For $W, Z \in \mathbf{C}^3$ we have the usual hermitian inner product defined by

$$\langle W, Z\rangle = \Sigma w_a \bar{z}_a, \qquad W = (w_0, w_1, w_2), \qquad Z = (z_0, z_1, z_2), \tag{4.1}$$

where here, as well as elsewhere throughout this paper, we use the index range $0 \leq a, b, c \leq 2$. The unitary group $U(3)$ is the group of all linear transformations on $\mathbf{C}^3$ leaving the inner product (4.1) invariant. $P_2(\mathbf{C})$ is the orbit space of $\mathbf{C}^3 - \{0\}$ under the action of the group $Z \to \eta Z$, where η is a complex number $\neq 0$. We thus have the projection map $\pi: \mathbf{C}^3 - \{0\} \to P_2(\mathbf{C})$. To a point $p \in P_2(\mathbf{C})$ a vector $Z \in \pi^{-1}(p)$ is called a homogeneous coordinate vector of p. We put

$$Z_0 = Z/\langle Z, Z\rangle^{1/2}$$

so that $\langle Z_0, Z_0\rangle = 1$. This relationship can be described by

$$\mathbf{C}^3 - \{0\} \xrightarrow{\lambda} S^5(1) \xrightarrow{h} P_2(\mathbf{C}), \qquad \pi = h \circ \lambda,$$

where $\lambda(Z) = Z_0$ and h is the Hopf fibering. The homogeneous coordinate vector Z_0 of a point $p \in P_2(\mathbf{C})$ is defined up to the change

$$Z_0 \to Z_0^* = e^{i\tau} Z_0, \qquad \tau \text{ real}.$$

Under this change we have

$$\langle dZ_0^*, Z_0^*\rangle = \langle dZ_0\rangle + i d\tau,$$

$$\langle dZ_0^*, Z_0^*\rangle = \langle dZ_0, dZ_0\rangle + i d\tau\{-\langle dZ_0, Z_0\rangle + \langle Z_0, dZ_0\rangle\} + d\tau^2.$$

The resulting invariant expression

$$ds^2 = \langle dZ_0, dZ_0\rangle - \langle dZ_0, Z_0\rangle\langle Z_0, dZ_0\rangle \tag{4.2}$$

is the Fubini-Study metric on $P_2(\mathbf{C})$.

Since Z_0 is defined by p up to a factor $\exp(i\tau)$, τ real, the unit tangent vectors to $P_2(\mathbf{C})$ at p can be identified with equivalence classes of pairs of unit vectors $\{Z_0, Z_1\}$ satisfying

$$\langle Z_0, Z_1 \rangle = 0,$$

with the equivalence relation given by

$$\{Z_0, Z_1\} \sim \{\exp(i\tau)Z_0, \exp(i\tau)Z_1\}.$$

Now let Z_a be a unitary frame in $\mathbf{C}^3$ so that

$$\langle Z_a, Z_b \rangle = \delta_{a\bar{b}}$$

In the bundle of all unitary frames on $\mathbf{C}^3$ we have

$$dZ_a = \sum_b \psi_{a\bar{b}} Z_b, \tag{4.3}$$

where $\psi_{a\bar{b}} = -\bar{\psi}_{b\bar{a}} = \langle dZ_a, Z_b \rangle$ is a 1-form. The $\psi_{a\bar{b}}$ are the Maurer-Cartan forms of the group $U(3)$ and so satisfy the Maurer-Cartan structure equations

$$d\psi_{a\bar{b}} = \sum_c \psi_{a\bar{c}} \wedge \psi_{c\bar{b}}. \tag{4.4}$$

These are obtained by exterior differentiation of (4.3). By (4.2) and (4.3) the Fubini-Study metric can be written

$$ds^2 = \sum_j \psi_{0\bar{j}} \bar{\psi}_{0\bar{j}}. \tag{4.5}$$

It is of the form (2.1) if we set

$$\omega_j = \psi_{0\bar{j}}. \tag{4.6}$$

If we choose

$$\omega_{i\bar{j}} = -(\psi_{j\bar{i}} - \delta_{ij}\psi_{0\bar{0}}), \tag{4.7}$$

then these forms satisfy the conditions (2.2) and (2.3). They are therefore the connection forms of the Fubini-Study metric and this metric is Kahlerian. Its curvature forms are

$$\Omega_{i\bar{j}} = -\omega_i \wedge \omega_{\bar{j}} - \delta_{ij} \Sigma \omega_k \wedge \omega_{\bar{k}}. \tag{4.8}$$

Thus the Fubini-Study metric has constant holomorphic sectional curvature.

Typically we want to apply these results to an immersed surface $x: M \to P_2(\mathbf{C})$. Over a neighbourhood $U \subseteq M$ we define a unitary frame Z_a along x as C^∞ maps

$$Z_a: U \subseteq M \to \mathbf{C}^3 - \{0\}$$

such that:

1) $\pi \circ Z_0: U \to P_2(\mathbf{C})$ is the immersion x;
2) $\{Z_0, Z_1, Z_2\}$ is a unitary frame in $\mathbf{C}^3$ for each point $p \in U$.

All of the above considerations remain valid for a frame field defined in this manner, where we must now regard (4.2) to (4.8) as equations restricted to M.

Let $x: M \to P_2(\mathbf{C})$ be a minimal surface which is neither holomorphic nor antiholomorphic. We assume in addition that $x(M)$ does not lie in a $P_1(\mathbf{C})$. Let $U \subseteq M$ be a neighbourhood and choose

$$Z: U \subseteq M \to \mathbf{C}^3 - \{0\}$$

to be a homogeneous coordinate vector for x. We are going to construct a unitary frame field along x. To this end let

$$Z_0 = Z/\langle Z, Z\rangle^{1/2}.$$

Recall from Section 2 that there is a unitary coframe ω_1, ω_2 on $P_2(\mathbf{C})$ such that

$$\omega_1 = s\phi, \qquad \omega_2 = t\bar{\phi}.$$

Let e_1, e_2 be the dual unitary frame. If $Z \in \pi^{-1}(p)$ then we can identify $T_p(P_2(\mathbf{C}))$ with $\{W \in \mathbf{C}^3 : \langle W, Z\rangle = 0\}$. Under this identification we let

$Z_1, Z_2 \in \mathbf{C}^3$ correspond to e_1, e_2, respectively. Then $\{Z_0, Z_1, Z_2\}$ is a unitary frame field along x. We have

$$
\begin{aligned}
dZ_0 &= \psi_{0\bar{0}} Z_0 + \psi_{0\bar{1}} Z_1 + \psi_{0\bar{2}} Z_2, \\
dZ_1 &= \psi_{1\bar{0}} Z_0 + \psi_{1\bar{1}} Z_1 + \psi_{1\bar{2}} Z_2, \qquad (4.9) \\
dZ_2 &= \psi_{2\bar{0}} Z_0 + \psi_{2\bar{1}} Z_1 + \psi_{2\bar{2}} Z_2,
\end{aligned}
$$

where

$$
\begin{aligned}
\psi_{0\bar{1}} &= \omega_1 = s\phi, \\
\psi_{0\bar{2}} &= \omega_2 = t\bar{\phi}, \qquad (4.10)
\end{aligned}
$$

Z_1 and Z_2 are defined over U. If $\tilde{Z}_1$ and $\tilde{Z}_2$ are similarly defined fields on $\tilde{U}$ and $U \cap \tilde{U} \neq \emptyset$ then on $U \cap \tilde{U}$, $\tilde{Z}_1 = \exp(i\tau_1)Z_1$ and $\tilde{Z}_2 = \exp(i\tau_2)Z_2$ where τ_1 and τ_2 are real valued functions. It follows that $\pi \circ Z_1$ and $\pi \circ Z_2$ are globally defined maps $M \to P_2(\mathbf{C})$.

Let us now consider the special case $M = S^2$. By Theorem 2 the cubic form $P = \omega_1 \omega_{\bar{1}2} \omega_{\bar{2}}$ is a holomorphic form on S^2. The Riemann-Roch Theorem then implies that $P = 0$ or equivalently that

$$s t \overline{c} = 0,$$

where $\omega_{1\bar{2}} = c\bar{\phi}$. As x is neither holomorphic nor antiholomorphic $s\bar{t}$ has only isolated zeroes. It follows that $c = 0$. By (4.7)

$$\omega_{1\bar{2}} = -\psi_{2\bar{1}}.$$

Hence $\psi_{2\bar{1}} = 0$. As $\psi_{2\bar{0}}$ is a form of type (1, 0) we get

$$dZ_2 \equiv \psi_{2\bar{2}} Z_2 \bmod \phi.$$

This means that the map $\pi \circ Z_2 : S^2 \to P_2(\mathbf{C})$ is holomorphic. We call it the directrix curve of the minimal surface $x : S^2 \to P_2(\mathbf{C})$ and denote it by $\Delta : S^2 \to P_2(\mathbf{C})$. Similar reasoning shows that the map $\pi \circ Z_1 : S^2 \to P_2(\mathbf{C})$ is antiholomorphic.

Adopting the terminology used for minimal surfaces in S^4 we say a minimal surface $x: M \to P_2(\mathbf{C})$ is superminimal if $P = 0$. Of course it follows that all minimal two-sheres in $P_2(\mathbf{C})$ are superminimal. The above considerations show that we can associate a directrix curve to a superminimal surface. Note that the construction of the directrix curve from the minimal surface is very simple; it involves basically only differentiations. We will call equations (4.9) with $\psi_{1\bar{2}} = 0$ the *structure equations* of a superminimal surface in $P_2(\mathbf{C})$.

From the directrix curve $\Delta: M \to P_2(\mathbf{C})$ we can recover the superminimal surface $x: M \to P_2(\mathbf{C})$. Let Z be a homogeneous coordinate vector for Δ. Set $Z_0 = Z/\langle Z, Z\rangle^{1/2}$. For each point $p \in U \subseteq M$ choose the field Z_1 such that Z_0, Z_1 is a unitary basis for the vector space spanned by Z, $\partial Z/\partial\zeta$ and choose the field Z_2 such that Z_0, Z_1, Z_2 is a unitary basis for the vector space spanned by Z, $\partial Z/\partial\zeta$, $\partial^2 Z/\partial\zeta^2$. Here ζ is a local complex coordinate on U. Then $\{Z_0, Z_1, Z_2\}$ is a unitary frame field along Δ. We have

$$
\begin{aligned}
dZ_0 &= \psi_{0\bar{0}}Z_0 + \psi_{0\bar{1}}Z_1, \\
dZ_1 &= \psi_{1\bar{0}}Z_0 + \psi_{1\bar{1}}Z_1 + \psi_{1\bar{2}}Z_2, \\
dZ_2 &= \qquad\qquad \psi_{2\bar{1}}Z_1 + \psi_{2\bar{2}}Z_2,
\end{aligned}
\tag{4.11}
$$

where $\psi_{0\bar{1}}$ and $\psi_{1\bar{2}}$ are forms of type $(1, 0)$. The frame $\{Z_0, Z_1, Z_2\}$ is classically known as the *Frenet frame* for the holomorphic curve Δ. The forms

$$
\omega_1 = \psi_{1\bar{2}}, \qquad \omega_2 = \psi_{1\bar{0}}
\tag{4.12}
$$

are a unitary coframe for the curve $\pi\circ Z_1$. This coframe is already "normalized" according to Section 2. Moreover, the connection form $\omega_{1\bar{2}}$ of this coframe is given by

$$\omega_{1\bar{2}} = -\psi_{0\bar{2}} = 0.$$

It follows that $\pi\circ Z_1: M \to P_2(\mathbf{C})$ is a superminimal surface. It is easy to see that $\pi\circ Z_2$ is an antiholomorphic curve. To compare equations (4.11) with the structure equations for the superminimal surface x we renumber the unitary frame $\{Z_0, Z_1, Z_2\}$ along x by cyclicly permuting $\{0, 1, 2\}$. The structure equations for x then become the equations (4.11). We conclude that the curve $\pi\circ Z_1$ is the superminimal surface $x: M \to P_2(\mathbf{C})$.

We have proved the following theorem:

THEOREM 4. *To each superminimal surface $x: M \to P_2(\mathbf{C})$ which is neither holomorphic nor antiholomorphic there is associated a unique holomorphic curve $\Delta: M \to P_2(\mathbf{C})$ and conversely. The construction of Δ from x and the construction of x from Δ involve only differentiations.*

Since every compact Riemann surface can be realized as an immersed algebraic curve in $P_2(\mathbf{C})$, we have the following analogue of Theorem 3:

THEOREM 5. *Every compact Riemann surface can be conformally immersed in $P_2(\mathbf{C})$ as a superminimal surface, which is neither a holomorphic nor an anti-holomorphic curve.*

REFERENCES

[1] Bryant, R. L. Every compact Riemann surface may be immersed conformally and minimally into S^4, preprint.

[2] Chern, S. S. On minimal spheres in the four spheres, Studies and Essays Presented to Y. W. Chen, Taiwan, (1970) 137–150, also *Selected Papers*, Springer-Verlag, New York (1978) 421–434.

[3] Chern, S. S. On the minimal immersions of the two-sphere in a space of constant curvature, *Problems in Analysis*, Princeton University Press (1970), 27–40.

[4] Din, A. M. and Zakrzewski, W. J., General classical solutions in the $\mathbf{C}P^{n-1}$ model, *Nucl. Phys. B* **174** (1980), 397–406.

[5] Eells, J. and Wood, J. C. Harmonic maps from surfaces to complex projective spaces, preprint.

[6] Wolfson, J. G. Minimal surfaces in complex manifolds, Ph.D. Thesis University of California, Berkeley (1982).

Reprinted from
Amer. J. Maths. **105** (1983) 59–83.

On Surfaces of Constant Mean Curvature in a Three-Dimensional Space of Constant Curvature

SHIING-SHEN CHERN*

Recently W. Y. Hsiang and his collaborators found examples of immersions of the three-sphere S^3 into the four-dimensional euclidean space E^4, which have constant mean curvature but are not round [2].

By a theorem of H. Hopf such examples do not exist in one lower dimension [1]. In this note we wish to give a proof of the corresponding theorem where the ambient space is of any constant curvature.

Let M be a three-dimensional Riemannian manifold of constant curvature c. Relative to orthonormal frames $xe_1e_2e_3$, $x \in M$, the structure equations are

$$\begin{aligned} d\theta_\alpha &= \sum \theta_\beta \wedge \theta_{\beta\alpha}, \\ d\theta_{\alpha\beta} &= \sum \theta_{\alpha\gamma} \wedge \theta_{\gamma\beta} + \Theta_{\alpha\beta}, \end{aligned} \tag{1}$$

where the indices have the range

$$1 \leq \alpha, \beta, \gamma \leqq 3, \tag{2}$$

and θ_α is an orthonormal coframe, $\theta_{\alpha\beta}(= -\theta_{\beta\alpha})$ are the connection forms, and

$$\Theta_{\alpha\beta} = -c\theta_\alpha \wedge \theta_\beta \tag{3}$$

are the curvature forms. Equations (3) express the fact that M is of constant curvature c.

Let $f: N \to M$ be an immersed two-dimensional surface. We restrict to frames $xe_1e_2e_3$, such that $x \in N$ and e_3 is the unit normal vector to N at x, supposing N to be oriented. Then

$$\theta_3 = 0, \tag{4}$$

and by (1),

$$\theta_{i3} = \sum h_{ik}\theta_k, \tag{5}$$

where

$$h_{ik} = h_{ki}. \tag{6}$$

and in what follows, we agree on the index range

$$1 \leqq i, j, k \leqq 2. \tag{7}$$

---

*Work done under partial support of NSF grant MCS-8023356.

The first and second fundamental forms are respectively

$$\text{(8)} \qquad \begin{aligned} \mathrm{I} &= \theta_1^2 + \theta_2^2 \\ \mathrm{II} &= h_{11}\theta_1^2 + 2h_{12}\theta_1\theta_2 + h_{22}\theta_2^2. \end{aligned}$$

The invariants

$$\text{(9)} \qquad \begin{aligned} H &= \tfrac{1}{2}(h_{11} + h_{22}) \\ K_l &= h_{11}h_{22} - h_{12}^2. \end{aligned}$$

which are the two elementary symmetric functions of the eigenvalues of II with respect to I, are called respectively the *mean curvature* and the *total curvature* of N. The latter has an induced Riemannian metric. By (1) its *Gaussian curvature* is

$$\text{(10)} \qquad K_i = K_l + c.$$

N is called *totally umbilical*, (resp. *totally geodesic*) if

$$\text{(11)} \qquad \mathrm{II} - H\mathrm{I} = 0 \text{ (resp. } \mathrm{II} = 0)$$

Exterior differentiation of (5) and use of (1) give,

$$\text{(12)} \qquad \sum Dh_{ik} \wedge \theta_k = 0,$$

where

$$\text{(13)} \qquad Dh_{ik} = dh_{ik} - \sum_j h_{ij}\theta_{kj} - \sum_j h_{kj}\theta_{ij}$$

By putting

$$\text{(14)} \qquad Dh_{ik} = \sum h_{ikj}\theta_j,$$

we get, from (12)

$$\text{(15)} \qquad h_{ikj} = h_{ijk}.$$

Thus h_{ikj} is symmetric in any two of its indices. A symmetric tensor h_{ik}, with this symmetry property of its covariant derivatives, is called by some mathematicians a *Codazzi tensor*.

The complex structure on N is defined by

$$\text{(16)} \qquad \phi = \theta_1 + i\theta_2.$$

By (1) its exterior derivative is given by

$$\text{(17)} \qquad d\phi = i\phi \wedge \theta_{12}$$

The form in (11),

$$\text{(18)} \qquad \mathrm{II} - H\mathrm{I} = \tfrac{1}{2}(h_{11} - h_{22})(\theta_1^2 - \theta_2^2) + 2h_{12}\theta_1\theta_2$$

has trace zero, and is the real part of the complex two-form

$$\text{(19)} \qquad \Lambda = \hat{H}\phi^2,$$

where

$$\hat{H} = \tfrac{1}{2}(h_{11} - h_{22}) - h_{12}i. \tag{20}$$

Clearly Λ is uniquely determined by II − HI. Hence both of them are forms intrinsically associated to N, independent of choice of frames.

Theorem 1. *If $H = const$, Λ is a holomorphic two-form on N.*

Proof. The hypothesis implies

$$h_{11i} + h_{22i} = 0. \tag{21}$$

By (13) and (14) we have

$$\begin{aligned} dh_{11} &= 2h_{12}\theta_{12} + h_{111}\theta_1 + h_{112}\theta_2, \\ dh_{12} &= -(h_{11} - h_{22})\theta_{12} + h_{121}\theta_1 + h_{122}\theta_2, \\ dh_{22} &= -2h_{12}\theta_{12} + h_{221}\theta_1 + h_{222}\theta_2, \end{aligned} \tag{22}$$

from which it follows that

$$d\hat{H} = 2i\hat{H}\theta_{12} + (h_{111} - ih_{112})\phi \tag{23}$$

Locally we write

$$\phi = \lambda dz, \tag{24}$$

so that

$$\Lambda = \hat{H}\lambda^2\, dz^2 \tag{25}$$

It suffices to show that the coefficient $\hat{H}\lambda^2$ of dz^2 in this expression is a holomorphic function of z.

By substituting (24) into (17), we get

$$d\lambda + i\lambda\theta_{12} \equiv 0, \text{ mod } dz,$$

while (23) implies

$$d\hat{H} - 2i\hat{H}\theta_{12} \equiv 0, \text{ mod } dz.$$

It follows that

$$d(\hat{H}\lambda^2) \equiv 0, \text{ mod } dz.$$

which proves the theorem.

As a corollary we have:

Theorem 2. *$f: S^2 \to M$ be an immersed two sphere with constant mean curvature, where M is a three-dimensional manifold of constant curvature. Then the surface is totally umbilical.*

Proof. This follows from the fact that $\Lambda = 0$.

REFERENCES

1. H. Hopf, Uber Flächen mit einer Relation zwischen den Hauptkrümmungen, Math. Nachr. 4 (1950–51), 232–249.
2. Wu-yi Hsiang, Zhen-huan Teng, and Wen-ci Yu, New examples of constant mean curvature immersions of 3-sphere into euclidean 4-space, in Proc. Nat. Acad. Sci, USA, vol 79, 3931–2, (1982).

Department of Mathematics
University of California
Berkeley, CA 94720
USA

Reprinted from
Geometric Dynamics, Springer Lecture Notes **1007** Springer–Verlag (1983) 104–108.

Deformation of Surfaces Preserving Principal Curvatures

By Shiing-shen Chern[1]

1. Introduction and Statement of Results

The isometric deformation of surfaces preserving the principal curvatures was first studied by O. Bonnet in 1867. Bonnet restricted himself to the complex case, so that his surfaces are analytic, and the results are different from the real case. After the works of a number of mathematicians, W. C. Graustein took up the real case in 1924 –, without completely settling the problem. An authoritative study of this problem was carried out by Elie Cartan in [2], using moving frames. Based on this work, we wish to prove the following:

Theorem: The non-trivial families of isometric surfaces having the same principal curvatures are the following:

1) a family of surfaces of constant mean curvature;
2) a family of surfaces of non-constant mean curvature. Such surfaces depend on six arbitrary constants, and have the properties:
   a) they are *W*-surfaces;
   b) the metric

$$d\hat{s}^2 = (\operatorname{grad} H)^2\, ds^2/(H^2 - K),$$

   where ds^2 is the metric of the surface and H and K are its mean curvature and Gaussian curvature respectively, has Gaussian curvature equal to -1.

By a non-trivial family of surfaces we mean surfaces which do not differ by rigid motions. The theorem is a local one and deals only with pieces of surfaces. We suppose that they do not contain umbilics and that they are C^5.

The analytic formulation of the problem leads to an over-determined system of partial differential equations. It must be the simple geometrical nature of the problem that the integrability conditions give the clear-cut conclusion stated in the theorem. The surfaces in class 2) are clearly of interest. An analogous problem is concerned with non-trivial families of isometric surfaces with lines of curvature preserved. They also have a simple description and are given by the molding surfaces; cf. [1, pp. 269–284].

I wish to thank Konrad Voss for calling my attention to this problem.

---

[1] Work done under partial support of NSF grant MCS 77-23579

2. Formulation of Problem

We consider in the euclidean space E^3 a piece of oriented surface M, of sufficient smoothness and containing no umbilics. Over M there is then a well-defined field of orthonormal frames $xe_1e_2e_3$, such that $x \in M$, e_3 is the unit normal vector at x, and e_1, e_2 are along the principal directions. We have then

$$\begin{aligned} dx &= \omega_1 e_1 + \omega_2 e_2 \\ de_1 &= \omega_{12} e_2 + \omega_{13} e_3, \\ de_2 &= -\omega_{12} e_1 + \omega_{23} e_3, \\ de_3 &= -\omega_{13} e_1 - \omega_{23} e_3, \end{aligned} \tag{1}$$

the ω's are one-forms on M. Our choice of the frames allows us to set

$$\begin{aligned} \omega_{12} &= h\omega_1 + k\omega_2 \\ \omega_{13} &= a\omega_1, \quad \omega_{23} = c\omega_2, \quad a > c. \end{aligned} \tag{2}$$

Then a and c are the two principal curvatures at x. As usual we denote the mean curvature and the Gaussian curvatures by

$$H = \tfrac{1}{2}(a + c), \quad K = ac. \tag{3}$$

The functions and forms satisfy the structure equations obtained by exterior differentiation of (1). They give

$$\begin{aligned} d\omega_1 &= \omega_{12} \wedge \omega_2, \quad d\omega_2 = \omega_1 \wedge \omega_{12}, \\ d\omega_{12} &= -K\omega_1 \wedge \omega_2, \\ d\omega_{13} &= \omega_{12} \wedge \omega_{23}, \quad d\omega_{23} = \omega_{13} \wedge \omega_{12}. \end{aligned} \tag{4}$$

The equation in the second line of (4) is called the Gauss equation and the equations in the last line of (4) are called the Codazzi equations.

Using (2), the Codazzi equations give

$$\begin{aligned} \{da - (a - c)h\omega_2\} \wedge \omega_1 = 0, \\ \{dc - (a - c)k\omega_1\} \wedge \omega_2 = 0. \end{aligned} \tag{5}$$

We introduce the functions u, v by

$$2dH = d(a + c) = (a - c)(u\omega_1 + v\omega_2). \tag{6}$$

Then we have

$$\begin{aligned} \frac{1}{a - c} da &= (u - k)\omega_1 + h\omega_2, \\ \frac{1}{a - c} dc &= k\omega_1 + (v - h)\omega_2, \end{aligned} \tag{7}$$

and

$$d \log(a - c) = (u - 2k)\omega_1 - (v - 2h)\omega_2. \tag{8}$$

We note also the relation

$$4(\operatorname{grad} H)^2 = (a - c)^2(u^2 + v^2). \tag{9}$$

For our treatment we introduce the forms

$$\theta_1 = u\omega_1 + v\omega_2, \quad \theta_2 = -v\omega_1 + u\omega_2, \tag{10}$$

$$\alpha_1 = u\omega_1 - v\omega_2, \quad \alpha_2 = v\omega_1 + u\omega_2. \tag{11}$$

Thus $\theta_1 = 0$ is tangent to the level curves $H = \text{const}$ and $\alpha_1 = 0$ is its symmetry with respect to the principal directions. If $H \neq \text{const}$, the quadratic differential form

$$\begin{aligned} d\hat{s}^2 = \theta_1^2 + \theta_2^2 = \alpha_1^2 + \alpha_2^2 &= (u^2 + v^2)(\omega_1^2 + \omega_2^2) \\ &= \frac{(\operatorname{grad} H)^2}{H^2 - K} ds^2 \end{aligned} \tag{12}$$

defines a conformal metric on M.

We find it convenient to make use of the Hodge $*$-operator, such that

$$\begin{aligned} *\omega_1 = \omega_2, \quad & *\omega_2 = -\omega_1, \\ *^2 = -1 \quad & \text{on one-forms.} \end{aligned} \tag{13}$$

Then we have

$$*\theta_1 = \theta_2, \quad *\theta_2 = -\theta_1, \tag{14}$$

$$*\alpha_1 = \alpha_2, \quad *\alpha_2 = -\alpha_1. \tag{15}$$

Using these notations Eq. (6) and (8) can be written

$$2dH = d(a + c) = (a - c)\theta_1, \tag{6a}$$

$$d\log(a - c) = \alpha_1 + 2*\omega_{12}. \tag{8a}$$

Suppose M^* is a surface which is isometric to M with preservation of the principal curvatures. We shall denote the quantities pertaining to M^* by the same symbols with asterisks, so that

$$a^* = a, \quad c^* = c. \tag{16}$$

As M and M^* are isometric, we have

$$\begin{aligned} \omega_1^* &= \cos\tau\,\omega_1 - \sin\tau\,\omega_2, \\ \omega_2^* &= \sin\tau\,\omega_1 + \cos\tau\,\omega_2. \end{aligned} \tag{17}$$

Exterior differentiation gives

$$\begin{aligned} d\omega_1^* &= (-d\tau + \omega_{12}) \wedge \omega_2^*, \\ d\omega_2^* &= \omega_1^* \wedge (-d\tau + \omega_{12}), \end{aligned}$$

so that

$$\omega_{12}^* = -d\tau + \omega_{12}. \tag{18}$$

By (8a) we get

$$\alpha_1 + 2*\omega_{12} = \alpha_1^* + 2*\omega_{12}^*.$$

Applying the $*$-operator to this equation, we find

$$\omega_{12}^* - \omega_{12} = \tfrac{1}{2}(\alpha_2^* - \alpha_2).$$

This gives

$$d\tau = \tfrac{1}{2}(\alpha_2 - \alpha_2^*). \tag{19}$$

We wish to simplify the last expression.

From (6a) we have

$$\theta_1^* = \theta_1,$$

i.e.

$$u^* \omega_1^* + v^* \omega_2^* = u\omega_1 + v\omega_2,$$

which gives, in view of (17),

$$\begin{aligned} u^* &= \cos\tau u - \sin\tau v \\ v^* &= \sin\tau u + \cos\tau v. \end{aligned} \tag{20}$$

It follows that

$$\alpha_2^* = \sin 2\tau \cdot \alpha_1 + \cos 2\tau \cdot \alpha_2.$$

Putting

$$t = \cot\tau, \tag{21}$$

we get from (19),

$$dt = t\alpha_1 - \alpha_2. \tag{22}$$

This is the total differential equation satisfied by the angle τ of rotation of the principal directions during the isometric deformation. In order that the deformation be non-trivial it is necessary and sufficient that the Eq. (22) be completely integrable. This is expressed by the conditions

$$\begin{aligned} d\alpha_1 &= 0, \\ d\alpha_2 &= \alpha_1 \wedge \alpha_2. \end{aligned} \tag{23}$$

When the mean curvature H is constant, we have

$$u = v = 0$$

and $t =$ const. This gives the theorem of Bonnet (cf. [3]):

Theorem (Bonnet): A surface of constant mean curvature can be isometrically deformed preserving the principal curvatures. During the deformation the principal directions rotate by a fixed angle.

3. Connection Form Associated to a Coframe

Given the linearly independent one-forms ω_1, ω_2, the first two equations in (4) uniquely determine the form ω_{12}. We call ω_1, ω_2 the (orthonormal) coframe of the metric

$$ds^2 = \omega_1^2 + \omega_2^2 \tag{24}$$

and ω_{12} the connection form associated to it. The discussions leading to (18) give the following lemma:

Lemma 1. When the coframe undergoes the transformation (17), the associated connection forms are related by (18).

We now consider a conformal transformation of the metric

$$d\hat{s}^2 = A^2\, ds^2 = A^2(\omega_1^2 + \omega_2^2), \tag{25}$$

where $A > 0$ is a function on M. Let

$$\omega_1^* = A\omega_1, \quad \omega_2^* = A\omega_2. \tag{26}$$

Then we have:

Lemma 2. Under the changes of coframe (26) the associated connection forms are related by

$$\omega_{12}^* = \omega_{12} - i(\partial - \bar{\partial}) \log A. \tag{27}$$

Here $\partial, \bar{\partial}$ are the differentiation operators relative to the complex structure $\omega = \omega_1 + i\omega_2$ of M. The proof is by straightforward calculation and will be omitted. We note, however, the useful formula

$$*(\partial - \bar{\partial}) f = -i df \tag{28}$$

where f is a function on M.

4. Surfaces of Non-Constant Mean Curvature

Suppose $H \neq$ const. Then

$$A = +(u^2 + v^2)^{1/2} > 0, \tag{29}$$

and we write

$$u + iv = A \exp(i\psi). \tag{30}$$

Let

$$\begin{aligned} \omega &= \omega_1 + i\omega_2, \\ \theta &= \theta_1 + i\theta_2, \\ \alpha &= \alpha_1 + i\alpha_2. \end{aligned} \tag{31}$$

Then

$$\begin{aligned} \theta &= A \exp(-i\psi)\omega, \\ \alpha &= A \exp(i\psi)\omega, \end{aligned} \tag{32}$$

so that

$$\alpha = \exp(2i\psi)\theta. \tag{33}$$

The forms ω, θ, α define the same complex structure on M and the operators $*, \partial, \bar{\partial}$ can be used without ambiguity.

Let $\omega_{12}, \theta_{12}, \alpha_{12}$ be the connection forms associated to the coframes ω_1, ω_2; θ_1, θ_2; α_1, α_2 respectively. By Lemmas 1 and 2, Sect. 3, we have the fundamental relation

$$\theta_{12} = \omega_{12} + d\psi - i(\partial - \bar{\partial}) \log A = 2d\psi + \alpha_{12}. \tag{34}$$

In addition, from (23) we have

$$\alpha_{12} = \alpha_2. \tag{35}$$

The second equation of (23) then implies that the metric $d\hat{s}^2$ on M has Gaussian curvature equal to -1. Moreover, the Eq. (35) shows that the curves $\alpha_2 = 0$ are geodesics and the curves $\alpha_1 = 0$ have geodesic curvatures equal to 1, i.e., are horocycles relative to the metric $d\hat{s}^2$.

From (8 a) and (23) we get

$$d * \omega_{12} = 0. \tag{36}$$

Applying $*$ to (34), we get, by using (28),

$$* \theta_{12} = * \omega_{12} + * d\psi - d \log A = 2 * d\psi - \alpha_1. \tag{37}$$

Exterior differentiation of the last equation gives, in view of (23), (36),

$$d * d\psi = 0, \tag{38}$$

which says that ψ *is a harmonic function*. Differentiation of (37) then gives

$$d * \theta_{12} = 0. \tag{39}$$

By differentiating (6 a) and using (8 a), we get

$$d\theta_1 + (\alpha_1 + 2 * \omega_{12}) \wedge \theta_1 = 0.$$

But

$$d\theta_1 = \theta_{12} \wedge \theta_2 = - * \theta_{12} \wedge \theta_1. \tag{40}$$

From (37) we find

$$- * \theta_{12} + \alpha_1 + 2 * \omega_{12} = 2d \log A.$$

It follows that

$$d \log A \wedge \theta_1 = 0, \tag{41}$$

and we set

$$d \log A = B\theta_1. \tag{42}$$

This is a differential equation in $\log A$. But $\partial\bar{\partial} \log A$ is related to the Gaussian curvature K of M. We wish to combine these facts to draw the remarkable conclusion that M is a W-surface.

This involves further computation of the integrability conditions. The simplest way is to make use of the coframe α_1, α_2, because their exterior derivatives satisfy the simple Eq. (23). For a function f on M we define

$$df = f_1 \alpha_1 + f_2 \alpha_2. \tag{43}$$

Its cross covariant derivatives satisfy the commutation formula

$$f_{21} - f_{12} + f_2 = 0. \tag{44}$$

Moreover, the condition for ψ to be a harmonic function is

$$\psi_{11} + \psi_{22} + \psi_1 = 0. \tag{45}$$

Note also that, by (37),

$$*\theta_{12} = -(2\psi_2 + 1)\alpha_1 + 2\psi_1\alpha_2. \tag{46}$$

By (6a) and (8a), the condition for M to be a W-surface is

$$(\alpha_1 + 2*\omega_{12}) \wedge \theta_1 = 0.$$

Using (37) and (42), this can be written

$$2\psi_1 \cos 2\psi + (2\psi_2 + 1)\sin 2\psi = 0. \tag{47}$$

From (42) we have

$$(\log A)_1 = B\cos 2\psi, \quad (\log A)_2 = B\sin 2\psi, \tag{48}$$

whose differentiations give

$$\begin{aligned}(\log A)_{1i} &= B_i \cos 2\psi - 2B\psi_i \sin 2\psi,\\ (\log A)_{2i} &= B_i \sin 2\psi + 2B\psi_i \cos 2\psi, \quad i = 1, 2.\end{aligned} \tag{49}$$

The commutation formula (44) applied to $\log A$ gives

$$B_1 \sin 2\psi - B_2 \cos 2\psi + B\{2\psi_1 \cos 2\psi + (2\psi_2 + 1)\sin 2\psi\} = 0. \tag{50}$$

But there is another equation between B_1, B_2, to be derived from the Gauss equation

$$d\omega_{12} = -ac\omega_1 \wedge \omega_2 = -acA^{-2}\alpha_1 \wedge \alpha_2, \tag{51}$$

as follows: From (34) we have

$$\omega_{12} = d\psi + \alpha_2 + (\log A)_2\alpha_1 - (\log A)_1\alpha_2. \tag{52}$$

Substituting into the above equation, we get

$$-(\log A)_{11} - (\log A)_{22} + \{-(\log A)_1 + 1\} + acA^{-2} = 0,$$

or, by (49),

$$\begin{aligned}&-B_1 \cos 2\psi - B_2 \sin 2\psi\\ &\quad + B\{2\psi_1 \sin 2\psi - (2\psi_2 + 1)\cos 2\psi\} + 1 + acA^{-2} = 0.\end{aligned} \tag{53}$$

Solving for B_1, B_2 from (50), (53),

$$\begin{aligned}B_1 + B(2\psi_2 + 1) - (1 + acA^{-2})\cos 2\psi &= 0,\\ B_2 - 2B\psi_1 - (1 + acA^{-2})\sin 2\psi &= 0.\end{aligned} \tag{54}$$

Differentiating the first equation with respect to the second index, the second equation with respect to the first index, subtracting, and using the Eq. (45) that ψ

is a harmonic function, we get

$$\begin{aligned} & -2(1+acA^{-2})\{2\psi_1 \cos 2\psi + (2\psi_2+1)\sin 2\psi\} \\ & \qquad + A^{-2}\{-(ac)_1 \sin 2\psi + (ac)_2 \cos 2\psi\} = 0. \end{aligned} \tag{55}$$

The expression in the last braces is the coefficient of $\alpha_1 \wedge \alpha_2$ in

$$-*d(ac) \wedge \theta_2.$$

Now

$$4ac = (a+c)^2 - (a-c)^2,$$

and its differential can be calculated, using (6 a) and (8 a). We get

$$\frac{2d(ac)}{a-c} = (a+c)\theta_1 - (a-c)(\alpha_1 + 2*\omega_{12})$$

and

$$\begin{aligned} -\frac{2}{(a-c)^2}(*d(ac)) \wedge \theta_2 &= (\alpha_2 - 2\omega_{12}) \wedge \theta_2 \\ &= -\{2\psi_1 \cos 2\psi + (2\psi_2+1)\sin 2\psi\}\alpha_1 \wedge \alpha_2. \end{aligned}$$

Hence (55) becomes

$$(1+H^2 A^{-2})\{2\psi_1 \cos 2\psi + (2\psi_2+1)\sin 2\psi\} = 0.$$

Since the first factor is non-zero, the second factor must vanish, which is the condition (47) for M to be a W-surface.

On M with the metric $d\hat{s}^2$ of Gaussian curvature -1 we search for a harmonic function ψ satisfying (47). We shall show that such a function depends on two constants. In fact, Eq. (47) allows us to put

$$2\psi_1 = C \sin 2\psi, \quad 2\psi_2 + 1 = -C\cos 2\psi. \tag{56}$$

Differentiation gives

$$\begin{aligned} 2\psi_{1i} &= C_i \sin 2\psi + 2C\psi_i \cos 2\psi, \\ 2\psi_{2i} &= -C_i \cos 2\psi + 2C\psi_i \sin 2\psi, \ i = 1, 2. \end{aligned} \tag{57}$$

The commutation formula for ψ and Eq. (45) give

$$\begin{aligned} -C_1 \cos 2\psi - C_2 \sin 2\psi + 2C\psi_1 \sin 2\psi - C(2\psi_2+1)\cos 2\psi - 1 &= 0, \\ C_1 \sin 2\psi - C_2 \cos 2\psi + 2C\psi_1 \cos 2\psi + C(2\psi_2+1)\sin 2\psi &= 0. \end{aligned} \tag{58}$$

Solving for C_1, C_2, we get

$$\begin{aligned} C_1 + C(2\psi_2+1) + \cos 2\psi &= 0, \\ C_2 - 2C\psi_1 + \sin 2\psi &= 0. \end{aligned} \tag{59}$$

It can be verified by differentiating (59) that the commutation relation for C is satisfied. Hence there exist harmonic functions ψ satisfying (47). The solution depends on two arbitrary constants, the values of ψ and C at an initial point.

From our discussion the differentials of the functions $\log A$, B, a, c are all determined. Hence our surfaces, e.g., the surfaces of non-constant mean curvature which can be isometrically deformed in a non-trivial way preserving the principal curvatures, depend on 6 arbitrary constants. This proves the main statement of our theorem in Sect. 1, the other statements being proved before.

Our derivation makes use of the 5th order jet of the surface M, which is therefore supposed to be of class 5.

References

[1] Bryant, R.; Chern, S.; Griffiths, P. A.: Exterior differential systems. Proceedings of 1980 Beijing DD-Symposium. Science Press, Beijing, China and Gordon and Breach, New York, 1982, vol. 1, pp. 219–338

[2] Cartan, E.: Sur les couples de surfaces applicables avec conservation des courbures principales. Bull. Sc. Math. 66 (1942), 1–30, or Oeuvres Complètes, Partie III, vol. 2, 1591–1620

[3] Darboux, G.: Théorie des surfaces, Partie 3. Paris 1894, p. 384

Reprinted from
Differential Geometry and Complex Analysis, Volume in Memory of H. Rauch, Springer–Verlag (1984) 155–163.

ON RIEMANNIAN METRICS ADAPTED TO THREE-DIMENSIONAL CONTACT MANIFOLDS

by

S.S. Chern
Mathematical Sciences Research Institute
1000 Centennial Drive
Berkeley, California 94720

and

R.S. Hamilton
Department of Mathematics
University of California, San Diego
La Jolla, California 92093

0. Introduction It was proved by R. Lutz and J. Martinet [8] that every compact orientable three-dimensional manifold M has a contact structure. The latter can be given by a one-form ω, the contact form, such that $\omega \wedge d\omega$ never vanishes; ω is defined up to a non-zero factor. A Riemannian metric on M is said to be adapted to the contact form ω if: 1) ω has the length 1; and 2) $d\omega = 2_* \omega$, $_*$ being the Hodge operator. The Webster curvature W, defined below in [9], is a linear combination of the sectional curvature of the plane ω and the Ricci curvature in the direction perpendicular to ω.

Adapted Riemannian metrics have interesting properties. The main result of the paper is the theorem:

Every contact structure on a compact orientable three-dimensional manifold has a contact form and an adapted Riemannian metric whose Webster curvature is either a constant ≤ 0 or is everywhere strictly positive.

The problem is analogous to Yamabe's problem on the conformal transformation of Riemannian manifolds Most recently, R. Schoen has proved Yamabe's conjecture in all cases, including that of positive scalar curvature [9]. It is thus an interesting question whether in the second case of our theorem the Webster curvature can be made a positive constant.

---

1),2) Research supported in part by NSF grants DMS84-03201 and DMS84-01959.

After our theorem was proved, we learned that a similar theorem on CR-manifolds of any odd dimension has been proved by Jerison and Lee. [7] As a result, our curvature was identified with the Webster curvature. We feel that our viewpoint is sufficiently different from Jerison-Lee and that the three-dimensional case has so many special features to merit a separate treatment.

In an appendix, Alan Weinstein gives a topological implication of the vanishing of the second fundamental form in (54). For an interesting account of three-dimensional contact manifolds, cf. [2].

1. Contact Structures. Let M be a manifold and B a subbundle of the tangent bundle TM. There is a naturally defined anti-symmetric bilinear form Λ on B with values in the quotient bundle TM/B

(1) $\Lambda: B \times B \longrightarrow TM/B$

defined by the Lie bracket;

(2) $\Lambda(V,W) \equiv [V,W] \bmod B.$

It is easy to verify that the value of $\Lambda(V,W)$ at a point $p \in M$ depends only on the values of V and W at p. The bundle B defines a foliation if and only if it satisfies the Frobenius integrability condition $\Lambda = 0$. Conversely, a <u>contact structure</u> on M is a subbundle B of the tangent bundle of codimension 1 such that Λ is non-singular at each point $p \in M$. This can only occur when the dimension of M is odd.

It is an interesting problem to find some geometric structure which can be put on every three-manifold, since this would be helpful in studying its topology. Along these lines we have the following remarkable theorem of Lutz and Martinent (see [8], [10]).

1.1 Theorem. *Every compact orientable three-manifold possesses a contact structure.*

There are many different contact structures possible, since the set of B with $\Lambda \neq 0$ is open. Even on S^3 there are contact structures for which the bundles B_1 and B_2 are topologically distinct. Nevertheless the notion of a contact structure is rather flabby, in the following sense. We say B is conjugate to B_* if there is a diffeomorphism $\varphi: M \longrightarrow M$ which has $\varphi(B) = B_*$. Then we have the following result due to Gray (see [4]).

1.2 Theorem. *Given a contact structure* B, *any other contact structure* B_* *close enough to* B *is conjugate to it.*

2. Metrics adapted to contact structures. A contact form ω is a 1-form on M which is nowhere zero and has the contact bundle B for its null space. In a three-manifold a non-zero 1-form ω is a contact form for the contact structure B = Null ω if and only if $\omega \wedge d\omega \neq 0$ at every point. The contact structure B determines the contact form up to a scalar multiple. The choice of a contact form ω also determines a vector field V in the following way.

2.1 Lemma. *There exists a unique vector field* V *such that* $\omega(V) = 1$ *and* $d\omega(V,W) = 0$ *for all* $W \in TM$.

Proof. Choose V_0 with $\omega(V_0) = 1$. Since $d\omega \wedge \omega \neq 0$, the form $d\omega$ is non-singular on B. Therefore there exists a unique $V_1 \in B$ with

$$d\omega(V_1,W) = d\omega(V_0,W)$$

for all $W \in B$. Let $V = V_0 - V_1$. Then $\omega(V) = \omega(V_0) - \omega(V_1) = 1$, and $d\omega(V,W) = 0$ for all $W \in B$. Since V is transverse to B and $d\omega(V,V) = 0$, we have $d\omega(V,W) = 0$ for all $W \in TM$.

Locally any two non-zero vector fields are conjugate by a diffeomorphism. However, this fails globally, since a vector field may have closed orbits while a nearby vector field does not. It is a classical result that locally any two contact forms are conjugate by a diffeomorphism. But globally two nearby contact forms may not be conjugate, since the vector fields they determine may not be.

A choice of a Riemannian metric on a contact manifold determines a choice of the contact form ω up to sign by the condition that ω have length 1. Let $*$ denote the Hodge star operator. We make the following definition.

2.2 Definition. A Riemannian metric on a contact three-manifold is said to be adapted to the contact form ω if ω is of length one and satisfies the structural equation

$$d\omega = 2 *\omega. \tag{3}$$

Such metrics have nice properties with respect to the contact structure. For example, we have the following results.

2.3 Lemma. *If the metric is adapted to the form ω, then the vector field V determined by ω is the unit vector field perpendicular to B.*

Proof. Let V be the unit vector field perpendicular to B. Then $\omega(V) = 1$, and for all vectors W in B we have $d\omega(V,W)=2*\omega(V,W)=0$. Hence V is the vector field determined by the contact form ω.

2.4 Lemma. *If the metric is adapted to the contact form ω, then the area form on B is given by $\frac{1}{2}\, d\omega$.*

Proof. The area form on B is $*\omega$.

A CR structure on a mainfold is a contact structure together with a complex structure on the contact bundle B; that is, an involution $J{:}B \longrightarrow B$ with $J^2 = -I$ where I is the identity. If M has dimension 3 then B has dimension 2, and a complex structure on B is equivalent to a conformal structure; that is knowing how to rotate by $90°$. Hence, a Riemannian metric on a contact three-manifold also produces a CR structure. CR structures have been extensively studied since they arise naturally on the boundaries of complex manifolds. The following observation will be basic to our study.

2.5 Theorem. *Let* M *be an oriented three-manifold with contact structure* B. *For every choice of contact form* ω *and a* CR *structure* J *there exists a unique Riemannian metric* g *adapted to the contact form* ω *and inducing the* CR *structure* J.

Proof. The form ω determines the unit vector field V perpendicular to B. The metric on B is determined by the conformal structure J and the volume form $*\omega\big|B = \frac{1}{2}\, d\omega\big|B$.

3. Structural equations. We begin with a review of the structural equations of Riemannian geometry. Let $\omega_\alpha, 1 \leq \alpha, \beta \leq \dim M$, be an orthonormal basis of 1-forms on a Riemannian manifold M. Then there exists a unique anti-symmetric matrix of 1-forms $\varphi_{\alpha\beta}$ such that the structural equations

(4) $$d\omega_\alpha + \varphi_{\alpha\beta} \wedge \omega_\beta = 0.$$

hold on M. The forms $\varphi_{\alpha\beta}$ describe the Levi-Civita connection of the metric in the moving frame ω_α. We can also view the ω_α as intrinsically defined 1-forms on the principal bundle of orthonormal bases. Then the $\varphi_{\alpha\beta}$ are also intrinsically defined as 1-forms on this principal bundle, and the collection $\{\omega_\alpha, \varphi_{\alpha\beta}\}$ forms an orthonormal basis of 1-forms in the induced metric on the principal bundle. The curvature tensor $R_{\alpha\beta\gamma\delta}$ is defined by the structural equation

(5) $$d\varphi_{\alpha\beta} = -\varphi_{\alpha\gamma} \wedge \varphi_{\gamma\beta} + R_{\alpha\beta\gamma\delta}\,\omega_\gamma \wedge \omega_\delta, 1 \leq \alpha, \beta, \gamma, \delta \leq \dim M,$$

where the summation convention applies.

In three-dimensions it is natural to replace a pair of indices in an anti-symmetric tensor by the third index. Thus we will write φ_{12}

$= \varphi_3$ and $R_{1212} = K_{33}$, etc. Here $K_{\alpha\beta}$ are the components of the Einstein tensor

(6) $$K_{\alpha\beta} = \frac{1}{2} Rg_{\alpha\beta} - R_{\alpha\beta},$$

which has the property that, for any unit vector V, K(V,V) is the Riemannian sectional curvature of the plane $V^{\perp}$. The structural equations then take the following form.

3.1 Structural equations in three dimensions.

$$d\omega_1 = \varphi_2 \wedge \omega_3 - \varphi_3 \wedge \omega_2,$$

(7) $$d\omega_2 = \varphi_3 \wedge \omega_1 - \varphi_1 \wedge \omega_3,$$

$$d\omega_3 = \varphi_1 \wedge \omega_2 - \varphi_2 \wedge \omega_1,$$

and

$$d\varphi_1 = \varphi_2 \wedge \varphi_3 + K_{11}\omega_2 \wedge \omega_3 + K_{12}\omega_3 \wedge \omega_1 + K_{13}\omega_1 \wedge \omega_2,$$

(8) $$d\varphi_2 = \varphi_3 \wedge \varphi_1 + K_{21}\omega_2 \wedge \omega_3 + K_{22}\omega_3 \wedge \omega_1 + K_{23}\omega_1 \wedge \omega_2,$$

$$d\varphi_3 = \varphi_1 \wedge \varphi_2 + K_{31}\omega_2 \wedge \omega_3 + K_{32}\omega_3 \wedge \omega_1 + K_{33}\omega_1 \wedge \omega_2, K_{\alpha\beta} = K_{\beta\alpha}.$$

If the metric is adapted to the contact from ω, we choose the frames such that $\omega_3 = \omega$. As a consequence K_{33} is the sectional curvature of the plane $V^{\perp}$ and $\frac{1}{2}(K_{11}+K_{22})$ is the Ricci curvature in the direction V. The Webster curvature is defined by

(9) $$W = \frac{1}{8}(K_{11}+K_{22}+2K_{33}+4)$$

and has remarkable properties.

We proceed to illustrate these equations with three examples which are very relevant to our discussion, the sphere S^3, the unit tangent bundle of a compact orientable surface of genus > 1, and the

Heisenberg group H^3.

3.2 Example. The sphere S^3 is defined by the equation

$$x^2+y^2+z^2+w^2 = 1 \tag{10}$$

in R^4. Differentiating we get

$$\omega_0 = xdx + y\,dy + z\,dz + w\,dw = 0. \tag{11}$$

A specific choice of an orthonormal basis in the induced metric is

$$\begin{aligned} \omega_1 &= x\,dy - y\,dx + z\,dw - w\,dz, \\ \omega_2 &= x\,dz - z\,dx + y\,dw - w\,dy, \\ \omega_3 &= x\,dw - w\,dx + y\,dz - z\,dy. \end{aligned} \tag{12}$$

The reader can verify that if $<dx,dx> = 1$, $<dx,dy> = 0$, etc., then $<\omega_1,\omega_1> = 1$, $<\omega_1,\omega_2> = 0$, etc., and that $<\omega_0,\omega_0> = 1$, $<\omega_0,\omega_1> = 0$, etc. Taking exterior derivative we have

$$d\omega_1 = 2\omega_2\wedge\omega_3, \quad d\omega_2 = 2\omega_3\wedge\omega_1, \quad d\omega_3 = 2\omega_1\wedge\omega_2, \tag{13}$$

and hence in this basis

$$\varphi_1 = \omega_1, \quad \varphi_2 = \omega_2, \quad \varphi_3 = \omega_3, \tag{14}$$

which makes

$$K_{11} = 1, \quad K_{22} = 1, \quad K_{33} = 1, \tag{15}$$

and the other entries are zero. The Webster curvature $W = 1$.

3.3 Example. The unit tangent bundle of a compact orientable surface of genus $\neq 1$.

Let N be a compact orientable surface of genus g. If N is equipped with a Riemannian metric, its orthonormal coframe θ_1, θ_2, and the connection form θ_{12} satisfy the structural equations

$$(16)\qquad d\theta_1 = \theta_{12}\wedge\theta_2,\ d\theta_2 = \theta_1\wedge\theta_{12},\ d\theta_{12} = -K\theta_1\wedge\theta_2,$$

where K is the Gaussian curvature. Suppose $g\neq 1$. We can choose the metric such that

$$(17)\qquad K = \epsilon = \begin{cases} +1, & \text{when } g=0, \\ -1, & \text{when } g>1. \end{cases}$$

The unit tangent bundle T_1N of N, as a three-dimensional manifold, has the metric

$$(18)\qquad \frac{1}{4}(\theta_1^2 + \theta_2^2 + \theta_{12}^2).$$

Putting

$$(19)\qquad \omega_1 = \frac{1}{2}\theta_1,\ \omega_2 = \frac{1}{2}\theta_2,\ \omega_3 = -\frac{1}{2}\epsilon\theta_{12},$$

we find

$$(20)\qquad d\omega_1 = 2\epsilon\omega_2\wedge\omega_3,\ d\omega_2 = 2\epsilon\omega_3\wedge\omega_1,\ d\omega_3 = 2\omega_1\wedge\omega_2,$$

and

$$(21)\qquad \varphi_1 = \omega_1,\ \varphi_2 = \omega_2,\ \varphi_3 = (2\epsilon-1)\omega_3.$$

It follows that

$$(22)\qquad K_{11} = K_{22} = 1,\quad K_{33} = 4\epsilon-3,$$

all other $K_{\alpha\beta}$'s being zero. By (9) we get

$$W = \epsilon.$$

This includes the example in §3.2 when $g = 0$, for the unit tangent bundle of S^2 is the real projective space RP^3, which is covered by S^3, and our calculation is local. On the other hand, T_1N, for $g > 1$, has a contact structure and an adapted Riemannian metric with $W = -1$.

3.4. Example. The Heisenberg group.

We can make C^2 into a Lie group by identifying (z,w) with the matrix

$$\begin{bmatrix} 1 & 0 & 0 \\ z & 1 & 0 \\ w & -\bar{z} & 1 \end{bmatrix}. \tag{23}$$

The subgroup given by the variety

$$z\bar{z} + w + \bar{w} = 0 \tag{24}$$

is the Heisenberg group H^3. The group acts on itself by the translations

$$\begin{aligned} z &\longrightarrow z + a, \\ w &\longrightarrow w - \bar{a}z + b, \end{aligned} \tag{25}$$

which leave invariant the complex forms

$$dz \text{ and } dw + \bar{z}\, dz. \tag{26}$$

Hence an invariant metric is given by

$$ds^2 = |dz|^2 + |dw + \bar{z}\, dz|^2. \tag{27}$$

Introduce the real coordinates

$$z = x + iy \qquad w = u + iv. \tag{28}$$

Then the variety (24) is

(29) $$x^2 + y^2 + 2u = 0$$

and differentiation gives

(30) $$du + x\,dx + y\,dy = 0.$$

Then an orthonormal basis of 1-forms in the metric above is given by

(31) $$\omega_1 = dx, \quad \omega_2 = dy, \quad \omega_3 = dv + x\,dy - y\,dx,$$

and we compute

(32) $$\begin{cases} d\omega_1 = 0, & d\omega_2 = 0, & d\omega_3 = 2\omega_1 \wedge \omega_2, \\ \varphi_1 = \omega_1, & \varphi_2 = \omega_2, & \varphi_3 = -\omega_3, \\ K_{11} = 1, & K_{22} = 1, & K_{33} = -3, \end{cases}$$

and the other entries are zero. By (9) we have W=0. All these examples give metrics adapted to a contact form $\omega = \omega_3$, since in an orthonormal basis $*\omega_3 = \omega_1 \wedge \omega_2$.

In general, given a metric adapted to a contact form ω, we shall restrict our attention to orthonormal bases of 1-forms ω_1, ω_2, ω_3 with $\omega_3 = \omega$. Considering the dual basis of vectors, we only need to choose a unit vector in B. These form a principal circle bundle, and all of our structural equations will live naturally on this circle bundle. It turns out to be advantageous to compare the general situation to that on the Heisenberg group. Therefore, we introduce the forms ψ_1, ψ_2, ψ_3 and the matrix $L_{11}, L_{12}, \ldots, L_{33}$ defined by

(33) $$\begin{cases} \varphi_1 = \psi_1 + \omega_1, & \varphi_2 = \psi_2 + \omega_2, & \varphi_3 = \psi_3 - \omega_3, \\ K_{11} = L_{11} + 1, & K_{22} = L_{22} + 1, & K_{33} = L_{33} - 3, \\ K_{12} = L_{12}, & K_{13} = L_{13}, & K_{23} = L_{23}. \end{cases}$$

Thus the ψ and L all vanish on the Heisenberg group. We then compute the following.

3.5. Structure equations for an adapted metric. They are:

$$(34)\qquad \begin{cases} d\omega_1 = \psi_2\wedge\omega_3-\psi_3\wedge\omega_2, \\ d\omega_2 = \psi_3\wedge\omega_1-\psi_1\wedge\omega_3, \\ d\omega_3 = 2\omega_1\wedge\omega_2, \end{cases}$$

and

$$(35)\qquad \begin{cases} \psi_1\wedge\omega_2 - \psi_2\wedge\omega_1 = 0, \\ \psi_1\wedge\omega_1 + \psi_2\wedge\omega_2 = 0, \end{cases}$$

and

$$(36)\qquad \begin{cases} d\psi_1 = \psi_2\wedge\psi_3+L_{11}\omega_2\wedge\omega_3+L_{12}\omega_3\wedge\omega_1+L_{13}\omega_1\wedge\omega_2, \\ d\psi_2 = \psi_3\wedge\psi_1+L_{21}\omega_2\wedge\omega_3+L_{22}\omega_3\wedge\omega_1+L_{23}\omega_1\wedge\omega_2, \\ d\psi_3 = \psi_1\wedge\psi_2+L_{31}\omega_2\wedge\omega_3+L_{32}\omega_3\wedge\omega_1+L_{33}\omega_1\wedge\omega_2. \end{cases}$$

Proof. **The equation $d\omega_3 = 2\omega_1\wedge\omega_2$ comes from the condition $d\omega = 2{*}\omega$ that the metric is adapted to the contact form ω. Then the corresponding structural equation yields $\psi_1\wedge\omega_2 - \psi_2\wedge\omega_1 = 0$. Using $dd\omega_3 = 0$ we compute $\psi_1\wedge\omega_1 + \psi_2\wedge\omega_2 = 0$ also.**

3.6. Corollary. We can find functions a and b on the principal circle bundle so that

$$(37)\qquad \begin{cases} \psi_1 = a\omega_1+b\omega_2, \\ \psi_2 = b\omega_1-a\omega_2. \end{cases}$$

Proof. This follows algebraically from the equations (35).

It is even more convenient to write these equations in complex form. We make the following substitutions.

3.7. Complex substitutions.

On account of the complex structure in B it is convenient to use the complex notation. We shall set:

$$
(38)\quad \begin{cases}
\Omega = \omega_1+i\omega_2, & \omega = \omega_3, \\
\Psi = \psi_1+i\psi_2, & \psi = \psi_3, \\
\iota = a+ib, & \\
p = \frac{1}{2}(L_{11}+L_{22}), \quad q = \frac{1}{2}(L_{11}-L_{22}), & r = L_{12}, \\
s = q+ir, & \\
z = \frac{1}{2}(L_{13}+iL_{23}), & \\
t = L_{33}, & \\
W = \frac{1}{4}(t-a^2-b^2), &
\end{cases}
$$

where W is the <u>Webster</u> <u>curvature</u>, to be verified below. Note that $\Psi = \iota\bar{\Omega}$. Thus Ω and ω give a basis for the 1-forms on M, while ι and ψ define the connection.

3.8 Complex structural equations.

$$
(39)\quad \begin{aligned}
d\Omega &= i(\psi\wedge\Omega-\iota\bar{\Omega}\wedge\omega), \\
d\omega &= i\Omega\wedge\bar{\Omega},
\end{aligned}
$$

and

$$
(40)\quad \begin{cases}
d\psi = i[2W\Omega\wedge\bar{\Omega}+(z\bar{\Omega}-\bar{z}\Omega)\wedge\omega], \\
d\iota \equiv i(2\iota\psi+z\Omega-s\omega) \mod \bar{\Omega}, \\
p+|\iota|^2 = 0.
\end{cases}
$$

Proof. This is a direct computation. Note that the real functions p,W and the complex functions z,s give the curvature of the metric.

The equation $p+\left|\iota\right|^2 = 0$ has the important consequence that we can compute the Webster curvature W from the $K_{\alpha\beta}$. The result is the expression for W in (9).

The following notation will be useful. If f is a function on a Riemannian manifold with frame ω_α, then

(41) $$df = D_\alpha f \cdot \omega_\alpha,$$

where $D_\alpha f$ is the derivative of f in the direction of the dual vector field V_α. If f is a function on the principal bundle then we can still define $D_\alpha f$ as the derivative in the direction of the horizontal lifting of V_α. In this case we will have

(42) $$df \equiv D_\alpha f \cdot \omega_\alpha \mod \varphi_{\alpha\beta}.$$

If the function f represents a tensor then $D_\alpha f$ are its covariant derivatives, and the extra terms in $\varphi_{\alpha\beta}$ depend on what kind of tensor is represented. In the example if T is a covariant 1-tensor and

(43) $$f = T(V_\gamma),$$

then,

(44) $$df = D_\alpha f \cdot \omega_\alpha + T(V_\beta)\, \varphi_{\beta\gamma},$$

while if T is a covariant 2-tensor and

(45) $$f = T(V_\gamma, V_\delta),$$

then

(46) $$df = D_\alpha f \cdot \omega_\alpha + T(V_\beta, V_\delta)\, \varphi_{\beta\gamma} + T(V_\gamma, V_\beta)\, \varphi_{\beta\delta},$$

and so on. In the complex notation we write

(47) $$df = \partial f \cdot \Omega + \bar{\partial} f \cdot \bar{\Omega} + D_V f \cdot \omega$$

as the definition of the differential operators ∂f, $\bar{\partial} f$, and $D_v f$. As usual

$$(48)\qquad \begin{cases} \partial f = \frac{1}{2}(D_1 f - iD_2 f), \\ \bar{\partial} f = \frac{1}{2}(D_1 f + iD_2 f), \\ D_v f = D_3 f, \end{cases}$$

reflecting the transition from real to complex notation. If f is a function on the principal circle bundle coming from a symmetric k-tensor on B then

$$(49)\qquad df = \partial f \cdot \Omega + \bar{\partial} f \cdot \bar{\Omega} + D_v f \cdot \omega + ikf\psi.$$

For example, the function ι represents a trace-free symmetric 2-form on B, and the structural equation for ι tells us

3.9. Lemma.

$$(50)\qquad \bar{\partial}\iota = iz \text{ and } D_v \iota = -is.$$

4. Change of basis. We start with the simplest change of basis, namely rotation through an angle θ. We take θ to be a function on M and study what happens on the principal circle bundle. The new basis ω_1^*, ω_2^*, ω_3^* is given by $\omega_3^* = \omega_3 = \omega$ and

$$(51)\qquad \begin{aligned} \omega_1^* &= \cos.\theta\ \omega_1 - \sin.\theta \omega_2, \\ \omega_2^* &= \sin.\theta\ \omega_1 + \cos.\theta\ \omega_2 \end{aligned}$$

or in complex terms $\omega^* = \omega$ and

$$(52)\qquad \Omega^* = e^{i\theta}\ \Omega.$$

Then from the structural equations we immediately find that

4.1. Lemma.

$$\psi^* = \psi + d\theta,$$

(53)

$$\iota^* = \iota\, e^{2i\theta}.$$

Now a function or tensor on the principal circle bundle comes from one on M by the pull-back if and only if it is invariant under rotation by θ. Thus we see that the curvature form $d\psi^* = d\psi$ is invariant and hence lives on M. The form $\Omega \wedge \bar{\Omega}$ is also invariant, so $W = W^*$ is invariant and W is a function on M. This W is the scalar curvature introduced by Webster (see [11]). Likewise $|\iota|^2$ is invariant and hence a function on M. The function ι defines a tensor $\iota\bar{\Omega}^2$ which is invariant. Hence its real and imaginary parts

$$a(\omega_1^2 - \omega_2^2) + 2b\ \omega_1\omega_2,$$

(54)

$$b(\omega_1^2 - \omega_2^2) - 2a\ \omega_1\omega_2$$

define trace-free symmetric bilinear forms on B (they differ by rotation). This form is called the torsion tensor by Webster (see [11]); it is analogous to the second fundamental form for a surface.

We now consider more interesting changes of basis. First we change the CR structure while leaving the contact form ω fixed. In order to keep the metric adapted to the contact form we must leave $\omega_1 \wedge \omega_2$ invariant. This gives a new basis

$$\omega_1^* = A\omega_1 + B\omega_2,$$

(55)

$$\omega_2^* = C\omega_1 + D\omega_2,$$

$$\omega_3^* = \omega_3$$

with $AD - BC = 1$. An infinitesimal change of basis is given by the tangent to a path at $t = 0$. Thus an infinitesimal change of the basis which changes CR structure but leaves the contact form invariant and keeps the metric adapted is given by

$$\omega_1' = g\omega_1 + h\omega_2,$$
$$\omega_2' = k\omega_1 + l\omega_2,$$
$$\omega_3' = 0$$

with $g+l = 0$. Since the rotations are trivial we may as well take $h=k$. This gives

$$\omega_1' = g\omega_1 + h\omega_2,$$
$$(56) \qquad \omega_2' = h\omega_1 - g\omega_2,$$
$$\omega_3' = 0.$$

In complex notation if $f = g+ih$ then

$$(57) \qquad \Omega' = f\bar{\Omega} \text{ and } \omega' = 0.$$

For future use we compute the infinitesimal change ψ' in ψ and ι' in ι from the structural equations (39), (40). We find that f transforms as a 2-tensor

$$(58) \qquad df = \partial f \cdot \Omega + \bar{\partial} f \cdot \bar{\Omega} + D_v f \cdot \omega + 2if\psi$$

and that

4.2. Lemma.

$$\iota' = -i\, D_v f,$$
$$(59)$$
$$\psi' = i(\partial f \cdot \bar{\Omega} - \bar{\partial}\bar{f} \cdot \Omega) - (\iota\bar{f} + \bar{\iota} f)\omega$$

using the fact that we know $\psi \wedge \Omega$ and ψ is real.

On the other hand we may wish to fix the CR structure and change the contact form while keeping the metric adapted. In this case let $\omega_3^* = f^2\omega_3$ where f is a positive real function. Excluding rotation we find that to keep the metric adapted we need

$$\omega_1^* = f\cdot\omega_1 - D_2 f\cdot\omega_3,$$
(60) $$\omega_2^* = f\cdot\omega_2 + D_1 f\cdot\omega_3,$$
$$\omega_3^* = f^2\omega_3.$$

In complex notation

$$\Omega^* = f\Omega + 2i\,\bar\partial f\cdot\omega,$$
(61)
$$\omega^* = f^2\omega.$$

For an infinitesimal variation we differentiate to obtain

$$\Omega' = f'\,\Omega + 2i\,\bar\partial f'\cdot\omega,$$
(62)
$$\omega' = 2f'\,\omega.$$

Hence changes of metric fixing the CR structure are given by a potential function f, much the same way as changes of metric fixing a conformal structure. The main difference is that the derivatives of f enter the formula for the new basis.

As a consequence of $ddf = 0$ we have

(63) $$\partial\bar\partial f - \bar\partial\partial f + iD_V f = 0.$$

We also define the sub-Laplace operator

(64) $$\square f = 2(\partial\bar\partial f + \bar\partial\partial f) = (D_1 D_1 f + D_2 D_2 f).$$

Then a straightforward computation substituting in the structural equations yields

(4.3. **Lemma.**)

4.3. Lemma.

(65)
$$\psi^* = \psi + 3i\left[\frac{\partial f}{f}\,\Omega - \frac{\bar{\partial} f}{f}\,\bar{\Omega}\right] - \left[\frac{\Box f}{2f} + 6\,\frac{\partial f \bar{\partial} f}{f^2}\right],$$
$$\iota^* = \iota - 2\,\frac{\bar{\partial}\bar{\partial} f}{f} - 6\left[\frac{\bar{\partial} f}{f}\right]^2.$$

Differentiating the first we get

$$d\psi^* \equiv d\psi - 2i\,\frac{\Box f}{f}\,\Omega\wedge\bar{\Omega} \mod \omega,$$

which shows the remarkable relation given by

4.4 Lemma.

(66)
$$f^3 W^* = fW - \Box f.$$

4.5. Corollary. In an infinitesimal variation

(67)
$$W' = -\Box f' - 2\,f'\,W.$$

5. Energies. Let μ be the measure on M

(68)
$$\mu = \omega_1\wedge\omega_2\wedge\omega_3 = \frac{i}{2}\,\Omega\wedge\bar{\Omega}\wedge\omega$$

induced by the metric. Here are two interesting energies which we may form. The first is

(69)
$$E_W = \int_M W\,\mu,$$

which is analogous to the energy

(70)
$$E = \int_M R\,\mu$$

in the Yamabe problem. The second is

(71) $$E_\iota = \int_M |\iota|^2 \mu,$$

which is a kind of Dirichlet energy.

In this section we shall study the critical points of these energies.

First we observe that for computational reasons it is easier to integrate over the principal circle bundle P. The measure there is

(72) $$\nu = \omega_1 \wedge \omega_2 \wedge \omega_3 \wedge \psi_3 = \frac{i}{2}\, \Omega \wedge \bar{\Omega} \wedge \omega \wedge \psi.$$

If f is a function on the base M then

(73) $$\int_P f\, \nu = 2\pi \int_M f\, \mu,$$

so nothing is lost.

Next we observe that we can integrate by parts.

5.1. Lemma. For any f on P

(74) $$\int_P \partial f \cdot \nu = 0 \text{ and } \int_P D_V f \cdot \nu = 0.$$

Proof. The first follows from

$$\int_P d(f \bar{\Omega} \wedge \omega \wedge \psi) = 0$$

and the second follows from

$$\int_P d(f \Omega \wedge \bar{\Omega} \wedge \psi) = 0,$$

since $d\Omega \equiv 0 \mod \bar{\Omega}, \omega$ and $d\omega \equiv 0 \mod \Omega, \bar{\Omega}$ and $d\psi \equiv 0 \mod \Omega, \bar{\Omega}, \omega$.

5.2. Theorem. *The energy E_W is critical over all contact forms with a fixed CR structure and fixed volume if and only if W is constant. It is critical* *over all CR structures with a fixed contact form if and only if* $\iota = 0$.

Proof. We compute the infinitesimal variation $E_W{}'$. Fixing the CR structure and varying the potential f of the contact form with $\omega^* = f^2\omega$ gives $\nu' = 4f'\nu$ and

$$E_W{}' = \int_P (-\Box f' + 2f' W)\, \nu = 2\int f' W\, \nu,$$

since $\Box$ integrates away. The volume is fixed when $\int f' \nu = 0$. Thus, $E_W' = 0$ precisely when W is constant.

Fixing the contact form and varying the CR structure we use the following.

5.3 Lemma.

(75)
$$E_W = \tfrac{1}{2}\int_P d\psi\wedge\omega\wedge\psi.$$

Proof. We use the structural equation to see

$$d\psi\wedge\omega = 2iW\Omega\wedge\bar{\Omega}\wedge\omega$$

and integrate by parts to get the result. Then we have

$$E_W{}' = \tfrac{1}{2}\int_P d\psi'\wedge\omega\wedge\psi + d\psi\wedge\omega\wedge\psi'$$

(using $\omega' = 0$), and this gives

$$E_W{}' = \tfrac{i}{2}\int_P \psi'\wedge\Omega\wedge\bar{\Omega}\wedge\psi.$$

Then using Lemma 4.2 we get

$$E_W{}' = -\tfrac{1}{2}\int_P (\iota\bar{f}+\bar{\iota}f)\, \nu,$$

so that the CR structure is critical for fixed ω precisely when $\iota=0$.

Next we consider the energy E_ι.

5.4. **Theorem.** *The energy* E_ι *is critical over all* CR *structures with fixed contact form if and only if* $D_V\iota = 0$, *which is equivalent to* $s = 0$, *or* $K_{11} = K_{22}$ *and* $K_{12} = 0$. *The energy* E_ι *is critical over all contact forms with fixed* CR *structure and fixed volume if and only if*

$$(76) \qquad 2i(\partial z - \bar{\partial}\bar{z}) + 3p = \text{constant}.$$

Proof. The energy E_ι is given by

$$E_\iota = \int_P |\iota|^2 \nu,$$

so its first variation is

$$E_\iota' = \int_P (\iota' \bar{\iota} + \iota \bar{\iota}')\nu + |\iota|^2 \nu'.$$

When ω is fixed, $\omega' = 0$ and $\nu' = 0$. By Lemma 4.2 we have the result that if $\Omega' = f\,\bar{\Omega}$ then $\iota' = -iD_V f$, and this gives

5.5. **Lemma.**

$$E_\iota' = 2\,\mathrm{Im} \int_P \bar{f}\, D_V \iota\ \nu.$$

Since f is any real function on M, we see $E_\iota' = 0$ when $D_V\iota = 0$. Then $s = 0$ by Lemma 3.9 and $K_{11} = K_{22}$ and $K_{12} = 0$ by substitution (38).

This condition says that, at each point of M, the sectional curvature of all planes perpendicular to the contact plane B are equal.

If, on the other hand, we fix the CR structure and vary the contact form by a potential f, we have from Lemma 4.3

$$\iota^* = \iota - 2\,\frac{\bar\partial\bar\partial f}{f} - 6\left[\frac{\bar\partial f}{f}\right]^2.$$

Taking an infinitesimal variation

$$\iota' = -2\bar\partial\bar\partial f',\quad \nu' = 3f'\,\nu$$

Then the variation in E_ι is

$$E_\iota{}' = \int_P \{-2(\iota\,\partial\partial f' + \bar\iota\,\bar\partial\bar\partial f') + 3|\iota|^2 f'\}\,\nu,$$

from which we see that $E_\iota{}' = 0$ precisely when

$$2(\partial\partial\iota + \bar\partial\bar\partial\bar\iota) - 3|\iota|^2$$

is constant. Since $\partial\iota = iz$ by Lemma 3.9, and $|\iota|^2+p = 0$, this gives the equation (76).

6. Changing Webster Scalar Curvature. The problem of fixing the CR structure and changing the Webster scalar curvature is precisely analogous to the Yamabe problem of fixing the conformal structure and changing the scalar curvature, except the problem is subelliptic, and the estimates and constants for the 3-dimensional CR case look like the 4-dimensional conformal case. The first result is the following.

6.1. Theorem. *Let M be a compact orientable three-manifold with fixed CR structure. Then we can change the contact form so that the Webster scalar curvature W of the adapted Riemannian metric is either positive or zero or negative everywhere.*

Proof. We have $f^3W^* = fW - \Box f$ from Lemma 4.4. We take f to be the eigenfunction of $W-\Box$ with lowest eigenvalue λ_1. By the strict maximum principle for subelliptic equations (see Bony [1]) we conclude that f is strictly positive. Since $Wf - \Box f = \lambda_1 f$ we have

$f^2W^* = \lambda_1$. Hence W^* always has the same sign as λ_1.

Next we show that in the negative curvature case we can make W whatever we want, in particular, a negative constant.

6.2. Theorem. *Let M be a compact orientable three-manifold with a fixed CR structure. If some contact form has negative Webster scalar curvature, then every negative function $W<0$ is the Webster scalar curvature of one and only one contact form ω.*

Proof. Let C be the space of all contact forms and let $\mathcal{F}$ be the space of functions. We define the operator P by

$$P{:}C \longrightarrow \mathcal{F}, \quad P(\omega) = W.$$

Let $\mathcal{F}^-$ be the space of negative functions and let C^- be the space of contact forms with negative Webster curvature. Then

$$P{:}C^- \longrightarrow \mathcal{F}^- \tag{77}$$

is also defined. We claim the P in (77) is a global diffeomorphism. This follows from the following observations.

a) C^- is not empty.

b) P is locally invertible.

c) P is proper (the inverse image of a compact set is compact).

d) $\mathcal{F}^-$ is simply connected.

We then argue that (a) allows us to start inverting somewhere, (b) allows us to continue the inverse along paths, (c) says that the inverse doesn't stop until we run out of $\mathcal{F}^-$, and (d) tells us that the inverse is independent of the path and hence unique.

Before we start the proof we remark on a few technical details. There are two possible approaches to the proof. One is to work with C^∞ functions and quote the Nash–Moser theorem (see [5] for an exposition) using the ideas in [6] to handle the subelliptic estimates. The other is to work with the Folland–Stein spaces S^p_k (see [3]) which measure k derivatives in the direction of the contact structure in L^p norm. We can take $\omega\epsilon S^p_{k+2}$ and $W\epsilon S^p_k$ provided pk>8 so that $W\epsilon C^\circ$ by the appropriate Sobolev inclusion. The easiest case analytically is to take p = 2, which necessitates $k \geqslant 5$.

We proceed with the proof. Observation (a) follows from the hypothesis. To see (b) we compute the derivative of P, and apply the inverse function theorem.

In fact, from Corollary 4.5 we write

$$\bar{\Box} f' + 2\bar{W}f' = -W',$$

by putting dashes on the original metric. The operator $\bar{\Box} + 2\bar{W}$ has zero null space by the maximum principle, since $\bar{W} < 0$. Since it is self-adjoint, it must also be onto and hence invertible. This proves that DP is invertible when $\bar{W} < 0$, and so P is locally invertible on all of C^-.

To see assertion (c) that P is proper, we apply the maximum principle to the equation

$$f^3W = f\bar{W} - \bar{\Box}f.$$

Where f is a maximum $\bar{\Box}f \leqslant 0$, and where f is a minimum $\bar{\Box}f \geqslant 0$. Since W and $\bar{W}$ are both negative we get the estimate

$$\text{(78)} \qquad \left[(\bar{W}/W)_{\min}\right]^{\frac{1}{2}} \leqslant f_{\min} \leqslant f_{\max} \leqslant \left[(\bar{W}/W)_{\max}\right]^{\frac{1}{2}}.$$

Notice that the estimate fails if W and $\bar{W}$ are positive. Having control of the maximum and minimum of f, it is easy to control the higher derivatives using the equation and the subelliptic Garding's inequality

(79) $$\|f\|_{S^p_{k+2}} \leq C\,(\|\bar{\Box} f\|_{S^p_k} + \|f\|_{L^p}).$$

In the C^∞ case this shows P is proper. For given any compact set of M, we have uniform bounds on W_{max} and W_{min} and all $\|W\|_{S^p_k}$. This gives bounds on f_{max} and f_{min} and all $\|f\|_{S^p_k}$ for all f in the preimage, so the preimage is compact since C^∞ is a Montel space. To work in the Banach space S^p_k we also need the following observation. Suppose we have a sequence of contact forms ω_n with $W_n \longrightarrow \bar{W} < 0$ in S^p_k. The previous estimates give bounds on ω_n in S^p_{k+2}, which implies convergence of a subsequence in S^p_k. Let $\omega_n \longrightarrow \bar{\omega}$, and write $\omega_n = f_n^2 \bar{\omega}$. The maximum principle estimate shows $f_n \longrightarrow 1$ in C°. Then using the equation we get the estimate

(80) $$\|f_n - 1\|_{S^p_{k+2}} \leq C\,\|W_n - \bar{W}\|_{S^p_k};$$

this shows $\omega_n \longrightarrow \bar{\omega}$ in S^p_{k+2}, and proves P is proper.

The assertion (d) that $\mathcal{T}^-$ is simply connected follows by shrinking along straight line paths to W = -1. This completes the proof of the theorem.

7. Minimizing Torsion. We consider finally the problem of minimizing the energy

(81) $$E_\iota = \int_P |\iota|^2 \nu$$

representing the L^2 norm of the torsion by the heat equation with the contact form ω fixed. From Lemma 5.5 we have the result that if we take a path of Ω's depending on t with $\Omega' = f\bar{\Omega}$ then

$$E_\iota' = 2\,\mathrm{Im} \int_P \bar{f}\, D_V \iota\, \nu.$$

Following the gradient flow of E_ι we let $f = i\, D_V \iota$. This gives heat equation for E_ι. Since $D_V \iota = -$ is by Lemma 3.9 we get the following results.

7.1. Heat Equation Formulas.

$$\Omega' = i\, D_V \iota \cdot \bar{\Omega},$$

(82) $$E_\iota' = -2\int_P |s|^2\, \nu,$$

$$\iota' = D_V^2 \iota.$$

These equations show that if the solution exists for all time then the energy E_ι decreases and the curvature $s \longrightarrow 0$. The equation $\iota' = D_V^2 \iota$ is a highly degenerate parabolic equation, since the right hand side involves only the second derivative in the one direction V. Nevertheless, it is not a bad equation, since the maximum principle applies. This shows that the maximum absolute value of ι decreases. The equation is in fact just the ordinary heat equation restricted to each orbit in the flow of V. Physically we can imagine the manifold P to be made of a bundle of wires insulated from each other, with the heat flowing only along the wires. When the orbits of V are closed, the analysis should be fairly straightforward. When the orbits of V are dense, things are much more complicated, and probably lead to small divisor problems.

A regular foliation is one where each leaf is compact and the space of leaves is Hausdorff. In this case we always have a Seifert foliation, one where each leaf has a neighborhood which is a finite quotient of a bundle. In three dimensions the Seifert foliated manifolds are well-understood by the topologists, and provide many of the nice examples. We conjecture the following result.

7.2. Conjecture. Let M be a compact three-manifold with a fixed contact form ω whose vector field V induces a Seifert foliation. There there exists a CR structure on M such that the associated metric has $s = 0$, i.e., the sectional curvature of all planes at a given point perpendicular to the contact bundle $B = \text{Null}\ \omega$ are equal. The metric is obtained as the limit of the heat equation flow as $t \longrightarrow \infty$.

REFERENCES

[1] J.M. Bony, Principe du maximum, inégalité de Harnack, et unicité du problème de Cauchy pour les opérateurs elliptiques dégénerés, Ann. Inst. Fourier 19(1969), 277-304.

[2] A. Douady, Noeuds et structures de contact en dimension 3, d'après Daniel Bennequin, Séminaire Bourbaki, 1982/83, no.° 604.

[3] G.B. Folland and E.M. Stein, Estimates for the $\bar{\partial}_b$-complex and analysis on the Heisenberg group, Comm. Pure and App. Math 27(1974), 429-522.

[4] J.W. Gray, Some global properties of contact structures, Annals of Math 69(1959), 421-450.

[5] R. Hamilton, The inverse function theorem of Nash and Moser, Bull. Amer. Math. Soc. 7(1982), 65-222.

[6] R. Hamilton, Three-manifolds with positive Ricci curvature, J. Diff. Geom. 17(1982), 255-306.

[7] D. Jerison and J. Lee, A subelliptic, non-linear eigenvalue problem and scalar curvature on CR manifolds, Microlocal Analysis, Amer. Math Soc. Contemporary Math Series, 27(1984), 57-63.

[8] J. Martinet, Formes de contact sur les variétés de dimension 3, Proc. Liverpool Singularities Symp II, Springer Lecture Notes in Math 209(1971), 142-163.

[9] R. Schoen, Conformal deformation of a Riemannian metric to constant scalar curvature, preprint 1984.

[10] W. Thurston and H.E. Winkelnkemper, On the existence of contact forms. Proc. Amer. Math. Soc. 52(1975), 345-347.

[11] S.M. Webster, Pseudohermitian structures on a real hypersurface, J. Diff. Geom. 13(1978), 25-41.

APPENDIX

by

Alan Weinstein

THREE-DIMENSIONAL CONTACT MANIFOLDS WITH VANISHING TORSION TENSOR

In a lecture on some of the material in the preceding paper, Professor Chern raised the question of determining those 3-manifolds admitting a contact structure and adpated Riemannian metric for which the torsion invariant $c^2=a^2+b^2$ is identically zero. (See §3. A variational characterization of such structures is given in Theorem 5.2.) The purpose of this note is to show that the class of manifolds in question consists of certain Seifert fiber manifolds over orientable surfaces, and that the first real Betti number $b_1(M)$ of each such manifold M is even. These results are not new; see our closing remarks.

By a simple computation, it may be seen that the matrix $\begin{bmatrix} a & b \\ b & -a \end{bmatrix}$ (see Corollary 3.5) represents the Lie derivative of the induced metric on the contact bundle B with respect to the contact vector field V. We thus have:

A.1. Lemma. *The invariant* c^2 *is identically zero if and only if* V *is a Killing vector field. (In other words,* M *is a "K-contact manifold"; see* [1].)

We would like the flow generated by V to be periodic. If this is not the case, we can make it so by changing the structures in the following way. Let G be the closure, in the automorphism group of M with its contact and metric structures, of the 1-parameter group generated by V. G must be a torus, so in its Lie algebra we can find Killing vector fields V' arbitrarily close to V and having periodic flow. Let ω' be the 1-form which annihilates the subbundle B' perpendicular to V' and which satisfies $\omega'(V') \equiv 1$. For V' sufficiently close to V, ω' will be so close to the original contact form ω that it is itself a contact form. Since the flow of V' leaves the metric invariant, it leaves the invariant the form ω', from which it follows that V' is the contact vector field associated with ω'.

Having made the changes described in the previous paragraph, we may revert to our original notation, dropping primes, and assume that the flow of V is periodic. A rescaling of ω will even permit us to assume that the least period of V is 1. (Note that, by Gray's theorem [2], we could actually assume that the new contact structure equals the one which was originally given.)

Suppose for the moment that the action of $S^1 = \mathbb{R}/\mathbb{Z}$ generated by V is free. Then M is a principal S^1 bundle over the surface M/S^1. The form ω is a connection on this bundle; since ω is a contact form, the corresponding curvature form on M/S^1 is nowhere vanishing. Thus M/S^1 is an orientable surface, and the Chern class of the fibration $M \longrightarrow M/S^1$ is non-zero. By the classification of surfaces, $b_1(M/S^1)$ is even; by the Gysin sequence, $b_1(M) = b_1(M/S^1)$ and is therefore even as well.

We are left to consider the case where the action of S^1, although locally free, is not free. The procedure which we will follow is that of [8]. Let $\Gamma \subsetneq S^1$ be the (finite) subgroup generated by the isotropy groups of all the elements of M. Then M is a branched cover of M/Γ, and M/Γ is a principal bundle over M/S^1 with fiber the circle S^1/Γ. The branched covering map $M \longrightarrow M/\Gamma$ induces isomorphisms on real cohomology, so it suffices to show that $b_1(M/\Gamma)$ is even. To see this, we consider the fibration $S^1/\Gamma \longrightarrow M/\Gamma \longrightarrow M/S^1$. The quotient spaces M/Γ and M/S^1 are V-manifolds in the sense of [4], and we have a fibre bundle in that category. The base M/S^1 is actually a topological surface which is orientable since it carries a nowhere-zero 2-form on the complement of its singular points. Now the contact form may once again be considered as a connection on our V-fibration, and so, just as in the preceding paragraph, we may conclude that $b_1(M/\Gamma)$ is even.

Remarks. A K-contact manifold is locally a 1-dimensional bundle over an almost-Kähler manifold. When the base is Kähler, the contact manifold is called <u>Sasakian.</u> Using harmonic forms, Tachibana

[5] is shown that the first Betti number of a compact Sasakian manifold is even. On the other hand, since every almost complex structure on a surface is integrable, every 3-dimensional K-contact mainfold is Sasakian, and hence our result follows from Tachibana's theorem. In higher dimensions, compact symplectic manifolds with odd Betti numbers in even dimension are known to exist [3] [7], and circle bundles over them will carry K-contact structures, while having odd Betti numbers in even dimension.

The paper [6] contains a study of which Seifert fiber manifolds over surfaces actually admit S^1-invariant contact structures.

Acknowledgments. This research was partially supported by NSF Grant DMS84-03201. I would like to thank Geoff Mess for his helpful advice.

REFERENCES

1. D. Blair, Contact Manifolds in Riemannian Geometry, Lecture Notes in Math., vol. 59 (1976).
2. J. Gray, Some global properties of contact structures, Ann. of Math. 69 (1959), 421-450.
3. D.McDuff, Examples of simply connected symplectic manifolds which are not Kähler, preprint, Stony Brook, 1984.
4. I. Satake, The Gauss-Bonnet theorem for V-manifolds, J. Math. Soc. Japan 9 (1957), 464-492.
5. S. Tachibana, On harmonic tensors in compact Sasakian spaces, Tohoku Math. J. 17 (1965), 271-284.
6. C.B. Thomas, Almost regular contact manifolds, J. Diff. Geom. 11 (1976), 521-533.
7. W. Thurston, Some simple examples of symplectic manifolds. Proc. A.M.S. 55 (1976), 467-468.
8. A. Weinstein, Symplectic V-manifolds, periodic orbits of hamiltonian systems, and the volume of certain riemannian manifolds, Comm. Pure Appl. Math. 30 (1977), 265-271.

Reprinted from
Arbeitstagung Bonn 1984 Springer Lecture Notes 1111 Springer–Verlag (1985) 279–308.

oc. Natl. Acad. Sci. USA
ol. 82, pp. 2217–2219, April 1985
athematics

Iarmonic maps of S^2 into a complex Grassmann manifold

(harmonic sequences/Frenet harmonic sequences/fundamental collineations/harmonic flags/"crossing" and "turning" constructions)

HIING-SHEN CHERN† AND JON WOLFSON‡

epartment of Mathematics, University of California at Berkeley, and Mathematical Sciences Research Institute, Berkeley, CA 94720; and ‡Department of athematics, Rice University, Houston, TX 77251

ontributed by Shiing-shen Chern, December 5, 1984

BSTRACT **Let G(k, n) be the Grassmann manifold of all in C_n, the complex spaces of dimensions k and n, respec-ely, or, what is the same, the manifold of all projective aces P_{k-1} in P_{n-1}, so that G(1, n) is the complex projective ace P_{n-1} itself. We study harmonic maps of the two-dimen-nal sphere S^2 into G(k, n). The case $k = 1$ has been the bject of investigation by several authors [see, for example, n, A. M. & Zakrzewski, W. J. (1980) *Nucl. Phys. B* 174, 7–406; Eells, J. & Wood, J. C. (1983) *Adv. Math.* 49, 217–3; and Wolfson, J. G. *Trans. Am. Math. Soc.*, in press]. The rmonic maps $S^2 \to G(2, 4)$ have been studied by Raman-han [Ramanathan, J. (1984) *J. Differ. Geom.* 19, 207–219]. e shall describe all harmonic maps $S^2 \to G(2, n)$. The meth-is based on several geometrical constructions, which lead om a given harmonic map to new harmonic maps, in which e image projective spaces are related by "fundamental col-eations." The key result is the degeneracy of some funda-ental collineations, which is a global consequence, following om the fact that the domain manifold is S^2. The method ex-ds to G(k, n).**

ometry of G(k, n)

e consider C_n equipped with the standard hermitian inner oduct. That is, for $\mathbf{Z}, \mathbf{W} \in C_n$,

$$\mathbf{Z} = (z_1, \ldots, z_n), \qquad \mathbf{W} = (w_1, \ldots, w_n), \tag{1.1}$$

have

$$(\mathbf{Z}, \mathbf{W}) = \sum z_A \overline{w}_A = \sum z_A w_{\overline{A}}. \tag{1.2}$$

hroughout this note we will agree on the following ranges of dices:

$$1 \leqq A, B, C, \ldots \leqq n, \qquad 1 \leqq \alpha, \beta, \gamma, \ldots \leqq k, \qquad k+1 \leqq i, j, \ldots \leqq n, \tag{1.3}$$

d we shall use the convention $\overline{z}_A = z_{\overline{A}}$ and also the summa-on convention. A frame consists of an ordered set of n vec-rs $\mathbf{Z}_A$, such that

$$\mathbf{Z}_1 \wedge \ldots \wedge \mathbf{Z}_n \neq 0. \tag{1.4}$$

is called *unitary*, if

$$(\mathbf{Z}_A, \mathbf{Z}_B) = \delta_{A\overline{B}}. \tag{1.5}$$

we write

$$d\mathbf{Z}_A = \omega_{A\overline{B}} \mathbf{Z}_B, \tag{1.6}$$

e publication costs of this article were defrayed in part by page charge yment. This article must therefore be hereby marked "*advertisement*" accordance with 18 U.S.C. §1734 solely to indicate this fact.

the coefficients $\omega_{A\overline{B}}$ are the Maurer–Cartan forms of the unitary group U(n). They are skew-hermitian; i.e.,

$$\omega_{A\overline{B}} + \omega_{\overline{B}A} = 0, \qquad \overline{\omega}_{B\overline{A}} = \omega_{\overline{B}A}. \tag{1.7}$$

Taking the exterior derivative of Eq. **1.6**, we get the Maurer–Cartan equations

$$d\omega_{A\overline{B}} = \omega_{A\overline{C}} \wedge \omega_{C\overline{B}}. \tag{1.8}$$

These equations contain all the local geometry of G(k, n).

An element C_k of G(k, n) can be defined by the multivector $\mathbf{Z}_1 \wedge \ldots \wedge \mathbf{Z}_k \neq 0$, defined up to a factor. This defines a G-structure on G(k, n), with $G = U(k) \times U(n-k)$. (We have called such a structure a *Segre* structure.) In particular, the form

$$ds^2 = \omega_{\alpha\overline{i}}\, \omega_{\overline{\alpha}i} \tag{1.9}$$

is a positive definite hermitain form on G(k, n) and defines an hermitian metric. Its Kahler form is

$$\Omega = \frac{i}{2}\, \omega_{\alpha\overline{i}} \wedge \omega_{\overline{\alpha}i}. \tag{1.10}$$

By using Eq. **1.8** it can be immediately verified that Ω is closed, so that the metric ds^2 is kahlerian.

By the expressions for $d\omega_{\alpha\overline{i}}$ we see that the connection forms are $\omega_{\alpha\overline{\beta}}$, $\omega_{i\overline{j}}$; their exterior derivatives give the curvature forms of the metric ds^2.

Surfaces in G(K, n)

Consider an oriented surface **M** immersed by a smooth map f into G(k, n). It acquires an induced riemannian metric and hence a complex structure. Using the latter, we write the induced metric as $ds^2_{\mathbf{M}} = \varphi\overline{\varphi}$, φ being a complex-valued one-form, defined up to a factor of absolute value 1. For $x \in \mathbf{M}$ the image $f(x) \in G(k, n)$ has an orthogonal space $f(x)^\perp \in G(n-k, n)$, which describes a surface $\mathbf{M}^\perp$. If $\mathbf{Z} \in f(x)$, then

$$d\mathbf{Z} \equiv \varphi \mathbf{X} + \overline{\varphi} \mathbf{Y}, \qquad \text{mod } f(x), \tag{2.1}$$

where $\mathbf{X}, \mathbf{Y} \in f(x)^\perp$. If $\mathbf{Z} \in C_n - \{0\}$, we denote by [**Z**] the point in P_{n-1} with **Z** as the homogeneous coordinate vector. Then

$$\partial : [\mathbf{Z}] \mapsto [\mathbf{X}], \qquad \overline{\partial} : [\mathbf{Z}] \mapsto [\mathbf{Y}] \tag{2.2}$$

define projective collineations of the projectivized space $[f(x)]$ into $[f(x)^\perp]$. We shall call these the *fundamental collineations*. Dually there are adjoint fundamental collineations from $[f(x)^\perp]$ to $[f(x)]$.

To express the situation analytically we choose, locally, a field of unitary frames $\mathbf{Z}_A$, so that $\mathbf{Z}_\alpha(x)$ span $f(x)$. Then

$$f^* \omega_{\alpha\overline{i}} = a_{\alpha\overline{i}} \varphi + b_{\alpha\overline{i}} \overline{\varphi}. \tag{2.3}$$

The fundamental collineations ∂ and $\bar\partial$ send $[\mathbf{Z}_\alpha]$ to $[\mathbf{X}_\alpha]$ and $[\mathbf{Y}_\alpha]$, respectively, where

$$\mathbf{X}_\alpha = a_{\alpha\bar{i}}\mathbf{Z}_i, \qquad \mathbf{Y}_\alpha = b_{\alpha\bar{i}}\mathbf{Z}_i. \qquad [2.4]$$

The metric $ds^2_{\mathbf{M}}$ has a connection form ρ, which is a real one-form satisfying the equation

$$d\varphi = -i\,\rho \wedge \varphi. \qquad [2.5]$$

Taking the exterior derivative of Eq. **2.3** and using Eqs. **1.8** and **2.5**, we get

$$Da_{\alpha\bar{i}} \wedge \varphi + Db_{\alpha\bar{i}} \wedge \bar\varphi = 0, \qquad [2.6]$$

where

$$\begin{aligned} Da_{\alpha\bar{i}} &= da_{\alpha\bar{i}} - a_{\beta\bar{i}}\,\omega_{\alpha\bar\beta} + a_{\alpha\bar{j}}\,\omega_{\bar{j}i} - i\,a_{\alpha\bar{i}}\rho, \\ Db_{\alpha\bar{i}} &= db_{\alpha\bar{i}} - b_{\beta\bar{i}}\,\omega_{\alpha\bar\beta} + b_{\alpha\bar{j}}\,\omega_{\bar{j}i} + i\,b_{\alpha\bar{i}}\rho. \end{aligned} \qquad [2.7]$$

Define

$$\begin{aligned} \mathbf{DX}_\alpha &= d\mathbf{X}_\alpha - (\omega_{\alpha\bar\beta} + i\,\rho\,\delta_{\alpha\bar\beta})\mathbf{X}_\beta, \\ \mathbf{DY}_\alpha &= d\mathbf{Y}_\alpha - (\omega_{\alpha\bar\beta} - i\,\rho\,\delta_{\alpha\bar\beta})\mathbf{Y}_\beta. \end{aligned} \qquad [2.8]$$

Then the condition for the map f to be harmonic (1) is given by *Theorem 1*.

THEOREM 1. *The property that* f *is a harmonic map is given by one of the following conditions, which are equivalent:*

(i) $\mathbf{D}a_{\alpha\bar{i}} \equiv 0, \quad \mathrm{mod}\ \varphi,$

(ii) $\mathbf{D}b_{\alpha\bar{i}} \equiv 0, \quad \mathrm{mod}\ \bar\varphi,$

(iii) $\mathbf{DX}_\alpha \equiv 0, \quad \mathrm{mod}\ \mathbf{Z}_\beta, \varphi,$

(iv) $\mathbf{DY}_\alpha \equiv 0, \quad \mathrm{mod}\ \mathbf{Z}_\beta, \bar\varphi.$

From this criterion we immediately draw the conclusion that a holomorphic or an anti-holomorphic map of $\mathbf{M}$ into $G(k, n)$ is harmonic. Thus we shall study harmonic maps that are not $\pm$ holomorphic. We have also the following theorem.

THEOREM 2. *Let* $f{:}\mathbf{M} \to G(k, n)$ *be a harmonic map. Then* (i) $f^\perp{:}\ \mathbf{M} \to G(n-k, n)$ *is harmonic, where* $f^\perp(x) = f(x)^\perp$, $x \in \mathbf{M}$.

(ii) *The images of* $[f(x)]$ *under the fundamental collineations* ∂, $\bar\partial$ *are of constant dimensions, say* $k_1 - 1$, $k_2 - 1$, *and the maps of* $\mathbf{M}$ *into* $G(k_1, n)$, $G(k_2, n)$ *so defined are harmonic. Denote the images by* $\partial[f(x)]$, $\bar\partial[f(x)]$, *respectively. If* $k_1 = k$ (*resp* $k_2 = k$), *then the image under* $\bar\partial$ (*resp* ∂) *of* $\partial[f(x)]$ (*resp* $\bar\partial[f(x)]$) *is* $[f(x)]$ *itself.*

(iii) *The kernels of the fundamental collineations* ∂, $\bar\partial$ *are of constant dimensions. If their orthogonal complements in* $[f(x)]$ *are of dimensions* $l_1 - 1$, $l_2 - 1$, *respectively, the maps so defined into* $G(l_1, n)$ *and* $G(l_2, n)$ *are harmonic.*

It is advantageous to use projective geometry, and we write $[f(x)] = \mathbf{L}_0$, being a projective space of dimension $k - 1$. Continuing the construction in part *ii* of *Theorem 2*, we get two sequences of harmonic maps

$$\begin{aligned} &\mathbf{L}_0 \xrightarrow{\partial} \mathbf{L}_1 \longrightarrow \ldots \longrightarrow \ldots \\ &\mathbf{L}_0 \xrightarrow{\bar\partial} \mathbf{L}_{-1} \longrightarrow \ldots \longrightarrow \ldots \end{aligned} \qquad [2.9]$$

connected by fundamental collineations. Each of the sequences in **2.9** will be called a *harmonic sequence*. By construction two consecutive projective spaces of a harmonic sequence are orthogonal. If the members of a harmonic sequence are of the same dimension, the sequence can be extended in both directions, giving

$$\begin{aligned} &\ldots \longrightarrow \mathbf{L}_{-1} \xrightarrow{\partial} \mathbf{L}_0 \xrightarrow{\partial} \mathbf{L}_1 \longrightarrow \quad \ldots \\ &\ldots \longleftarrow \mathbf{L}_{-1} \xleftarrow{\bar\partial} \mathbf{L}_0 \xleftarrow{\bar\partial} \mathbf{L}_1 \longleftarrow \quad \ldots \end{aligned} \qquad [2.10]$$

A harmonic sequence such that any two members are orthogonal will be called a *Frenet harmonic sequence*. A harmonic sequence is called *full* if its members span the whole $\mathbf{P}_{n-1}$.

An example of a Frenet harmonic sequence is given by the vertices $\mathbf{L}_0, \mathbf{L}_1, \ldots, \mathbf{L}_{n-1}$ of the Frenet frame of a holomorphic curve $\mathbf{L}_0(x)$, $x \in \mathbf{M}$, in $G(1, n) = \mathbf{P}_{n-1}$. The Frenet formula can be expressed as a Frenet harmonic sequence

$$\mathbf{L}_0 \xrightarrow{\partial} \mathbf{L}_1 \longrightarrow \ldots \xrightarrow{\partial} \mathbf{L}_{n-1}. \qquad [2.11]$$

In fact, the fundamental theorem on harmonic maps $\mathbf{S}^2 \to \mathbf{P}_{n-1}$ is that any such map is a member of a Frenet harmonic sequence for then it can be obtained from a holomorphic curve through fundamental collineations.

For $k = 2$ our process can be briefly described as follows. If the harmonic sequence degenerates, we are reduced to a harmonic map into $\mathbf{P}_{n-1}$. If it does not, we will "straighten" it into a full Frenet harmonic sequence. This is done through two constructions, which we call *crossing* and *turning*, respectively. Their success depends on the vanishing of certain holomorphic differential forms and thus on the fact that the domain manifold is $\mathbf{S}^2$. Their inverse processes depend on choices of holomorphic sections of $\mathbf{P}_1$-bundles.

Vanishing Theorems

The restriction on the harmonic maps whose domain manifold is $\mathbf{S}^2$ arises from the fact that $\mathbf{S}^2$ has no holomorphic differential forms of positive degree except zero. From our analytical data we are able to construct such forms and obtain in this way strong conclusions. Our first "vanishing theorem" is *Theorem 3*.

THEOREM 3. *Consider a harmonic map* $f{:}\mathbf{S}^2 \to G(k, n)$. *Let*

$$c_{\alpha\bar\beta} = a_{\alpha\bar{i}}\,b_{\bar\beta i}. \qquad [3.1]$$

Then

$$\det(c_{\alpha\bar\beta} + t\,\delta_{\alpha\bar\beta}) = t^k \text{ identically in } t. \qquad [3.2]$$

This follows from the fact that, with t as a parameter, $\det(c_{\alpha\bar\beta}\varphi^2 + t\,\delta_{\alpha\bar\beta})$ is a holomorphic differential form. This theorem, in different but equivalent formulations, was known to Ramanathan (2), Uhlenbeck (personal communication), and others.

Our next vanishing theorem is concerned with a Frenet harmonic sequence, as follows.

THEOREM 4. *Let* $\mathbf{L}_i{:}\mathbf{S}^2 \to G(k, n)$, $0 \leqq i \leqq s - 1$, $n \geqq ks$, *be harmonic maps that form a Frenet harmonic sequence*

$$\mathbf{L}_0 \xrightarrow{\partial} \mathbf{L}_1 \longrightarrow \ldots \xrightarrow{\partial} \mathbf{L}_{s-1}. \qquad [3.3]$$

Let $\pi{:}\mathbf{L}^\perp_{s-1} \to \mathbf{L}_0$ *be the standard projection. Then the map*

$$\pi \circ \partial{:}\mathbf{L}_{s-1} \longrightarrow \mathbf{L}_0 \qquad [3.4]$$

is degenerate. When $n = ks$, $\partial{:}\mathbf{L}_{s-1} \to \mathbf{L}_0$, *so in this case the fundamental collineation is degenerate.*

The proof of this theorem again relies on the vanishing of a lomorphic differential. More vanishing theorems will be ed.

rmonic Maps of S^2 into $G(2, n)$

e shall describe our crossing and turning constructions. To press the situation vividly we shall call the image, under a rmonic map, of a point $x \in M$ in $G(1, n)$ (resp $G(2, n)$) a rmonic point (resp a harmonic line). Similarly, a *harmonic g* consists of a harmonic point with a harmonic line ough it. The crossing construction passes from a harmonline to a harmonic flag and the turning construction passes m a harmonic flag to a double harmonic flag—i.e., a harnic line containing two orthogonal harmonic points.

We shall describe crossing for $n = 6$, which generalizes to y n. In P_5 consider the harmonic sequence

$$L_{-1} \xrightarrow{\partial} L_0 \xrightarrow{\partial} L_1, \qquad [4.1]$$

ich is full. By *Theorem 3* we have $\det(c_{\alpha\bar\beta}) = 0$. If $c_{\alpha\bar\beta} = 0$, is a Frenet harmonic sequence and the end lines contain rmonic points.

Consider the remaining case $rk(c_{\alpha\bar\beta}) = 1$. $L_0^\perp$ is three-dinsional (as a projective space) and L_{-1}, L_1 are nonintercting and nonorthogonal lines in it. The harmonic map S^2 $G(2, 3)$ defined by

$$x \mapsto L_0(x)^\perp, \qquad x \in S^2$$

s fundamental collineations that we denote by

$$'\partial, '\bar\partial: L_0(x)^\perp \longrightarrow L_0(x). \qquad [4.2]$$

r hypothesis says that their restrictions $'\partial|_{L_{-1}}$, $'\bar\partial|_{L_1}$ are of nk 1. We therefore get four geometrically defined points, eir images on L_0, and $\ker('\partial|_{L_{-1}})$ on L_{-1}, and $\ker('\bar\partial|_{L_1})$ on . The first two points are orthogonal, and the same is true the lines

$$\lambda_0 = \mathrm{im}('\partial|_{L_{-1}}) \wedge \ker('\bar\partial|_{L_1}),$$
$$\lambda_{-1} = \mathrm{im}('\bar\partial|_{L_1}) \wedge \ker('\partial|_{L_{-1}}). \qquad [4.3]$$

t λ_1 be the line, uniquely determined, that is orthogonal to th λ_0 and λ_{-1}. Then we have *Theorem 5*.

THEOREM 5. *In the above construction the lines* $\lambda_{-1} \to \lambda_0$ λ_1 *form a Frenet harmonic sequence.*

This is a global theorem on S^2. Its proof depends on the nishing of a holomorphic differential form of degree 5. The sociation of the Frenet harmonic sequence in *Theorem 5* to 1 is called crossing.

We must understand "recrossing," which is to recover L_0 om the Frenet harmonic sequence in *Theorem 5*. It involves the choice of a holomorphic section of a P_1-bundle over S^2—i.e., that of a meromorphic function over S^2.

To explain turning consider a harmonic flag. Its harmonic point is a vertex of the Frenet frame of some holomorphic curve in P_{n-1}. In our treatment harmonic flags always arise from degenerate fundamental collineations. Consider therefore the Frenet harmonic sequence

$$\lambda_{-1} \xrightarrow{\partial} \lambda_0 \xrightarrow{\partial} p_0, \qquad [4.4]$$

where $p_0 = \mathrm{im}\partial$ and $p_1 = \ker\partial \in \lambda_0$ are points and (p_1, λ_0) is a harmonic flag. Denote by q_{-1} the point on λ_{-1} satisfying $\partial(q_{-1}) = p_1$ and denote by q_1 the point on λ_0 orthogonal to p_1. Let μ_0 be the line spanned by q_{-1} and q_1. The vanishing of a holomorphic differential implies that the sequence

$$\mu_{-1} \xrightarrow{\partial} \mu_0 \xrightarrow{\partial} p_1 \xrightarrow{\partial} p_0$$

is a Frenet harmonic sequence. This construction is called turning. By iterating turnings we will come to a Frenet harmonic sequence

$$\nu_{-1} \xrightarrow{\partial} \nu_0 \xrightarrow{\partial} p_{k-1} \longrightarrow \dots \longrightarrow p_0,$$

with $p_k = \ker\partial \in \nu_0$ but $p_k \,\overline{\in}\, \partial(\nu_{-1})$. The line ν_0 can then be shown to be a double harmonic flag containing two orthogonal harmonic points.

The inverse procedure to turning, to be called returning, can be achieved by choosing a number of holomorphic sections of P_1-bundles.

In these constructions some fundamental collineations may vanish identically, in which case we get a holomorphic curve of $G(2, n)$. As a result we may state our theorem as follows.

THEOREM 6. *To every harmonic map* $L: S^2 \to G(2, n)$ *there is associated either* (i) *a holomorphic curve* $\wedge: S^2 \to G(2, n)$ *or* (ii) *a holomorphic curve* $\Delta: S^2 \to P_{n-1}$ *and a harmonic line* λ *joining two adjacent vertices of the Frenet frame of* Δ. *The map* L *can be constructed from* $\wedge$ *(in case* i) *or* λ *(in case* ii) *by choosing a number of holomorphic sections of* P_1*-bundles.*

Further references on our topic can be found in refs. 3 and 4.

S.-s.C. is partially supported by National Science Foundation Grants DMS 84-03201 and DMA 84-01959 and J.W. is partially supported by National Science Foundation Grant DMS 84-05186.

1. Eells, J. & Lemaire, L. (1978) *Bull. Lond. Math. Soc.* **10,** 1–68.
2. Ramanathan, J. (1984) *J. Differ. Geom.* **19,** 207–219.
3. Chern, S. S. & Wolfson, J. G. (1983) *Am. J. Math.* **105,** 59–83.
4. Din, A. M. & Zakrzewski, W. J. (1981) *Lett. Math. Phys.* **5** (6), 553–561.

Société Mathématique de France
Astérisque, hors série, 1985, p. 67-77

MOVING FRAMES

BY

Shiing-shen CHERN[1]

Introduction

The method of moving frames has a long history. It is at the heart of kinematics and its conscious application to differential geometry could be traced back at least to SERRET and FRENET[2]. In his famous "Leçons sur la théorie des surfaces," based on lectures at the Sorbonne 1882–1885, DARBOUX had the revolutionary idea using frames depending on two parameters and integrability conditions turn up. DARBOUX's method was generalized to arbitrary Lie groups by Émile COTTON [14] for the search of differential invariants, whose work should be considered a forerunner of the general method of moving frames.

CARTAN's first paper on moving frames was published in 1910 [7]. He immediately observed that DARBOUX's partial derivatives should be combined into the "Maurer-Cartan forms" and DARBOUX's integrability conditions are essentially the Maurer-Cartan equations. This extends the method to the case where the ambient space is acted on by any Lie group and, more generally, by an infinite pseudogroup in CARTAN's sense, such as the pseudogroup of all complex analytic automorphisms in n variables. The emphasis

[1] Work done under partial support of NSF grant MCS 77-23579.

[2] The so-called Serret-Frenet formulas were published by J.A. SERRET in MONGE's "Application de l'Analyse à la Géométrie", 5th edition (1850), p. 566, and by F. FRENET in 1847

on the Maurer-Cartan forms and their counterparts in infinite pseudo-groups cannot be over-estimated. For in this way we dispense with the local coordinates and these forms even in the case of the infinite pseudo-groups, are defined up to a transformation of a finite-dimensional Lie group, the gauge group, as it is called nowadays.

In almost all his works on differential geometry CARTAN used moving frames as the local tool. His general theory was developed in [8] and [9], the latter being based on notes edited by J. LERAY of a course given at the Sorbonne in 1931–1932. While we gather here today to honor Élie CARTAN, it is my feeling that his method of moving frames has been bypassed. As a principal reason I will quote Hermann WEYL [24] in his review of [9] :

"All of the author's books, the present one not excepted, are highly stimulating, full of original viewpoints, and profuse in interesting geometric details. CARTAN is undoubtedly the greatest living master of differential geometry... Nevertheless, I must admit that I found the book, like most of CARTAN's papers, hard reading..."

In this paper I wish to give a review of this method. I believe it is best appreciated through examples, and I will begin with some applications, old and new.

1. Riemannian geometry of two dimensions

Let M be a two-dimensional oriented Riemannian manifold. Let SM be its unit tangent bundle. A unit tangent vector can be denoted by xe_1, x being the origin. It determines uniquely the orthonormal frame xe_1e_2 consistent with the orientation of M. Thus SM can be identified with the (orthonormal) frame bundle of M, and we have the projection

$$\pi : SM \longrightarrow M, \tag{1}$$

sending xe_1e_2 to x. If $x\omega_1\omega_2$ is the dual coframe of xe_1e_2, the metric on M can be written

$$ds^2 = \omega_1^2 + \omega_2^2. \tag{2}$$

The space SM can also be interpreted as the space of all decompositions (2) of ds^2 as a sum of squares, with $\omega_1 \wedge \omega_2 > 0$.

The fundamental theorem of local Riemannian geometry says that there is a uniquely defined form ω_{12} in SM, the connection form, satisfying the equations

$$\begin{aligned} d\omega_1 &= \omega_{12} \wedge \omega_2, \\ d\omega_2 &= \omega_1 \wedge \omega_{12}. \end{aligned} \tag{3}$$

Classically this is expressed as the unique determination of the Levi-Civita connection without torsion. Taking the exterior derivatives of (3), one derives

$$d\omega_{12} = -K\omega_1 \wedge \omega_2, \tag{4}$$

where K is the Gaussian curvature. Formulas (3) and (4) contain all the information on local Riemannian geometry in two dimensions.

They give global consequences as well. A little meditation convinces one that (4) must be the formal basis of the Gauss-Bonnet formula, and this is indeed the case. It turns out that the proof of the n-dimensional Gauss-Bonnet formula can be based on this idea and can be reduced to a problem of the exterior differential calculus [11]. Interpreted in another way, (4) says that the form ω_{12} in SM has its exterior derivative in the base manifold M. This notion is called a transgression. It seems to me that this concept was not known to GAUSS.

The formalism also adapts well with variational problems. In CARTAN's original introduction of the exterior differential calculus, as a generalization of earlier works of DARBOUX and FROBENIUS, he introduced two differentials d and δ, which are commutative:

$$d\delta = \delta d. \tag{5}$$

Then the exterior product of the one-forms α and β is defined by

$$(\alpha \wedge \beta)(d, \delta) = \alpha(d)\beta(\delta) - \beta(d)\alpha(\delta) \tag{6}$$

and the exterior derivative of α was called the "bilinear covariant" and was defined by

$$d\alpha(d, \delta) = d\alpha(\delta) - \delta\alpha(d). \tag{7}$$

If we have a family of curves on M, we can let d be the variation along the curves and choose the frames so that $\omega_2(d) = 0$, (which means that e_1 are the unit tangent vectors and $\omega_1(d)$ is the element of arc), and we let δ be the variation among the curves.Then the first equation of (3) gives

$$d\omega_1(\delta) - \delta\omega_1(d) = \omega_{12}(d)\omega_2(\delta). \tag{8}$$

The second term in (8) is the first variation of the arc length, from which it is easily derived that the equations of the geodesics are

$$\omega_2 = \omega_{12} = 0. \tag{9}$$

Pursuing this method further, we can find the formula for the second variation of arc length and JACOBI's equation, etc.

2. Isometric deformation of surfaces

The isometry or isometric deformation of surfaces is a fundamental problem in classical surface theory. In 1867 BONNET studied these problems under the following additional conditions [4].

A) preservation of the lines of curvature;

B) preservation of the principal curvatures,

assuming in each case that the surfaces are free from umbilics. Since the Gaussian curvature is invariant under an isometry, condition B) is equivalent to :

B') preservation of the mean curvature.

BONNET solved Problem A), but not B). On B) he gave, however, an interesting theorem (see below).

A more interesting problem is a family of isometric surfaces, such that the isometries satisfy the condition A) or B). Such a family is called non-trivial, if it is not generated by one of its surfaces through rigid motions.

When these problems are formulated analytically, they give over-determined systems of partial differential equations, where the number of equations exceeds the number of unknown functions. Significant conclusions can be drawn only through repeated prolongations. In both problems fifth-order jets have to be used and the calculations are long. We will state the theorems, and refer the proofs to [5, pp. 269-284] and [12] respectively.

THEOREM 2.1 (Bonnet). — *A non-trivial family of isometric surfaces of non-zero Gaussian curvature, preserving the lines of curvature, is a family of cylindrical molding surfaces.*

The cylindrical molding surfaces can be kinematically described as follows : Take a cylinder Z and a curve C on one of its tangent planes. A molding surface is the locus described by C as a tangent plane rolls about Z.

THEOREM 2.2. — *A non-trivial family of isometric surfaces preserving the principal curvatures is one of the following :*

α) (the general case) a family of surfaces of constant mean curvature;

β) (the exceptional case) a family of surfaces of non-constant mean curvature.

They depend on six arbitrary constants and have the properties :

β1) they are W-surfaces;

β2) the metric

$$d\hat{s}^2 = (\mathrm{grad}\, H)^2 ds^2 / (H^2 - K),$$

where ds^2 is the metric of the surface and H and K are its mean curvature and Gaussian curvature respectively, has Gaussian curvature equal to -1.

The case α) was proved by BONNET, using a Lie transformation. In [12] we gave a complete solution of Problem B, an outstanding unsolved problem in classical surface theory, to which CARTAN made important contributions. We made use of the method of moving frames, with the connection form ω_{12} playing a decisive role.

3. Minimal two-spheres in a complex Grassmann manifold

There has been a lot of recent works on harmonic mappings of surfaces or minimal surfaces in the complex projective space $P_n(\mathbf{C})$ or the Grassmann manifold $G_{2,n+1}$, which we interpret as the space of all lines in $P_n(\mathbf{C})$. I shall try to imagine how CARTAN would approach the problem, using projective geometry [10]. The points of $P_n(\mathbf{C})$ will be identified with their homogeneous coordinate vectors $Z \in C_{n+1} - \{0\}$, and the conditions we shall deal with are to be invariant under the change $Z \to \lambda Z$. The space $P_n(\mathbf{C})$ is provided with the Study-Fubini metric, and the metric notions are valid.

Given a point $Z \in P_n(\mathbf{C})$, all the points of $P_n(\mathbf{C})$ orthogonal to Z form a hyperplane $Z^\perp \in P_n^*(\mathbf{C})$, which is the dual space of all hyperplanes of $P_n(\mathbf{C})$. A surface $Z : M \to P_n(\mathbf{C})$ gives rise to a dual surface $Z^\perp : M \to P_n^*(\mathbf{C})$, and vice versa. It is easily seen that Z is minimal if and only if $Z^\perp$ is minimal. We make use of the complex structure on M induced by its metric. The orthogonal projection of ∂Z (resp $\overline{\partial} Z$) into $Z^\perp$ is a form of type (1,0) (resp. (0,1)) with value in $Z^\perp$ and defines a point A (resp. B) $\in Z^\perp \subset P_n(\mathbf{C})$. It is easily verified that A and B are orthogonal points. The fundamental theorem on minimal two-spheres in $P_n(\mathbf{C})$ says that *if $M = S^2$ is not a holomorphic or an anti-holomorphic curve, then both A and B describe minimal two-spheres.*

Applying this theorem, one gets a complete description of the minimal S^2 in $P_n(\mathbf{C})$ as follows : *Let C be a rational curve in $P_n(\mathbf{C})$, which does not belong to a lower-dimensional space, and let $Z_0 Z_1 \ldots Z_n$, $Z_0 \in C$, be its Frenet frame. Then every Z_i, $0 \le i \le n$, describes a minimal two-sphere. Conversely every minimal two-sphere in $P_n(\mathbf{C})$ which does not belong to $P_m(\mathbf{C})$, $m < n$, is obtained in this way.*

Using the method of moving frames, we immediately stumble on invariant complex-valued forms of type $(k, 0)$, $k \ge 3$. The structure equations imply that they are holomorphic. On a two-sphere they must be zero. This is the key step of the proof. For details *cf.* [25]. The theorem was first stated by A.M. DIN and W.J. ZAKRZEWSKI [15], while the first mathemacian's proof was given by J. EELLS and J. WOOD [16].

The corresponding problem of minimal S^2 in $G_{2,4}$ was solved by J. RAMANATHAN [23]. J. WOLFSON and I have extended the above ideas to the more general case of $G_{2,n+1}$, and I wish to give a brief description of our results [13].

Consider a minimal surface

$$L : M \longrightarrow G_{2,n+1}, \tag{10}$$

which we regard as a manifold of lines in $P_n(\mathbf{C})$ of two real dimensions. The space $L(x)^\perp$ orthogonal to $L(x)$, $x \in M$, is of complex dimension $n-2$. It is an element of $G_{n-1,n+1}$, the Grassmann manifold of all projective spaces of dimension $n-2$ in $P_n(\mathbf{C})$. $L(x)^\perp$, $x \in M$, defines a surface in $G_{n-1,n+1}$, and L is minimal if and only if $L^\perp$ is minimal; they are called dual minimal surfaces.

As above, if $Z \in L$, ∂Z (resp. $\bar\partial Z$) defines a point A (resp.B) $\in L^\perp$. This defines mappings

$$A, B : L \longrightarrow L^\perp, \tag{11}$$

which are projective collineations. Reversing the role of L and $L^\perp$ (and of ∂ and $\bar\partial$), we have also the collineations

$$A^\perp, B^\perp : L^\perp \longrightarrow L. \tag{11a}$$

We shall call A, B, $A^\perp$, $B^\perp$ the *fundamental collineations*. In particular, we have

$$B^\perp \circ A : L \longrightarrow L. \tag{11b}$$

The determinant $\det(B^\perp \circ A)$ gives a form of type (4,0) on M, which is holomorphic by the structure equations. Thus, if $M = S^2$, we have

$$\det(B^\perp \circ A) = 0. \tag{12}$$

This is the first global result, using the fact that the surface is S^2 and has no non-zero holomorphic differential forms of degree > 0.

This result is sufficient to give a solution of our problem for $n = 3$. For in this case $\dim L^\perp = 1$ and condition (12) becomes

$$(\det B)(\det A) = 0. \tag{13}$$

Hence one of the determinants is zero. Suppose $\det A = 0$, $A \neq 0$. Then the collineations A and $A^\perp$ are degenerate, and the images $A(L)$ and $A^\perp(L^\perp)$ are points.

From the fact that L in (10) is a minimal surface it follows by standard calculations that $A(L)$ and $A^{\perp}(L^{\perp})$ describe minimal surfaces in $P_3(\mathbf{C})$. The line joining $A(L)$ and $A^{\perp}(L^{\perp})$ has an orthogonal line λ. There are points $Z_0, Z_3 \in \lambda$ such that Z_0 describes a holomorphic curve and Z_3 an antiholomorphic curve and Z_0, $A(L)$, $A^{\perp}(L^{\perp})$, Z_3 are the vertices of the Frenet frame of Z_0.

For $n = 4$ we have $\dim L^{\perp} = 2$. In the generic case the lines $A(L)$ and $B(L)$ meet in a point P; P describes a minimal surface in $P_4(\mathbf{C})$.

Our main idea is to make use of constructions of projective geometry to derive from the minimal surfaces (10) in $G_{2,n+1}$, minimal surfaces of $P_n(\mathbf{C})$, thus reducing to a problem already solved. This requires several geometric constructions whose details we refer to [13]. It suffices to remark that we need to consider the osculating spaces of higher order to exhaust the ambient space, which involve the jets of higher order of the mapping.

4. General theory*

The general problem to be treated is the following : Let G be a Lie group and $H \subset G$ a closed subgroup. Let $N = G/H$ be the homogeneous space of left cosets, acted on by G by left multiplication. The fundamental problem is to find all the local invariants of a smooth submanifold

$$f : M \longrightarrow N, \tag{14}$$

i.e., if

$$f^* : M^* \longrightarrow N \tag{15}$$

is another submanifold, to find the necessary and sufficient conditions that there exist a local diffeomorphism

$$T : M \longrightarrow M^* \tag{16}$$

and an element $g \in G$ such that

$$f^* \circ T = g \circ f. \tag{17}$$

If these conditions are satisfied, M and M^* are said to be *congruent*. A related problem, the so-called fixed parametrization problem, is to take $M = M^*$ and $T =$ identity, and to ask for the existence of $g \in G$ such that (17) holds.

---

* I wish to thank D. BERNARD and G. JENSEN for their comments on this section.

We denote by f_k the k-th order jet defined by f. If T exists, and if for each point $x \in M$ there exists $g(x) \in G$ such that

$$(f^* \circ T)_k(x) = (g(x) \circ f)_k(x), \tag{18}$$

we say that f and f^* are G-deformable of order k.

Consider the diagram

$$\begin{array}{ccc} & & G \\ & & \big\downarrow \pi \\ M & \xrightarrow{f} & G/H, \end{array} \tag{19}$$

when π sends an element $g \in G$ to the coset gH. Let G and H be of dimensions r and s respectively, and let $\omega^1, \ldots, \omega^r$ be the left-invariant Maurer-Cartan forms in G, such that

$$\omega^{s+1} = \cdots = \omega^r = 0 \tag{20}$$

defines the foliation of G by the left cosets of H. A moving frame is a local (smooth) map

$$U \xrightarrow{F} G, \tag{21}$$

where U is a neighborhood in M, such that $f = \pi \circ F$, i.e., it assigns to each point $x \in U$ a point of the coset $f(x) \in G/H$. If $\dim M = m$ and if $\alpha^1, \ldots, \alpha^m$ are linearly independent one-forms on M, we have

$$F^*\omega^\lambda = \sum_i a_i^\lambda \alpha^i, \qquad 1 \leq i \leq m, s+1 \leq \lambda \leq r. \tag{22}$$

A new moving frame is given by $\widetilde{F} = Fh$, the latter being the right multiplication of F by $h : U \to H$. Under such a change $F^*\omega^\lambda$ undergoes a transformation of the adjoint group $\mathrm{ad}(H)$, and we have the change

$$A = (a_i^\lambda) \longrightarrow \widetilde{A} = SAT, \tag{23}$$

where S is the transformation induced on ω^λ by $\mathrm{ad}(h)$, $h \in H$, and T is any non-singular $(n \times n)$-matrix. The invariants of A under this change are the first-order invariants of the submanifold M.

The simplest case, and this is the case in most applications, is when $\mathrm{ad}(H)$ acts transitively on the Grassmann bundle of m-planes of the tangent bundle of G/H. Then the Maurer-Cartan forms of G can be so chosen that

$$F^*\omega^\lambda = 0, \qquad m+s+1 \leq \lambda \leq r, \tag{24}$$

and that $F^*\omega^{s+1},\ldots,F^*\omega^{s+m}$ are linearly independent. The main step is to take the exterior derivative of (24) and make use of the Maurer-Cartan structure equations. This will lead to the second-order invariants of M.

The analysis of the general case depends on the behavior of the invariants and has generally to be divided into many cases. As CARTAN himself remarked, the method provides a *practical and rapid* mechanism leading to the invariants, which could be functions or differential forms, exterior or ordinary. It is effective in both classical problems and problems where higher-order jets are involved. A characteristic feature is the exclusive use of differential forms, so that local coordinates do not appear. This makes the method useful in both local and global problems, as illustrated in the applications in the last two sections. Personally I find the effectiveness and superiority of the method in classical problems just as noteworthy as the general theory.

CARTAN tried to reduce the congruence problem to that of contact. Let $N^{(0)} = N$, $N^{(q+1)} =$ the bundle of tangent m-planes to $N^{(q)}$, $q \geq 0$. For the submanifold (14) let $f^{(0)} = f$ and let $f^{(q+1)}$ be the prolongation of $f^{(q)}$, i.e., $f^{(q+1)} : M \to N^{(q+1)}$ is defined by

$$f^{(q+1)}(x) = f^{(q)}_*(T_x M), \quad x \in M. \tag{25}$$

Then the submanifolds (14) and (15) are said to have a k-th order contact at $x \in M$ and $x^* \in M^*$, if and only if

$$f^{(k)}(x) = f^{*(k)}(x^*). \tag{26}$$

They are said to have a G-contact of order k at x and x^*, if there exists a transformation $g \in G$ such that f and $g \circ f^*$ have k-th order contact at x and $g(x^*)$.

In [21] JENSEN proved the following theorem :

THEOREM. — *Suppose f and f^* have the same contact type. Then there exists a k with the following property : if there is a diffeomorphism* (16) *such that f and f^* have G-contact of order k at x and $T(x)$ for every $x \in M$, then f and f^* are congruent, i.e.,* (17) *is satisfied.*

Recently in [20] S. HÜCKEL proved a generalized congruence theorem, which is even valid for non-transitive actions of G and from which JENSEN's theorem may be deduced. The condition of "same contact type" is replaced by the condition that certain osculating spaces of order $k-1$ (relative to the action of G) have a constant dimension, and the a priori existence of a diffeomorphism T is replaced by the "continuity of G-contact of order k", a topological property of the space $\Gamma_k \subset M \times M^* \times G$ of triples (x, x^*, g) such that $g \in G$ realizes the G-contact of order k between M at x and M^* at x^*.

The importance of the method depends to a certain extent on the significance of structures with a higher order of smoothness. The question whether the conclusions in § 2 are true with a lower degree of smoothness is a valid one, but perhaps not very interesting. In the study of partial differential equations the step usually goes backwards, which is to consider generalized solutions with less smoothness.

There is strong reason why differential forms furnish the right tool for the study of submanifolds. This is because in the last analysis the group G plays the fundamental role; its Maurer-Cartan forms and Maurer-Cartan equations are pulled back in a natural way by the moving frame.

REFERENCES

[1] BERNARD (D.). — *Espaces homogènes et repère mobile.* — IV Coll. Int. de Géom. Diff. Santiago de Compostela, 1978, p. 55–63.

[2] BERNARD (D.). — *Immersions et repères mobiles.* — Symposium E.B. Christoffel, Basel, Birkhäuser, 1981.

[3] BERNARD (D.). — Congruence, contact et repères de Frenet, *Differential Geometry*, Peniscola 1982, p. 21–35. — Berlin, Springer-Verlag, 1984, (*Lecture Notes in Math.*, **1045**).

[4] BONNET (O.). — Mémoire sur la théorie des surfaces applicables sur une surface donnée, *J. de l'École Polytechnique*, t. **25**, 1867, p. Cahier 41, p. 209–230, Cahier 42, p. 1–151.

[5] BRYANT (R.), CHERN (S.) and GRIFFITHS (P.). — Exterior differential systems, *Proc. of the 1980 Beijing Symposium on Differential Geometry and Differential Equations*, Science Press, Beijing, China and Gordon and Breach, New York, 1982, p. 219–338.

[6] BURNS (D.). — Harmonic maps from CP^1 to CP^n, in *Harmonic maps*, p. 48–56. Berlin, Springer-Verlag, 1982, (*Lecture Notes in Math.*, **949**).

[7] CARTAN (É). — La structure des groupes de transformations continus et la théorie du trièdre mobile, *Bull. Soc. Math. France*, t. **34**, 1910, p. 250–284, or *Oeuvres complètes*, Partie III, vol. 1, p. 145–178.

[8] CARTAN (É). — La méthode du repère mobile, la théorie des groupes continus et les espaces généralisés, *Exposés de Géométrie V*, Paris, Hermann, 1935 or *Oeuvres complètes*, Partie III, vol. 2, p. 1259–1320.

[9] CARTAN (É). — *La théorie des groupes finis et continus et la géométrie diférentielle traitées par la méthode du repère mobile*, (rédigée par J. LERAY). — Paris, Gauthier-Villars, 1937.

[10] CARTAN (É). — *Leçons sur la géométrie projective complexe.* — Paris, Gauthier-Villars, 1931.

[11] CHERN (S.). — A simple intrinsic proof of the Gauss-Bonnet formula for closed Riemannian manifolds, *Annals of Math.*,, t. **45**, 1944, p. 747–752, or *Selected Papers*, Berlin, Springer-Verlag, 1978, p. 83–88.

[12] CHERN (S.). — Deformation of surfaces preserving principal curvatures, *Differential Geometry*, edited by I. Chavel, Berlin, Springer-Verlag, 1984, p. 116–122, preprint MSRI 010-83, Berkeley.

[13] CHERN (S.) and WOLFSON (J.). — Harmonic maps of S^2 into a complex Grassmann manifold, *Proc. Nat. Acad. Sci. USA*, t. **82**, 1985, p. 2217–2219.

[14] COTTON (É). — Généralisation de la théorie du trièdre mobile, *Bull. Soc. Math. France*, t. **33**, 1905, p. 1–23.

[15] DIN (A.M.) and ZAKRZEWSKI (W.J.). — General classical solutions in the CP^{n-1} model, *Nucl. Physics B*, t. **174**, 1980, p. 397–406.

[16] EELLS (J.) and WOOD (J.). — Harmonic maps from surfaces to complex projective spaces, *Advances in Math.*, t. **49**, 1983, p. 217–263.

[17] EELLS (J.). — *Bibliography for harmonic maps.* — Warwick, 1984.

[18] GREEN (M.). — The moving frame, differential invariants and rigidity theorems for curves in homogeneous spaces, *Duke Math. J.*, t. **45**, 1978, p. 735–779.

[19] GRIFFITHS (P.). — On Cartan's method of Lie groups and moving frames as applied to existence and uniqueness questions in differential geometry, *Duke Math. J.*, t. **41**, 1974, p. 775–814.

[20] HÜCKEL (S.). — Invariants différentiels, *G*-contact, théorème de congruence, Thèse Strasbourg, 1981.

[21] JENSEN (G. R.). — *Higher Order Contact of Submanifolds of Homogeneous Spaces.* Berlin, Springer-Verlag, 1977, (*Lecture Notes in Math.*, **610**).

[22] JENSEN (G.R.). — Deformation of submanifolds of homogeneous spaces, *J. Differential Geometry*, t. **16**, 1981, p. 213-246.

[23] RAMANATHAN (J.). — Harmonic maps from S^2 to $G_{2,4}$, *J. Differential Geometry*, t. **19**, 1984, p. 207-219.

[24] WEYL (H). — Cartan on groups and differential geometry, *Bull. Amer. Math. Soc.*, t. **44**, 1938, p. 598–601.

[25] WOLFSON (J.). — Minimal surfaces in complex manifolds, Ph. D. Thesis, University of California, Berkeley, 1982.

Shiing-shen CHERN,
Mathematical Sciences Research Institute, and
University of California, Berkeley,
Berkeley, California 94720, U.S.A.

Pseudospherical Surfaces and Evolution Equations

By S. S. Chern and K. Tenenblat*

We consider evolution equations, mainly of type $u_t = F(u, u_x, \ldots, \partial^k u/\partial x^k)$, which describe pseudo-spherical surfaces. We obtain a systematic procedure to determine a linear problem for which a given equation is the integrability condition. Moreover, we investigate how the geometrical properties of surfaces provide analytic information for such equations.

Introduction

A major contribution to the development of the study of nonlinear evolution equations was the inverse-scattering method introduced by Gardner, Greene, Kruskal, and Miura [5] and by Zakharov and Shabat [10], which was generalized by Ablowitz, Kaup, Newell, and Segur [1].

Sasaki [7] gave a geometrical interpretation for the inverse-scattering problem, considered by Ablowitz et al., in terms of pseudospherical surfaces (p.s.s.). Based on this interpretation (see Section 1), we consider the following definition.

Let M^2 be a two-dimensional differentiable (C^∞) manifold, with coordinates (x, t). We say that a differential equation for a real function $u(x, t)$ describes a p.s.s. if it is a necessary and sufficient condition for the existence of differentiable functions $f_{\alpha\beta}$, $1 \leq \alpha \leq 3$, $1 \leq \beta \leq 2$, depending on u and its derivatives, such that the 1-forms

$$\omega_\alpha = f_{\alpha 1}\, dx + f_{\alpha 2}\, dt$$

satisfy the structure equations of a p.s.s. (ω_1, ω_2 are the forms which determine the metric, and ω_3 is the connection form). This is equivalent to saying that the differential equation for u is the integrability condition for the problem

$$dv = \Omega v, \qquad (*)$$

Address for correspondence: K. Tenenblat, Department of Mathematics, Universidade de Brasília, Brasília, Brazil.

*The first author is partially supported by NSF grants DMS 84-03201 and DMS 84-01939, and the second author by CNPq.

STUDIES IN APPLIED MATHEMATICS 74, 55–83(1986) 55

Published by Elsevier Science Publishing Co., Inc. 0022-2526/86/$3.50

where v is a vector and Ω is a traceless 2×2 matrix of 1-forms given by

$$\Omega = \frac{1}{2}\begin{pmatrix} w_2 & w_1 - w_3 \\ w_1 + w_3 & -w_2 \end{pmatrix}.$$

We observe that whenever $f_{21} = \eta$ is a parameter, and the functions f_{11} and f_{31} do not depend on the parameter η, this is the problem considered by Ablowitz et al. [1].

Examples of differential equations which describe a p.s.s. are sine-Gordon, sinh-Gordon, KdV, MKdV, Burgers, etc. (see Sections 1, 2). In this paper we are concerned mainly with the following question: Is there any systematic way to obtain the functions $f_{\alpha\beta}$ associated with a differential equation which describes a p.s.s.? Equivalently, is there any systematic way to determine the problem ($*$) for which a given equation is the integrability condition? The importance of this question is due to the general observation that all of these nonlinear partial differential equations, which are exactly solvable, are also integrability conditions for certain first-order systems such as ($*$). Moreover, we investigate how the geometrical properties of a p.s.s. may provide analytic information for such equations.

In Section 2 we consider the class of differential equations of type

$$u_t = F\left(u, \frac{\partial u}{\partial x}, \ldots, \frac{\partial^k u}{\partial x^k}\right)$$

which describe p.s.s., where $f_{21} = \eta$ is a parameter and the functions $f_{\alpha\beta}$ depend on $u, \partial u/\partial x, \ldots, \partial^k u/\partial x^k$. We obtain an algorithm which characterizes all such equations (which may eventually depend on the parameter η). Moreover, the results are constructive (see Section 2.7) in the sense that they provide the problem ($*$) associated to a given equation, whenever it is in the class of equations considered above. Among the applications, (see Example 2.7) we point out that for $k = 2$ we obtain a new family of equations,[1] which describe p.s.s., given by

$$h_t = \pm l_{xx} + (hl)_x,$$

where h and l are any differentiable functions of $u(x,t)$ such that $h_u \neq 0$ and $l_u \neq 0$ [see Example 2.7(b)]. An interesting problem that has to be investigated is to characterize those equations in our classification which do not depend on the parameter η. An example of such a result is obtained in Proposition 2.8.

In Section 3 we characterize the sine-Gordon and sinh-Gordon equations as nongeneric cases of equations of type

$$u_{xt} = F\left(u, \frac{\partial u}{\partial x}, \ldots, \frac{\partial^k u}{\partial x^k}\right)$$

which describe p.s.s.

[1] This equation can be transformed into the Burgers equation or the one solved by Fokas and Yortsos in *SIAM J. Appl. Math.* 42, No. 2 (1982).

In Section 4 we describe a geometrical result which, under certain conditions, provides an infinite number of conservation laws and Backlund transformations for differential equations which describe p.s.s. A generalization of Proposition 4.2 and an algorithm which explicitly provides the conservation laws are given in [9].

1. Inverse scattering problem and differential equations which describe pseudospherical surfaces

The inverse scattering problem, due to Ablowitz et al. [1], considers

$$\begin{aligned} v_{1,x} &= -i\xi v_1 + q(x,t)v_2, \\ v_{2,x} &= i\xi v_2 + r(x,t)v_1, \end{aligned} \tag{1.1}$$

and the time dependence

$$\begin{aligned} v_{1,t} &= Av_1 + Bv_2, \\ v_{2,t} &= Cv_1 + Dv_2. \end{aligned} \tag{1.2}$$

The compatibility condition for (1.1) and (1.2), assuming the eigenvalues ξ are invariants, provides a certain set of conditions on $A,\dots,D$ which must be satisfied. Without loss of generality one can take $D = -A$ and the conditions are

$$\begin{aligned} A_x &= qC - rB, \\ B_x + 2i\xi B &= q_t - 2Aq, \\ C_x - 2i\xi C &= r_t + 2Ar. \end{aligned} \tag{1.3}$$

In order to solve (1.3) for A, B, C, in general, one finds that still another condition has to be satisfied. This latter condition is the evolution equation.

In terms of exterior differential forms the inverse-scattering problem can be formulated as follows: (1.1) and (1.2) are given by

$$dv = \Omega v, \tag{1.4}$$

where v is a vector whose coordinates are v_1, v_2, and Ω is a traceless 2×2 matrix of 1-forms given by

$$\Omega = \begin{pmatrix} -i\xi\,dx + A\,dt & q\,dx + B\,dt \\ r\,dx + C\,dt & i\xi\,dx - A\,dt \end{pmatrix}.$$

The integrability condition for (1.4) is given by

$$d\Omega - \Omega \wedge \Omega = 0. \tag{1.5}$$

Moreover, an evolution equation must be satisfied for the existence of a solution A, B, C for (1.5).

Whenever the functions are real, Sasaki [7] gave a geometrical interpretation for the problem. Consider the 1-forms defined by

$$\begin{aligned} \omega_1 &= (r+q)\,dx + (C+B)\,dt, \\ \omega_2 &= \eta\,dx + 2A\,dt, \\ \omega_3 &= (r-q)\,dx + (C-B)\,dt, \end{aligned} \tag{1.6}$$

where $\eta = -2i\xi$. Then (1.5) is equivalent to saying that ω_1, ω_2, and ω_3 satisfy the following relations:

$$\begin{aligned} d\omega_1 &= \omega_3 \wedge \omega_2, \\ d\omega_2 &= \omega_1 \wedge \omega_3, \\ d\omega_3 &= \omega_1 \wedge \omega_2. \end{aligned} \tag{1.7}$$

Let M be a two-dimensional differentiable manifold parametrized by coordinates x, t. We consider a metric on M defined by ω_1 and ω_2. The two first equations in (1.7) are the structure equations which determine the connection form ω_3, and the last equation in (1.7), the Gauss equation, determines that the Gaussian curvature of M is -1, i.e. M is a pseudospherical surface. Moreover, an evolution equation must be satisfied for the existence of forms (1.6) satisfying (1.7). This justifies the definition of a differential equation which describes a p.s.s. that we considered in the introduction.

In this paper, we will restrict ourselves to the case where $f_{21} = \eta$ is a parameter. More precisely, we say that a differential equation for $u(x,t)$ *describes a pseudospherical surface* if it is a necessary and sufficient condition for the existence of functions $f_{\alpha\beta}$, $1 \le \alpha \le 3$, $1 \le \beta \le 2$, depending on u and its derivatives, $f_{21} = \eta$, such that the 1-forms

$$\begin{aligned} \omega_1 &= f_{11}\,dx + f_{12}\,dt, \\ \omega_2 &= \eta\,dx + f_{22}\,dt, \\ \omega_3 &= f_{31}\,dx + f_{32}\,dt, \end{aligned} \tag{1.8}$$

satisfy the structure equations (1.7) of a pseudospherical surface. It follows from this definition that for each nontrivial solution u of the differential equation, one gets a metric defined on M, whose Gaussian curvature is -1.

It has been known, for a long time, that the sine-Gordon (SG) equation describes a pseudospherical surface. More recently, other equations, such as the Korteweg-deVries (KdV) and modified Korteweg-deVries (MKdV), were also shown to describe such surfaces [7].

Examples: Let M be a differentiable surface, parametrized by coordinates x, t.

(a) Consider

$$
\begin{aligned}
\omega_1 &= \frac{1}{\eta}\sin u\, dt,\\
\omega_2 &= \eta\, dx + \frac{1}{\eta}\cos u\, dt,\\
\omega_3 &= u_x\, dx.
\end{aligned}
\tag{1.9}
$$

Then, M is a pseudospherical surface iff u satisfies the SG equation

$$u_{xt} = \sin u.$$

(b) For

$$
\begin{aligned}
\omega_1 &= -\eta u_x\, dt,\\
\omega_2 &= \eta\, dx + \left(\tfrac{1}{2}\eta u^2 + \eta^3\right) dt,\\
\omega_3 &= u\, dx + \left(u_{xx} + \tfrac{1}{2}u^3 + \eta^2 u\right) dt,
\end{aligned}
\tag{1.10}
$$

M is a pseudospherical surface iff u satisfies the MKdV equation

$$u_t = u_{xxx} + \tfrac{3}{2}u^2 u_x.$$

(c) Consider

$$
\begin{aligned}
\omega_1 &= (1-u)\, dx + \left(-u_{xx} + \eta u_x - \eta^2 u - 2u^2 + \eta^2 + 2u\right) dt,\\
\omega_2 &= \eta\, dx + \left(\eta^3 + 2\eta u - 2u_x\right) dt,\\
\omega_3 &= -(1+u)\, dx + \left(-u_{xx} + \eta u_x - \eta^2 u - 2u^2 - \eta^2 - 2u\right) dt.
\end{aligned}
\tag{1.11}
$$

M is a pseudospherical surface iff u satisfies the KdV equation

$$u_t = u_{xxx} + 6uu_x.$$

(d) Let

$$
\begin{aligned}
\omega_1 &= u_x\, dx,\\
\omega_2 &= \eta\, dx + \frac{1}{\eta}\cosh u\, dt,\\
\omega_3 &= \frac{1}{\eta}\sinh u\, dt.
\end{aligned}
\tag{1.12}
$$

M is a pseudospherical surface iff u satisfies the ShG equation

$$u_{xt} = \sinh u.$$

Other examples are given in Example 2.7 and Proposition 2.8.

2. A characterization of differential equations of the type $u_t = F(u, \partial u/\partial x, \ldots, \partial^k u/\partial x^k)$ which describe pseudospherical surfaces

In this section, we consider differential equations for $u(x,t)$ of the form

$$u_t = F\left(u, \frac{\partial u}{\partial x}, \ldots, \frac{\partial^k u}{\partial x^k}\right) \tag{2.1}$$

which describe p.s.s., where the functions $f_{\alpha\beta}$ depend on $u, \partial u/\partial x, \ldots, \partial^k u/\partial x^k$, $1 \le \alpha \le 3$, $1 \le \beta \le 2$, and $f_{21} = \eta$ is a parameter. Unless explicitly stated, we allow the equation (2.1) and the functions $f_{\alpha\beta}$ to involve the parameter η.

We first obtain necessary conditions on the functions $f_{\alpha\beta}$ (Lemma 2.1). In particular we obtain that f_{11} and f_{31} depend only on u. By assuming a generic condition on f_{11} and f_{31} we obtain the generic case Theorem 2.2. The nongeneric cases are given by Theorems 2.3, 2.4, and 2.5. Except for the nongeneric case of Theorem 2.5, the differential equations (2.1) are characterized by the fact that F and the forms ω_α are algebraically determined by f_{11}, f_{31}, f_{22} (or f_{12}), and their derivatives. Moreover, in the generic case these three functions have to satisfy certain differential equations. The family of equations given by

$$h_t = \pm l_{xx} + (hl)_x,$$

where h and l are any differentiable functions of $u(x,t)$ with $h_u \neq 0$ and $l_u \neq 0$, provides examples for the generic case when $k = 2$ [see Example 2.7(b)]. The KdV is also in the generic case. The MKdV is included in a family of equations which describe p.s.s. in a nongeneric case, when $k = 3$ and the coefficient of $\partial^3 u/\partial x^3$ in F is equal to one (see Proposition 2.8). An example for the nongeneric case of Theorem 2.5 is given in [9], by the equation

$$u_t = \left(u_x^{-1/2}\right)_{xx} + u_x^{3/2}.$$

We introduce the following notation:

$$z_0 = u, \qquad z_1 = \frac{\partial u}{\partial x}, \ldots, \qquad z_k = \frac{\partial^k u}{\partial x^k}. \tag{2.2}$$

Moreover, we consider the functions defined on the space of variables $x, t, z_0, \ldots, z_k$. We assume that the functions do not depend explicitly on x and t.

The following lemma gives necessary conditions on the functions $f_{\alpha\beta}$ for the existence of a differential equation (2.1) which describes a p.s.s. From now on we will be using the notation (2.2), and the variables appearing in lower indices will denote partial differentiation.

LEMMA 2.1. *Let*

$$z_{0,t} = F(z_0, \ldots, z_k) \tag{2.3}$$

be a differential equation which describes a p.s.s., with associated 1-forms $\omega_\alpha = f_{\alpha 1}\,dx + f_{\alpha 2}\,dt$, *where* $f_{21} = \eta$ *is a parameter. If* $f_{\alpha\beta}$ *are functions of* $z_0, \ldots, z_k$, *then*

$$\begin{gathered} f_{11,z_i} = f_{31,z_i} = 0, \qquad 1 \le i \le k, \\ f_{12,z_k} = f_{22,z_k} = f_{32,z_k} = 0, \\ f_{22,z_{k-1}} = 0, \\ f_{11,z_0}^2 + f_{31,z_0}^2 \neq 0. \end{gathered} \tag{2.4}$$

Moreover,

$$\begin{gathered} -Ff_{11,z_0} + \sum_{i=0}^{k-1} z_{i+1} f_{12,z_i} + \eta f_{32} - f_{22} f_{31} = 0, \\ \sum_{i=0}^{k-2} z_{i+1} f_{22,z_i} - f_{11} f_{32} + f_{12} f_{31} = 0, \\ -Ff_{31,z_0} + \sum_{i=0}^{k-1} z_{i+1} f_{32,z_i} + \eta f_{12} - f_{22} f_{11} = 0. \end{gathered} \tag{2.5}$$

Proof: From (2.2) and (2.3) it follows that

$$\begin{aligned} dz_i \wedge dt &= z_{i+1}\, dx \wedge dt, \qquad 0 \le i \le k-1, \\ dz_0 \wedge dx &= -F dx \wedge dt. \end{aligned} \tag{2.6}$$

Since the 1-forms ω_α satisfy the structure equations (1.7), we have that

$$\sum_{i=0}^{k} f_{11,z_i}\, dz_i \wedge dx + \sum_{i=0}^{k} f_{12,z_i}\, dz_i \wedge dt + (\eta f_{32} - f_{22} f_{31})\, dx \wedge dt = 0,$$

$$\sum_{i=0}^{k} f_{22,z_i}\, dz_i \wedge dt + (-f_{11} f_{32} + f_{12} f_{31})\, dx \wedge dt = 0,$$

$$\sum_{i=0}^{k} f_{31,z_i}\, dz_i \wedge dx + \sum_{i=0}^{k} f_{32,z_i}\, dz_i \wedge dt + (\eta f_{12} - f_{22} f_{11})\, dx \wedge dt = 0.$$

Using (2.6) in the above equations, we get $f_{11,z_i} = f_{31,z_i} = 0$ for $1 \le i \le k$, $f_{12,z_k} = f_{22,z_k} = 0$, $f_{32,z_k} = 0$, and

$$-Ff_{11,z_0} + \sum_{i=0}^{k-1} z_{i+1} f_{12,z_i} + \eta f_{32} - f_{22} f_{31} = 0,$$

$$\sum_{i=0}^{k-1} z_{i+1} f_{22,z_i} - f_{11} f_{32} + f_{12} f_{31} = 0,$$

$$-Ff_{31,z_0} + \sum_{i=0}^{k-1} z_{i+1} f_{32,z_i} + \eta f_{12} - f_{22} f_{11} = 0.$$

Taking the z_k derivative of the second equation above, we obtain $f_{22,z_{k-1}} = 0$. Hence we have obtained (2.5). Finally, we observe that if f_{11,z_0} and f_{31,z_0} vanish simultaneously, then it follows from (2.5) that the equation (2.3) cannot be the necessary and sufficient condition for the ω_α to satisfy the structure equation of a p.s.s. □

A necessary condition for the existence of an equation $z_{0,t} = F(z_0, \ldots, z_k)$ describing a p.s.s., as in Lemma 2.1, is that $f_{\alpha\beta}$ satisfies (2.4). Under a generic assumption on f_{11} and f_{31}, the following result shows that such equations are algebraically determined by f_{11}, f_{31}, f_{22}, and their derivatives. In order to state the theorem we introduce the notation

$$L = \begin{vmatrix} f_{11} & f_{31} \\ f_{11,z_0} & f_{31,z_0} \end{vmatrix}, \qquad H = \begin{vmatrix} f_{11} & f_{31} \\ f_{31,z_0} & f_{11,z_0} \end{vmatrix},$$
$$P = \begin{vmatrix} f_{11,z_0} & f_{31,z_0} \\ f_{11,z_0z_0} & f_{31,z_0z_0} \end{vmatrix}, \qquad M = f_{31,z_0}^2 - f_{11,z_0}^2, \tag{2.7}$$

and

$$B = \sum_{i=0}^{k-2} z_{i+1} f_{22,z_i}. \tag{2.8}$$

Moreover, whenever $L \neq 0$ we define A^j recursively as follows:

$$A^{k-1} = 0,$$
$$A^j = -\sum_{i=0}^{k-1} z_{i+1} A^{j+1}_{z_i} + \frac{1}{L}\left(z_1 L_{z_0} + \eta H\right) A^{j+1}$$
$$+ \frac{1}{L}(-z_1 P + \eta M) B_{z_{j+1}} + f_{22,z_{j+1}} H, \qquad 0 \le j \le k-2. \tag{2.9}$$

THEOREM 2.2. *Let $f_{\alpha\beta}$, $1 \le \alpha \le 3$, $1 \le \beta \le 2$, be differentiable functions of $z_0, \dots, z_k$ such that (2.4) holds, and $f_{21} = \eta$ a nonzero parameter. Suppose $HL \ne 0$. Then $z_{0,t} = F(z_0, \dots, z_k)$ describes a p.s.s., with associated 1-forms $\omega_\alpha = f_{\alpha 1}\,dx + f_{\alpha 2}\,dt$, if and only if*

$$F = \frac{1}{L}\sum_{i=0}^{k-1} z_{i+1}B_{z_i} + \frac{1}{HL}\left(-z_1\frac{L}{\eta} + f_{31}^2 - f_{11}^2\right)\sum_{i=0}^{k-2} z_{i+1}A^i$$

$$+\frac{B}{HL}(z_1M + \eta L) + z_1\frac{f_{22}}{\eta}, \tag{2.10}$$

and

$$f_{12} = \frac{f_{11}f_{22}}{\eta} + \frac{1}{H}\left(-\frac{f_{11}}{\eta}\sum_{i=0}^{k-2} z_{i+1}A^i + f_{31,z_0}B\right),$$

$$f_{32} = \frac{f_{31}f_{22}}{\eta} + \frac{1}{H}\left(-\frac{f_{31}}{\eta}\sum_{i=0}^{k-2} z_{i+1}A^i + f_{11,z_0}B\right), \tag{2.11}$$

where f_{11}, f_{31}, f_{22} satisfy the following differential equations:

$$\frac{L}{\eta}f_{22,z_j} - \frac{L}{\eta}\sum_{i=0}^{k-2}\left(z_{i+1}\frac{A^i}{H}\right)_{z_j} + A^j + \frac{M}{H}B_{z_j} + \frac{B}{H^2}(LP + M^2)\delta_{j0} = 0, \tag{2.12}$$

$$0 \le j \le k-1,$$

where $\delta_{j0} = 0$ if $j \ne 0$ and $\delta_{00} = 1$.

Proof: Suppose $z_{0,t} = F(z_0, \dots, z_k)$ describes a p.s.s. Then it follows from Lemma 2.1 that (2.5) is satisfied. By hypothesis f_{11,z_0} and f_{31,z_0} do not vanish simultaneously; therefore (2.5) is equivalent to the following system of equations:

$$\sum_{i=0}^{k-1} z_{i+1}\left(f_{12,z_i}f_{31,z_0} - f_{32,z_i}f_{11,z_0}\right) + \eta\left(f_{32}f_{31,z_0} - f_{12}f_{11,z_0}\right) + f_{22}H = 0, \tag{2.13}$$

$$B - f_{11}f_{32} + f_{12}f_{31} = 0, \tag{2.14}$$

$$FL + \sum_{i=0}^{k-1} z_{i+1}\left(f_{12,z_i}f_{31} - f_{32,z_i}f_{11}\right)$$

$$+\eta\left(f_{32}f_{31} - f_{12}f_{11}\right) - f_{22}\left(f_{31}^2 - f_{11}^2\right) = 0. \tag{2.15}$$

Taking the z_k derivative of (2.13) and the z_{k-1} derivative of (2.14), it follows from (2.4) that

$$f_{12,z_{k-1}} f_{31,z_0} - f_{32,z_{k-1}} f_{11,z_0} = 0,$$

$$f_{12,z_{k-1}} f_{31} - f_{32,z_{k-1}} f_{11} + B_{z_{k-1}} = 0.$$

Therefore

$$f_{12,z_{k-1}} = \frac{f_{11,z_0}}{L} B_{z_{k-1}},$$

$$f_{32,z_{k-1}} = \frac{f_{31,z_0}}{L} B_{z_{k-1}}.$$

Recursively, taking the z_{j+1} derivative of (2.13) and the z_j derivative of (2.14) for $1 \leq j \leq k-1$, we obtain

$$f_{12,z_j} = -\frac{1}{L}\left(f_{11}A^j - f_{11,z_0}B_{z_j}\right), \tag{2.16}$$

$$f_{32,z_j} = -\frac{1}{L}\left(f_{31}A^j - f_{31,z_0}B_{z_j}\right), \qquad 1 \leq j \leq k-1,$$

where A^j is given by (2.9). Taking the z_1 derivative of (2.13), we obtain

$$f_{12,z_0} f_{31,z_0} - f_{32,z_0} f_{11,z_0} + A^0 = 0. \tag{2.17}$$

Hence, it follows from (2.13) and (2.14) that

$$f_{32} f_{31,z_0} - f_{12} f_{11,z_0} - \frac{1}{\eta} \sum_{i=0}^{k-2} z_{i+1} A^i + \frac{f_{22}}{\eta} H = 0,$$

$$- f_{32} f_{11} + f_{12} f_{31} + B = 0,$$

where we have used (2.16) and (2.17). From the last two equations we obtain f_{12} and f_{32} given by (2.11). Hence, it follows from (2.16), (2.11), and (2.15) that F is given by (2.10). Moreover, since f_{12} and f_{32} must satisfy (2.16) and (2.17), it follows from (2.11) that f_{11}, f_{31}, and f_{22} must satisfy the differential equations (2.12).

Conversely, if F, f_{12}, and f_{32} are given by (2.10), (2.11) and if f_{11}, f_{31}, f_{22} satisfy (2.8), it follows from a straightforward computation that the 1-forms ω_α satisfy the structure equations of a p.s.s. if and only if $z_{0,t} = F$. □

In the preceding theorem we made the generic assumption $HL \neq 0$. Therefore we need to consider the nongeneric cases $L = 0$ and $H = 0$. Observe that $L = 0$ is equivalent to saying that f_{11} and f_{31}, which are functions of z_0, are multiples of each other, i.e., $f_{11}=0$ or $f_{31}=0$ or $f_{31}=\lambda f_{11}$, where λ does not depend on z_0. $H=0$ with $L\neq 0$ is equivalent to considering $f_{31}^2-f_{11}^2=c$, where c does not depend on z_0. The nongeneric cases are treated in the following Theorems 2.3–2.5. As before, B will denote the function defined by (2.8).

THEOREM 2.3. *Let $f_{\alpha\beta}$, $1\leq\alpha\leq 3$, $1\leq\beta\leq 2$, be differentiable functions of $z_0,\ldots,z_k$ such that (2.4) holds and $f_{21}=\eta$ is a nonzero parameter. Suppose that $f_{11}=0$, or that $f_{31}=0$. Then $z_{0,t}=F(z_0,\ldots,z_k)$ describes a p.s.s., with associated 1-forms $\omega_\alpha=f_{\alpha 1}\,dx+f_{\alpha 2}\,dt$, if and only if $f_{22,z_{k-2}}=0$ and*

$$F=\frac{1}{\eta f_{31,z_0}}\sum_{i=0}^{k-1} z_{i+1}\left\{\left[\sum_{j=0}^{k-2} z_{j+1}\left(\frac{B}{f_{31}}\right)_{z_j}\right]_{z_i}+(f_{22}f_{31})_{z_i}\right\}-\frac{\eta B}{f_{31}f_{31,z_0}}, \tag{2.18}$$

$$f_{12}=-\frac{B}{f_{31}},$$
$$f_{32}=\frac{1}{\eta}\left[\sum_{i=0}^{k-2} z_{i+1}\left(\frac{B}{f_{31}}\right)\zeta_1+f_{22}f_{31}\right], \tag{2.19}$$

or

$$F=\frac{1}{\eta f_{11,z_0}}\sum_{i=0}^{k-1} z_{i+1}\left\{\left[-\sum_{j=0}^{k-2} z_{j+1}\left(\frac{B}{f_{11}}\right)_{z_j}\right]_{z_i}+(f_{22}f_{11})_{z_i}\right\}+\frac{\eta B}{f_{11}f_{11,z_0}}, \tag{2.20}$$

$$f_{12}=\frac{1}{\eta}\left[-\sum_{i=0}^{k-2} z_{i+1}\left(\frac{B}{f_{11}}\right)_{z_i}+f_{22}f_{11}\right],$$
$$f_{32}=\frac{B}{f_{11}}. \tag{2.21}$$

Proof: If $z_{0,t}=F(z_0,\ldots,z_k)$ describes a p.s.s., then it follows from Lemma 2.1 that (2.5) is satisfied. By hypothesis $f_{11}=0$; therefore (2.5) is given by

$$\sum_{i=0}^{k-1} z_{i+1}f_{12,z_i}+\eta f_{32}-f_{22}f_{31}=0, \tag{2.22}$$

$$B+f_{12}f_{31}=0, \tag{2.23}$$

$$-Ff_{31,z_0}+\sum_{i=0}^{k-1} z_{i+1}f_{32,z_i}+\eta f_{12}=0. \tag{2.24}$$

Taking the z_k derivative of (2.22) and the z_{k-1} derivative of (2.23), it follows from (2.4) that $f_{12,z_{k-1}} = 0$ and

$$B_{z_{k-1}} = f_{22,z_{k-2}} = 0.$$

From (2.23) and (2.22) we get f_{12} and f_{32} given by (2.19). Therefore, it follows from (2.24) that F is given by (2.18).

Similarly, we obtain (2.20) and (2.21) whenever $f_{31} = 0$. The converse is a straightforward computation. □

THEOREM 2.4. *Let $f_{\alpha\beta}$, $1 \le \alpha \le 3$, $1 \le \beta \le 2$, be differentiable functions of $z_0, \ldots, z_k$ such that (2.4) holds and $f_{21} = \eta$ is a nonzero parameter. Suppose $f_{31} = \lambda f_{11} \neq 0$, where λ does not depend on z_0. Then $z_{0,t} = F(z_0, \ldots, z_k)$ describes a p.s.s. with associated 1-forms $\omega_\alpha = f_{\alpha 1}\, dx + f_{\alpha 2}\, dt$, if and only if*

(a) *f_{22} does not depend on z_i, $0 \le i \le k$; $f_{32} = \lambda f_{12}$; and*

$$F = \frac{1}{f_{11,z_0}} \left[\sum_{i=0}^{k-1} z_{i+1} f_{12,z_i} + \lambda(\eta f_{12} - f_{11} f_{22}) \right] \tag{2.25}$$

whenever $\lambda^2 - 1 = 0$;

(b) *$f_{22,z_{k-2}} = 0$, and*

$$F = \frac{1}{(\lambda^2-1) f_{11,z_0}} \left\{ \sum_{i=0}^{k-1} \frac{z_{i+1}}{\eta} \left[\left(\sum_{j=0}^{k-2} z_{j+1} \left(\frac{B}{f_{11}} \right)_{z_j} \right)_{z_i} + (\lambda^2 - 1)(f_{11} f_{22})_{z_i} \right] - \frac{\eta B}{f_{11}} \right\}, \tag{2.26}$$

$$\begin{aligned} f_{12} &= \frac{1}{\lambda^2 - 1} \left[\frac{1}{\eta} \sum_{i=0}^{k-2} z_{i+1} \left(\frac{B}{f_{11}} \right)_{z_i} - \lambda \frac{B}{f_{11}} \right] + \frac{f_{11} f_{22}}{\eta}, \\ f_{32} &= \frac{1}{\lambda^2 - 1} \left[\frac{\lambda}{\eta} \sum_{i=0}^{k-2} z_{i+1} \left(\frac{B}{f_{11}} \right)_{z_i} - \frac{B}{f_{11}} \right] + \frac{\lambda}{\eta} f_{11} f_{22} \end{aligned} \tag{2.27}$$

whenever $\lambda^2 - 1 \neq 0$.

Proof: Suppose $z_{0,t} = F(z_0, \ldots, z_k)$ describes a p.s.s. Since $f_{31} = \lambda f_{11}$, it follows from Lemma 2.1 that

$$\begin{aligned} -F f_{11,z_0} + \sum_{i=0}^{k-1} z_{i+1} f_{12,z_i} + \eta f_{32} - \lambda f_{22} f_{11} &= 0, \\ B - f_{11}(f_{32} - \lambda f_{12}) &= 0, \\ -\lambda F f_{11,z_0} + \sum_{i=0}^{k-1} z_{i+1} f_{32,z_i} + \eta f_{12} - f_{11} f_{22} &= 0. \end{aligned} \tag{2.28}$$

(a): If $\lambda^2 - 1 = 0$, then (2.28) is equivalent to

$$- Ff_{11,z_0} + \sum_{i=0}^{k-1} z_{i+1} f_{12,z_i} + \eta f_{32} - \lambda f_{11} f_{22} = 0, \tag{2.29}$$

$$B - f_{11}(f_{32} - \lambda f_{12}) = 0, \tag{2.30}$$

$$\sum_{i=0}^{k-1} z_{i+1}(\lambda f_{12,z_i} - f_{32,z_i}) - \eta\lambda(\lambda f_{12} - f_{32}) = 0. \tag{2.31}$$

Taking successive derivatives of (2.31) with respect to $z_k, z_{k-1}, \ldots, z_1$, we obtain

$$f_{32} = \lambda f_{12}.$$

Therefore, from (2.30) we get that f_{22} does not depend on z_i, $0 \le i \le k$. It follows from (2.29) that F is given by (2.25).

(b): If $\lambda^2 - 1 \neq 0$, then (2.28) is equivalent to

$$\sum_{i=0}^{k-1} z_{i+1}(\lambda f_{12,z_i} - f_{32,z_i}) + \eta(\lambda f_{32} - f_{12}) - (\lambda^2 - 1) f_{11} f_{22} = 0, \tag{2.32}$$

$$B - f_{11}(f_{32} - \lambda f_{12}) = 0, \tag{2.33}$$

$$- F(\lambda^2 - 1) f_{11,z_0} + \sum_{i=0}^{k-1} z_{i+1}(\lambda f_{32,z_i} - f_{12,z_i}) + \eta(\lambda f_{12} - f_{32}) = 0. \tag{2.34}$$

Taking the z_k derivative of (2.32) and the z_{k-1} derivative of (2.33), we obtain respectively

$$\lambda f_{12,z_{k-1}} - f_{32,z_{k-1}} = 0,$$

$$f_{22,z_{k-2}} = 0.$$

It follows from (2.33) that

$$\lambda f_{12} - f_{32} = -\frac{B}{f_{11}},$$

and hence from (2.32)

$$\eta(\lambda f_{32} - f_{12}) = \sum_{i=0}^{k-2} z_{i+1}\left(\frac{B}{f_{11}}\right)_{z_i} + (\lambda^2 - 1) f_{11} f_{22}.$$

From the last two equations we obtain f_{12} and f_{32} given by (2.27), and from (2.34) we obtain (2.26).

The converse is a straightforward computation. □

In order to consider the nongeneric case $f_{31}^2 - f_{11}^2 = c$, where c does not depend on z_i, we need to introduce the following notation. Let

$$\begin{aligned} E^{k-1} &= 0, \\ E^j &= -\sum_{i=0}^{k-1} z_{i+1} E_{z_i}^{j+1} + \left(-z_1 \frac{L}{c} + \eta\right) B_{z_{j+1}}, \qquad 0 \le j \le k-2. \end{aligned} \tag{2.35}$$

THEOREM 2.5. *Let $f_{\alpha\beta}$, $1 \le \alpha \le 3$, $1 \le \beta \le 2$, be differentiable functions of $z_0, \ldots, z_k$ such that* (2.4) *holds, and $f_{21} = \eta$ a nonzero parameter. Suppose $f_{31}^2 - f_{11}^2 = c$, where c does not depend on z_i. Then, $z_{0,t} = F(z_0, \ldots, z_k)$ describes a p.s.s., with associated 1-forms $\omega_\alpha = f_{\alpha 1}\,dx + f_{\alpha 2}\,dt$, if and only if*

$$F = \frac{1}{L}\left[\sum_{i=1}^{k-1} z_{i+1} B_{z_i} - z_1\left(f_{12,z_0} f_{31} - f_{32,z_0} f_{11}\right) - \eta\left(f_{32} f_{31} - f_{12} f_{11}\right) + f_{22} c\right], \tag{2.36}$$

where f_{12} and f_{32} are functions of f_{11}, f_{31}, f_{22} which satisfy, for $1 \le j \le k-1$,

$$\begin{aligned} f_{12,z_j} &= \frac{1}{c}\left(f_{11} E^j - f_{31} B_{z_j}\right), \\ f_{32,z_j} &= \frac{1}{c}\left(f_{31} E^j - f_{11} B_{z_j}\right), \end{aligned} \tag{2.37}$$

$$\begin{aligned} f_{12,z_0} f_{11} - f_{32,z_0} f_{31} + E^0 &= 0, \\ -f_{11} f_{32} + f_{12} f_{31} + B &= 0, \end{aligned} \tag{2.38}$$

and f_{11}, f_{31}, f_{22} satisfy the differential equation

$$-\sum_{i=0}^{k-2} z_{i+1} E^i + \eta B = 0. \tag{2.39}$$

Proof. Suppose $z_{0,t} = F(z_0, \ldots, z_k)$ describes a p.s.s. Since $f_{31}^2 - f_{11}^2 = c \neq 0$, it follows that $L \neq 0$. From Lemma 2.1 we have

$$\sum_{i=0}^{k-1} z_{i+1}\left(f_{12,z_i} f_{11} - f_{32,z_i} f_{31}\right) + \eta B = 0, \tag{2.40}$$

$$B - f_{11} f_{32} + f_{12} f_{31} = 0, \tag{2.41}$$

$$LF + \sum_{i=0}^{k-1} z_{i+1}\left(f_{12,z_i} f_{31} - f_{32,z_i} f_{11}\right) + \eta\left(f_{32} f_{31} - f_{12} f_{11}\right) - f_{22} c = 0. \tag{2.42}$$

Taking the z_{j+1} derivative of (2.40) and the z_j derivative of (2.41) for $j = k-1, k-2,\ldots,1$, we obtain (2.37). Taking the z_1 derivative of (2.40) and considering (2.41), we obtain (2.38). Therefore, substituting (2.37) and (2.38) in (2.40), we get that f_{11}, f_{31}, f_{22} must satisfy the differential equation (2.39). Equation (2.36) follows from (2.42) and (2.38).

The converse is a straightforward computation. □

We observe that, except for the nongeneric case of Theorem 2.5, the above results provide a constructive way to obtain the functions $f_{\alpha\beta}$ for a differential equation which describes a p.s.s. as in Theorems 2.2–2.4. We conclude this section by applying the procedure to specific equations. Moreover, we provide new examples of equations which describe a p.s.s.

Remark 2.6: A special class of equations, included in the generic case (Theorem 2.2), is obtained whenever $f_{11,z_0} = f_{31,z_0} \neq 0$. In this case, the expressions (2.10)–(2.12) are simpler, since

$$\begin{aligned} M &= P = 0, \\ H &= L = -af_{11,z_0}, \\ f_{31} - f_{11} &= a \neq 0, \end{aligned} \tag{2.43}$$

where a is independent of z_i.

Example 2.7 (Applications):

(a) Consider the Burgers equation, given by

$$z_{0,t} = z_2 + z_1 z_0.$$

Equating the right-hand side of this equation with F given by (2.10), where $k = 2$, we obtain $f_{22,z_0} = L$, $P = 0$, $L_{z_0} = 0$, $f_{31,z_0z_0} = f_{11,z_0z_0} = 0$, $H = -z_0M + D$, $B = z_1L$, $A^0 = \eta M$, where M, D, and L are constants. The differential equations (2.12) imply $M = -\frac{1}{4}$, $L = \eta/2$, and $f_{22} = (\eta/2)z_0 - 2\eta D$.

Therefore, taking $D = 0$, we obtain, up to a choice of constants,

$$f_{11} = \frac{z_0}{2}, \qquad f_{31} = -\eta, \qquad f_{22} = \frac{\eta}{2}z_0.$$

Hence from (2.11) we get

$$f_{12} = \frac{z_0^2}{4} + \frac{z_1}{2}, \qquad f_{32} = -\frac{\eta}{2}z_0.$$

Therefore the Burgers equation describes a p.s.s. with associated forms

$$\omega_1 = \frac{z_0}{2}\,dx + \left(\frac{z_0^2}{4} + \frac{z_1}{2}\right)dt,$$

$$\omega_2 = \eta\,dx + \frac{\eta}{2}z_0\,dt,$$

$$\omega_3 = -\eta\,dx - \frac{\eta}{2}z_0\,dt.$$

(b) The Burgers equation is a special case of a class of equations which describe p.s.s. In fact, let $h(u(x,t))$ and $l(u(x,t))$ be differentiable functions of $u(x,t)$ such that $h_u \neq 0$ and $l_u \neq 0$. Then

$$h_t = \pm l_{xx} + (hl)_x$$

describes a p.s.s. with associated 1-forms

$$\omega_1 = h\,dx + (hl \pm l_x)\,dt,$$

$$\omega_2 = \eta\,dx + \eta l\,dt,$$

$$\omega_3 = \mp \eta\,dx \mp \eta l\,dt.$$

This class of equations and associated 1-forms are obtained from Theorem 2.2 by considering $k = 2$.

(c) Consider the KdV equation, which is given by

$$z_{0,t} = z_3 + 6z_0 z_1.$$

Equating the right-hand side of this equation with F given by (2.10), where we are considering (2.43) and $k = 3$, we obtain

$$f_{11} = \frac{2}{a}z_0 + \gamma,$$

$$f_{22} = -2z_1 + 2\eta z_0 - 2a\eta\gamma - a^2\eta + \eta^3,$$

$$f_{31} = \frac{2}{a}z_0 + \gamma + a,$$

where γ is any constant. The differential equations (2.12) are trivially verified and (2.11) determine f_{12} and f_{32}. Choosing $a = -2$ and $\gamma = 1$, we get

$$f_{12} = -z_2 + \eta z_1 - 2z_0^2 - \eta^2 z_0 + 2z_0 + \eta^2,$$

$$f_{32} = -z_2 + \eta z_1 - 2z_0^2 - \eta^2 z_0 - 2z_0 - \eta^2.$$

Taking $\omega_\alpha = f_{\alpha 1}\,dx + f_{\alpha 2}\,dt$, $1 \leq \alpha \leq 3$, with $f_{21} = \eta$, we obtain the forms ω_α (1.11) for the KdV equation.

(d) As an application of Theorem 2.3 we obtain the forms ω_α (1.10) for the MKdV equation. Considering

$$z_{0,t} = z_3 + \tfrac{3}{2}z_0^2 z_1,$$

we equate the right-hand side of this equation with F given by (2.18), where

$k = 3$. We obtain

$$f_{22} = \tfrac{1}{2}\eta z_0^2 + \eta^3,$$

$$f_{31} = z_0.$$

Therefore, it follows from (2.19) that

$$f_{12} = -\eta z_1,$$

$$f_{32} = z_2 + \tfrac{1}{2}z_0^3 + \eta^2 z_0.$$

Taking $\omega_\alpha = f_{\alpha 1}\,dx + f_{\alpha 2}\,dt$, $1 \le \alpha \le 3$, with $f_{11} = 0$, $f_{21} = \eta$, we get (1.10).

Similarly, for the MKdV equation

$$z_{0,t} = z_3 - \tfrac{3}{2}z_0^2 z_1,$$

using (2.20) and (2.21), we obtain

$$f_{11} = z_0$$

$$f_{22} = -\frac{\eta}{2}z_0^2 + \eta^3,$$

$$f_{12} = z_2 - \frac{z_0^3}{2} + \eta^2 z_0,$$

$$f_{32} = -\eta z_1.$$

Therefore $\omega_\alpha = f_{\alpha 1}\,dx + f_{\alpha 2}\,dt$, $1 \le \alpha \le 3$, $f_{31} = 0$, and $f_{21} = \eta$ satisfy the structure equations of a p.s.s.

(e) The equation

$$z_{0,t} = z_2 + z_1^2$$

is an example for Theorem 2.4. In fact, it describes a p.s.s. with associated forms given by

$$w_1 = e^{z_0}\,dx + e^{z_0}(z_1 - \eta)\,dt,$$

$$w_2 = \eta\,dx - \eta^2\,dt,$$

$$w_3 = e^{z_0}\,dx + e^{z_0}(z_1 - \eta)\,dt.$$

(f) An example for Theorem 2.5 is given in [9], with the equation

$$u_t = \left(u_x^{-1/2}\right)_{xx} + u_x^{3/2}.$$

We observe that the equations of the type $z_{0,t} = F(z_0,\ldots,z_k)$ obtained in the preceding theorems, in general, may involve the parameter η. One would like to characterize all such equations which do not involve η. The following proposition considers this problem for the nongeneric case of Theorem 2.3, when $k = 3$ and the coefficient of z_3 in F is equal to one.

PROPOSITION 2.8. *Let*

$$z_{0,t} = F(z_0,\ldots,z_3)$$

be a differential equation which describes a p.s.s. as in Theorem 2.3. Suppose $F_{z_3} = 1$. *Then F is independent of* η *if and only if it is of the form*

$$\tilde{z}_{0,t} = \tilde{z}_3 \pm \tilde{z}_1 \tilde{z}_0^2 + a\tilde{z}_1,$$

where a is a constant, $\tilde{z}_0 = g(z_0)$, *and g is any differentiable function of* z_0 *independent of the parameter* η, *with* $g_{z_0} \neq 0$.

Proof: From Theorem 2.3 the equation is characterized by (2.18) and (2.19), where $k = 3$, i.e.

$$F = \frac{1}{f_{31,z_0}}\left\{ z_3 \frac{f_{22,z_0}}{\eta f_{31}} + \frac{3z_2 z_1}{\eta}\left(\frac{f_{22,z_0}}{f_{31}}\right)_{z_0} \right.$$

$$\left. + \frac{z_1^3}{\eta}\left(\frac{f_{22,z_0}}{f_{31}}\right)_{z_0 z_0} + z_1\left[\frac{(f_{22}f_{31})_{z_0}}{\eta} - \eta\frac{f_{22,z_0}}{f_{31}}\right]\right\}, \tag{2.44}$$

where f_{31} and f_{22} are functions of z_0. It follows from the hypothesis on the coefficient of z_3 that

$$f_{22,z_0} = \eta f_{31} f_{31,z_0}. \tag{2.45}$$

Substitute (2.45) into (2.44). Since F is independent of η, we get that the coefficients of $z_2 z_1$, z_1^3, and z_1 in (2.44) must be independent of η, i.e.

$$\frac{\partial}{\partial\eta}\left(\frac{f_{31,z_0z_0}}{f_{31,z_0}}\right) = 0, \tag{2.46}$$

and

$$\frac{\partial}{\partial\eta}\left(f_{31}^2 + \frac{f_{22}}{\eta} - \eta^2\right) = 0. \tag{2.47}$$

From (2.46), we get that f_{31} is of the form

$$f_{31}(z_0,\eta) = \alpha(\eta)h(z_0) + \beta(\eta), \tag{2.48}$$

where $\alpha(\eta) \neq 0$ and $h_{z_0} \neq 0$, since $f_{31,z_0} \neq 0$. It follows from (2.48) and (2.47) that

$$f_{22} = \eta^3 - (\alpha h + \beta)^2 \eta + l(z_0)\eta, \tag{2.49}$$

where l is independent of η. Substituting (2.48) and (2.49) into (2.45), we obtain

$$l_{z_0} = \frac{3}{2} \frac{\partial}{\partial z_0} \left[(\alpha h + \beta)^2 \right].$$

Since l is independent of η, it follows from the last equality that α and β are constants and

$$l = \tfrac{3}{2}(\alpha h + \beta)^2 + a,$$

where a is also a constant. Therefore, using (2.48), (2.49), and (2.44), we get

$$\begin{aligned} f_{31} &= \alpha h + \beta, \\ f_{22} &= \eta^3 + \tfrac{1}{2}(\alpha h + \beta)^2 \eta + a\eta, \end{aligned} \tag{2.50}$$

and

$$z_{0,t} = z_3 + 3z_1 z_2 \frac{h_{z_0 z_0}}{h_{z_0}} + z_1^3 \frac{h_{z_0 z_0 z_0}}{h_{z_0}} + z_1\left[\tfrac{3}{2}(\alpha h + \beta)^2 + a \right].$$

Considering

$$\tilde{z}_0 = \sqrt{\tfrac{3}{2}}\,(\alpha h + \beta), \tag{2.51}$$

the above equation reduces to

$$\tilde{z}_{0,t} = \tilde{z}_3 + \tilde{z}_1 \tilde{z}_0^2 + a\tilde{z}_1, \tag{2.52}$$

where as before $\tilde{z}_k = \partial^k \tilde{z}_0 / \partial x^k$. Moreover, it follows from (2.50), (2.51), and (2.19) that the 1-forms associated with the equation (2.52) are

$$\begin{aligned} \omega_1 &= -\eta\sqrt{\tfrac{2}{3}}\,\tilde{z}_1\,dt, \\ \omega_2 &= \eta\,dx + \left(\eta^3 + \frac{\eta}{3}\tilde{z}_0^2 + a\eta \right) dt, \\ \omega_3 &= \sqrt{\tfrac{2}{3}}\,\tilde{z}_0\,dx + \sqrt{\tfrac{2}{3}}\left(\tilde{z}_2 + \tfrac{1}{3}\tilde{z}_0^3 + \eta^2 \tilde{z}_0 + a\tilde{z}_0 \right) dt. \end{aligned} \tag{2.53}$$

Similarly, for the case $f_{31} = 0$, one gets the equation

$$\tilde{z}_{0,t} = \tilde{z}_3 - \tilde{z}_1 \tilde{z}_0^2 + a\tilde{z}_1. \quad \square$$

3. Some results on nongeneric cases for differential equations of type $u_{xt} = F(u, u_x, \ldots, \partial^k u/\partial x^k)$ which describe pseudospherical surfaces

In this section we characterize the SG and ShG equations as nongeneric cases of equations of the type $u_{xt} = F(u, u_x, \ldots, \partial^k u/\partial x^k)$ which describe p.s.s. With the notation introduced in (2.2), we will consider equations of the form

$$z_{1,t} = F(z_0, z_1, \ldots, z_k).$$

Moreover, the functions will be defined in the space of variables $x, t, z_0, \ldots, z_k$.

THEOREM 3.1. *Let $f_{\alpha\beta}$ be differentiable functions of $z_0, \ldots, z_k$, $1 \le \alpha \le 3$, $1 \le \beta \le 2$, such that $f_{11} = 0$, $f_{32} = 0$, and $f_{21} = \eta$. Then $z_{1,t} = F(z_0, \ldots, z_k)$ describes a p.s.s., with associated 1-forms $\omega_\alpha = f_{\alpha 1}\,dx + f_{\alpha 2}\,dt$, if and only if*

$$F = \frac{\eta}{a}(b \sin az_0 + c \cos az_0), \tag{3.1}$$

and

$$\begin{aligned} f_{12} &= b \sin az_0 + c \cos az_0, \\ f_{22} &= b \cos az_0 - c \sin az_0, \\ f_{31} &= az_1, \end{aligned} \tag{3.2}$$

where $a \ne 0$, b, c do not depend on z_i, $0 \le i \le k$.

Proof: Suppose $z_{1,t} = F(z_0, \ldots, z_k)$ describes a p.s.s. with ω_α satisfying the structure equations (1.7). Then

$$\begin{aligned} &\sum_{i=0}^{k} f_{12,z_i}\,dz_i \wedge dt - f_{31} f_{22}\,dx \wedge dt = 0, \\ &\sum_{i=0}^{k} f_{22,z_i}\,dz_i \wedge dt + f_{12} f_{31}\,dx \wedge dt = 0, \\ &\sum_{i=0}^{k} f_{31,z_i}\,dz_i \wedge dx + \eta f_{12}\,dx \wedge dt = 0. \end{aligned} \tag{3.3}$$

Since

$$\begin{aligned} dz_i \wedge dt &= z_{i+1}\,dx \wedge dt, \qquad 0 \le i \le k-1, \\ dz_1 \wedge dx &= -F\,dx \wedge dt, \end{aligned}$$

it follows from (3.3) that

$$\begin{aligned} f_{31,z_i} &= 0, \qquad i \ne 1, \\ f_{12,z_k} &= f_{32,z_k} = 0, \end{aligned} \tag{3.4}$$

and

$$\sum_{i=0}^{k-1} z_{i+1} f_{12, z_i} - f_{31} f_{22} = 0, \tag{3.5}$$

$$\sum_{i=0}^{k-1} z_{i+1} f_{22, z_i} + f_{12} f_{31} = 0, \tag{3.6}$$

$$-F f_{31, z_1} + \eta f_{12} = 0. \tag{3.7}$$

Taking derivatives of (3.5) and (3.6) with respect to $z_k, \ldots, z_2$ successively, it follows from (3.4) that

$$f_{12, z_i} = f_{22, z_i} = 0, \qquad 1 \leq i \leq k.$$

Therefore, taking double derivative of (3.5) with respect to z_1, we obtain

$$f_{31, z_1 z_1} = 0.$$

Hence,

$$f_{31} = a z_1 + e.$$

Inserting f_{31} in (3.5) and (3.6), it follows that

$$e = 0,$$

$$f_{12, z_0} - a f_{22} = 0,$$

$$f_{22, z_0} + a f_{12} = 0.$$

From these equations we get that

$$f_{12, z_0 z_0} + a^2 f_{12} = 0,$$

and f_{12}, f_{22}, and f_{31} are given by (3.2). Observe that $a \neq 0$; otherwise $\omega_3 = 0$ contradicts the fact that ω_3 is the connection form of a p.s.s. Now, it follows from (3.7) that F is given by (3.1).

The converse is a straightforward computation. □

In the preceding result, choosing $b = \bar{b}/\eta$ and $c = \bar{c}/\eta$, where $\bar{b}, \bar{c}$ are constants, we obtain F independent of η. For $a = 1$, $b = 1/\eta$, and $c = 0$ one gets the SG equation and the 1-forms ω_α as in (1.9).

Using arguments similar to the above theorem, one proves

THEOREM 3.2. *Let $f_{\alpha\beta}$ be differentiable functions of $z_0, \ldots, z_k$, $1 \leq \alpha \leq 3$, $1 \leq \beta \leq 2$ such that $f_{12} = 0$, $f_{31} = 0$, and $f_{21} = \eta$. Then $z_{1,t} = F(z_0, \ldots, z_k)$ describes a p.s.s.*

with associated 1*-forms* $\omega_\alpha = f_{\alpha 1}\,dx + f_{\alpha 2}\,dt$ *if and only if*

$$F = \frac{\eta}{a}(b\cosh az_0 + c\sinh az_0),$$

$$f_{11} = az_1,$$

$$f_{22} = b\sinh az_0 + c\cosh az_0,$$

$$f_{32} = b\cosh az_0 + c\sinh az_0,$$

where $a \neq 0$, b, c, *do not depend on* z_i, $0 \le i \le k$.

Taking $a = 1$, $b = 0$, and $c = 1/\eta$, one gets the ShG equation and the forms ω_α as in (1.12).

4. A geometric method which provides conservation laws and Bäcklund transformations

In this section we show how the geometric properties of a p.s.s. may be applied to obtain analytic results for equations which describe p.s.s.

Given a nonlinear differential equation for a function $u(x,t)$, suppose there exist operators D and F, defined on an appropriate space of functions, such that

$$\frac{\partial}{\partial t}D(u(x,t)) + \frac{\partial}{\partial x}F(u(x,t)) = 0$$

for all solutions u of the initial equation. Then the above relation is called *a conservation* law of the given differential equation. Moreover, the functional

$$I(u) = \int_{-\infty}^{\infty} D(u(x,t))\,dx$$

is called a *conserved quantity*, since $I(u)$ is independent of t for each solution u which satisfies appropriate conditions as $x \to \pm\infty$.

The classical Bäcklund theorem originated in the study of p.s.s., relating solutions of the SG equation. Recently other transformations have been found relating solutions of specific equations (see [6] and [8] for references). Such transformations are called *Bäcklund transformations* after the classical one. A Bäcklund transformation which relates solutions of the same equation is called a *self-Bäcklund transformation*. An interesting fact which has been observed is that differential equations which have self-Bäcklund transformation also admit a superposition formula. The importance of such formulas is due to the following: if u_0 is a solution of the differential equation and u_1, u_2 are solutions of the same equation obtained by the self-Bäcklund transformation, then the superposition formula provides a new solution $\bar{u}$ algebraically. By this procedure one obtains the soliton solutions of the differential equation.

In what follows we show that geometrical properties of p.s.s. provide a systematic method to obtain an infinite number of conservation laws and, under certain conditions, Bäcklund transformations.

Let M be a surface endowed with a C^∞ Riemannian metric. Consider a local orthonormal frame field e_1, e_2; let ω_1, ω_2 be its dual coframe, and ω_{12} the connection form. Then the structure equations are

$$d\omega_1 = \omega_{12} \wedge \omega_2,$$
$$d\omega_2 = \omega_1 \wedge \omega_{12}, \tag{4.1}$$
$$d\omega_{12} = -K\omega_1 \wedge \omega_2,$$

where K is the Gaussian curvature of M. M is a pseudospherical surface whenever $K \equiv -1$.

PROPOSITION 4.1. *Let M be a C^∞ Riemannian surface. M is pseudospherical iff given any unit vector v_0 tangent to M at $p_0 \in M$, there exists an orthonormal frame field v_1, v_2, locally defined, such that $v_1(p_0) = v_0$ and the associated 1-forms $\theta_1, \theta_2, \theta_{12}$ satisfy*

$$\theta_{12} + \theta_2 = 0. \tag{4.2}$$

In this case, θ_1 is a closed form.

Proof: We have to show that the differential-form equation (4.2) is completely integrable iff M is a pseudospherical surface. Let $\mathscr{I}$ be the ideal generated by the form $\gamma = \theta_{12} + \theta_2$. Then it follows from (4.1) that

$$\begin{aligned} d\gamma &= d\theta_{12} + \theta_1 \wedge \theta_{12} \\ &= d\theta_{12} + \theta_1 \wedge \gamma - \theta_1 \wedge \theta_2 \\ &\equiv d\theta_{12} - \theta_1 \wedge \theta_2 \pmod{\mathscr{I}}. \end{aligned}$$

Therefore, $\mathscr{I}$ is closed under exterior differentiation iff M is a pseudospherical surface. Hence the first part of the proposition follows from the Frobenius theorem. The fact that θ_1 is closed follows from (4.2) and the structure equations (4.1). □

We observe that the above result is the intrinsic version of the integrability theorem for the classical Bäcklund theorem [4, Theorem 2]. Moreover, the integral curves of the vector fields v_1, v_2 are respectively geodesics and horocycles of the pseudospherical surface.

In order to give the analytical interpretation of Proposition 4.1, we need the relations between the 1-forms associated to different orthonormal frames defined on a surface. Let M be a Riemannian surface; let e_1, e_2 and v_1, v_2 be two orthonormal frame fields with $\omega_1, \omega_2, \omega_{12}$ and $\theta_1, \theta_2, \theta_{12}$ the associated 1-forms respectively. We may consider both frames with the same orientation; then

$$e_1 = \cos\phi\, v_1 + \sin\phi\, v_2,$$
$$e_2 = -\sin\phi\, v_1 + \cos\phi\, v_2.$$

Therefore

$$\begin{aligned} \omega_1 &= \cos\phi\,\theta_1 + \sin\phi\,\theta_2, \\ \omega_2 &= -\sin\phi\,\theta_1 + \cos\phi\,\theta_2, \\ \omega_{12} &= \theta_{12} + d\phi, \end{aligned} \tag{4.3}$$

where ϕ is the rotation angle of the frames.

From now on, we will denote by (E) a differential equation for $u(x,t)$ which describes a p.s.s. with associated 1-forms

$$\begin{aligned} \omega_1 &= f_{11}\,dx + f_{12}\,dt, \\ \omega_2 &= \eta\,dx + f_{22}\,dt, \\ \omega_{12} &= f_{31}\,dx + f_{32}\,dt, \end{aligned} \tag{4.4}$$

where $f_{\alpha\beta}$ are functions of $u(x,t)$ and its derivatives. (Observe that we are denoting ω_3 by ω_{12}, which is the classical notation for the connection form.)

The analytic interpretation of Proposition 4.1 is the following.

PROPOSITION 4.2. *Let* (E) *be a differential equation which describes a p.s.s. with associated* 1-*forms* (4.4). *Then, for each solution u of* (E), *the system of equations for* $\phi(x,t)$

$$\begin{aligned} &\phi_x - f_{31} - f_{11}\sin\phi - \eta\cos\phi = 0, \\ &\phi_t - f_{32} - f_{12}\sin\phi - f_{22}\cos\phi = 0 \end{aligned} \tag{4.5}$$

is completely integrable. Moreover, for each solution u of (E) *and corresponding solution* ϕ,

$$(f_{11}\cos\phi - \eta\sin\phi)\,dx + (f_{12}\cos\phi - f_{22}\sin\phi)\,dt \tag{4.6}$$

is a closed form.

Proof: It follows from (4.3) and Proposition 4.1 that u is a solution of (E) iff

$$\omega_{12} - d\phi + \sin\phi\,\omega_1 + \cos\phi\,\omega_2 = 0 \tag{4.7}$$

is completely integrable for ϕ. In this case,

$$\cos\phi\,\omega_1 - \sin\phi\,\omega_2 \tag{4.8}$$

is a closed form. Hence, inserting (4.4) into (4.7) and (4.8), we obtain respectively the system of equations (4.5) whose integrability condition is (E) and the closed form (4.6). □

Whenever (E) does not involve the parameter η, the closed form (4.6) may provide an infinite number of conservation laws. In fact, when the solution $\phi(x,t,\eta)$ of (4.5) and the functions $f_{\alpha\beta}$ are analytic in η, we get an infinite number of conservation laws for E by equating like powers of η in (4.6). A generalization of this result and an algorithm which explicitly provides the conservation laws are obtained in [9].

Under certain conditions, (4.5) provides Bäcklund transformations for (E). Suppose we can eliminate u from (4.5); then we get

$$u = G(\phi) \tag{4.9}$$

and a differential equation for ϕ,

$$L(\phi) = 0. \tag{H}$$

These two last equations are equivalent to (4.5). Therefore, from the proof of Proposition 4.2, it follows that:

Proposition 4.3. *Let* (E) *be a differential equation which describes a p.s.s., with associated* 1*-forms* (4.4). *Suppose* (4.5) *is equivalent to a system of equations* (4.9) *and* (H). *Given a solution u of* (E), *then the system of equations* (4.5) *is completely integrable and* ϕ *is a solution of* (H). *Conversely, if* ϕ *is a solution of* (H), *then u defined by* (4.9) *is a solution of* (E).

Example 4.4 (Applications): In what follows, we apply the above result to obtain Bäcklund transformations.

(a) We have seen that the SG equation

$$u_{xt} = \sin u \tag{4.10}$$

describes a p.s.s. with associated 1-forms given by (1.9). In this case (4.5) is given by

$$\begin{aligned} &\phi_x - u_x - \eta\cos\phi = 0, \\ &\phi_t - \frac{1}{\eta}\cos(u-\phi) = 0, \end{aligned} \tag{4.11}$$

which is equivalent to the system of equations

$$u = \phi + \arccos(\eta\phi_t), \tag{4.12}$$

$$\phi_{tx} - \sqrt{1-\eta^2\phi_t^2}\cos\phi = 0. \tag{HSG}$$

It follows from Proposition 4.3 that given a solution u of SG, then for each constant $\eta \neq 0$, the system of equations (4.11) is completely integrable and ϕ is a solution of (HSG). Conversely, if ϕ is a solution of (HSG), where η is an arbitrary constant, then u defined by (4.12) satisfies the SG equation. The system of equations (4.11) is a Bäcklund transformation between the SG equation and (HSG).

In order to obtain the self-Bäcklund transformation for the SG equation we observe that (HSG) is invariant under the transformation $(\phi, \eta) \to (\pi - \phi, -\eta)$. If u is a solution of SG and (ϕ, η) satisfies (4.11), then (4.12) holds and ϕ is a solution for (HSG). It follows from the invariance of the latter equation and the preceding considerations that $\bar{u}$ defined by

$$\bar{u} = \pi - \phi + \arccos(\eta\phi_t) \tag{4.13}$$

is another solution of SG. From (4.12) and (4.13) we obtain

$$\phi = \tfrac{1}{2}(u - \bar{u} + \pi).$$

We substitute this relation into (4.11), and we get

$$(u + \bar{u})_x = 2\eta \sin \frac{u - \bar{u}}{2},$$

$$(u - \bar{u})_t = \frac{2}{\eta} \sin \frac{\bar{u} + u}{2},$$

which is the self-Backlund transformation for the sine-Gordon equation (4.10).

(b) In Section 1 we have seen that the MKdV equation

$$u_t = u_{xxx} + \tfrac{3}{2}u^2 u_x \tag{4.14}$$

describes a p.s.s. with associated 1-forms given by (1.10). In this case (4.5) is given by

$$\begin{aligned} &\phi_x - u - \eta \cos\phi = 0 \\ &\phi_t - u_{xx} - \tfrac{1}{2}u^3 - \eta^2 u + \eta u_x \sin\phi - \left(\tfrac{1}{2}\eta u^2 + \eta^3\right)\cos\phi = 0. \end{aligned} \tag{4.15}$$

Observe that we can eliminate u, obtaining an equivalent system of equations

$$u = \phi_x - \eta\cos\phi, \tag{4.16}$$

$$\phi_t = \phi_{xxx} + \tfrac{1}{2}\phi_x^3 + \tfrac{3}{2}\eta^2\phi_x\cos^2\phi. \tag{HMKdV}$$

It follows from Proposition 4.3 that given a solution of the MKdV (4.14), then for each constant η, the system of equations (4.15) is completely integrable and ϕ is a solution of (HMKdV). Conversely, if ϕ is a solution of (HMKdV), where η is an arbitrary constant, then u defined by (4.16) satisfies the MKdV equation.

In order to obtain the self-Bäcklund transformation for the MKdV equation, we observe that (HMKdV) is invariant under the transformation $(\phi, \eta) \to (\phi, -\eta)$. If u is a solution of MKdV and (ϕ, η) satisfies (4.15), then it follows from the invariance of (HMKdV) and the preceding considerations that

$$\bar{u} = 2\phi_x + \eta \cos\phi \tag{4.17}$$

is also a solution of MKdV. Therefore, from (4.16) and (4.17) we obtain

$$\phi_x = \tfrac{1}{2}(u + \bar{u}).$$

Consider $u = w_x$, $\bar{u} = \bar{w}_x$. We may take

$$\phi = \tfrac{1}{2}(w + \bar{w}).$$

Inserting ϕ into (4.15), we obtain the self-Bäcklund transformation for the MKdV equation (4.14), which is given by

$$(\bar{w} - w)_x = 2\eta \cos\frac{\bar{w} + w}{2},$$

$$(\bar{w} + w)_t = (\eta w_x^2 + 2\eta^3)\cos\frac{\bar{w} + w}{2} - 2\eta w_{xx} \sin\frac{\bar{w} + w}{2} + 2w_{xxx} + w_x^3 + 2\eta^2 w_x.$$

(c) The KdV equation

$$u_t = u_{xxx} + 6uu_x$$

describes a p.s.s. with associated 1-forms given by (1.11). In this case, the system of equations (4.5) is given by

$$\begin{aligned} \phi_x &= -(1+u) + (1-u)\sin\phi + \eta\cos\phi, \\ \phi_t &= (1+\sin\phi)(-u_{xx} + \eta u_x - \eta^2 u - 2u^2) \\ &\quad + (1-\sin\phi)(-\eta^2 - 2u) \\ &\quad + \cos\phi\,(\eta^3 + 2\eta u - 2u_x). \end{aligned} \tag{4.18}$$

Since u can be eliminated from (4.18), we obtain the following equivalent system of equations:

$$u = \frac{1}{1+\sin\phi}(\sin\phi + \eta\cos\phi - \phi_x - 1), \tag{4.19}$$

$$\phi_t = \phi_{xxx} + \frac{\phi_x}{1+\sin\phi}\left[(2-\sin\phi)\phi_x^2 - 3\phi_{xx}\cos\phi + 6(\eta\cos\phi + \sin\phi - 1)\right]. \tag{HKdV}$$

Therefore, it follows from Proposition 4.3 that if u is a solution of the KdV equation, then for each constant η, the system of equations (4.18) is completely integrable and ϕ is a solution of (HKdV). Conversely, if ϕ is a solution of (HKdV), where η is an arbitrary constant, then u defined by (4.19) satisfies the KdV equation.

As a consequence we can obtain the self-Bäcklund transformation for the KdV equation. In fact, if u is a solution of KdV and (ϕ, η) satisfies (4.18), then, since (HKdV) is invariant under the transformation $(\phi, \eta) \to (\pi - \phi, -\eta)$, it follows that

$$\bar{u} = \frac{1}{1+\sin\phi}(\sin\phi + \eta\cos\phi + \phi_x - 1) \tag{4.20}$$

is also a solution of KdV. From (4.19) and (4.20) we obtain

$$\bar{u} - u = \frac{2\phi_x}{1+\sin\phi}.$$

Considering $u = w_x$ and $\bar{u} = \bar{w}_x$,

$$w - \bar{w} - \lambda = \frac{4}{1+\tan\frac{1}{2}\phi}, \tag{4.21}$$

where λ is a constant. Inserting (4.21) into (4.18), we get

$$\begin{aligned}(w+\bar{w})_x &= \frac{\eta^2}{2} - \tfrac{1}{2}(w - \bar{w} - \lambda - 2 + \eta)^2,\\ (w-\bar{w})_t &= 2w_{xxx} + 2(w - \bar{w} - \lambda - \eta)w_{xx}\\ &\quad + \left(\frac{\eta^2}{2} + w_x\right)\left[(w - \bar{w} - \lambda - 2 - \eta)^2 + 4w_x - \eta^2\right].\end{aligned}$$

Choosing $\lambda = -\eta$, we obtain

$$\begin{aligned}(w+\bar{w})_x &= \frac{\eta^2}{2} - \tfrac{1}{2}(w - \bar{w} - 2 + 2\eta)^2\\ (w-\bar{w})_t &= 2w_{xxx} + 2(w - \bar{w} - 2)w_{xx}\\ &\quad + \left(\frac{\eta^2}{2} + w_x\right)\left[(w - \bar{w} - 2)^2 + 4w_x - \eta^2\right],\end{aligned}$$

which is the known self-Bäcklund transformation for the KdV equation.

We conclude this section by observing that the systematic procedure to obtain self-Bäcklund transformations for a differential equation (E) which describe a pseudospherical surface is based on Proposition 4.3 and on the fact that the associated equation (H) is invariant under a certain transformation. It would be interesting to have a deeper understanding of the invariance property. As for the assumptions in Proposition 4.3. they certainly occur for equations of the type $u_t = F(u, u_x, \ldots, \partial^k u/\partial x^k)$ considered in Section 2. In fact, from Lemma 2.1, we have that f_{11} and f_{31} depend only on u, and $f_{11,u}, f_{31,u}$ do not vanish simultaneously. This allows one to obtain u in terms of ϕ (locally) from the first equation of (4.5).

References

1. M. J. Ablowitz, D. J. Kaup, A. C. Newell, and H. Segur, The inverse scattering transform—Fourier analysis for nonlinear problems, *Stud. Appl. Math.* 53:249–315 (1974).
2. S. S. Chern and C. K. Peng, Lie groups and KdV equations, *Manuscripta Math.* 28:207–217 (1979).
3. S. S. Chern and K. Tenenblat, Foliations on a surface of constant curvature and the modified Korteweg-deVries equations, *J. Differential Geom.* 16:347–349 (1981).
4. S. S. Chern and C. L. Terng, An analogue of Bäcklund's theorem in affine geometry, *Rocky Mountain J. Math.* 10:105–124 (1980).
5. C. S. Gardner, J. M. Greene, M. D. Kruskal, and R. M. Miura, *Phys. Rev. Lett.* 19:1095–1097 (1967).
6. R. M. Miura (Ed.), *Bäcklund Transformations, The Inverse Scattering Method, Solitons and Their Applications*, Lecture Notes in Mathematics, Vol. 515, Springer, New York, 1976.
7. R. Sasaki, Soliton equations and pseudospherical surfaces, *Nucl. Phys. B* 154:343–357 (1979).
8. A. C. Scott, F. Y. F. Chu, and D. W. McLaughlin, The soliton. A new concept in applied science, *Proc. IEEE* 61:1443–1483 (1973).
9. K. Tenenblat and J. A. Cavalcante, Conservation laws for nonlinear evolution equations, to appear.
10. V. E. Zakharov, and A. B. Shabat, Exact theory of two-dimensional self-focusing and one-dimensional self-modulation of waves in nonlinear media, *Soviet Phys. JETP* 34:62–69 (1972).

University of California at Berkeley

Mathematical Sciences Research Institute, Universidade de Brasília

(Received May 16, 1985)

J.A. BARROSO editor, Aspects of Mathematics and its Applications

ON A CONFORMAL INVARIANT OF THREE-DIMENSIONAL MANIFOLDS

Shiing-shen CHERN[1]

Mathematical Sciences Research Institute, Berkeley, California 94720, U.S.A.

Dedicated to Leopoldo Nachbin

1. Introduction

In [3] James Simons and I introduced an invariant for a compact orientable three-dimensional Riemannian manifold as follows: Let M denote the manifold, oriented, and let P be the bundle of its orthonormal frames, so that we have

$$\pi : P \to M, \tag{1.1}$$

where π is the projection, mapping a frame $xe_1e_2e_3$ to its origin x. A section $s : M \to P$ of the bundle satisfies the condition $\pi \circ s =$ identity and can be viewed as a field of frames. It is well known that in our case such a section always exists; we say that M is *parallelizable*. (All our manifolds and maps are $\mathscr{C}^\infty$.)

To such a frame field the *Levi-Civita connection* of the metric is given by an antisymmetric matrix of one-forms:

$$\omega = (\omega_{ij}), \qquad 1 \leq i, j \leq 3, \tag{1.2}$$

and its *curvature* by an antisymmetric matrix of two-forms:

$$\Omega = (\Omega_{ij}), \qquad 1 \leq i, j \leq 3. \tag{1.3}$$

Throughout this paper our small Latin indices will run from 1 to 3. We have

[1] Work done under partial support of National Science Foundation grants MCS 77-23579 and MCS 8120790.

$$d\omega_{ik} = \sum_j \omega_{ij} \wedge \omega_{jk} + \Omega_{ik} . \tag{1.4}$$

We introduce the three-form

$$T = \frac{1}{8\pi^2} \sum_{i,j,k} (\omega_{ij} \wedge \Omega_{ij} - \tfrac{1}{3}\omega_{ij} \wedge \omega_{jk} \wedge \omega_{ki}) , \tag{1.5}$$

and consider the integral

$$\Phi(s) = \int_M \tfrac{1}{2} T . \tag{1.6}$$

It will be proved below that for another section $t : M \to P$, $\Phi(t) - \Phi(s)$ is an integer, so that $\Phi(s)$ mod 1 is independent of s. This defines an invariant $\Phi(M) \in \mathbb{R}/\mathbb{Z}$, where $\Phi(M) = \Phi(s)$ mod 1.

In [3] we proved the theorems:

Theorem (1.1). *$\Phi(M)$ is a conformal invariant, i.e. it remains unchanged under a conformal transformation of the metric.*

Theorem (1.2). *$\Phi(M)$ has a critical value at M if and only if M is locally conformally flat.*

In this paper we will give direct proofs of these theorems, both because the three-dimensional case has special features and because the invariant $\Phi(M)$ has come up in various other connections. In fact, up to an additive constant it is the eta invariant of M, which was introduced by Atiyah, Patodi, and Singer via spectral theory [1]. It has also been used by W. Thurston in his theory of hyperbolic manifolds [6], and Robert Meyerhoff has shown that, for certain hyperbolic manifolds it takes values which are dense on the unit circle [5]. It is also related to the concept of 'anomaly' in two-dimensional gauge field theory in physics.

2. Family of Connections

We shall use arbitrary frame fields to develop the Riemannian geometry on M. Let $xe_1e_2e_3$ be a frame, and ω^1, ω^2, ω^3, its dual coframe.

Let the inner product be

$$(2.1)\qquad \langle e_i, e_j\rangle = g_{ij}\,.$$

We introduce g^{ij} through the equations

$$(2.2)\qquad \sum g_{ij}g^{jk} = \delta_i^k$$

and use the g's to raise or lower indices, as in classical tensor analysis. Then the *connection forms* ω_i^j or ω_{ij} are determined, uniquely, by the equations

$$(2.3)\qquad \mathrm{d}\omega^i = \sum \omega^j \wedge \omega_j^i\,, \qquad \omega_{ij} + \omega_{ji} = \mathrm{d}g_{ij}\,.$$

The *curvature forms* are defined by

$$(2.4)\qquad \Omega_i^j = \mathrm{d}\omega_i^j - \sum \omega_i^k \wedge \omega_k^j\,.$$

By exterior differentiation of (2.4) we get the *Bianchi identity*

$$(2.5)\qquad \mathrm{d}\Omega_i^j = \sum \omega_i^k \wedge \Omega_k^j - \sum \Omega_i^k \wedge \omega_k^j\,.$$

To avoid confusion notice our convention that the upper index in ω_i^j, Ω_i^j is the second index. Thus $\sum_j \omega_i^j g_{jk} = \omega_{ik}$ ($\neq \omega_{ki}$ in general).

We introduce the cubic form

$$(2.6)\qquad 8\pi^2 T = -\tfrac{1}{3}\sum \omega_i^j \wedge \omega_j^k \wedge \omega_k^i + \sum \omega_i^j \wedge \Omega_j^i - \sum \omega_i^j \wedge \Omega_j^i\,.$$

On our three-manifold M, T is of course closed. But the basic reason for its importance and interesting properties is that formally by (2.4), (2.5), we have

$$(2.7)\qquad \begin{aligned} 8\pi^2\,\mathrm{d}T &= \sum (\Omega_i^i \wedge \Omega_j^j - \Omega_i^j \wedge \Omega_j^i) \\ &= \sum \delta_{i_1 i_2}^{j_1 j_2}\,\Omega_{j_1}^{i_1} \wedge \Omega_{j_2}^{i_2}\,, \end{aligned}$$

which is the *first Pontrjagin form*.

Consider a family of connections on M, depending on a parameter τ. Then ω_i^j, Ω_i^j, T all involve τ, and we have the fundamental formula

$$8\pi^2 \frac{\partial T}{\partial \tau} = -\mathrm{d}\Big\{\sum \omega^i_i \wedge \frac{\partial \omega^j_j}{\partial \tau} - \sum \omega^j_i \frac{\partial \omega^i_j}{\partial \tau}\Big\} \tag{2.8}$$
$$+ 2\sum \Big(\frac{\partial \omega^i_i}{\partial \tau} \wedge \Omega^j_j - \frac{\partial \omega^j_i}{\partial \tau} \wedge \Omega^i_j\Big).$$

The proof of (2.8) is straightforward. It follows by differentiation of (2.6), and using the formulas obtained by differentiation of (2.3), (2.4) with respect to τ. It is useful to observe that the last term is a polarization of the Pontrjagin form.

Let P' be the bundle of all frames of M, so that we have

$$\begin{array}{ccc} P & \xrightarrow{\ i\ } & P' \\ & {}_{\pi}\searrow \quad \swarrow_{\pi'} & \\ & M & \end{array} , \tag{2.9}$$

where i is the inclusion. Then T in (2.6) can be considered as a form in P' and its pull-back i^*T is the T given by (1.5). We will make such identifications when there is no danger of confusion.

We consider $\Phi(s)$ defined by (1.6). When t is another section, then $t(M) - s(M)$, as a three-dimensional cycle is homologous, modulo torsion, to an integral multiple of the fiber P_x, $x \in M$. But P_x is topologically $SO(3)$ and ω_{ij} reduces on it to its Mauer–Cartan forms. If $j : P_x = SO(3) \to P$ is the inclusion,

$$\pm\tfrac{1}{2} j^* T = \frac{1}{8\pi^2}\, \omega_{12} \wedge \omega_{23} \wedge \omega_{31},$$

whose integral over P_x is 1. Hence $\Phi(s)$ mod 1 is independent of s.

We wish to clarify the relation between orthonormal frame fields and arbitrary frame fields. By the Schmidt orthogonalization process P is a retract of P', under which the origin of the frame is fixed. The retraction we denote by $r : P' \to P$. We consider the form T defined in (2.6) to be in P'. If $s' : M \to P'$ is a section, then $s = r \circ s'$ is also a section and they are homotopic through sections. Since $\mathrm{d}T = 0$, we have

$$\int_{s'M} T = \int_{sM} i^* T. \tag{2.10}$$

Hence $\Phi(M)$ can be computed through an arbitrary frame field by the left-hand side of the last equation.

3. Proofs of the Theorems

In view of the above remark we can, for local considerations, use a local coordinate system u^i and the resulting natural frame field $\partial/\partial u^i$. We shall summarize the well-known formulas, which are

$$\omega_i^j = \sum \Gamma_{ik}^j \, \mathrm{d}u^k,$$

$$\Gamma_{ik}^j = \sum g^{jl}\Gamma_{ilk},$$

(3.1a)

$$\Gamma_{ilk} = \tfrac{1}{2}\left(\frac{\partial g_{il}}{\partial u^k} + \frac{\partial g_{kl}}{\partial u^i} - \frac{\partial g_{ik}}{\partial u^l}\right),$$

$$\Omega_i^j = \tfrac{1}{2}\sum R_{ikl}^j \, \mathrm{d}u^k \wedge \mathrm{d}u^l .$$

The R_{ijkl} satisfy the symmetry relations

(3.1b)

$$R_{ijkl} = -R_{jikl} = -R_{ijlk}, \qquad R_{ijkl} = R_{klij},$$

$$R_{ijkl} + R_{iklj} + R_{iljk} = 0, \qquad R_{ijkl,h} + R_{ijlh,k} + R_{ijhk,l} = 0,$$

where the comma denotes covariant differentiation. They imply

(3.2) $$\Omega_{ij} + \Omega_{ji} = 0, \qquad \sum \Omega_{ij} \wedge \mathrm{d}x^j = 0,$$

and

(3.3) $$\sum \Omega_i^i = \sum g^{ik}\Omega_{ik} = 0 .$$

We also introduce the *Ricci curvature* and the *scalar curvature by*

(3.4) $$R_k^i = \sum_j R_{kj}^{ij}, \qquad R = \sum R_i^i .$$

For treatment of the conformal geometry we define

$$(3.5)\qquad \begin{aligned} C^j_{kl} &= R^j_{k,l} - R^j_{l,k} - \tfrac{1}{4}(\delta^j_k R_{,l} - \delta^j_l R_{,k}), \\ C_{jkl} &= R_{jk,l} - R_{jl,k} - \tfrac{1}{4}(g_{jk} R_{,l} - g_{jl} R_{,k}). \end{aligned}$$

Then the Bianchi identities give

$$(3.6)\qquad \sum C^j_{kj} = 0, \qquad C_{jkl} + C_{klj} + C_{ljk} = 0.$$

These relations imply that the matrix

$$(3.7)\qquad C = \begin{bmatrix} C^1_{23} & C^1_{31} & C^1_{12} \\ C^2_{23} & C^2_{31} & C^2_{12} \\ C^3_{23} & C^3_{31} & C^3_{12} \end{bmatrix}$$

is symmetric and that the matrix GC, where $G = (g_{ij})$, has trace zero.

Schouten proved [4, p. 92] that the three-dimensional manifold M is conformally flat, if and only if $C = 0$, i.e. $C_{ijk} = 0$.

By integrating (2.8), we get

$$(3.8)\qquad 8\pi^2 \frac{\partial}{\partial \tau} \int_M T = 2 \int_M \Delta,$$

where

$$(3.9)\qquad \Delta = \sum \left(\frac{\partial \omega^i_i}{\partial \tau} \wedge \Omega^j_j - \frac{\partial \omega^j_i}{\partial \tau} \wedge \Omega^i_j \right) = -\sum \frac{\partial \omega^j_i}{\partial \tau} \wedge \Omega^i_j,$$

by (3.3).

We consider a family of metrics $g_{ij}(\tau)$ and put

$$(3.10)\qquad v_{ij} = \frac{\partial g_{ij}}{\partial \tau}.$$

To prove Theorem (1.1) we suppose this is a conformal family of metrics, i.e.

(3.11) $$v_{ij} = \sigma g_{ij}\,.$$

From (3.1a) we find

(3.12) $$\frac{\partial \omega^j_i}{\partial \tau} = \tfrac{1}{2} \sum (\delta^j_i \sigma_k + \delta^j_k \sigma_i - g_{ik} g^{jl} \sigma_l)\,\mathrm{d}x^k\,,$$

where $\sigma_k = \partial\sigma/\partial x^k$. By the second equation of (3.2) and the equation (3.3) we find $\Delta = 0$. This proves that $\int_M T$ is independent of τ, and hence Theorem (1.1).

To prove Theorem (1.2) we consider v_{ij} such that the trace $\sum v^i_i = 0$. Geometrically this means that we consider the tangent space of the space of conformal structures on M. From (3.1a) we find

(3.13) $$\frac{\partial \omega^j_i}{\partial \tau} = \sum g^{jl}\Big\{-\sum v_{lk}\omega^k_i + \tfrac{1}{2}\sum \Big(\frac{\partial v_{il}}{\partial u^k} + \frac{\partial v_{kl}}{\partial u^i} - \frac{\partial v_{ik}}{\partial u^l}\Big)\,\mathrm{d}u^k\Big\}.$$

It follows that

(3.14) $$\Delta = \sum_{i,j} \Omega^{ij}\Big\{-\sum v_{jk}\omega^k_i + \tfrac{1}{2}\mathrm{d}v_{ij} + \tfrac{1}{2}\sum_k \Big(\frac{\partial v_{kj}}{\partial u^i} - \frac{\partial v_{ki}}{\partial u^j}\Big)\,\mathrm{d}u^k\Big\}.$$

The term in the middle is zero, because Ω^{ij} is antisymmetric and $\mathrm{d}v_{ij}$ is symmetric in i, j. To the integral of the last term we apply Stokes theorem to reduce it to an integral involving only the v_{ij}, and not their derivatives. We will omit the details of this computation. The result is that the condition

$$\frac{\partial}{\partial \tau}\int_M T = 0$$

is equivalent to

(3.15) $$\int_M \mathrm{Tr}(VC)\,\mathrm{d}u^1 \wedge \mathrm{d}u^2 \wedge \mathrm{d}u^3 = 0\,.$$

If the metric is conformally flat, we have $C = 0$ and hence the vanishing of the above integral.

Conversely, at a critical point of Φ we must have $\mathrm{Tr}(VC) = 0$ for all symmetric V satisfying $\mathrm{Tr}(VG^{-1}) = 0$. Hence C is a multiple of G^{-1} or GC is a multiple of the unit matrix. But GC has trace zero. Hence it must itself be zero and we have $C = 0$. This proves Theorem (1.2).

References

[1] M.F. Atiyah, V.K. Patodi and I.M. Singer, Spectral asymmetry and Riemannian geometry II, Math. Proc. Cambridge Philos. Soc. 78 (1975) 405–432.

[2] S.S. Chern, Complex manifolds without potential theory, 2nd Edition (Springer, Berlin, 1979) Appendix 97–150.

[3] S.S. Chern and J. Simons, Characteristic forms and geometrical invariants, Ann. of Math 99 (1974) 48–69 or S.S. Chern, Selected papers (Springer, Berlin, 1978) 444–465.

[4] L.P. Eisenhart, Riemannian Geometry (Princeton Univ. Press, Princeton, 1949).

[5] R. Meyerhoff, The Chern–Simons invariant for hyperbolic 3-manifolds, thesis (Princeton University, Princeton, 1981).

[6] W. Thurston, Three-dimensional manifolds, Kleinian groups, and hyperbolic geometry, The Mathematical Heritage of Henri Poincaré (Amer. Math. Soc., Providence, RI, 1983) 87–111.

Annals of Mathematics, **125** (1987), 301–335

Harmonic maps of the two-sphere into a complex Grassmann manifold II*

By Shiing-Shen Chern** and Jon G. Wolfson***

Introduction

Let $G(k, n)$ be the Grassmann manifold of all k-dimensional subspaces $\mathbf{C}^k$ in a complex number space $\mathbf{C}^n$ of dimension n. The manifold can also be regarded as that of all P_{k-1} in P_{n-1}, the projective spaces of dimensions $k-1$ and $n-1$, respectively. In particular, we have $G(1, n) = P_{n-1}$. $G(k, n)$ has a canonical Kählerian metric. We will study the minimal immersions or harmonic maps of the two-sphere S^2 into $G(k, n)$. The harmonic maps $S^2 \to G(1, n) = P_{n-1}$ were first determined by Din and Zakrzewski ([5], cf. also [7], [10]) and the harmonic maps $S^2 \to G(2, 4)$ were determined by Ramanathan [8].

A fundamental role is played by the transforms of a harmonic map of a surface. To define the ∂-transform (or $\bar{\partial}$-transform) consider an immersion $f\colon M \to G(k, n)$, when M is an oriented surface. M acquires an induced Riemannian metric, which we write as

$$ds_M^2 = \varphi\bar{\varphi},$$

where φ is a complex-valued one-form, defined up to a factor of absolute value 1. This form φ defines a complex structure on M. For $x \in M$ the space $f(x)$ has an orthogonal space $f(x)^{\perp}$ of dimension $n-k$. We denote by $[f(x)]$ and $[f(x)^{\perp}]$ their corresponding projective spaces, of dimensions $k-1$ and $n-k-1$, respectively. For a vector $Z(x) \in f(x)$ the orthogonal projection of ∂Z in $f(x)^{\perp}$ is a multiple of φ, and hence, by cancelling out φ, defines a point of $f(x)^{\perp}$. This defines a projective collineation $\partial\colon [f(x)] \to [f(x)^{\perp}]$, to be called a *fundamental collineation*. The mapping defined by sending $x \in M$ to the image of $[f(x)]$ under ∂ is called the ∂-transform. Similarly, we define the $\bar{\partial}$-transform.

Concerning this transform the main result is that, if f is harmonic, then its ∂-transform and $\bar{\partial}$-transform are harmonic. Notice that the fundamental collinea-

*The paper bearing the same title was an announcement of the main results, which appeared in Proc. Natl. Acad. Sci. U.S.A. 82 (1985), 2217–2219.

**Work done under partial support of NSF Grants DMS 84-03201 and DMA 84-01939.

tions may degenerate, as a consequence of which the transforms will be of lower dimension or zero. In the latter case the map f is holomorphic. (We will use "holomorphic" to mean "holomorphic or anti-holomorphic".)

Successive applications of the ∂-transform (or $\bar{\partial}$-transform) give rise to a sequence of harmonic maps

$$[f(x)] \xrightarrow{\partial} \partial[f(x)] \xrightarrow{\partial} \partial^2[f(x)] \to \cdots,$$

of which two consecutive ones are by construction orthogonal. The "crossing" construction, similar to orthogonalization in linear algebra, but of course more sophisticated, associates to it a harmonic sequence of which any two members are orthogonal. Such a harmonic sequence will be called *Frenet* and plays an important role. The construction is based on projective geometry. Its possibility is a global property, due to the fact that the domain manifold is S^2. Its inverse operation, the *recrossing*, depends on the choice of a holomorphic section of a P_1-bundle. In Section 7 we will describe the crossing construction for $k = 2$.

Our strategy is to establish the existence of a degenerate fundamental collineation in a harmonic sequence and, if there is none, in the Frenet harmonic sequence obtained from crossings. The latter always has a degenerate fundamental collineation, a consequence of a vanishing theorem on S^2 (cf. §3).

We consider $G(2, n)$. In Section 8 we will describe a construction, to be called "turning", whose successive applications lead from a degenerate fundamental collineation of rank one to a holomorphic curve in P_{n-1}. Its inverse operation, the "returning", depends on the choice of a holomorphic section of a P_1-bundle.

A holomorphic map $M \to G(2, n)$ is harmonic. We will call it a *classical harmonic map of the first kind*. Another harmonic map is obtained by taking a holomorphic curve $M \to G(1, n)$ and taking the line joining two consecutive vertices of its Frenet frame or by doing a similar construction with a holomorphic and an antiholomorphic curve $M \to G(1, n)$ (see §8). We will call these *classical harmonic maps of the second kind*.

Our main theorem is the following:

THEOREM. *Let f: $S^2 \to G(2, n)$ be a harmonic map. It can be obtained in one of the following ways*: 1) *from a classical harmonic map of the first kind through ∂ and $\bar{\partial}$-transforms and recrossings*; 2) *from a classical harmonic map of the second kind through ∂- and $\bar{\partial}$-transforms and recrossings and returnings.*

From this viewpoint the fundamental theorem of harmonic maps $S^2 \to G(1, n) = P_{n-1}$ can be stated as follows: *Any harmonic map $S^2 \to P_{n-1}$ can be obtained from a holomorphic map $S^2 \to P_{n-1}$ through ∂ and $\bar{\partial}$-transforms.*

After one of us (J. W.) reported on our work at a conference in August 1984, the same problem was studied by F. Burstall and John Wood, using their twistor method. A recent preprint by Karen K. Uhlenbeck treats a more general problem and contains many interesting results. It is desirable to study harmonic maps of S^2 into a symmetric Riemannian or Hermitian manifold or into $G(k, \infty)$. In this sense our paper gives only the first step. But there are enough remarkable features to warrant such a study. We will even give separate treatments of the cases $G(2,4)$, $G(2,5)$, $G(2,6)$, because we find their understanding is basic for that of the general case. By different methods the case $G(2,4)$ was studied by J. Ramanathan and the case $G(2,5)$ by Aithal [1]. We will use the standard method of moving frames, which is effective in such problems and is in principle elementary.

It is a pleasure to thank Robert Bryant for some useful and interesting discussions and to thank John Wood for his careful reading of our first manuscript. Part of the research reported on in this paper was conducted at MSRI, Berkeley. The second author would like to express his gratitude to MSRI for their hospitality.

1. Geometry of $G(k, n)$

We equip $\mathbf{C}^n$ with the standard Hermitian inner product, so that, for Z, $W \in \mathbf{C}^n$,

(1.1) $$Z = (z_1, \ldots, z_n), \qquad W = (w_1, \ldots, w_n),$$

we have

(1.2) $$(Z, W) = \sum z_A \overline{w}_A = \sum z_A w_{\bar{A}}.$$

Throughout this paper we will agree on the following ranges of indices:

(1.3) $1 \leq A, B, C, \ldots \leq n, 1 \leq \alpha, \beta, \gamma, \ldots \leq k, k+1 \leq i, j, h, \ldots \leq n.$

We shall use the summation convention, and the convention

(1.4) $$\bar{z}_A = z_{\bar{A}}, \qquad \bar{t}_{A\bar{B}} = t_{\bar{A}B}, \text{ etc.}$$

A frame consists of an ordered set of n linearly independent vectors Z_A, so that

(1.5) $$Z_1 \wedge \cdots \wedge Z_n \neq 0.$$

It is called *unitary*, if

(1.6) $$(Z_A, Z_B) = \delta_{A\bar{B}}.$$

The space of unitary frames can be identified with the unitary group $U(n)$. With

(1.7) $$dZ_A = \omega_{A\bar{B}} Z_B,$$

the $\omega_{A\bar{B}}$ are the Maurer-Cartan forms of $U(n)$. They are skew-Hermitian; i.e., we have

(1.8) $$\omega_{A\bar{B}} + \omega_{\bar{B}A} = 0.$$

Taking the exterior derivative of (1.7), we get the Maurer-Cartan equations of $U(n)$:

(1.9) $$d\omega_{A\bar{B}} = \omega_{A\bar{C}} \wedge \omega_{C\bar{B}}.$$

An element $\mathbf{C}^k$ of $G(k, n)$ can be defined by the multivector $Z_1 \wedge \cdots \wedge Z_k \neq 0$, defined up to a factor. The vectors Z_α and their orthogonal vectors Z_i are defined up to a transformation of $U(k)$ and $U(n-k)$, respectively, so that $G(k, n)$ has a G-structure, with $G = U(k) \times U(n-k)$. In particular, the form

(1.10) $$ds^2 = \omega_{\alpha\bar{i}}\omega_{\bar{\alpha}i}$$

is a positive definite Hermitian form on $G(k, n)$, and defines a Hermitian metric. Its Kähler form is

(1.11) $$\Omega = \sqrt{-1}/2\omega_{\alpha\bar{i}} \wedge \omega_{\bar{\alpha}i}.$$

By (1.9), it can be immediately verified that Ω is closed, so that the metric ds^2 is Kählerian.

2. Harmonic maps of surfaces

Let M be an oriented Riemannian surface and let f: $M \to G(k, n)$ be a non-constant harmonic map. Denote the Riemannian metric on M by $ds_M^2 = \varphi\bar{\varphi}$, where φ is a complex-valued one-form; φ is defined up to a complex factor of absolute value 1. For $x \in M$ the image $f(x) \in G(k, n)$ has an orthogonal space $f(x)^\perp \in G(n-k, n)$. If $Z \in f(x)$, we can write

(2.1) $$dZ \equiv \varphi X + \bar{\varphi} Y, \qquad \operatorname{mod} f(x),$$

where $X, Y \in f(x)^\perp$. If $Z \in \mathbf{C}^n - \{0\}$, we denote by $[Z]$ the point in P_{n-1} with Z as the homogeneous coordinate vector. Then

(2.2) $$\partial\colon [Z] \mapsto [X], \quad \bar{\partial}\colon [Z] \mapsto [Y],$$

if not zero, are well-defined projective collineations of the projectivized space $[f(x)]$ into $[f(x)^\perp]$. We shall call these the *fundamental collineations*. Dually there are adjoint fundamental collineations from $[f(x)^\perp]$ to $[f(x)]$. Clearly the fundamental collineation $\bar{\partial}$ (resp. ∂) is zero, if and only if f is holomorphic (resp. anti-holomorphic).

To express the situation analytically we choose, locally, a field of unitary frames Z_A, so that Z_α span $f(x)$. Then we have

(2.3) $$f^*\omega_{\alpha\bar{i}} = a_{\alpha\bar{i}}\varphi + b_{\alpha\bar{i}}\bar{\varphi}.$$

By (1.7) the fundamental collineations ∂ and $\bar\partial$ send $[Z_\alpha]$ to $[X_\alpha]$ and $[Y_\alpha]$ respectively, where

$$X_\alpha = a_{\alpha\bar i} Z_i, \qquad Y_\alpha = b_{\alpha\bar i} Z_i. \tag{2.4}$$

If f is an isometry

$$f^*(ds^2) = ds_M^2 = \varphi\bar\varphi, \tag{2.5}$$

we have, by (1.10),

$$\begin{aligned} a_{\alpha\bar i} b_{\bar\alpha i} &= 0, \\ a_{\alpha\bar i} a_{\bar\alpha i} + b_{\alpha\bar i} b_{\bar\alpha i} &= 1. \end{aligned} \tag{2.6}$$

When M is the two-sphere we can, without loss of generality, assume that f is an isometry (except, of course, at branch points). However, for higher genus M we do not make this assumption.

The metric ds_M^2 has a connection form ρ, which is a real one-form satisfying the equation

$$d\varphi = -\sqrt{-1}\,\rho \wedge \varphi. \tag{2.7}$$

Its exterior derivative gives the Gaussian curvature K as follows:

$$d\rho = -\sqrt{-1}\,/2K\varphi \wedge \bar\varphi. \tag{2.8}$$

Taking the exterior derivative of (2.3) and using (1.9), (2.7), we get

$$Da_{\alpha\bar i} \wedge \varphi + Db_{\alpha\bar i} \wedge \bar\varphi = 0, \tag{2.9}$$

where

$$\begin{aligned} Da_{\alpha\bar i} &= da_{\alpha\bar i} - a_{\beta\bar i}\omega_{\alpha\bar\beta} + a_{\alpha\bar j}\omega_{j\bar i} - \sqrt{-1}\,a_{\alpha\bar i}\rho, \\ Db_{\alpha\bar i} &= db_{\alpha\bar i} - b_{\beta\bar i}\omega_{\alpha\bar\beta} + b_{\alpha\bar j}\omega_{j\bar i} + \sqrt{-1}\,b_{\alpha\bar i}\rho. \end{aligned} \tag{2.10}$$

From (2.9) it follows that

$$\begin{aligned} Da_{\alpha\bar i} &= p_{\alpha\bar i}\varphi + q_{\alpha\bar i}\bar\varphi, \\ Db_{\alpha\bar i} &= q_{\alpha\bar i}\varphi + r_{\alpha\bar i}\bar\varphi. \end{aligned} \tag{2.11}$$

The quadratic differential form

$$Da_{\alpha\bar i}\varphi + Db_{\alpha\bar i}\bar\varphi = p_{\alpha\bar i}\varphi^2 + 2q_{\alpha\bar i}\varphi\bar\varphi + r_{\alpha\bar i}\bar\varphi^2 \tag{2.12}$$

is the "second fundamental form" of the map f. It is well-known that the vanishing of its trace is the condition that f be harmonic, so that $q_{\alpha\bar i} = 0$; cf. [3]. We define also

$$\begin{aligned} DX_\alpha &= dX_\alpha - \left(\omega_{\alpha\bar\beta} + \sqrt{-1}\,\rho\delta_{\alpha\bar\beta}\right)X_\beta, \\ DY_\alpha &= dY_\alpha - \left(\omega_{\alpha\bar\beta} - \sqrt{-1}\,\rho\delta_{\alpha\bar\beta}\right)Y_\beta. \end{aligned} \tag{2.13}$$

We get therefore the following criterion for the harmonicity of f, which we will apply repeatedly:

THEOREM 2.1. *The property that f is a harmonic map is expressed by one of the following conditions, which are equivalent*:

(a) $Da_{\alpha\bar{i}} \equiv 0, \bmod \varphi,$

(b) $Db_{\alpha\bar{i}} \equiv 0, \bmod \bar{\varphi},$

(c) $DX_{\alpha} \equiv 0, \bmod Z_{\beta}, \varphi,$

(d) $DY_{\alpha} \equiv 0, \bmod Z_{\beta}, \bar{\varphi}.$

Conditions (c) and (d) follow by the differentiation of (2.4) and the use of (1.7). The condition for a harmonic map is generally given by a second-order differential system. This theorem expresses it as a first-order system. It is like replacing the Laplace equation by the Cauchy-Riemann equations.

It is easy to see, using the criteria of Theorem 2.1, that a holomorphic map $M \to G(k, n)$ is harmonic. Another simple class of harmonic maps $M \to G(k, n)$ can be described as follows. Let h: $M \to P_{n-1}$ be a holomorphic curve. The span of any k distinct elements of the Frenet frame of h describes a map $M \to G(k, n)$. It was first shown by Din and Zakrzewski [6] that these maps are conformal and harmonic. This fact also follows easily from Theorem 2.1.

Theorem 2.1 allows us to study the global behavior of the maps, ∂, $\bar{\partial}$ when f is harmonic. In particular, we will show that the matrices $(a_{\alpha\bar{i}})$, $(b_{\alpha\bar{i}})$ have constant ranks except at isolated points.

The map f: $M \to G(k, n)$ induces over M the tautological k-dimensional complex vector bundle $\mathscr{S}$, with fibers $f(x)$, $x \in M$. In terms of our frames Z_A, a vector $Z \in f(x)$ can be written

$$Z = \xi^{\alpha} Z_{\alpha}, \tag{2.14}$$

and we have the natural connection defined by

$$DZ = \left(d\xi^{\alpha} + \xi^{\beta}\omega_{\beta\bar{\alpha}}\right) Z_{\alpha}. \tag{2.15}$$

On $\mathscr{S}$, which is of real dimension $2k + 2$, there is an almost complex structure defined by the forms

$$\theta^{\alpha} = d\xi^{\alpha} + \xi^{\beta}\omega_{\beta\bar{\alpha}}, \quad \text{and} \quad \varphi. \tag{2.16}$$

By (1.9) and (2.7) it can be immediately verified that these satisfy the Frobenius condition. Hence, by the Newlander-Nirenberg theorem there is a complex structure on $\mathscr{S}$ and $\mathscr{S}$ is a holomorphic bundle over M. Similarly, its orthogonal bundle $\mathscr{T}$, with fibers $f(x)^{\perp}$, $x \in M$, is also a homomorphic bundle over M. In fact, if $Z = \eta^i Z_i \in f(x)^{\perp}$, the forms defining the complex structure on $\mathscr{T}$ are

$$\psi^i = d\eta^i + \eta^j \omega_{j\bar{i}}, \quad \text{and} \quad \varphi. \tag{2.17}$$

Let $T^{1,0}$ be the cotangent bundle of M of type $(1,0)$, so that its sections can be written as $f\varphi$, f being a function. A section of the tensor product $\mathcal{T} \otimes T^{1,0}$ can be written $\eta^i Z_j \otimes \varphi$, and its covariant differential is given by

$$D\eta^j = d\eta^j + \eta^k \omega_{k\bar{j}} - \sqrt{-1}\,\eta^j \rho. \tag{2.18}$$

On $\mathcal{T} \otimes T^{1,0}$ there is a complex structure defined by the forms

$$\tilde{\psi}^j = d\eta^j + \eta^k \omega_{k\bar{j}} - \sqrt{-1}\,\eta^j \rho, \quad \text{and} \quad \varphi. \tag{2.19}$$

We define the mapping

$$\mathcal{D}\colon \mathcal{S} \to \mathcal{T} \otimes T^{1,0} \tag{2.20}$$

by

$$\mathcal{D}(\xi^\alpha Z_\alpha) = \xi^\alpha a_{\alpha\bar{i}} Z_i \otimes \varphi \tag{2.21}$$

keeping M pointwise fixed. Both sides of (2.20) being holomorphic bundles, we will prove that $\mathcal{D}$ is a holomorphic bundle map if f is harmonic. In fact, substituting

$$\eta^j = \xi^\alpha a_{\alpha\bar{j}}$$

into $\tilde{\psi}^j$ in (2.19), we find

$$\tilde{\psi}^j \equiv 0 \qquad \text{mod } \theta^\alpha, \varphi.$$

It follows that except at isolated points the map $\mathcal{D}$ and so the matrix $(a_{\alpha\bar{i}})$ have constant rank. The holomorphicity of $\mathcal{D}$ also implies that its image extends continuously and smoothly over the isolated singularities of $\mathcal{D}$ and so the image of $\mathcal{D}$ defines a bundle over M. Note that the fundamental collineation ∂ is simply the projectivization of $\mathcal{D}$. Thus the image of ∂, $\partial[f(x)]$, is a projective bundle over M. Denoting dim $\partial[f(x)] = k_1 - 1$, we define the ∂-transform of f:

$$\partial f\colon M \to G(k_1, n) \tag{2.22}$$

by $(\partial f)(x) = \partial[f(x)]$, $x \in M$. Similarly $\bar{\partial}[f(x)]$ is a projective bundle and the fundamental collineation $\bar{\partial}$ is a projective bundle map. Also we have the $\bar{\partial}$-transform

$$\bar{\partial} f\colon M \to G(k_2, n) \tag{2.23}$$

defined by $(\bar{\partial} f)(x) = \bar{\partial}[f(x)]$, $x \in M$, where dim $\bar{\partial}[f(x)] = k_2 - 1$.

Consider the vectors $Z \in f(x)$, such that $Y = 0$ in (2.1). They form a subspace $\ker \bar{\partial} \subset f(x)$. If f is harmonic, the above argument shows that $\ker \bar{\partial}$ is of constant dimension. We define

$$\delta_1 f\colon M \to G(l_1, n), \tag{2.24}$$

which sends $x \in M$ to the orthogonal complement of $\ker \bar{\partial}$ in $f(x)$. Similarly, we define $\delta_2 f$, using the operator ∂.

THEOREM 2.2. *Let f: $M \to G(k, n)$ be a harmonic map. Then*
(a) *The map $f^{\perp}: M \to G(n-k, n)$, defined by*

$$f^{\perp}(x) = f(x)^{\perp}, \qquad x \in M,$$

is harmonic
(b) *The maps $\partial f, \bar{\partial} f, \delta_1 f, \delta_2 f$ are harmonic.*
(c) *If $k_1 = k$, $\bar{\partial}\partial f$ is f itself.*

Proof. Using the criteria in Theorem 2.1, the proof of (a) is immediate.

To prove the first statement in (b) choose frames so that Z_α span $f(x)$ and Z_u span $\partial f(x)$, when the indices have the ranges

$$k+1 \leq u, v \leq k+k_1, k+k_1+1 \leq \lambda, \mu \leq n.$$

Then $a_{\alpha\bar{\lambda}} = 0$, and the matrix $(a_{\alpha\bar{u}})$ has rank k_1. Since f is harmonic, it follows from Theorem 2.1 and (2.10) that

$$a_{\alpha\bar{u}}\omega_{u\bar{\lambda}} \equiv 0, \qquad \text{mod } \varphi, \tag{2.25}$$

which implies $\omega_{u\bar{\lambda}} \equiv 0$, mod φ.

We now apply to the map ∂f the criterion of harmonicity in Theorem 2.1. The space $(\partial f)(x)$ is spanned by Z_u and its orthogonal space by Z_α, Z_λ. We have

$$\begin{aligned} \omega_{u\bar{\alpha}} &= -\omega_{\bar{\alpha}u} = -b_{\bar{\alpha}u}\varphi - a_{\bar{\alpha}u}\bar{\varphi}, \\ \omega_{u\bar{\lambda}} &\equiv 0, \qquad \text{mod } \varphi. \end{aligned} \tag{2.26}$$

By condition (b) of Theorem 2.1 we see readily that ∂f is harmonic.

In the same way we prove the other statements in (b).

The most interesting case is when $k_1 = k$. From (2.26) we see immediately that the $\bar{\partial}$-transform of $\partial f(x)$ is $f(x)$ itself. This completes the proof of Theorem 2.2.

Repeating the constructions of Theorem 2.2 we get two sequences of harmonic maps

$$f_0(=f) \xrightarrow{\partial} f_1 \xrightarrow{\partial} f_2 \to \cdots, \tag{2.27}$$

$$f_0 \xrightarrow{\bar{\partial}} f_{-1} \xrightarrow{\bar{\partial}} f_{-2} \to \cdots, \tag{2.27a}$$

whose image spaces are connected by fundamental collineations. Such sequences will be called *harmonic sequences*.

The most interesting case is when the k_i's are equal. Then we can combine the sequences into one:

$$\cdots f_{-2} \underset{\bar{\partial}}{\overset{\partial}{\rightleftarrows}} f_{-1} \underset{\bar{\partial}}{\overset{\partial}{\rightleftarrows}} f_0 \underset{\bar{\partial}}{\overset{\partial}{\rightleftarrows}} f_1 \cdots. \tag{2.28}$$

By construction two consecutive spaces $[f_i(x)]$ and $[f_{i+1}(x)]$, $x \in M$, of a harmonic sequence are orthogonal. If any two such members of a harmonic sequence (2.28) are orthogonal, the latter is called a *Frenet harmonic sequence.* A Frenet harmonic sequence whose members span the ambient space will be called a *full Frenet harmonic sequence.*

We saw above that if $f\colon M \to G(k, n)$ is a map, not necessarily harmonic, then the tautological bundle inherits the structure of a holomorphic rank k bundle. The same construction shows that any rank k subbundle of the trivial n-bundle over M inherits a holomorphic structure. Projectivizing, we see that any P_{k-1} subbundle of the trivial P_{n-1} bundle over M has a natural holomorphic structure. With respect to this structure, a section s of a P_{k-1} subbundle B is holomorphic (resp. antiholomorphic) if

$$(ds, t) \equiv 0 \qquad \text{mod } \varphi \text{ (resp. mod } \bar{\varphi}) \tag{2.29}$$

for all sections t orthogonal in B to s. (Two sections s and t of B are orthogonal if any vectors Z and W representing s and t respectively satisfy

$$(Z, W) = 0.$$

This condition is clearly independent of the choice of representatives.)

3. Vanishing theorems

Harmonic maps whose domain manifold is S^2 have many special properties. These arise from the fact that S^2 has no holomorphic differential forms of positive degree except zero. From our analytical data we are able to construct such forms and obtain in this way strong conclusions. Our first "vanishing theorem" is:

THEOREM 3.1. *Consider a harmonic map $f\colon S^2 \to G(k, n)$. Let*

$$c_{\alpha\bar{\beta}} = \alpha_{\alpha\bar{i}} b_{\bar{\beta}i}. \tag{3.1}$$

Then

$$\det(c_{\alpha\bar{\beta}} + t\delta_{\alpha\bar{\beta}}) = t^k \text{ identically in } t. \tag{3.2}$$

Proof. Using conditions (a), (b) of Theorem 2.1, we find

$$dc_{\alpha\bar{\beta}} \equiv 2\sqrt{-1}\,\rho c_{\alpha\bar{\beta}} + c_{\gamma\bar{\beta}}\omega_{\alpha\bar{\gamma}} - c_{\alpha\bar{\gamma}}\omega_{\gamma\bar{\beta}}, \qquad \text{mod } \varphi. \tag{3.3}$$

If ζ is a local complex coordinate then

$$\varphi = \lambda\, d\zeta \tag{3.4}$$

where λ is a positive function. Taking its exterior derivative, we get

$$d\lambda \equiv -\sqrt{-1}\,\rho\lambda \qquad \text{mod } \varphi. \tag{3.5}$$

Thus we have

$$d\left(c_{\alpha\bar{\beta}}\lambda^2\right) \equiv c_{\gamma\bar{\beta}}\omega_{\alpha\bar{\gamma}} - c_{\alpha\bar{\gamma}}\omega_{\gamma\bar{\beta}}, \qquad \text{mod } \varphi. \tag{3.6}$$

Consider the function $\det(c_{\alpha\bar{\beta}}\lambda^2 + t\delta_{\alpha\bar{\beta}})$, t being a parameter. We wish to show that its differential is zero mod φ. From (3.6) we get

$$\begin{aligned} d\left(c_{\alpha\bar{\beta}}\lambda^2 + t\delta_{\alpha\bar{\beta}}\right) \equiv 1/\lambda^2 \Big\{ &\left(c_{\gamma\bar{\beta}}\lambda^2 + t\delta_{\gamma\bar{\beta}}\right)\omega_{\alpha\bar{\gamma}} \\ &- \left(c_{\alpha\bar{\gamma}}\lambda^2 + t\delta_{\alpha\bar{\gamma}}\right)\omega_{\gamma\bar{\beta}}\Big\}, \qquad \text{mod } \varphi. \end{aligned} \tag{3.7}$$

The conclusion

$$d \det\left(c_{\alpha\bar{\beta}}\lambda^2 + t\delta_{\alpha\bar{\beta}}\right) \equiv 0, \qquad \text{mod } \varphi,$$

follows by direct calculation. Thus $\det(c_{\alpha\bar{\beta}}\lambda^2 + t\delta_{\alpha\bar{\beta}})$ is a holomorphic function. This implies that, with t as a parameter, $\det(c_{\alpha\bar{\beta}}\varphi^2 + t\delta_{\alpha\bar{\beta}})$ is a holomorphic differential form. The theorem follows from the fact that S^2 has no holomorphic differential forms of positive degree except zero. □

This theorem, in different but equivalent formulations, was known to J. Ramanathan, K. Uhlenbeck, and others. We remark that if we denote the adjoint (with respect to the hermitian inner product) of $\bar{\partial}$ by $\bar{\partial}^*$ then $(c_{\alpha\bar{\beta}})$ is the matrix representation of the composition $\bar{\partial}^* \circ \partial$.

Our next vanishing theorem is concerned with a Frenet harmonic sequence, as follows:

THEOREM 3.2. *Let* L_λ: $S^2 \to G(k, n)$, $0 \le \lambda \le s - 1$, $n \ge ks$ *be harmonic maps which form a Frenet harmonic sequence*

$$L_0 \xrightarrow{\partial} L_1 \xrightarrow{\partial} \cdots \xrightarrow{\partial} L_{s-1}. \tag{3.8}$$

Let π: $L_{s-1}^{\perp} \to L_0$ *be the orthogonal projection. Then the map*

$$\pi \circ \partial : L_{s-1} \to L_0 \tag{3.9}$$

is degenerate. When $n = ks$, ∂: $L_{s-1} \to L_0$; *so in this case the fundamental collineation is degenerate.*

Proof. For each λ, $0 \le \lambda \le s - 1$, choose a unitary frame $(Z_{\lambda_1}, \ldots, Z_{\lambda_k})$ spanning the k-dimensional space L_λ. Let A_λ, $0 \le \lambda \le s - 2$, denote the matrix representation of the map ∂: $L_\lambda \to L_{\lambda+1}$ with respect to the bases determined by this frame. Similarly, let B denote the matrix representation of the map $\pi \circ \partial$: $L_{s-1} \to L_0$. Our convention is that the A_λ and B act on the right. Note that since each L_λ has dimension k the A_λ and B are $k \times k$ matrices, of which the A_λ are non-singular. We will show that B is singular.

We complete the partial unitary frame

$$\{Z_{0_1},\ldots,Z_{0_k},Z_{1_1},\ldots,Z_{1_k},\ldots,Z_{s-1_1},\ldots,Z_{s-1_k}\}$$

to a unitary framing of $\mathbf{C}^n$

$$\{Z_{0_1},\ldots,Z_{s-1_k},X_1,\ldots,X_t\} \tag{3.10}$$

where $n = ks + t$. Write

$$(Z_\lambda) = \begin{pmatrix} Z_{\lambda_1} \\ \vdots \\ Z_{\lambda_k} \end{pmatrix}, \qquad 0 \le \lambda \le s-1$$

(3.11)

$$(X) = \begin{pmatrix} X_1 \\ \vdots \\ X_t \end{pmatrix}.$$

The following structure equations are consequences of (3.8) and of the fact that the Maurer-Cartan matrix associated to the frame (3.10) is skew-Hermitian:

$$\begin{aligned} d(Z_0) &= \pi_0(Z_0) + A_0\varphi(Z_1) \qquad\qquad -{}^t\bar{B}\bar{\varphi}(Z_{s-1}) - {}^t\bar{F}\bar{\varphi}(X), \\ d(Z_\tau) &= -{}^t\bar{A}_{\tau-1}\bar{\varphi}(Z_{\tau-1}) + \pi_\tau(Z_\tau) + A_\tau\varphi(Z_{\tau+1}), \qquad 1 \le \tau \le s-2 \\ d(Z_{s-1}) &= B\varphi(Z_0) + {}^t\bar{A}_{s-2}\bar{\varphi}(Z_{s-2}) + \pi_{s-1}(Z_{s-1}) + G\varphi(X) \end{aligned} \tag{3.12}$$

where the π_λ, $0 \le \lambda \le s-1$, are $k \times k$ skew-Hermitian matrices of 1-forms and F and G are $k \times t$ matrices of functions.

The Maurer-Cartan equations of $U(n)$ are obtained by taking the exterior derivative of (3.12). They give

$$\begin{aligned} dA_\sigma &\equiv \pi_\sigma A_\sigma - A_\sigma\pi_{\sigma+1} + \sqrt{-1}\,\rho A_\sigma, \quad \bmod \varphi, \qquad 0 \le \sigma \le s-2, \\ dB &\equiv \pi_{s-1}B - B\pi_0 + \sqrt{-1}\,\rho B, \qquad \bmod \varphi. \end{aligned} \tag{3.13}$$

Consider the map $L_0 \to L_0$ given by

$$L_0 \xrightarrow{\partial} L_1 \xrightarrow{\partial} \cdots \xrightarrow{\partial} L_{s-1} \xrightarrow{\pi\circ\partial} L_0.$$

In matrix notation this map is represented by

$$A_0A_1 \cdots A_{s-2}B\colon L_0 \to L_0.$$

(Remember that these matrices act on the right.) We have

$$
\begin{aligned}
d(A_0A_1 \cdots A_{s-2}B)& \\
&= \sum_{\sigma=0}^{s-2} A_0 \cdots A_{\sigma-1} dA_\sigma A_{\sigma+1} \cdots A_{s-2}B + A_0 \cdots A_{s-2}\,dB \\
&\equiv \sum_{\sigma=0}^{s-2} A_0 \cdots A_{\sigma-1}\big(\pi_\sigma A_\sigma - A_\sigma \pi_{\sigma+1} + \sqrt{-1}\,\rho A_\sigma\big) A_{\sigma+1} \cdots B \\
&\quad + A_0 \cdots A_{s-2}\big(\pi_{s-1}B - B\pi_0 + \sqrt{-1}\,\rho B\big), \qquad \operatorname{mod} \varphi, \\
&\equiv \pi_0 A_0 \cdots A_{s-1}B - A_0 \cdots A_{s-1}B\pi_0 + \sqrt{-1}\,s\rho A_0 \cdots A_{s-2}B, \quad \operatorname{mod} \varphi.
\end{aligned}
$$

Hence

$$\det(A_0A_1 \cdots A_{s-2}B\varphi^s + tI), \qquad I = \text{identity matrix},$$

is a holomorphic differential form (with t as a parameter). Since the domain manifold is S^2, we get

$$\det(A_0A_1 \cdots A_{s-2}B + tI) = t^k$$

identically in t. In particular

$$\det(A_0A_1 \cdots A_{s-2}B) = 0.$$

Since $A_0, \ldots, A_{s-2}$ and B are $k \times k$ matrices and the A_σ, $0 \le \sigma \le s-2$, are nonsingular we have

$$\det(B) = 0.$$

Therefore the map $\pi \circ \partial\colon L_{s-1} \to L_0$ is degenerate.

More vanishing theorems will be proved and used in the following sections.

4. Harmonic maps of S^2 to $G(2, 4)$

Following our general notation we choose frames $Z_1, \ldots, Z_4$, so that Z_1, Z_2 span $f(x)$ and Z_3, Z_4 span $f(x)^\perp$, $x \in S^2$. The fundamental collineations ∂, $\bar\partial$ send Z_α to $a_{\alpha\bar i} Z_i$ and $b_{\alpha\bar i} Z_i$, respectively. We write

$$A = (a_{\alpha\bar i}), \qquad B = (b_{\alpha\bar i}). \tag{4.1}$$

Our basic fact is

$$\det(c_{\alpha\bar\beta}) = \det A \cdot \det \bar B = 0, \tag{4.2}$$

so that one of the determinants in the product is zero. Suppose

$$\det(A) = 0. \tag{4.3}$$

If $(a_{\alpha\bar i})$ is identically zero, the harmonic map is an anti-holomorphic curve in $G(2, 4)$. Discarding this case, the fundamental collineation ∂ is of rank one. We choose the frames so that Z_1 is the kernel (i.e., $\partial Z_1 = 0$) and Z_3 is the image of ∂. Then

$$A = \begin{pmatrix} 0 & 0 \\ a & 0 \end{pmatrix}, \qquad a \neq 0. \tag{4.4}$$

Taking account of the relations (2.6), we write

$$B = \begin{pmatrix} \bar{b}_1 & \bar{b}_2 \\ 0 & \bar{b}_3 \end{pmatrix}. \tag{4.5}$$

For definiteness we suppose $\det B = \bar{b}_1 \bar{b}_3 \neq 0$. The frames Z_A are defined up to the change $Z_A \to Z_A^* = \exp(\tau_A \sqrt{-1}) Z_A$, τ_A real, i.e., to a transformation of the group $U(1) \times \cdots \times U(1)$. Thus the points $[Z_A]$ in P_3 are well-defined.

We apply the condition that f is harmonic. By Theorem 2.1 it includes

$$Da_{1\bar{3}} \equiv 0, \qquad Da_{2\bar{4}} \equiv 0, \qquad \operatorname{mod} \varphi,$$

and they give

$$\omega_{1\bar{2}} = p\varphi, \qquad \omega_{3\bar{4}} = q\varphi. \tag{4.6}$$

Therefore we have

$$\begin{aligned} dZ_1 &= \omega_{1\bar{1}} Z_1 + p\varphi Z_2 + \bar{b}_1 \bar{\varphi} Z_3 + \bar{b}_2 \bar{\varphi} Z_4, \\ dZ_2 &= -\bar{p}\bar{\varphi} Z_1 + \omega_{2\bar{2}} Z_2 + a\varphi Z_3 + \bar{b}_3 \bar{\varphi} Z_4, \\ dZ_3 &= -b_1 \varphi Z_1 - \bar{a}\bar{\varphi} Z_2 + \omega_{3\bar{3}} Z_3 + q\varphi Z_4, \\ dZ_4 &= -b_2 \varphi Z_1 - b_3 \varphi Z_2 - \bar{q}\bar{\varphi} Z_3 + \omega_{4\bar{4}} Z_4. \end{aligned} \tag{4.7}$$

It is of interest to observe that under our choice of frames all the non-diagonal terms are pure forms, i.e., multiples of φ or $\bar{\varphi}$.

By Theorem 2.2 the maps $S^2 \to P_3$ defined by

$$x \mapsto Z_\lambda(x), \qquad \lambda = 2, 3, \quad x \in S^2 \tag{4.8}$$

are harmonic. This can also be seen immediately from the equations (4.7) by application of the criteria of Theorem 2.1.

To simplify the equations (4.7) we introduce

$$\begin{aligned} Z_1' &= \frac{-\bar{p} Z_1 + \bar{b}_3 Z_4}{\| -\bar{p} Z_1 + \bar{b}_3 Z_4 \|}, \\ Z_4' &= \frac{-b_1 Z_1 + q Z_4}{\| -b_1 Z_1 + q Z_4 \|}, \end{aligned} \tag{4.9}$$

These are unit vectors, and we have:

LEMMA 4.1. *Z_1' and Z_4' are perpendicular.*

Proof. Consider the symmetric differential form

$$P = \omega_{1\bar{2}}\omega_{2\bar{3}}\omega_{3\bar{1}} + \omega_{4\bar{2}}\omega_{2\bar{3}}\omega_{3\bar{4}} \tag{4.10}$$
$$= -a(pb_1 + qb_3)\varphi^3,$$

which is independent of the choice of the frames and is well-defined on S^2. Using a complex coordinate ζ on S^2, so that $\varphi = \lambda\, d\zeta$ (locally), we have

$$P = -a(pb_1 + qb_3)\lambda^3\, d\zeta^3.$$

We shall prove that the coefficient of $d\zeta^3$ is a holomorphic function of ζ. For this purpose we write

$$\omega_{1\bar{2}} = u\,d\zeta, \qquad \omega_{2\bar{3}} = v\,d\zeta, \tag{4.11}$$
$$\omega_{3\bar{1}} = x\,d\zeta, \qquad \omega_{4\bar{2}} = y\,d\zeta, \qquad \omega_{3\bar{4}} = z\,d\zeta,$$

so that

$$P = v(ux + yz)\,d\zeta^3. \tag{4.12}$$

By differentiating (4.11) and using the structure equations, we get

$$\begin{aligned} du &\equiv y\omega_{1\bar{4}} + u(\omega_{1\bar{1}} - \omega_{2\bar{2}}),\\ dv &\equiv v(\omega_{2\bar{2}} - \omega_{3\bar{3}}),\\ dx &\equiv x(\omega_{3\bar{3}} - \omega_{1\bar{1}}),\\ dy &\equiv y(\omega_{4\bar{4}} - \omega_{2\bar{2}}),\\ dz &\equiv -x\omega_{1\bar{4}} + z(\omega_{3\bar{3}} - \omega_{4\bar{4}}), \qquad \mod d\zeta, \end{aligned} \tag{4.13}$$

from which we conclude

$$d(vux + vyz) \equiv 0, \qquad \mod d\zeta.$$

On S^2 such a form P is identically zero, and the lemma follows.

Relative to the frames Z_1', Z_2, Z_3, Z_4' we have

$$\begin{aligned} dZ_1' &= \omega'_{1\bar{1}}Z_1' + r\varphi Z_2 + \omega'_{1\bar{4}}Z_4',\\ dZ_2 &= -\bar{r}\bar{\varphi}Z_1' + \omega_{2\bar{2}}Z_2 + a\varphi Z_3,\\ dZ_3 &= -\bar{a}\bar{\varphi}Z_2 + \omega_{3\bar{3}}Z_3 + s\varphi Z_4',\\ dZ_4' &= \omega'_{4\bar{1}}Z_1' - \bar{s}\bar{\varphi}Z_3 + \omega'_{4\bar{4}}Z_4', \end{aligned} \tag{4.14}$$

and our genericity hypotheses imply $ras \neq 0$. Denoting by $\omega'_{A\bar{B}}$ the coefficients in the above equations, we have

$$\omega'_{1\bar{3}} = 0,$$

and, by exterior differentiation,

$$\omega'_{1\bar{4}} \wedge \bar{s}\bar{\varphi} = 0.$$

Hence we can write $\omega'_{1\bar{4}} = -\bar{t}\bar{\varphi}$ or $\omega'_{4\bar{1}} = t\varphi$.

We wish to prove that $t = 0$. In fact, consider the quartic form

$$P_1 = \omega'_{1\bar{2}}\omega'_{2\bar{3}}\omega'_{3\bar{4}}\omega'_{4\bar{1}} = arst\,\varphi^4,$$

which is well-defined on S^2. As in the above lemma we can prove that this form is holomorphic (i.e., locally of the form $h(\zeta)\,d\zeta^4$, where $h(\zeta)$ is a holomorphic function), and is hence zero. It follows that $t = 0$ or $\omega'_{1\bar{4}} = 0$.

Thus $[Z'_1]$ describes a holomorphic curve in P_3, with $[Z'_4]$ as its dual antiholomorphic curve and with $Z'_1 Z_2 Z_3 Z'_4$ as its Frenet frame. We will call $[Z'_1]$ the *directrix curve* of the harmonic map f. See Figure 1.

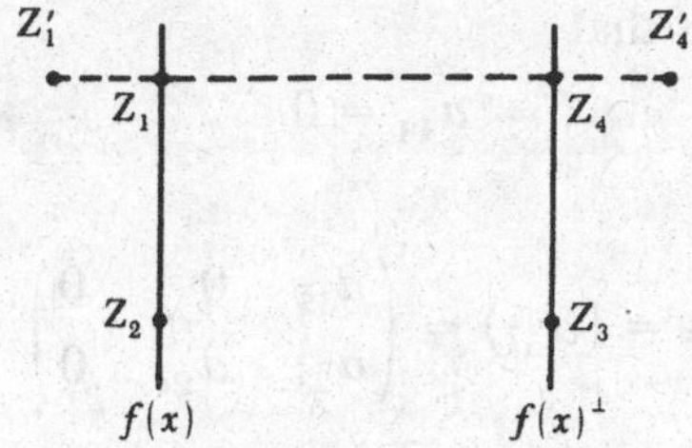

FIGURE 1

To recover the line $f(x) = Z_1 Z_2$ we set

$$(4.15)\qquad \begin{aligned} Z_1 &= (1 + |w|^2)^{-1/2}(Z'_1 + wZ'_4), \\ Z_4 &= (1 + |w|^2)^{-1/2}(-\bar{w}Z'_1 + Z'_4). \end{aligned}$$

The point Z_1 satisfies, by (4.7), the condition

$$(4.16)\qquad (dZ_1, Z_4) \equiv 0, \qquad \text{mod } \bar{\varphi}.$$

This can be written

$$(4.17)\qquad dw + w(\omega'_{4\bar{4}} - \omega'_{1\bar{1}}) \equiv 0, \qquad \text{mod } \bar{\varphi}.$$

Thus the map f involves a directrix curve $[Z'_1]$ and the choice of an anti-holomorphic section of the bundle $Z'_1 Z'_4$ over M.

5. Harmonic maps of S^2 to $G(2, 5)$

This case, like the case of harmonic maps of S^2 to $G(2, 4)$, exhibits many special features not present in the general case. Let $f\colon S^2 \to G(2, 5)$ be a harmonic map and write $L_0 = [f(x)]$, $x \in S^2$. To keep our discussion reasonably brief we will only consider the case where both fundamental collineations ∂ and $\bar{\partial}$ have rank two, so that the ∂ and $\bar{\partial}$-transforms of L_0 are lines. We leave the other cases to the reader.

Denote the ∂ and $\bar{\partial}$-transforms of L_0 by L_1 and L_{-1} respectively. L_1 and L_{-1} are projective lines in P_2 which are not equal since if they are the condition $\det(c_{\alpha\bar{\beta}}) = 0$ of Theorem 3.1 implies that one of the fundamental collineations must be degenerate. It follows that the lines L_1 and L_{-1} intersect in a point. We choose frames $Z_1, \ldots, Z_5$ so that Z_1, Z_2 span L_0 and Z_3, Z_4, Z_5 span $L_0^{\perp}$. The fundamental collineations ∂ and $\bar{\partial}$ send Z_α to $a_{\alpha\bar{i}}Z_i$ and $b_{\alpha\bar{i}}Z_i$, respectively. By choosing Z_3, Z_4 to span L_1 and Z_4, Z_5 to span L_{-1} (so that the point $[Z_4]$ is the point of intersection of L_1 and L_{-1}) we have

$$a_{\alpha\bar{5}} = 0, \quad b_{\alpha\bar{3}} = 0, \quad \alpha = 1, 2. \tag{5.1}$$

We can further specify the frame by demanding that ∂ take the point $[Z_1]$ to the point $[Z_3]$. This means that

$$a_{1\bar{4}} = 0. \tag{5.2}$$

Thus

$$A = (a_{\alpha\bar{i}}) = \begin{pmatrix} a_{1\bar{3}} & 0 & 0 \\ a_{2\bar{3}} & a_{2\bar{4}} & 0 \end{pmatrix}$$

and

$$B = (b_{\alpha\bar{i}}) = \begin{pmatrix} 0 & b_{1\bar{4}} & b_{1\bar{5}} \\ 0 & b_{2\bar{4}} & b_{2\bar{5}} \end{pmatrix}.$$

Then since $c_{\alpha\bar{\beta}} = a_{\alpha\bar{i}} b_{\bar{\beta}i}$, $\operatorname{tr}(c_{\alpha\bar{\beta}}) = a_{2\bar{4}} b_{\bar{2}4}$. As $\operatorname{tr}(c_{\alpha\bar{\beta}}) = 0$ by Theorem 3.1 and $a_{2\bar{4}} \neq 0$, we have

$$b_{2\bar{4}} = 0. \tag{5.3}$$

It follows from our choices that the frame $Z_1, \ldots, Z_5$ is determined up to a transformation by the group $U(1) \times \cdots \times U(1)$ and thus that the points $[Z_A]$ in P_4 are well-defined.

Applying Theorem 2.1 to the conditions (5.1)–(5.3) and using the harmonicity of f we have

$$\omega_{1\bar{2}} = p\varphi, \quad \omega_{3\bar{4}} = q\varphi, \quad \omega_{3\bar{5}} = r\varphi, \quad \omega_{4\bar{5}} = s\varphi$$

and therefore

$$\begin{aligned}
dZ_1 &= \omega_{1\bar{1}}Z_1 + p\varphi Z_2 + a_{1\bar{3}}\varphi Z_3 + b_{1\bar{4}}\bar{\varphi} Z_4 + b_{1\bar{5}}\bar{\varphi} Z_5, \\
dZ_2 &= -\bar{p}\bar{\varphi} Z_1 + \omega_{2\bar{2}} Z_2 + a_{2\bar{3}}\varphi Z_3 + a_{2\bar{4}}\varphi Z_4 + b_{2\bar{5}}\bar{\varphi} Z_5, \\
dZ_3 &= -\bar{a}_{1\bar{3}}\bar{\varphi} Z_1 - \bar{a}_{2\bar{3}}\bar{\varphi} Z_2 + \omega_{3\bar{3}} Z_3 + q\varphi Z_4 + r\varphi Z_5, \\
dZ_4 &= -\bar{b}_{1\bar{4}}\varphi Z_1 - \bar{a}_{2\bar{4}}\bar{\varphi} Z_2 - \bar{q}\bar{\varphi} Z_3 + \omega_{4\bar{4}} Z_4 + s\varphi Z_5, \\
dZ_5 &= -\bar{b}_{1\bar{5}}\varphi Z_1 - \bar{b}_{2\bar{5}}\varphi Z_2 - \bar{r}\bar{\varphi} Z_3 - \bar{s}\bar{\varphi} Z_4 + \omega_{5\bar{5}} Z_5.
\end{aligned} \tag{5.4}$$

As in the case of $G(2,4)$ our choice of frame has the consequence that all non-diagonal terms in (5.4) are 1-forms of pure type. This makes the computations to follow particularly easy and elegant.

Consider the (5, 0) form

$$\begin{aligned} Q &= \omega_{1\bar{3}}\omega_{3\bar{5}}\omega_{5\bar{2}}\omega_{2\bar{4}}\omega_{4\bar{1}} \\ &= (a_{1\bar{3}} r \bar{b}_{2\bar{5}} a_{2\bar{4}} \bar{b}_{1\bar{4}})\varphi^5. \end{aligned}$$

Since the frame is defined up to a torus it is easy to verify that Q is invariantly, and hence, globally defined on S^2.

LEMMA 5.1. *Q is a holomorphic (5, 0) form.*

Proof. Let ζ be a local complex coordinate on S^2 and write

$$\begin{aligned} &\omega_{1\bar{3}} = t\,d\zeta, \quad \omega_{3\bar{5}} = u\,d\zeta, \quad \omega_{5\bar{2}} = v\,d\zeta, \\ &\omega_{2\bar{4}} = w\,d\zeta, \quad \omega_{4\bar{1}} = x\,d\zeta. \end{aligned}$$

Then we must show that the function $(tuvwx)$ is a holomorphic function of ζ or equivalently that

$$d(tuvwx) \equiv 0, \qquad \text{mod } d\zeta.$$

The structure equations applied to (5.4) imply that

$$\begin{aligned} dt \wedge d\zeta &= d\omega_{1\bar{3}} \\ &= \omega_{1\bar{1}} \wedge \omega_{13} + \omega_{1\bar{3}} \wedge \omega_{3\bar{3}} \\ &= \omega_{1\bar{1}} \wedge t\varphi + t\varphi \wedge \omega_{3\bar{3}}, \end{aligned}$$

so that

$$dt \equiv t(\omega_{1\bar{1}} - \omega_{3\bar{3}}), \qquad \text{mod } \varphi.$$

Similarly,

$$\begin{aligned} du &\equiv u(\omega_{3\bar{3}} - \omega_{5\bar{5}}), &&\text{mod } \varphi, \\ dv &\equiv v(\omega_{5\bar{5}} - \omega_{2\bar{2}}), &&\text{mod } \varphi, \\ dw &\equiv w(\omega_{2\bar{2}} - \omega_{4\bar{4}}), &&\text{mod } \varphi, \\ dx &\equiv x(\omega_{4\bar{4}} - \omega_{1\bar{1}}), &&\text{mod } \varphi, \end{aligned}$$

and the result follows.

COROLLARY 5.1. $Q = 0$.

Proof. Q is a holomorphic form on S^2.

As we are assuming that both fundamental collineations ∂ and $\bar{\partial}$ have rank two it follows that $a_{1\bar{3}}a_{2\bar{4}} \neq 0$ and $b_{2\bar{5}}b_{1\bar{4}} \neq 0$. Thus the corollary implies that $r = 0$.

Theorem 5.2. *The map $S^2 \to P_4$ given by $[Z_4]$ is harmonic.*

Proof. Using $r = 0$, apply Theorem 2.1. That is, the point of intersection of L_1 and L_{-1} defines a harmonic map $S^2 \to P_4$.

From the Din-Zakrzewski theory of harmonic maps $S^2 \to P_4$, this map is an element of the Frenet frame of some holomorphic curve $S^2 \to P_4$. The remainder of this discussion centers around finding this holomorphic curve and then showing how to reconstruct the map f from this holomorphic curve.

From (5.4) we have

$$dZ_4 \equiv (-\bar{a}_{2\bar{4}}Z_2 - \bar{q}Z_3)\bar{\varphi} + (-\bar{b}_{1\bar{4}}Z_1 + sZ_5)\varphi, \qquad \text{mod } Z_4.$$

This motivates the following choices:

$$Z_3' = \frac{-\bar{a}_{2\bar{4}}Z_2 - \bar{q}Z_3}{\|\bar{a}_{2\bar{4}}Z_2 + \bar{q}Z_3\|}, \qquad Z_5' = \frac{-\bar{b}_{1\bar{4}}Z_1 + sZ_5}{\|-\bar{b}_{1\bar{4}}Z_1 + sZ_5\|},$$

$$Z_2' = \frac{qZ_2 - a_{2\bar{4}}Z_3}{\|qZ_2 - a_{2\bar{4}}Z_3\|}, \qquad Z_1' = \frac{\bar{s}Z_1 + b_{1\bar{4}}Z_5}{\|\bar{s}Z_1 + b_{1\bar{4}}Z_5\|}.$$

It is clear that Z_3' and Z_5' define harmonic maps $S^2 \to P_4$. In fact Z_3', Z_4, Z_5' are adjacent elements of the Frenet frame of some holomorphic curve. Note that the harmonic map $[Z_4]$ is never holomorphic or antiholomorphic. It is possible however that the map $[Z_3']$ is holomorphic or that the map $[Z_5']$ is antiholomorphic. In any case the remaining two elements of the Frenet frame lie on the line $Z_1' \wedge Z_2'$.

The converse to the problem we have just been discussing is to reconstruct the harmonic map $f\colon S^2 \to G(2,5)$ from a holomorphic curve $S^2 \to P_4$. Suppose that $\tilde{Z}_1, \tilde{Z}_2, \ldots, \tilde{Z}_5$ is the Frenet frame of the holomorphic curve $[\tilde{Z}_1]\colon S^2 \to P_4$. Consider one of the lines $\lambda_0 = \tilde{Z}_1 \wedge \tilde{Z}_5$, $\mu_0 = \tilde{Z}_1 \wedge \tilde{Z}_2$ or $\nu_0 = \tilde{Z}_4 \wedge \tilde{Z}_5$, say $\lambda_0 = \tilde{Z}_1 \wedge \tilde{Z}_5$, for definiteness. Now λ_0 defines a P_1 bundle over S^2 equipped with a natural holomorphic structure, as discussed at the end of Section 2. Using this structure we choose a holomorphic section $\tilde{Z}_1'$ of λ_0 and let $\tilde{Z}_5'$ be the point on λ_0 orthogonal to $\tilde{Z}_1'$. Now consider the lines

$$\lambda_1 = \tilde{Z}_1' \wedge \tilde{Z}_2, \quad \lambda_{-1} = \tilde{Z}_5' \wedge \tilde{Z}_4$$

as defining holomorphic P_1-bundles now S^2. Choose a holomorphic section $\tilde{Z}_1''$ of λ_1 and let $\tilde{Z}_2'$ be the point on λ_1 orthogonal to $\tilde{Z}_1''$. This choice determines an antiholomorphic section $\tilde{Z}_5''$ of λ_{-1} as follows. The lines λ_1 and λ_{-1} determine maps $S^2 \to G(2,5)$. It is a simple matter to verify that both of these maps are harmonic, that the ∂ and $\bar{\partial}$ transforms of λ_1 are $[\tilde{Z}_3]$ and λ_{-1}, resp., and that the ∂ and $\bar{\partial}$ transforms of λ_{-1} are λ_1 and $[\tilde{Z}_3]$ resp. Thus we have

$$\begin{array}{ccc} & \overset{\partial}{\swarrow} & \lambda_1 \\ [\tilde{Z}_3] & & \bar{\partial}\downarrow\uparrow\partial \\ & \underset{\bar{\partial}}{\nwarrow} & \lambda_{-1}. \end{array}$$

Set $[\tilde{Z}_5''] = \bar{\partial}([\tilde{Z}_2'])$ and choose $\tilde{Z}_4'$ to be the point on λ_{-1} orthogonal to $\tilde{Z}_5''$. It is not difficult to verify that $\tilde{Z}_5''$ is an antiholomorphic section of λ_{-1} (and in fact $\tilde{Z}_4'$ is a holomorphic section). The line $\tilde{Z}_1'' \wedge \tilde{Z}_5''$ determines a map $S^2 \to G(2,5)$. This map is harmonic. For suitable choice of directrix curve $[\tilde{Z}_1]$, of line λ_0, μ_0, or ν_0 and of holomorphic sections it is our given map f: $S^2 \to G(2,5)$(see Figure 2).

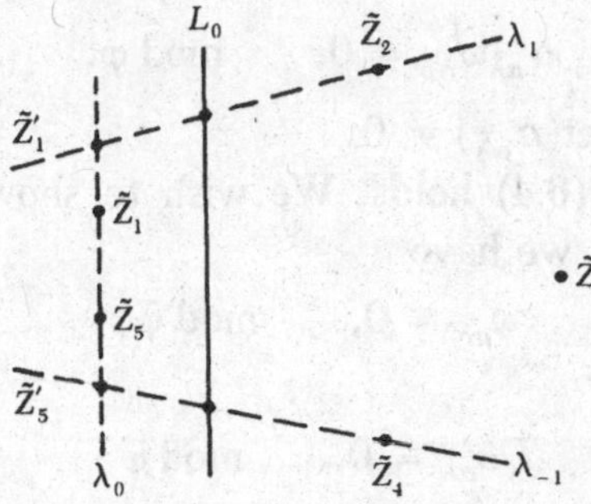

FIGURE 2

6. Harmonic maps of S^2 into $G(2,6)$

This is a typical case, containing all the features for general $G(2, n)$. We suppose that

$$A = (a_{\alpha i}), \qquad B = (b_{\alpha \bar{i}})$$

are both of rank two, so that the ∂ and $\bar{\partial}$-transforms of $[f(x)]$ are lines. We write $L_0 = [f(x)]$ and denote its ∂ and $\bar{\partial}$-transforms by L_1 and L_{-1} respectively. They form a sequence

$$L_{-1} \underset{\bar{\partial}}{\overset{\partial}{\rightleftarrows}} L_0 \underset{\bar{\partial}}{\overset{\partial}{\rightleftarrows}} L_1 \tag{6.1}$$

connected by fundamental collineations. By Theorem 3.1 the matrix $(c_{\alpha\bar{\beta}})$ is of rank < 2. It is of rank zero, if and only if L_{-1} and L_1 are perpendicular. We shall first briefly discuss this case. The condition for f to be harmonic can be put in a simple form as follows: Choose the frame Z_A, so that L_0, L_1, L_{-1} are spanned by Z_1, Z_2; Z_3, Z_4; Z_5, Z_6, respectively. Fix the following ranges of indices:

$$\alpha, \beta = 1, 2; \quad \lambda, \mu = 3, 4, \quad \rho, \sigma = 5, 6; \tag{6.2}$$

then we have

$$a_{\alpha\bar{\rho}} = 0, \quad b_{\alpha\bar{\lambda}} = 0. \tag{6.3}$$

LEMMA 6.1. *In the above notation the necessary and sufficient conditions for f to be harmonic are*

$$\omega_{\lambda\bar{\rho}} \equiv 0, \qquad \text{mod } \varphi. \tag{6.4}$$

Proof. If f is harmonic, we have, by Theorem 2.1,

$$Da_{\alpha\bar{i}} \equiv 0, \qquad \text{mod } \varphi. \tag{6.5}$$

For $i = \rho$ this gives

$$a_{\alpha\bar{\lambda}}\omega_{\lambda\bar{\rho}} \equiv 0, \qquad \text{mod } \varphi.$$

This implies (6.4), since $\det(a_{\alpha\bar{\lambda}}) \neq 0$.

Conversely, suppose (6.4) holds. We wish to show that (6.5), for $i = \lambda$, follows. In fact, from (6.4) we have

$$\omega_{\rho\bar{\lambda}} \equiv 0, \qquad \text{mod } \bar{\varphi},$$

which gives, by (6.3),

$$Db_{\alpha\bar{\lambda}} \equiv 0, \qquad \text{mod } \bar{\varphi}.$$

Our desired conclusion follows from (2.11).

The situation can also be described as a triad of harmonic maps

(6.6)

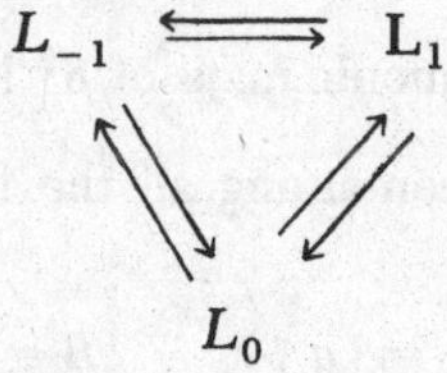

connected by fundamental collineations, with the lines L_0, L_{-1}, L_1 mutually perpendicular. It is important to remark that by our vanishing Theorem 3.2 one of the fundamental collineations must be degenerate. This leads to the construction of a holomorphic curve, as we will show below.

Consider now the case rank $(c_{\alpha\bar{\beta}}) = 1$. We shall define a geometrical construction, to be called *crossing*, which derives from (6.1) a Frenet harmonic sequence. We shall also describe its inverse operation, the *recrossing*.

The lines L_0, L_1, and L_{-1} are spanned by the points $[Z_\alpha]$, $[a_{\alpha\bar{i}}Z_i]$, and $[b_{\alpha\bar{i}}Z_i]$ respectively. Under the adjoint of $\bar{\partial}$, $\bar{\partial}^*$, $[a_{\alpha\bar{i}}Z_i]$ goes to $[c_{\alpha\bar{\beta}}Z_\beta]$, while under the adjoint of ∂, ∂^*, $[b_{\alpha\bar{i}}Z_i]$ goes to $[c_{\bar{\beta}\alpha}Z_\beta]$. Here adjoint is taken with respect to the hermitian inner product on $\mathbf{C}^6$. By hypothesis each pair consists of coincident points. We choose our frames so that the first point is $[Z_1]$. Then

(6.7) $$c_{1\bar{2}} = c_{2\bar{2}} = 0,$$

and by (2.6), we have also $c_{1\bar{1}} = 0$. Thus the second point is $[Z_2]$, and the two points are orthogonal. The kernel of the map $\bar{\partial}^*|_{L_1}$ is $[a_{1\bar{i}}Z_i]$, since $c_{1\bar{1}} = 0$. The kernel of the map $\partial^*|_{L_{-1}}$ is $[b_{2\bar{i}}Z_i]$, since $c_{2\bar{2}} = 0$. These two points are orthogonal since $c_{1\bar{2}} = 0$. We choose them to be Z_3 and Z_5 respectively, so that

(6.8) $$\begin{gathered} a_{1\bar{4}} = a_{1\bar{5}} = a_{1\bar{6}} = 0, \\ b_{2\bar{3}} = b_{2\bar{4}} = b_{2\bar{6}} = 0, \quad a_{1\bar{3}} \neq 0, \quad b_{2\bar{5}} \neq 0. \end{gathered}$$

By the condition on $c_{\alpha\bar{\beta}}$ we find

(6.9) $$a_{2\bar{5}} = b_{1\bar{3}} = 0.$$

We express now, according to Theorem 2.1, the conditions that f is harmonic. By (6.8), and (2.10), we get

(6.10) $$\begin{aligned} -a_{2\bar{s}}\omega_{1\bar{2}} + a_{1\bar{3}}\omega_{3\bar{s}} &\equiv 0, \quad \text{mod } \varphi, \quad s = 4,5,6, \\ -b_{1\bar{t}}\omega_{2\bar{1}} + b_{2\bar{5}}\omega_{5\bar{t}} &\equiv 0, \quad \text{mod } \bar{\varphi}, \quad t = 3,4,6. \end{aligned}$$

Using (6.9) and the fact that $a_{1\bar{3}} \neq 0$, $b_{2\bar{5}} \neq 0$, we get

(6.11) $$\omega_{3\bar{5}} \equiv 0, \quad \text{mod } \varphi.$$

The other equations of (6.10) give

(6.12) $$\begin{aligned} \omega_{3\bar{4}} &\equiv a_{2\bar{4}}/a_{1\bar{3}}\omega_{1\bar{2}}, \quad \omega_{3\bar{6}} \equiv a_{2\bar{6}}/a_{1\bar{3}}\omega_{1\bar{2}}, \quad \text{mod } \varphi, \\ \omega_{5\bar{4}} &\equiv b_{1\bar{4}}/b_{2\bar{5}}\omega_{2\bar{1}}, \quad \omega_{5\bar{6}} \equiv b_{1\bar{6}}/b_{2\bar{5}}\omega_{2\bar{1}}, \quad \text{mod } \bar{\varphi}. \end{aligned}$$

In the same way the harmonicity of f gives, from (6.9).

(6.13) $$\begin{aligned} a_{2\bar{4}}\omega_{4\bar{5}} + a_{2\bar{6}}\omega_{6\bar{5}} &\equiv 0, \quad \text{mod } \varphi, \\ b_{1\bar{4}}\omega_{4\bar{3}} + b_{1\bar{6}}\omega_{6\bar{3}} &\equiv 0, \quad \text{mod } \bar{\varphi}. \end{aligned}$$

From the first equation of (6.12) and the second equation of (6.13) we get

$$(b_{\bar{1}4}a_{2\bar{4}} + b_{\bar{1}6}a_{2\bar{6}})\omega_{1\bar{2}} \equiv 0, \quad \text{mod } \varphi.$$

The coefficient is $c_{2\bar{1}} \neq 0$. Hence

(6.14) $$\omega_{1\bar{2}} \equiv 0, \quad \text{mod } \varphi,$$

from which it follows that

(6.15) $$\begin{aligned} \omega_{3\bar{4}} &\equiv 0, \quad \omega_{3\bar{6}} \equiv 0, \quad \text{mod } \varphi, \\ \omega_{4\bar{5}} &\equiv 0, \quad \omega_{6\bar{5}} \equiv 0, \quad \text{mod } \varphi. \end{aligned}$$

LEMMA 6.2. *The map f considered above is harmonic if and only if the equations* (6.14) *and* (6.15) *are satisfied.*

Proof. It remains to prove the sufficiency of the condition, i.e., to prove

$$Da_{13} \equiv 0, \quad Da_{2\bar{t}} \equiv 0, \quad \text{mod } \varphi, \quad t = 3,4,6.$$

As in the proof of Lemma 6.1 we use the relations

$$db_{1\bar{3}} \equiv 0, \quad db_{2\bar{i}} \equiv 0, \qquad \mathrm{mod}\ \bar{\varphi},$$

and (2.11). Q.E.D.

Summarizing, we have the equations

(6.16)

$$d\begin{pmatrix} Z_1 \\ \cdot \\ \cdot \\ \cdot \\ \cdot \\ Z_6 \end{pmatrix} = \begin{pmatrix} \omega_{1\bar{1}} & a_{1\bar{2}}\varphi & a_{1\bar{3}}\varphi & b_{1\bar{4}}\bar{\varphi} & b_{1\bar{5}}\bar{\varphi} & b_{1\bar{6}}\bar{\varphi} \\ -a_{\bar{1}2}\bar{\varphi} & \omega_{2\bar{2}} & a_{2\bar{3}}\varphi & a_{2\bar{4}}\varphi & b_{2\bar{5}}\bar{\varphi} & a_{2\bar{6}}\varphi \\ -a_{\bar{1}3}\bar{\varphi} & -a_{\bar{2}3}\bar{\varphi} & \omega_{3\bar{3}} & a_{3\bar{4}}\varphi & a_{3\bar{5}}\varphi & a_{3\bar{6}}\varphi \\ -b_{\bar{1}4}\varphi & -a_{\bar{2}4}\bar{\varphi} & -a_{\bar{3}4}\bar{\varphi} & \omega_{4\bar{4}} & a_{4\bar{5}}\varphi & \omega_{4\bar{6}} \\ -b_{\bar{1}5}\varphi & -b_{\bar{2}5}\varphi & -a_{\bar{3}5}\bar{\varphi} & -a_{\bar{4}5}\bar{\varphi} & \omega_{5\bar{5}} & b_{5\bar{6}}\bar{\varphi} \\ -b_{\bar{1}6}\varphi & -a_{\bar{2}6}\bar{\varphi} & -a_{\bar{3}6}\bar{\varphi} & \omega_{6\bar{4}} & -b_{\bar{5}6}\varphi & \omega_{6\bar{6}} \end{pmatrix} \begin{pmatrix} Z_1 \\ \cdot \\ \cdot \\ \cdot \\ \cdot \\ Z_6 \end{pmatrix}.$$

LEMMA 6.3. $a_{3\bar{5}} = 0$ *in the equations* (6.16).

Proof. The vectors Z_1, Z_2, Z_3, Z_5 are each defined up to a factor of absolute value 1, while Z_4 and Z_6 are defined up to a unitary transformation. Hence the form

(6.17)
$$\begin{aligned} &\omega_{1\bar{3}}\omega_{3\bar{5}}\omega_{5\bar{2}}(\omega_{2\bar{4}}\omega_{4\bar{1}} + \omega_{2\bar{6}}\omega_{6\bar{1}}) \\ &= a_{1\bar{3}}a_{3\bar{5}}b_{\bar{2}5}c_{2\bar{1}}\varphi^5 \end{aligned}$$

is independent of the choice of the frame. It is of type (5, 0). By the structure equations we can show that it is holomorphic. Hence, it is zero on S^2. Since $a_{1\bar{3}}b_{\bar{2}5}c_{2\bar{1}} \neq 0$, the lemma follows. Q.E.D.

We put

(6.18)
$$\begin{aligned} &Z_1' = Z_1, \quad Z_2' = Z_5, \quad Z_3' = Z_2, \\ &Z_4' = Z_3, \quad Z_5' = Z_4, \ Z_6' = Z_6. \end{aligned}$$

Then equations (6.16) can be written

(6.19)

$$d\begin{pmatrix} Z_1' \\ \cdot \\ \cdot \\ \cdot \\ \cdot \\ Z_6' \end{pmatrix} = \begin{pmatrix} & & a_{1\bar{2}}\varphi & a_{1\bar{3}}\varphi & b_{1\bar{4}}\bar{\varphi} & b_{1\bar{6}}\bar{\varphi} \\ & & -b_{\bar{2}5}\varphi & 0 & -a_{\bar{4}5}\bar{\varphi} & b_{5\bar{6}}\bar{\varphi} \\ -a_{\bar{1}2}\bar{\varphi} & b_{2\bar{5}}\bar{\varphi} & & & a_{2\bar{4}}\varphi & a_{2\bar{6}}\varphi \\ -a_{\bar{1}3}\bar{\varphi} & 0 & & & a_{3\bar{4}}\varphi & a_{2\bar{6}}\varphi \\ -b_{\bar{1}4}\varphi & a_{4\bar{5}}\varphi & & & & \\ -b_{\bar{1}6}\varphi & -b_{\bar{5}6}\varphi & & & & \end{pmatrix} \begin{pmatrix} Z_1' \\ \cdot \\ \cdot \\ \cdot \\ \cdot \\ Z_6' \end{pmatrix},$$

where the diagonal blocks in the matrix are immaterial and the other elements are inferred from the fact that the matrix is skew-Hermitian. Notice the remarkable fact that the non-diagonal (2×2)-blocks are multiples of φ or $\bar{\varphi}$. By Lemma 6.1 this has the following geometrical meaning: Denote by $\lambda_0, \lambda_1, \lambda_{-1}$ the lines defined by $Z_1' \wedge Z_2'$, $Z_3' \wedge Z_4'$, $Z_5' \wedge Z_6'$ respectively. Then

$$\lambda_{-1} \xrightarrow{\partial} \lambda_0 \xrightarrow{\partial} \lambda_1 \tag{6.20}$$

is a Frenet harmonic sequence. The passage from (6.1) to (6.20) is called *crossing*. It is essentially an "orthogonalization process," but in a more sophisticated situation.

We should describe its inverse process, the *recrossing*, which is to find all harmonic sequences (6.1) with (6.20) as its crossing. Given (6.20), we choose $[Z_1'](x) \in \lambda_0$, $x \in S^2$, and let $[Z_2']$ be its orthogonal point on λ_0. By choosing $[Z_3'] = \partial[Z_2'] \in \lambda_1$, we wish to find the condition that the line L_0 determined by $Z_1' \wedge Z_3'$ gives rise to a harmonic sequence (6.1) with (6.20) as its crossing. Let

$$dZ_A' = \omega_{A\bar{B}}' Z_B'. \tag{6.21}$$

By Lemma 6.1 we have

$$\begin{aligned} \omega_{\alpha\bar{\lambda}}' &= a_{\alpha\bar{\lambda}}'\varphi, \quad \omega_{\alpha\bar{\rho}}' = b_{\alpha\bar{\rho}}'\bar{\varphi}, \\ \omega_{\lambda\bar{\rho}}' &= e_{\lambda\bar{\rho}}'\varphi. \end{aligned} \tag{6.22}$$

Our choice of Z_3 implies

$$\omega_{2\bar{4}}' = 0. \tag{6.23}$$

Its exterior differentiation gives

$$(a_{1\bar{4}}'\omega_{1\bar{2}}' + a_{2\bar{3}}'\omega_{3\bar{4}}') \wedge \varphi = 0. \tag{6.24}$$

We write

$$\omega_{1\bar{2}}' = a_{1\bar{2}}'\varphi + b_{1\bar{2}}'\bar{\varphi}, \quad \omega_{3\bar{4}}' = a_{3\bar{4}}'\varphi + b_{3\bar{4}}'\bar{\varphi}, \tag{6.25}$$

and carry out our crossing construction on the line $Z_1'Z_3'$. In order that $[Z_2'] \in \ker(\partial^*)$, $[Z_4'] \in \ker(\bar{\partial}^*)$, we must have respectively

$$a_{1\bar{2}}' = 0, \quad b_{3\bar{4}}' = 0. \tag{6.26}$$

By (6.24) these conditions are equivalent, since $a_{1\bar{4}}'a_{2\bar{3}}' \neq 0$. Thus the condition can be written

$$(dZ_1', Z_2') \equiv 0, \qquad \text{mod } \bar{\varphi}. \tag{6.27}$$

By Lemma 6.2 the line L_0 is harmonic. Hence *the recrossing is obtained by an anti-holomorphic section* $[Z_1'](x) \in \lambda_0(x)$, $x \in S^2$, *of the* P_1 bundle λ_0.

It remains to describe all Frenet harmonic sequences or harmonic triads. If the map is not of the first kind, i.e., if none of the fundamental collineations is identically zero, we will, by a geometrical construction to be called *turning*, construct a holomorphic curve in P_5, from which the map can be obtained by the inverse processes, the *returnings*.

We use the notation at the beginning of this section, and consider the harmonic triad (6.6), with L_0, L_1, L_{-1} spanned by Z_1, Z_2; Z_3, Z_4; Z_5, Z_6 respectively. By (6.4) we write

(6.28) $$\omega_{\lambda\bar{\rho}} = e_{\lambda\bar{\rho}}\varphi.$$

The fundamental collineation ∂: $L_1 \to L_{-1}$ being of rank one, we choose the frame so that $[Z_5]$ is the image and $[Z_4]$ is the kernel. By Theorem 2.2 it follows that

(6.29) $$[Z_3], [Z_5]: S^2 \to P_5$$

are harmonic. Analytically this means

(6.30) $$(e_{\lambda\bar{\rho}}) = \begin{pmatrix} e & 0 \\ 0 & 0 \end{pmatrix}, e \neq 0,$$

or

(6.30a) $$\omega_{4\bar{5}} = \omega_{3\bar{6}} = 0, \quad \omega_{4\bar{6}} = 0.$$

Exterior differentiation of the first two equations gives

(6.31) $$\omega_{5\bar{6}} = r\varphi, \quad \omega_{4\bar{3}} = s\varphi,$$

while the exterior differentiation of the last equation does not lead to any new relation.

Since the fundamental collineations $\partial, \bar{\partial}$ on L_0 are non-degenerate, we choose the frames, so that $\bar{\partial}[Z_1] = [Z_6]$. Then

(6.32) $$b_{1\bar{5}} = 0,$$

and

(6.33) $$b_{2\bar{5}}b_{1\bar{6}} \neq 0.$$

The frame Z_A is now determined up to the multiplication of each Z_A by a complex number of absolute value 1.

Lemma 6.4. $a_{2\bar{3}} = 0$.

Proof. The cubic form

$$\omega_{2\bar{3}}\omega_{3\bar{5}}\omega_{5\bar{2}} = - a_{2\bar{3}}eb_{\bar{2}5}\varphi^3$$

is independent of the choice of frame and is of type (3,0). By the structure equations it can be shown to be holomorphic and therefore to vanish on S^2. Since $eb_{\bar{2}5} \neq 0$, we have $a_{2\bar{3}} = 0$.

From the last condition we have $a_{1\bar{3}}a_{2\bar{4}} \neq 0$. The harmonicity condition $Da_{2\bar{3}} \equiv 0$, mod φ, then gives

(6.34)
$$\omega_{1\bar{2}} = -\bar{t}\bar{\varphi}.$$

We have therefore the equations

(6.35)
$$d\begin{pmatrix} Z_1 \\ \cdot \\ \cdot \\ \cdot \\ \cdot \\ Z_6 \end{pmatrix} = \begin{pmatrix} \omega_{1\bar{1}} & -\overline{t\varphi} & a_{1\bar{3}}\varphi & a_{2\bar{3}}\varphi & 0 & b_{1\bar{6}}\bar{\varphi} \\ t\varphi & \omega_{2\bar{2}} & 0 & a_{2\bar{4}}\varphi & b_{2\bar{5}}\bar{\varphi} & b_{2\bar{6}}\bar{\varphi} \\ -a_{\bar{1}3}\bar{\varphi} & 0 & \omega_{3\bar{3}} & -\overline{s\varphi} & e\varphi & 0 \\ -a_{\bar{2}3}\bar{\varphi} & -a_{\bar{2}4}\bar{\varphi} & s\varphi & \omega_{4\bar{4}} & 0 & 0 \\ 0 & -b_{\bar{2}5}\varphi & -e\varphi & 0 & \omega_{5\bar{5}} & r\varphi \\ -b_{\bar{1}6}\varphi & -b_{\bar{2}6}\varphi & 0 & 0 & -\overline{r\varphi} & \omega_{6\bar{6}} \end{pmatrix} \begin{pmatrix} Z_1 \\ \cdot \\ \cdot \\ \cdot \\ \cdot \\ Z_6 \end{pmatrix},$$

where

(6.36)
$$ea_{1\bar{3}}a_{2\bar{4}}b_{1\bar{6}}b_{2\bar{5}} \neq 0.$$

Setting

(6.37)
$$Z_4' = \frac{-a_{\bar{1}3}Z_1 - \bar{s}Z_4}{\|-a_{\bar{1}3}Z_1 - \bar{s}Z_4\|}, \qquad Z_1' = \frac{sZ_1 - a_{1\bar{3}}Z_4}{\|sZ_1 - a_{1\bar{3}}Z_4\|},$$
$$Z_6' = \frac{-b_{\bar{2}5}Z_2 + rZ_6}{\|-b_{\bar{2}5}Z_2 + rZ_6\|}, \qquad Z_2' = \frac{\bar{r}Z_2 + b_{2\bar{5}}Z_6}{\|\bar{r}Z_2 + b_{2\bar{5}}Z_6\|},$$

we see that

$$Z_4' \overset{\partial}{\to} Z_3 \overset{\partial}{\to} Z_5 \overset{\partial}{\to} Z_6'$$

is part of a Frenet frame.

We restrict ourselves to the general case that

(6.38)
$$Z_1'' = \bar{\partial}Z_4' \neq 0, \quad Z_2'' = \partial Z_6' \neq 0.$$

Then it follows from the general theory of holomorphic maps of S^2 to P_{n-1} that $[Z_1'']$ is a holomorphic curve and $[Z_2'']$ is its dual antiholomorphic curve. This construction is called *turning*.

The inverse process, *returning*, is to construct the harmonic line $Z_3 \wedge Z_4$ from the holomorphic curve $[Z_1'']$. The latter determines the Frenet harmonic sequence

(6.39)
$$Z_1'' \to Z_4' \to Z_3 \to Z_5 \to Z_6' \to Z_2''.$$

The vectors Z_1' and Z_2' are orthogonal to Z_4', Z_3, Z_5, Z_6', and are hence linear combinations of Z_1'', Z_2''. Moreover, by (6.35) and (6.37), we get

$$(dZ_1', Z_2') \equiv 0, \qquad \text{mod } \bar{\varphi}. \tag{6.40}$$

Hence $[Z_1']$ is an antiholomorphic section of the bundle $Z_1'' \wedge Z_2''$.

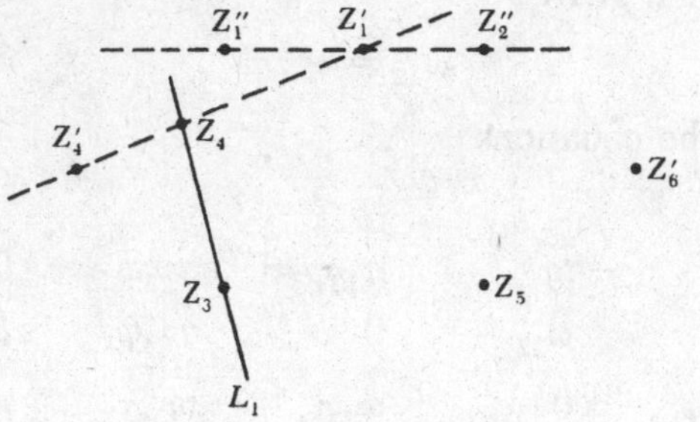

FIGURE 3

The point Z_4 satisfies the condition

$$(dZ_4, Z_1) \equiv 0, \qquad \text{mod } \bar{\varphi},$$

and is thus an antiholomorphic section of the line $Z_4'Z_1'$. See Figure 3.

The construction, by returning of the line $Z_4 \wedge Z_3$, and so the resulting harmonic triad, from the holomorphic curve $[Z_1'']$, involves two holomorphic sections of P_1-bundles.

In conclusion the harmonic maps $S^2 \to G(2, 6)$ are determined by a holomorphic curve in P_5 and three holomorphic sections of P_1-bundles (one for recrossing and two for returnings).

7. Crossing

Consider a harmonic map f: $S^2 \to G(2, n)$. It gives rise to a harmonic sequence

$$L_0 \xrightarrow{\partial} L_1 \xrightarrow{\partial} L_2 \to \cdots, \quad L_0 = [f(x)], \quad x \in S^2. \tag{7.1}$$

If the fundamental collineations do not degenerate, we will associate to f a Frenet harmonic sequence. This process is called *crossing*, which we developed in the last section for $n = 6$ and which we will define for any n. We will also study its inverse process, the *recrossing*, whose definition involves an arbitrary holomorphic section of a P_1-bundle. In this way the problem is reduced to the description of Frenet harmonic sequences.

We proceed by induction. Suppose that

$$L_0 \xrightarrow{\partial} L_1 \xrightarrow{\partial} \cdots \xrightarrow{\partial} L_s \tag{7.1a}$$

is a Frenet harmonic sequence and that the sequence obtained by adding $\partial\colon L_s \to L_{s+1}$ is not. We choose the unitary frame Z_A, $1 \le A \le n$, so that L_ρ is spanned by $Z_{2\rho+1}$, $Z_{2\rho+2}$, $0 \le \rho \le s$. Set

(7.2)
$$E_\rho = \begin{pmatrix} Z_{2\rho+1} \\ Z_{2\rho+2} \end{pmatrix}, \quad 0 \le \rho \le s,$$
$$X = \begin{pmatrix} Z_{2s+3} \\ \vdots \\ Z_n \end{pmatrix}.$$

Since (7.1a) is Frenet, we have

(7.3)
$$\begin{aligned} dE_0 &= \pi_0 E_0 + A_0 \varphi E_1 - {}^t\overline{C}\overline{\varphi} E_s + \overline{P}\overline{\varphi} X, \\ dE_\sigma &= -{}^t\overline{A}_{\sigma-1}\overline{\varphi} E_{\sigma-1} + \pi_\sigma E_\sigma + A_\sigma \varphi E_{\sigma+1}, \qquad 1 \le \sigma \le s-1, \\ dE_s &= C\varphi E_0 - {}^t\overline{A}_{s-1}\overline{\varphi} E_{s-1} + \pi_s E_s + Q\varphi X, \end{aligned}$$

where $A_0, A_1, \ldots, A_{s-1}$ and C are (2×2)-matrices of complex-valued functions, π_ρ are skew-Hermitian (2×2)-matrices of 1-forms, and P and Q are $2 \times (n - 2s - 2)$ matrices of complex-valued functions. A_τ, $0 \le \tau \le s-1$, is the matrix representation of the fundamental collineation

$$\partial\colon L_\tau \to L_{\tau+1}$$

with respect to the bases $\{Z_{2\tau+1}, Z_{2\tau+2}\}$ and $\{Z_{2\tau+3}, Z_{2\tau+4}\}$ of L_τ and $L_{\tau+1}$, respectively. It is by assumption non-singular, except perhaps at isolated points. From (7.3) we see that L_{s+1} is orthogonal to $L_1, \ldots, L_s$, but not to L_0. The latter means that $C \ne 0$. Now, C is the matrix representation of the collineation $\pi \circ \partial\colon L_s \to L_0$, where π is the orthogonal projection of L_{s+1} into L_0. By the vanishing Theorem 3.2, C is degenerate, so that C has rank one.

We choose our frame so that

(7.4)
$$Z_1 = \operatorname{Im}(\pi \circ \partial) \in L_0$$

and

(7.5)
$$Z_{2\rho+1} = \partial^\rho(Z_1) \in L_\rho, \qquad 0 \le \rho \le s \quad \Big(\partial^\rho = \underbrace{\partial \circ \cdots \circ \partial}_{\rho}\Big),$$

with $Z_{2\rho+2}$ orthogonal to $Z_{2\rho+1}$ in L_ρ. Since

(7.6)
$$A_\tau = \begin{pmatrix} a_{2\tau+1,\overline{2\tau+3}} & a_{2\tau+1,\overline{2\tau+4}} \\ a_{2\tau+2,\overline{2\tau+3}} & a_{2\tau+2,\overline{2\tau+4}} \end{pmatrix}$$

our choice of frame implies

$$a_{2\tau+1,\overline{2\tau+4}} = 0. \tag{7.7}$$

From the Vanishing Theorem 3.2 the collineation

$$(\pi \circ \partial) \circ \underbrace{\partial \circ \cdots \circ \partial}_{s} : L_0 \to L_0 \tag{7.8}$$

has trace zero. On applying it to Z_1, we see that the image must be a multiple of Z_1 by (7.4), and hence must be zero by the zero-trace property. This means

$$Z_{2s+1} \in \ker(\pi \circ \partial);$$

i.e., $\partial[Z_{2s+1}]$ is orthogonal to L_0. Since $\partial[Z_{2s+1}] \in L_{s+1}$ and is thus orthogonal to $Z_3, \dots, Z_{2s+1}$, we can choose

$$Z_{2s+3} = \partial Z_{2s+1} \tag{7.9}$$

and complete the Z's to form a unitary frame Z_A, of which each of $Z_1, \dots, Z_{2s+3}$ is defined up to a factor of absolute value 1.

The matrix

$$C = \begin{pmatrix} c_{2s+1,\bar{1}} & c_{2s+1,\bar{2}} \\ c_{2s+2,\bar{1}} & c_{2s+2,\bar{2}} \end{pmatrix} \tag{7.10}$$

represents the collineation

$$\pi \circ \partial : L_s \to L_0$$

relative to the bases Z_{2s+1}, Z_{2s+2} and Z_1, Z_2 of L_s and L_0 respectively. Our choice of the Z's implies

$$c_{2s+1,\bar{1}} = c_{2s+1,\bar{2}} = c_{2s+2,\bar{2}} = 0, \quad c_{2s+2,\bar{1}} \neq 0. \tag{7.11}$$

Applying the structure equations to (7.3) (i.e., taking their exterior derivatives), we obtain

$$\pi_\rho = \begin{pmatrix} \omega_{2\rho+1,\overline{2\rho+1}} & p_{\rho+1}\varphi \\ -\bar{p}_\rho \bar{\varphi} & \omega_{2\rho+2,\overline{2\rho+2}} \end{pmatrix}, \qquad 0 \le \rho \le s, \tag{7.12}$$

where p_ρ is a complex-valued function. Finally, the choice (7.9) implies that the matrix Q has the form

$$Q = \begin{pmatrix} * & 0 \cdots 0 \\ * & * \dots * \end{pmatrix}. \tag{7.13}$$

LEMMA 7.1. *Let $\omega_{A\bar{B}}$ be the coefficient of Z_B in dZ_A. The $(2s+3,0)$-form*

(7.14)
$$\begin{aligned}\Phi &= \omega_{1\bar{3}}\omega_{3\bar{5}} \cdots \omega_{2s-1,\overline{2s+1}}\omega_{2s+1,\overline{2s+3}}\omega_{2s+3,\bar{2}}\omega_{2\bar{4}}\omega_{4\bar{6}} \cdots \cdots \omega_{2s,\overline{2s+2}}\omega_{2s+2,\bar{1}} \\ &= \prod_{0\le\tau\le s-1} \det A_\tau \varphi^{2s}\omega_{2s+1,\overline{2s+3}}\omega_{2s+3,\bar{2}}\omega_{2s+2,\bar{1}}\end{aligned}$$

is globally defined on S^2 and is holomorphic.

Proof. The first statement is obvious and the second follows from the structure equations.

As a corollary we have $\Phi = 0$. In the product (7.14) we have $\det A_\tau \neq 0$. By (7.11), $\omega_{2s+2,\bar{1}} \neq 0$. Since $\partial\colon L_s \to L_{s+1}$ is non-degenerate, we have $\omega_{2s+1,\overline{2s+3}} \neq 0$. It follows that $\omega_{2s+3,\bar{2}} = -\bar{\omega}_{2,\overline{2s+3}} = 0$. Thus we can choose Z_{2s+4} of our frame such that

$$[Z_{2s+4}] = \bar{\partial}[Z_2]. \tag{7.15}$$

By this choice we have

$$P = \begin{pmatrix} * & * & * & \cdots & * \\ 0 & * & 0 & \cdots & 0 \end{pmatrix}. \tag{7.16}$$

This leads us to set

$$\begin{aligned}\lambda_0 &= Z_{2s+4} \wedge Z_1, \\ \lambda_\gamma &= Z_{2\gamma} \wedge Z_{2\gamma+1}, \quad 1 \le \gamma \le s+1,\end{aligned} \tag{7.17}$$

and we will identify these bivectors with the corresponding projective lines.

From the equations dE_0 and dE_1 in (7.3) we see that $\lambda_1 = Z_2 \wedge Z_3$ is a harmonic line. By using ∂' and $\bar{\partial}'$ to denote the transforms pertaining to λ_1, we see that λ_2 and λ_0 are its ∂'- and $\bar{\partial}'$-transforms. Writing out the equations (7.3) and making use of (7.7), (7.11), (7.12), (7.13), (7.16), we get the theorem:

THEOREM 7.2. *The sequence*

$$\lambda_0 \xrightarrow{\partial'} \lambda_1 \xrightarrow{\partial'} \cdots \xrightarrow{\partial'} \lambda_{s+1} \tag{7.18}$$

is a Frenet harmonic sequence.

Moreover, we have

$$Z_{2\gamma} = \partial'^{\gamma}(Z_1), \quad 1 \le \gamma \le s+1. \tag{7.19}$$

$Z_{2\gamma}$ is the intersection of λ_γ and $L_{\gamma-1}$.

The passage from (7.1a) to (7.18) is called *crossing*.

Its inverse process, the *recrossing*, involves an arbitrariness, as it passes from a Frenet harmonic sequence to a shorter one. Given (7.18), the point

$[Z_1] \in \lambda_0$ satisfies

$$(dZ_1, Z_{2s+4}) \equiv 0, \qquad \text{mod } \bar{\varphi}; \tag{7.20}$$

i.e., it defines an anti-holomorphic section of the P_1-bundle λ_0 over S^2. When $[Z_1]$ is chosen, we define $Z_{2\gamma}$ by (7.19) and $[Z_{2\gamma+1}]$ as the orthogonal point of $Z_{2\gamma}$ on λ_γ. Then L_ρ is spanned by $Z_{2\rho+1}$ and $Z_{2\rho+2}$, $0 \leq \rho \leq s$. The same argument as above shows that the sequence (7.1a) is harmonic.

8. Turning

Consider a degenerate harmonic map $f\colon S^2 \to G(2, n)$, that is, a harmonic map one of whose fundamental collineations is degenerate. Without loss of generality we can suppose that the ∂-collineation of f is degenerate and we can restrict our attention to the case where rank ∂ is one. Now f gives rise to the harmonic sequence

$$L_{-1} \overset{\bar{\partial}}{\leftarrow} L_0 \overset{\partial}{\rightarrow} p_0 \tag{8.1}$$

where $[f] = L_0$ is a line and p_0 is a projective point. We will associate to (8.1) a degenerate Frenet harmonic sequence, namely a sequence of the form

$$\bar{\partial}(\lambda_0) \overset{\bar{\partial}}{\leftarrow} \lambda_0 \overset{\partial}{\rightarrow} p_s \overset{\partial}{\rightarrow} \cdots \overset{\partial}{\rightarrow} p_0 \tag{8.2}$$

where λ_0 is a (projective) line, $\bar{\partial}(\lambda_0)$ is either a (projective) line or point, p_σ, $0 \leq \sigma \leq s$, is a (projective) point and the elements of the sequence are mutually orthogonal. The elements $\{p_s, \ldots, p_0\}$ of the sequence (8.2) are a part of the Frenet frame of some holomorphic curve $h\colon S^2 \to P_{n-1}$. We will show, moreover, that the sequence (8.2) can be taken so that the line λ_0 satisfies

$$\lambda_0 = p_{s+2} \wedge p_{s+1} \tag{8.3}$$

where either: (a) p_{s+2} and p_{s+1} are the two elements of the Frenet frame of h occurring adjacent to p_s, or (b) p_{s+1} is a holomorphic curve with Frenet frame $\{p_{s+1}, p_s, \ldots, p_0, \ldots\}$ and p_{s+2} is an antiholomorphic curve. The process of associating (8.2) and (8.3) to the degenerate harmonic map f is called *turning*. We will also study its inverse process, called *returning*. In this way the study of the degenerate harmonic maps is reduced to the well-known description of the Frenet frames of holomorphic curves in P_{n-1}.

We proceed by induction. Suppose that we begin with the degenerate Frenet harmonic sequence (8.2). We adapt a unitary frame Z_A, $1 \leq A \leq n$, to (8.2) by setting

$$\begin{aligned} Z_2 &= \ker \partial \in \lambda_0, \\ Z_{(s-\sigma)+4} &= p_\sigma, \qquad 0 \leq \sigma \leq s, \end{aligned} \tag{8.4}$$

and letting Z_3 be the point on λ_0 orthogonal to Z_2. Set

$$Z_1 = \bar{\partial}(Z_3) \tag{8.5}$$

where $\bar{\partial}$: $\lambda_0 \to \bar{\partial}(\lambda_0)$.

(8.5) depends on $Z_3 \notin \ker \bar{\partial}$. Suppose $Z_3 \in \ker \bar{\partial}$. By Theorem 2.2(b), $[Z_2]$ describes a harmonic map $S^2 \to P_{n-1}$. As $Z_2 = \ker \partial$, Theorem 2.2(b) implies that $[Z_3]$ also describes a harmonic map $S^2 \to P_{n-1}$. In fact either $[Z_2]$ and $[Z_3]$ are adjacent elements of the Frenet frame of a holomorphic curve $S^2 \to P_{n-1}$ (the next element of this frame is $[Z_4] = p_s$) or $[Z_3]$ is a holomorphic curve with Frenet frame $\{[Z_3], [Z_4], \ldots, [Z_{s+4}], \ldots\}$ and $[Z_2]$ is an antiholomorphic curve. We will discuss this phenomenon in detail, below. Thus if $Z_3 \in \ker \bar{\partial}$ then λ_0 satisfies (8.3) and we are done.

Complete $\{Z_1, \ldots, Z_{s+4}\}$ to a unitary frame $\{Z_1, \ldots, Z_n\}$ of $\mathbf{C}^n$. Since (8.2) is a Frenet harmonic sequence we have

$$\begin{aligned}
dZ_1 &= \omega_{1\bar{1}} Z_1 + a_1 \varphi Z_2 + a_2 \varphi Z_3 + \bar{b} \bar{\varphi} Z_{s+4} + \sum_{\tau = s+5}^{n} \bar{b}_\tau \bar{\varphi} Z_\tau \\
dZ_2 &= -\bar{a}_1 \bar{\varphi} Z_1 + \omega_{2\bar{2}} Z_2 + a_3 \varphi Z_3 + \sum_{\tau = s+5}^{n} \bar{c}_\tau \bar{\varphi} Z_\tau \\
dZ_3 &= -\bar{a}_2 \bar{\varphi} Z_1 - \bar{a}_3 \bar{\varphi} Z_2 + \omega_{3\bar{3}} Z_3 + a_4 \varphi Z_4 \\
&\ \vdots \\
dZ_\sigma &= -\bar{a}_\sigma \bar{\varphi} Z_{\sigma-1} + \omega_{\sigma\bar{\sigma}} Z_\sigma + a_{\sigma+1} \varphi Z_{\sigma+1} \qquad 4 \le \sigma \le s+3 \\
&\ \vdots \\
dZ_{s+4} &= -b\varphi Z_1 - \bar{a}_{s+4} \bar{\varphi} Z_{s+3} + \omega_{s+4, \overline{s+4}} Z_{s+4} + \sum_{\tau = s+5}^{n} a_\tau \varphi Z_\tau
\end{aligned} \tag{8.6}$$

where the a_A, b_τ, c_τ and b are complex valued functions. In particular the Maurer-Cartan 1-form $\omega_{s+4,\bar{1}} = -b\varphi$ is of type (1, 0).

Consider the $(s+3, 0)$ form

$$\Delta_s = \omega_{s+4,\bar{1}} \omega_{1\bar{3}} \omega_{3\bar{4}} \cdots \omega_{\sigma, \overline{\sigma+1}} \cdots \omega_{s+3, \overline{s+4}}.$$

Δ_s is obviously invariantly defined.

THEOREM 8.1. *Δ_s is a holomorphic $(s+3, 0)$-form on S^2.*

Proof. This follows from the structure equations.

COROLLARY 8.2. $\Delta_s = 0$.

Because $Z_3 \notin \ker \bar{\partial}$, the 1-form $\omega_{3\bar{1}} = -\bar{a}_2 \bar{\varphi}$ does not vanish. The 1-form $\omega_{3\bar{4}} = a_4 \varphi$ is non-zero because ∂: $\lambda_0 \to p_s$ has rank one. The 1-forms $\omega_{\sigma, \overline{\sigma+1}} = a_{\sigma+1}\varphi$, $4 \le \sigma \le s+3$, are non-zero because the fundamental collineations ∂: $p_\sigma \to p_{\sigma-1}$ are nonsingular. Thus the corollary implies that

$$\omega_{s+4,\bar{1}} = -b\varphi = 0. \tag{8.7}$$

Define

$$p_{s+1} = Z_3 = (\ker \partial)^{\perp}$$

THEOREM 8.3. *With the above notation, the line μ_0 describes a harmonic map $S^2 \to G(2, n)$, the point p_{s+1} describes a harmonic map $S^2 \to P_{n-1}$ and the sequence*

$$\bar{\partial}(\mu_0) \overset{\partial}{\leftarrow} \mu_0 \overset{\partial}{\rightarrow} p_{s+1} \overset{\partial}{\rightarrow} p_s \overset{\partial}{\rightarrow} \cdots \overset{\partial}{\rightarrow} p_0 \tag{8.8}$$

is a degenerate Frenet harmonic sequence.

Proof. This follows from (8.7) and the structure equations applied to (8.6).

The construction of the sequence (8.8) from a sequence (8.2) is called turning.

Consider again the case $Z_3 \in \ker \bar{\partial}$. Analytically this means that the 1-form $\omega_{3\bar{1}} = -\bar{a}_2\bar{\varphi}$ in (8.6) vanishes. Consider the function a_3 in (8.6). Because $\omega_{3\bar{1}} = 0$ it follows from the Maurer-Cartan equations and a simple lemma of Bers that either (a) a_3 has only isolated zeroes or (b) a_3 vanishes identically. In case (a) it is clear from (8.6) that $[Z_2]$ and $[Z_3]$ are adjacent elements of a Frenet frame. In case (b) it is clear that $[Z_3]$ is a holomorphic curve with Frenet frame $\{[Z_3], [Z_4], \ldots, [Z_{s+4}], \ldots\}$ and that $[Z_2]$ is an antiholomorphic curve. Moreover the osculating space of $[Z_2]$ is orthogonal to the $(s+1)$-th osculating space of $[Z_3]$. There is no global reason for a_3 to vanish so that case (a) is the "generic" case. In fact, formally, cases (a) and (b) are the same. For this reason we will call a pair Z_2 and Z_3 satisfying (a) or (b) *formally adjacent elements of a Frenet frame*. (We are indebted to J. Wood for pointing out case (b) to us.)

Let h: $M^2 \to P_{n-1}$ be a holomorphic curve and let $\{X_1, \ldots, X_n\}$ be its Frenet frame. It is a result of Din and Zakrzewski [5] that the span of any k distinct elements of this Frenet frame defines a conformal harmonic map $M^2 \to G(k, n)$. This and the above discussion motivate the following definition:

Definition. A harmonic map $M^2 \to G(2, n)$ is called a *classical harmonic map of the second kind* if it is the span of two formally adjacent elements of a Frenet frame.

Returning to the sequence (8.8), we denote by p_{s+2} the point of μ_0 orthogonal to the kernel of ∂: $\mu_0 \to p_{s+1}$. We can apply the operation of turning to (8.8) if and only if $p_{s+2} \notin \ker \bar{\partial}$. However, as we observed above, if $p_{s+1} \in \ker \bar{\partial}$ then the line μ_0 is spanned by two formally adjacent elements of a Frenet frame. That is, if $p_{s+2} \in \ker \bar{\partial}$ then μ_0 is a classical harmonic map of the second kind. By iterating turning we must eventually construct a classical harmonic map of the second kind since after sufficiently many iterations the ambient $\mathbf{C}^n$ will be exhausted.

THEOREM 8.4. *The operation of turning associates to every degenerate harmonic map $S^2 \to G(2, n)$ (not holomorphic or antiholomorphic) a classical harmonic map of the second kind.*

Remark. The classical harmonic map of the second kind in Theorem 8.4 can also be of the first kind; i.e., it can also be holomorphic. This occurs when in (8.8) the fundamental collineation $\bar{\partial}: \mu_0 \to \bar{\partial}(\mu_0)$ is zero.

Proof of Theorem 8.4. We must establish the first step of the induction. Let $f: S^2 \to G(2, n)$ be a degenerate harmonic map. Without loss of generality we can assume that the ∂-fundamental collineation has rank one. Let $L_0 = [f]$ and set $p_0 = \partial(L_0)$. The point p_0 is not necessarily orthogonal to $\bar{\partial}(L_0)$ which is either a point or a line. Set

$$p_2 = \ker \partial$$

and let p_1 be the point on L_0 orthogonal to p_2. By Theorem 2.2(b), p_1 describes a harmonic map $S^2 \to P_{n-1}$. Put

$$p_3 = \bar{\partial}(p_1).$$

As above if $p_1 \in \ker \bar{\partial}$ then by Theorem 2.2(b) L_0 is a classical harmonic map of the second kind and the theorem is proved. Thus we assume $p_1 \notin \ker \bar{\partial}$. Set

$$\lambda_0 = p_3 \wedge p_2.$$

It follows easily from the structure equations of f that λ_0 describes a degenerate harmonic map satisfying

$$\bar{\partial}(\lambda_0) \overset{\bar{\partial}}{\leftarrow} \lambda_0 \overset{\partial}{\rightarrow} p_1$$

where p_1 is orthogonal to $\bar{\partial}(\lambda_0)$. (Note that the point p_0 is *not* orthogonal to $\bar{\partial}(\lambda_0)$.) This construction does not use the vanishing of a holomorphic differential. The theorem now follows by induction using turning.

The inverse process of turning, called *returning*, involves an aribitrariness as it passes from a degenerate Frenet sequence to a shorter degenerate Frenet sequence. Let

$$\bar{\partial}(\nu_0) \overset{\bar{\partial}}{\leftarrow} \nu_0 \overset{\partial}{\rightarrow} p_{s+1} \overset{\partial}{\rightarrow} p_s \cdots \overset{\partial}{\rightarrow} p_0 \tag{8.9}$$

be a degenerate Frenet sequence. Consider the harmonic map ν_0 as a holomorphic P_1-bundle over S^2 and choose an antiholomorphic section τ of this bundle. Set

$$\mu_0(\tau) = \tau \wedge p_{s+1}$$

THEOREM 8.5. *The map $S^2 \to G(2, n)$ defined by $\mu_0(\tau)$ is harmonic. Moreover $\mu_0(\tau)$ belongs to the degenerate Frenet harmonic sequence*

$$\bar{\partial}(\mu_0(\tau)) \overset{\bar{\partial}}{\leftarrow} \mu_0(\tau) \overset{\partial}{\rightarrow} p_s \overset{\partial}{\rightarrow} \cdots \overset{\partial}{\rightarrow} p_0. \tag{8.10}$$

The proof is left to the reader.

The construction of (8.10) from (8.9) is called returning. It is obviously the inverse operation of turning.

9. Proof of the main theorem

After the above discussions, a proof follows immediately for the main theorem stated in the introduction. Given a harmonic map $S^2 \to G(2, n)$, there are two cases: 1) both ∂- and $\bar{\partial}$-transforms are nondegenerate; 2) at least one of them degenerates. Consider the first case. If the resulting harmonic sequence is not Frenet, we use crossing to construct a Frenet harmonic sequence. Continuing this construction as in Section 7, we arrive at a long Frenet harmonic sequence. By Theorem 3.2, one of the fundamental collineations at the end degenerates. The original harmonic map can be obtained from it through ∂- and $\bar{\partial}$-transforms and a number of recrossings. Each recrossing involves the choice of a holomorphic section of a P_1-bundle.

There remains the case of a Frenet harmonic sequence with a degenerate fundamental collineation at one end. If the collineation is of rank zero, the corresponding harmonic map is holomorphic, and we get a classical harmonic map of the first kind. If the collineation is of rank one, the turning construction leads to a holomorphic curve in P_{n-1} and the original harmonic map can be obtained from a classical harmonic map of the second kind through returnings. This completes the proof of the main theorem.

Thus the space of moduli of all harmonic maps $S^2 \to G(2, n)$ involves a holomorphic curve in $G(2, n)$ or one (or possibly two) holomorphic curve(s) in $G(1, n) = P_{n-1}$ and a number of holomorphic sections of P_1-bundles, which arise from the recrossings or returnings. Clearly the maximum number of such recrossings is $\leq [n/2] - 2$, where $[n/2]$ is the largest integer $\leq n/2$. When turnings are necessary, the maximum number is, by Section 8, $\leq n - 3$.

Added August 1986: In the year following the submission of this paper there has been much progress in the study of harmonic maps of surfaces into Grassmann manifolds. The second author, using the idea of the harmonic sequence, has given a description of the harmonic maps $f\colon M \to G(k, n)$, when M has genus zero or when M has genus one and $\deg f \neq 0$ (cf. J. Wolfson, Harmonic sequences and harmonic maps of surfaces into complex Grassmann manifolds). Also Burnstall and Salamon (cf. Tournaments, flags and harmonic maps) have given a description of the harmonic maps $S^2 \to G(k, n)$. Together with Uhlenbeck's work referred to in the introduction, this means there are three different descriptions of the harmonic maps $S^2 \to G(k, n)$ available at this writing.

MATHEMATICAL SCIENCES RESEARCH INSTITUTE, BERKELEY, CALIFORNIA

TULANE UNIVERSITY, NEW ORLEANS, LOUISIANA

BIBLIOGRAPHY

[1] A. R. AITHAL, Harmonic maps from S^2 to $G_2(\mathbf{C}^5)$, preprint.

[2] F. BURSTALL and J. C. WOOD, On the construction of harmonic maps from surfaces to complex Grassmanns, preliminary manuscript, January 1985.

[3] S. S. CHERN and S. I. GOLDBERG, On the volume-decreasing property of a class of real harmonic mappings, Amer. J. Math. **97** (1975), 133–147.

[4] S. S. CHERN and J. WOLFSON, Minimal surfaces by moving frames, Amer. J. Math. **105** (1983), 59–83.

[5] A. M. DIN and W. J. ZAKRZEWSKI, General classical solutions in the CP^{n-1} model, Nucl. Phys. B **174** (1980), 397–406.

[6] A. M. DIN and W. J. ZAKRZEWSKI, Classical solutions in Grassmannian σ-model, Letters in Math. Physics, **5**(6) (1981), 553–561.

[7] J. EELLS and J. C. WOOD, Harmonic maps from surfaces to complex projective spaces, Adv. in Math. **49** (1983), 217–263.

[8] J. RAMANATHAN, Harmonic maps from S^2 to $G(2,4)$, J. Diff. Geom. **19** (1984), 207–219.

[9] K. UHLENBECK, Harmonic maps into Lie groups (classical solutions of the chiral model), preprint.

[10] J. G. WOLFSON, On minimal two-spheres in Kähler manifolds of constant holomorphic sectional curvature, Trans. A. M. S. **290** (1985), 627–646.

(Received May, 28, 1985)

Math. Ann. 278, 381–399 (1987)

Mathematische Annalen

Tautness and Lie Sphere Geometry

Thomas E. Cecil[1,*] and Shiing-Shen Chern[2,**]

[1] Department of Mathematics, College of the Holy Cross, Worcester, MA 01610, USA
[2] Department of Mathematics, University of California, Berkeley, CA 94720, USA
and Mathematical Sciences Research Institute, 1000 Centennial Drive,
Berkeley, CA 94720, USA

Dedicated to Friedrich Hirzebruch on the occasion of his sixtieth birthday

An immersion f of a smooth compact connected manifold M into Euclidean space E^n is said to be *taut* if every Morse function of the form $L_p(x)=|p-f(x)|^2$, $p\in\mathbb{R}^n$, has the minimum number of critical points required by the Morse inequalities on M. Carter and West [3] showed that tautness is invariant under Moebius (conformal) transformations of $\mathbb{R}^n$ and under stereographic projection from $\mathbb{R}^n$ into the unit sphere S^n in $\mathbb{R}^{n+1}$. They showed further that a taut immersion must, in fact, be an embedding. Pinkall [13] showed that for the theory of taut embeddings, it is sufficient to study hypersurfaces in the following sense. Let $M\subset\mathbb{R}^n$ (or S^n) be a compact submanifold of codimension greater than one, and let M_ε be a tube of sufficiently small radius ε about M so that M_ε is an embedded hypersurface. Then M is taut if and only if M_ε is taut. Cecil and Ryan [6] showed that any isoparametric hypersurface in S^n is taut, and recently, Hsiang, Palais and Terng [8] have proven that any isoparametric submanifold of any codimension is taut.

The main result of this paper is that within the class of immersions, tautness is invariant under the group of Lie sphere transformations. This group is considerably larger than the group of Moebius transformations, as we will now explain.

In his work on contact transformations, Lie [9] developed his geometry of oriented spheres (see also Blaschke [2]). Lie established a bijective correspondence between the set of all oriented hyperspheres and point spheres in S^n and the points on the quadric hypersurface Q^{n+1} in real projective space P^{n+2} given by the equation $\langle x,x\rangle=0$, where $\langle\,,\rangle$ is an indefinite metric with signature $(n+1,2)$ on $\mathbb{R}^{n+3}$. Q^{n+1} contains projective lines but no linear subspaces of P^{n+2} of higher dimension. The 1-parameter family of oriented spheres, called a *parabolic pencil*, corresponding to the points of a projective line lying on Q^{n+1} consists of all oriented hyperspheres which are in oriented contact at a certain contact element on S^n. In this way, Lie established a diffeomorphism between the manifold of contact elements on S^n and the space Λ^{2n-1} of projective lines which lie on Q^{n+1}. A

* This work was done while the author was on sabbatical at the University of California, Berkeley in 1985–86

** Work done under support of NSF Grant No. DMS 84-03201

Lie sphere transformation is a projective transformation of P^{n+2} which takes Q^{n+1} to itself. Since a projective transformation takes lines to lines, a Lie transformation induces a contact transformation, i.e., a diffeomorphism of Λ^{2n-1}. The group G of Lie transformations is isomorphic to $O(n+1,2)/\{\pm I\}$, where $O(n+1,2)$ is the group of orthogonal transformations for the inner product $\langle , \rangle$ mentioned above. Moebius transformations are those Lie transformations which take point spheres to point spheres. The group of Moebius transformations is isomorphic to $O(n+1,1)/\{\pm I\}$.

The manifold Λ^{2n-1} has a contact structure, i.e., a 1-form ω such that $\omega \wedge (d\omega)^{n-1}$ does not vanish on Λ^{2n-1}. The condition $\omega = 0$ defines a $(2n-2)$-dimensional distribution D on Λ^{2n-1} which has integral submanifolds of dimension $n-1$ but none of higher dimension (see, for example, Arnold [1, p. 349ff.]). An immersion $\lambda: M^{n-1} \to \Lambda^{2n-1}$ of an $(n-1)$-manifold is called a *Legendre submanifold* if $\lambda^*\omega = 0$ on M^{n-1}.

An immersion f of a manifold M^{n-1} into S^n naturally induces a Legendre immersion $\lambda: M^{n-1} \to \Lambda^{2n-1}$. More generally, an immersion $f: M^k \to S^n$ of codimension greater than one induces a Legendre immersion $\lambda: B^{n-1} \to \Lambda^{2n-1}$, where B^{n-1} is the unit normal bundle to $f(M^k)$ in S^n.

If $\lambda: M^{n-1} \to \Lambda^{2n-1}$ is an arbitrary Legendre submanifold, then for each $x \in M^{n-1}$, the parabolic pencil of spheres determined by the line $\lambda(x)$ contains exactly one point sphere, which we denote by $Y_1(x)$. The map Y_1 from M^{n-1} into the Moebius space of point spheres is smooth but is not, in general, an immersion. It is called the *Moebius projection* of M^{n-1}. A Lie transformation α induces a new Legendre submanifold $\bar{\lambda}: M^{n-1} \to \Lambda^{2n-1}$ defined by $\bar{\lambda}(x) = \alpha(\lambda(x))$. We call this new submanifold αM.

We can now state our main result precisely. Let λ be a Legendre immersion of a compact, connected manifold M^{n-1} whose Moebius projection Y_1 is a taut immersion. Suppose that α is a Lie transformation such that the Moebius projection Z_1 of αM is an immersion. Then Z_1 is taut.

In the concluding section Sect. 3.D, we discuss the close relationship between taut and Dupin submanifolds and some remaining problems in these areas.

1. Lie Geometry of Oriented Spheres

A. Moebius Geometry of Unoriented Spheres

We will briefly recall the Moebius (conformal) geometry of spheres in $\mathbb{R}^n (n \geqq 2)$. To do this, one first introduces stereographic projection from $\mathbb{R}^n$ to the unit sphere S^n in $\mathbb{R}^{n+1}$. We denote the Euclidean inner product of two vectors u and v in $\mathbb{R}^n$ by $u \cdot v$. Recall the formula for stereographic projection $\sigma: \mathbb{R}^n \to S^n \subset \mathbb{R} \times \mathbb{R}^n$ given by

$$\sigma(u) = \left[\frac{1 - u \cdot u}{1 + u \cdot u}, \frac{2u}{1 + u \cdot u}\right].$$

The image of σ is $S^n - \{(-1, 0, \ldots)\}$. We follow σ with the canonical embedding of $\mathbb{R}^{n+1}$ into P^{n+1} as the complete of the hyperplane $(z_1 = 0)$ at infinity and obtain the map $\phi\sigma: \mathbb{R}^n \to P^{n+1}$,

$$\phi\sigma(u) = \left[\left(1, \frac{1 - u \cdot u}{1 + u \cdot u}, \frac{2u}{1 + u \cdot u}\right)\right], \tag{1.1}$$

where $[z]$ denotes the projective equivalence class of the vector $z \in \mathbb{R}^{n+2}$. Note that if $y \in S^n$ and $[z] = \phi(y) = [(1, y)]$, then $z = (z_1, \ldots, z_{n+2})$ satisfies the equation

$$b(z, z) = -z_1^2 + z_2^2 + \ldots + z_{n+2}^2 = 0. \tag{1.2}$$

The subset Σ^n in P^{n+1} determined by (1.2) is diffeomorphic to S^n and is called *Moebius space* or the *Moebius sphere.*

Let R_1^{n+2} denote the vector space with inner product b determined by the quadratic form in (1.2). A nonzero vector x in $\mathbb{R}_1^{n+2}$, is called *spacelike, timelike, lightlike*, respectively, depending on whether $b(x,x)$ is positive, negative or zero. The polar hyperplane of a spacelike point $[\xi]$ in P^{n+1} intersects Σ^n in an $(n-1)$-sphere S^{n-1}, which is the image under $\phi\sigma$ of an $(n-1)$-sphere in $\mathbb{R}^n$, unless it contains $[(+1, -1, 0, \ldots)]$, in which case it is the image under $\phi\sigma$ of a hyperplane in $\mathbb{R}^n$. Specifically, the sphere in $\mathbb{R}^n$ with center p and radius $r > 0$ is given by the equation

$$(u-p) \cdot (u-p) = r^2. \tag{1.3}$$

A direct computation shows that a point $u \in \mathbb{R}^n$ satisfies (1.3) if and only if $\phi\sigma(u)$ lies on the polar hyperplane of the point

$$[\xi] = \left[\left(\frac{1 + p \cdot p - r^2}{2}, \frac{1 - p \cdot p + r^2}{2}, p\right)\right]. \tag{1.4}$$

Note that the sum of the first two coordinates on the right is not zero. The hyperplane given by $u \cdot N = h$, for $[N] = 1$, corresponds to the polar hyperplane in P^{n+1} of the spacelike point

$$[\eta] = [(h, -h, N)]. \tag{1.5}$$

Thus, the spacelike points representing hyperplanes are precisely those points $[z]$ satisfying $z_1 + z_2 = 0$.

Conformality is naturally built into the structure. If S_1 and S_2 are two spheres in $\mathbb{R}^n$ represented by $[\xi_1]$ and $[\xi_2]$ as in (1.4), then S_1 and S_2 intersect orthogonally if and only if $|p_1 - p_2|^2 = r_1^2 + r_2^2$. This Euclidean condition is equivalent to the Moebius condition

$$b(\xi_1, \xi_2) = 0. \tag{1.6}$$

The same holds if either of S_1 and S_2 are planes represented by a point $[\eta]$ as in (1.5). Thus, one defines a *Moebius transformation* to be a projective transformation of P^{n+1} which preserves the polarity (1.2). For $A \in GL(n+2)$, let $\tilde{A}$ denote the projective transformation of P^{n+1} defined by $\tilde{A}[x] = [Ax]$. It is well known that if α is a projective transformation which preserves (1.2), then $\alpha = \tilde{A}$ for some $A \in O(n+1, 1)$, the orthogonal group on $\mathbb{R}_1^{n+2}$. The map $A \to \tilde{A}$ for $A \in O(n+1, 1)$ is thus a homomorphism of $O(n+1, 1)$ onto the group of Moebius transformations with kernel $\{\pm I\}$. Thus, the Moebius group is isomorphic to $O(n+1, 1)/\{\pm I\}$.

B. Lie Geometry of Oriented Spheres

The manifold K^{n+1} of unit spacelike vectors in $\mathbb{R}_1^{n+2}$, is a hyperboloid of one sheet. If P is a spacelike point in P^{n+1}, there are precisely two vectors $\pm\zeta$ in K^{n+1} which project to P. These can be taken to correspond to the two orientations of the sphere or plane represented by P through the expressions (1.4) or (1.5). To make this correspondence precise, one introduces one more homogeneous coordinate. Embed $\mathbb{R}_1^{n+2}$ into P^{n+2} by

$$(z_1, \ldots, z_{n+2}) \to [(z_1, \ldots, z_{n+2}, 1)].$$

If $b(\zeta, \zeta) = 1$, then the point $[(\zeta, 1)]$ in P^{n+2} lies on the quadric Q^{n+1} in P^{n+2} given in homogeneous coordinates by

$$\langle x, x\rangle = -x_1^2 + x_2^2 + \ldots + x_{n+2}^2 - x_{n+3}^2 = 0. \tag{1.7}$$

Q^{n+1} is called the *Lie quadric*. We now briefly outline the bijective correspondence between the points of Q^{n+1} and the set of all oriented hyperspheres, oriented hyperplanes and point spheres in $\mathbb{R}^n \cup \{\infty\}$.

Suppose that $[x] \in Q^{n+1}$ and $x_{n+3} \neq 0$. Then $[x]$ can be represented by a vector of the form $(\zeta, 1)$, where $b(\zeta, \zeta) = 1$. Suppose first that $\zeta_1 + \zeta_2 \neq 0$, so that $[\zeta]$ represents a sphere in Moebius geometry. If $[\zeta]$ is represented by a vector ξ as in (1.4), then $b(\xi, \xi) = r^2$. So ζ must be one of the vectors $\pm\xi/r$. In P^{n+2},

$$[(\zeta, 1)] = [(\pm\xi/r, 1)] = [(\xi, \pm r)].$$

One interprets the last coordinate as a *signed radius* of the unoriented sphere with center p and radius $r > 0$. Geometrically, one can adopt the convention that a positive radius corresponds to the orientation determined by the outward normal to the sphere, and negative radius corresponds to the orientation determined by the inward normal.

If $\zeta_1 + \zeta_2 = 0$, then $[\zeta]$ represents a hyperplane in $\mathbb{R}^n$ via (1.5) in Moebius geometry. For $\zeta = (h, -h, N)$ with $|N| = 1$, we have $b(\zeta, \zeta) = 1$. The two projective points $[(h, -h, N, \pm 1)]$ represent the two orientations of the hyperplane in $\mathbb{R}^n$ given by the equation $u \cdot N = h$. The usual convention is that $[(h, -h, N, 1)]$ represents the orientation determined by the unit normal N.

Finally, if $x_{n+3} = 0$, then $[x] = [(z, 0)]$ and $0 = \langle x, x\rangle = b(z, z)$. Then $[z]$ in P^{n+1} lies on the Moebius sphere Σ^n, and $[x]$ is a point sphere. Point spheres are not assigned an orientation.

Thus, we have established the following correspondence between the points of Q^{n+1} and the set of oriented spheres, oriented planes and points in $\mathbb{R}^n \cup \{\infty\}$.

Euclidean	*Lie*
Points: $u \in \mathbb{R}^n$	$\left[\left(\frac{1+u\cdot u}{2}, \frac{1-u\cdot u}{2}, u, 0\right)\right]$
∞	$[(1, -1, 0, 0)]$ (1.8)
Spheres: Center p, signed radius r	$\left[\left(\frac{1+p\cdot p-r^2}{2}, \frac{1-p\cdot p+r^2}{2}, p, r\right)\right]$
Planes: $u \cdot N = h$, unit normal N	$[(h, -h, N, 1)]$.

At times, it is useful to have formulas to convert the Lie coordinates back into Cartesian equations in $\mathbb{R}^n$. Suppose that $[x] \in Q^{n+1}$, $x = (x_1, \ldots, x_{n+3})$ and $x_1 + x_2 \neq 0$. If $x_{n+3} = 0$, then $[x]$ represents the point $u = (u_3, \ldots, u_{n+2})$ with $u_i = x_i/(x_1 + x_2)$, $3 \leq i \leq n+2$. If $x_{n+3} \neq 0$, then $[x]$ represents the sphere with center

$p=(p_3,\ldots,p_{n+2})$ and signed radius r given by $p_i=x_i/(x_1+x_2)$ and $r=x_{n+3}/(x_1+x_2)$. These formulas are clear after one notices that in the standard forms (1.8) above, the sum of the first two coordinates for points and spheres is one. If $x_1+x_2=0$, then $[x]$ represents a hyperplane or the improper point. If $x_{n+3}\neq 0$, then $[x]$ corresponds to the oriented hyperplane $u\cdot N=h$ with unit normal $N=(N_3,\ldots,N_{n+2})$ and height h given by $N_i=x_i/x_{n+3}$ and $h=x_1/x_{n+3}$. Finally, if $x_1+x_2=0$ and $x_{n+3}=0$, then the condition $\langle x,x\rangle=0$ forces $x_i=0$ for $3\leqq i\leqq n+2$, and $[x]=[(1,-1,0,\ldots)]$ corresponds to the improper point.

Two oriented spheres S_1 and S_2 in $\mathbb{R}^n$ are in *oriented contact* if they are tangent and their orientations agree at the point of tangency. If p_1 and p_2 are the respective centers of S_1 and S_2, r_1 and r_2 the respective signed radii, then the condition of oriented contact can be expressed analytically by

$$|p_1-p_2|=|r_1-r_2|. \tag{1.9}$$

If S_1 and S_2 are represented in the standard form (1.8) by $[k_1]$ and $[k_2]$, then (1.9) is equivalent to the condition

$$\langle k_1,k_2\rangle=0. \tag{1.10}$$

In the case where S_1 and/or S_2 is a plane or a point in $\mathbb{R}^n$, oriented contact has the logical meaning. That is, a sphere S and a plane π are in oriented contact if π is tangent to S and their orientations agree at the point of contact. Two oriented planes are in oriented contact if their unit normals are the same. They are in oriented contact at the improper point. A point sphere is in oriented contact with a sphere or plane S if it lies on S, and thus, the improper point is in oriented contact with every plane. In each case, the analytic condition for oriented contact is equivalent to (1.10).

C. Geometry of Oriented Spheres in S^n and H^n

In our treatment of tautness, we will also need to consider oriented hyperspheres in the sphere S^n and in hyperbolic space H^n. First we find the Moebius equation for the unoriented hypersphere in S^n with centers $\pm p\in S^n$ and spherical radius ϱ from p, $0\leqq\varrho\leqq\frac{\pi}{2}$. This hypersphere is the intersection of S^n with the hyperplane in $\mathbb{R}^{n+1}$ given by the equation $p\cdot y=\cos\varrho$. Suppose that $[z]=[(1,y)]$ for $y\in S^n$. Then

$$p\cdot y=\frac{-b(z,(0,p))}{b(z,e_1)}.$$

Hence, the equation $p\cdot y=\cos\varrho$ is equivalent to

$$b(z,(\cos\varrho,p))=0.$$

Therefore, a point $y\in S^n$ lies on the hypersphere determined by the equation $p\cdot y=\cos\varrho$ if and only if $[(1,y)]$ lies on the polar hyperplane in P^{n+1} to

$$[\xi]=[(\cos\varrho,p)]. \tag{1.11}$$

As in the Euclidean case, to obtain the two oriented spheres determined by the equation $p\cdot y=\cos\varrho$, note that $b(\xi,\xi)=\sin^2\varrho$ for ξ as in (1.11). If $\sin\varrho\neq 0$, let

$\zeta = \pm\xi/\sin\varrho$. Then the point $[(\zeta, 1)]$ is on Q^{n+1} and

$$[(\zeta, 1)] = [(\xi, \pm\sin\varrho)] = [(\cos\varrho, p, \pm\sin\varrho)].$$

We can incorporate the sign of the last coordinate into the radius and get that the oriented hypersphere with signed radius ϱ, $-\frac{\pi}{2} \leqq \varrho \leqq \frac{\pi}{2}$, and center p corresponds to the point

$$[(\cos\varrho, p, \sin\varrho)] \tag{1.12}$$

in Q^{n+1}. Point spheres in S^n correspond to those points in Q^{n+1} with last coordinate $\sin\varrho = 0$.

To treat oriented hyperspheres in H^n, let $\mathbb{R}_1^{n+1}$ be the Lorentz subspace of $\mathbb{R}_1^{n+2}$, spanned by the orthonormal basis $\{e_1, \dots, e_{n+1}\}$. Then, H^n is the hypersurface

$$\{y \in \mathbb{R}_1^{n+1} \mid b(y, y) = -1, y_1 \geqq 1\}$$

on which the restriction of b is a positive definite metric of constant curvature -1. For p, q in H^n, the distance $d(p, q) = \cosh^{-1}(-b(p, q))$. Thus, the equation for the sphere in H^n with center p and radius ϱ is

$$b(p, y) = -\cosh\varrho. \tag{1.13}$$

As before with S^n, we first embed $\mathbb{R}_1^{n+1}$ into P^{n+1} by setting $\phi(y) = [(y, 1)]$, where the last coordinate is in the spacelike direction e_{n+2}. If $[z] = [(y, 1)]$ for $y \in H^n$, then

$$b(p, y) = \frac{b(z, (p, 0))}{b(z, e_{n+2})}.$$

Hence, Eq. (1.13) is equivalent to the condition that $[(y, 1)]$ lie on the polar hyperplane in P^{n+1} to $[\xi] = [(p, \cosh\varrho)]$. Following exactly the same procedure as in the spherical case, we find that the oriented hypersphere in H^n with center p and signed radius ϱ corresponds to the point in Q^{n+1} given by

$$[(p, \cosh\varrho, \sinh\varrho)]. \tag{1.14}$$

D. Lie Sphere Geometry as the Basic Geometry

The Lie sphere geometry is the projective geometry in P^{n+2}, given the Lie quadric (1.7). Its group, the Lie sphere group, is isomorphic to $O(n+1, 2)/\{\pm I\}$. A point $[x] \in P^{n+2}$ is called *spacelike*, *timelike*, or *lightlike*, according as $\langle x, x \rangle$ is positive, negative, or zero. Note that a lightlike point is a point on Q^{n+1}, and thus corresponds to an oriented sphere in $\mathbb{R}^n$.

The quadric Q^{n+1}, or simply Q, carries lines, but not linear spaces of dimension $\geqq 2$. A line on Q is determined by two points $[x]$, $[y] \in Q$, satisfying

$$\langle x, y \rangle = 0.$$

Hence, all the lines on Q form a manifold of dimension $2n-1$, to be denoted by Λ^{2n-1}, or simply Λ. In $\mathbb{R}^n$, a line on Q corresponds to a family of ∞^1 oriented spheres such that any two of them are in oriented contact, i.e., all the oriented spheres tangent to an oriented hyperplane at a given point, i.e., an oriented element of contact.

From Lie sphere geometry one can build interesting subgeometries. For example, Moebius geometry is the Lie geometry relative to a fixed timelike point H, the Moebius space being the intersection of Q with the polar hyperplane $H^\perp$. By a Lie transformation we can suppose H to be the point $(0, \ldots, 0, 1)$, as in Sect. 1.B. Then $H^\perp$ has the equation $x_{n+3}=0$ and (1.7) reduces to (1.2). If we further fix a point $G \in H^\perp \cap Q$, to be the improper point, we get Euclidean geometry as the Lie geometry relative to the points H, G, where H and G are respectively timelike and lightlike, and $\langle G, H \rangle = 0$. It may be remarked that if only the lightlike point G is fixed, the resulting geometry is the *Laguerre geometry*, a point of Q orthogonal to G being a hyperplane of $\mathbb{R}^n$. In Laguerre geometry there are hyperplanes, but points have no meaning. See [2] for more detail on the geometries of Moebius, Laguerre and Lie.

A line ℓ in P^{n+2} not lying on Q either does not meet Q, is tangent to Q at a point p or meets Q in two distinct points, p_1, p_2. In the first case, ℓ consists entirely of timelike (resp. spacelike) points. In the second case $\ell - p$ has the same property. In the third case $\ell - p_1 - p_2$ is the union of two segments, consisting respectively of timelike and spacelike points.

From this discussion we can describe the behavior of the timelike points of P^{n+2} by the theorem:

Proposition 1.1. *The set of all timelike points of P^{n+2} forms a star-shaped domain such that, if any timelike point H is taken as the origin, any other timelike point can be joined to H by a segment consisting entirely of timelike points.*

We will call the line ℓ *timelike*, if it consists entirely of timelike points. A timelike line ℓ can be considered to be spanned by two orthogonal points, which can be supposed to be $[(1, 0, \ldots, 0)]$ and $[(0, \ldots, 0, 1)]$. The polar space $\ell^\perp$ of ℓ has then the equations

$$x_1 = 0, \quad x_{n+3} = 0.$$

By (1.7), we see that $\ell^\perp \cap Q$ is empty. This condition characterizes the timelike lines. Timelike lines are also characterized by the condition that the 2-plane in $\mathbb{R}_2^{n+3}$ determined by ℓ has signature $(-, -)$.

A two-dimensional plane E in P^{n+2} is called *timelike* if it contains a timelike line ℓ. Since $E^\perp \subset \ell^\perp$, the intersection of $E^\perp$ and Q is empty. This condition is also sufficient. For when it is satisfied, we can find $\ell^\perp \supset E^\perp$ such that $\ell^\perp$ does not intersect Q. Then $\ell = (\ell^\perp)^\perp \subset E$. Clearly no timelike plane contains a line of Q. For otherwise its timelike line would meet Q. Timelike planes are also characterized by the condition that the 3-plane in $\mathbb{R}_2^{n+3}$ determined by E has signature $(-, -, +)$.

To see more clearly the geometrical significance of a timelike plane, we choose coordinates and suppose E be spanned by the timelike points $[(1, 0, \ldots, 0)]$, $[(0, \ldots, 1)]$, and the spacelike point $[(0, 1, 0, \ldots, 0)]$. Then E has the equations

$$x_3 = \ldots = x_{n+2} = 0, \tag{1.15}$$

and $E \cap Q = C$ is the conic determined by (1.15) and the equation

$$-x_1^2 + x_2^2 - x_{n+3}^2 = 0.$$

By (1.8) the points of C, when interpreted as spheres of $\mathbb{R}^n$, are a family of concentric spheres with center at $p = 0$.

We list some of the correspondences in the following table:

Lie Geometry	*Geometry in* $\mathbb{R}^n$
Point on Q^{n+1}	Oriented sphere
Orthogonal points	Tangent spheres
Line on Q^{n+1}	Oriented element of contact
Timelike point $H \in P^{n+2}$	$H^{\perp} \cap Q$ as points of a Moebius geometry
Conic $C = E \cap Q$, E a timelike plane	Family of concentric spheres in a Moebius geometry

2. The Legendre Map

A. Structure Equations of Lie Sphere Geometry: Contact Structure on Λ

As discussed in Sect. 1.B, Sect. 1.D, the Lie sphere group is the group of all projective transformations of P^{n+2} leaving invariant the hyperquadric Q^{n+1} defined by the equation (1.7). By a slight change of notation we will write the equation of Q^{n+1} as

$$\langle y, y \rangle \equiv 2(y_1 y_2 + y_{n+2} y_{n+3}) + y_3^2 + \ldots + y_{n+1}^2, \tag{2.1}$$

where $y_1, \ldots, y_{n+3}$ are the homogeneous coordinates of P^{n+2}. We write the corresponding symmetric bilinear form as $\langle y, z \rangle$, for $y, z \in \mathbb{R}_2^{n+3}$. Its coefficients form the matrix

$$g = (g_{AB}) = \begin{pmatrix} J & 0 & 0 \\ 0 & I_{n-1} & 0 \\ 0 & 0 & J \end{pmatrix}, \tag{2.2}$$

where I_{n-1} is the unit $(n-1)\,x(n-1)$-matrix and

$$J = \begin{pmatrix} 0 & 1 \\ 1 & 0 \end{pmatrix}. \tag{2.3}$$

In this section, we agree on the following ranges of indices

$$\begin{gathered} 1 \leqq A, B, C \leqq n+3, \\ 3 \leqq i, j, k \leqq n+1. \end{gathered} \tag{2.4}$$

A *Lie frame* is an ordered set of $n+3$ vectors Y_A in $\mathbb{R}_2^{n+3}$ satisfying the relations

$$\langle Y_A, Y_B \rangle = g_{AB}. \tag{2.5}$$

The space of all Lie frames can be identified with the group $O(n+1, 2)$, of which the Lie sphere group, being isomorphic to $O(n+1, 2)/\{\pm I\}$, is a quotient group. In this space, we introduce the Maurer-Cartan forms ω_A^B by

$$dY_A = \Sigma \omega_A^B Y_B. \tag{2.6}$$

Differentiating (2.5), we get

$$\omega_{AB} + \omega_{BA} = 0, \tag{2.7}$$

where

$$\omega_{AB} = \sum g_{BC}\omega_A^C. \tag{2.8}$$

The Equation (2.7) says that the matrix

$$(\omega_{AB}) = \begin{pmatrix} \omega_1^2 & \omega_1^1 & \omega_1^i & \omega_1^{n+3} & \omega_1^{n+2} \\ \omega_2^2 & \omega_2^1 & \omega_2^i & \omega_2^{n+3} & \omega_2^{n+2} \\ \omega_j^2 & \omega_j^1 & \omega_j^i & \omega_j^{n+3} & \omega_j^{n+2} \\ \omega_{n+2}^2 & \omega_{n+2}^1 & \omega_{n+2}^i & \omega_{n+2}^{n+3} & \omega_{n+2}^{n+2} \\ \omega_{n+3}^2 & \omega_{n+3}^1 & \omega_{n+3}^i & \omega_{n+3}^{n+3} & \omega_{n+3}^{n+2} \end{pmatrix} \tag{2.9}$$

is anti-symmetric, a fact of which we shall make constant use in subsequent developments.

Taking the exterior derivative of (2.6), we get the Maurer-Cartan equations

$$d\omega_A^B = \sum_C \omega_A^C \wedge \omega_C^B, \tag{2.10}$$

which play a fundamental role in Lie sphere geometry.

The lines on Q^{n+1} form a manifold Λ of dimension $2n-1$. We can take a line on Q to be the line $Y_1 Y_{n+3}$ of a Lie frame. (From here on we identify a non-zero vector of $\mathbb{R}_2^{n+3}$ with the point of P^{n+2} of which it is the homogeneous coordinate vector.) On Λ, the form

$$\langle dY_1, Y_{n+3}\rangle = \omega_1^{n+2} \tag{2.11}$$

is defined up to a non-zero factor. Moreover, by using (2.10), we find

$$\omega_1^{n+2} \wedge (d\omega_1^{n+2})^{n-1} = \omega_1^{n+2} \wedge \left(\sum_i \omega_1^i \wedge \omega_i^{n+2}\right)^{n-1}$$

$$= (-1)^{n-1}(n-1)!\, \omega_1^{n+2} \wedge \omega_1^3 \wedge \omega_{n+3}^3 \wedge \ldots \wedge \omega_1^{n+1} \wedge \omega_{n+3}^{n+1} \neq 0.$$

Thus the form (2.11) defines a contact structure on Λ, which is then a contact manifold.

A submanifold

$$f: N \to \Lambda \tag{2.12}$$

is called a *Legendre submanifold*, if $f^*\omega_1^{n+2} = 0$.

B. The Legendre Map

Consider the diagram

$$\begin{array}{ccc} & B^{n-1} & \\ & \pi\downarrow & \\ & M^k \xrightarrow{i} \mathbb{R}^n, & \end{array} \tag{2.13}$$

where M^k is an immersed submanifold of the Euclidean space $\mathbb{R}^n$ and $B^{n-1} = B$ is the unit normal bundle. A point $b \in B$ determines an oriented hyperplane α which is perpendicular to b [$=$unit normal vector at $x = \pi(b)$] and tangent to M^k at x. The oriented spheres tangent to α at x define a line on Q, i.e., a point of Λ. The mapping

$$\lambda: B \to \Lambda \tag{2.14}$$

so defined is called the *Legendre map* of the submanifold M^k. The definition reminds one of that of a Gauss map, although they are different. We believe that the Legendre map, which is defined for a submanifold of $\mathbb{R}^n$ of any codimension, will play an important role in the study of submanifolds.

Proposition 2.1. *The Legendre map annihilates the contact form* ω_1^{n+2} *in* (2.11).

Proof. In the expression (2.11) for ω_1^{n+2}, we take Y_1 to be the point sphere x and Y_{n+3} to be the oriented hyperplane α. Their Lie sphere coordinates are given by (1.8), with the fundamental form (1.7). The vanishing of $\langle dY_1, Y_{n+3}\rangle$ follows immediately.

C. Second Fundamental Form of a Legendre Submanifold

The Legendre map λ defines a *Legendre submanifold* if for a generic choice of Y_1, the forms ω_1^i are linearly independent, i.e., if

$$\omega_1^3 \wedge \dots \wedge \omega_1^{n+1} \neq 0. \tag{2.15}$$

Here and later we pull back the forms to B^{n-1}, and we shall omit the symbols of such pull-backs for simplicity.

By Proposition 2.1, we have $\omega_1^{n+2} = 0$. By exterior differentiation and using (2.9), (2.10), we get

$$\sum_i \omega_1^i \wedge \omega_{n+3}^i = 0.$$

Because of (2.15) it follows by the well-known Cartan lemma that

$$\omega_{n+3}^i = \sum h^{ij} \omega_1^j, \tag{2.16}$$

where

$$h^{ij} = h^{ji}. \tag{2.17}$$

The quadric differential form

$$II(Y_1) = \sum_{i,j} h^{ij} \omega_1^i \omega_1^j, \tag{2.18}$$

defined up to a non-zero factor and depending on the choice of Y_1, is called the *second fundamental form.*

Consider a curve $\gamma(t)$ on B. The set of points in Q^{n+1} lying on the lines $\lambda(\gamma(t))$ form a ruled surface in Q^{n+1}. We look for the conditions that the ruled surface be a developable, i.e., consisting of lines which are tangent to a curve. Let $Y_{n+3} + rY_1$ be the point of contact. By (2.6), we have

$$d(rY_1 + Y_{n+3}) \equiv \sum_i (r\omega_1^i + \omega_{n+3}^i) Y_i, \bmod Y_1, Y_{n+3}. \tag{2.19}$$

It follows that the tangent direction of $\gamma(t)$ must satisfy the equations

$$\sum_j (r\delta^{ij} + h^{ij}) \omega_1^j = 0, \tag{2.20}$$

and r is a root of the equation

$$\det(r\delta^{ij} + h^{ij}) = 0. \tag{2.21}$$

By (2.17) all the roots of (2.21) are real. Denote them by $r_1, \ldots, r_{n-1}$. The point $r_\alpha Y_1 + Y_{n+3}$, $1 \leqq \alpha \leqq n-1$, is called a *curvature sphere* of λ at b. If r is a root of (2.21) of multiplicity m, the Eq. (2.20) define a subspace T_r of dimension m of $T_b B$, the tangent space to B at b. T_r is called a *principal space* of $T_b B$, the latter being thus decomposed into a direct sum of its principal spaces.

The Legendre submanifolds are an important object of study in the differential geometry of Lie sphere geometry. The simplest ones are those which satisfy a Dupin condition, as follows: A connected submanifold $N \subset B$ is called a *curvature submanifold*, if its tangent space is everywhere a principal space. A Legendre submanifold is called a *Dupin submanifold*, if, for every curvature submanifold $N \subset B$, the lines $\lambda(b)$, $b \in N$, pass through a fixed point. This definition of "Dupin" is the same as that used by Pinkall [12]. It is weaker than the definition of "Dupin" used by Thorbergsson [15] and Cecil-Ryan [7] who assumed that the curvature spheres have constant multiplicities on B.

We wish to prove the following local theorem on Legendre submanifolds:

Proposition 2.2. *Let $\lambda: B^{n-1} \to \Lambda$ be a Legendre submanifold. Suppose that the multiplicity m of a curvature sphere is constant on an open subset U of B^{n-1}. Then the distribution of the corresponding principal spaces is integrable on U.*

Proof. We can suppose Y_{n+3} to be the curvature sphere, so that $r=0$ is a root of multiplicity m of (2.21). We also suppose $m \geqq 2$, as otherwise there is nothing to prove. Among the Eq. (2.20)

$$\sum_j h^{ij} \omega_1^j = 0 \quad \text{or} \quad \omega_{n+3}^i = 0,$$

there are exactly $n-1-m$ independent ones, while the others are their linear combinations. We can choose Y_i so that ω_{n+3}^α, $3 \leqq \alpha, \beta \leqq n-m+1$, are orthogonal to the distribution, while $\omega_{n+3}^\varrho = 0$, $n-m+2 \leqq \varrho \leqq n+1$. Since Y_1 is generic, we suppose that ω_{n+3}^α, ω_1^ϱ are linearly independent. Differentiating the equations $\omega_{n+3}^\varrho = 0$ and using (2.10), we get

$$\omega_{n+3}^1 \wedge \omega_1^\varrho + \sum_\alpha \omega_{n+3}^\alpha \wedge \omega_\alpha^\varrho = 0.$$

It follows that

$$\omega_{n+3}^1 \equiv 0, \operatorname{mod} \omega_{n+3}^\alpha, \omega_1^\varrho \text{ (each } \varrho).$$

Since $m \geqq 2$, this holds only when

$$\omega_{n+3}^1 \equiv 0, \operatorname{mod} \omega_{n+3}^\alpha.$$

The proposition is proved, if we show

$$d\omega_{n+3}^\alpha \equiv 0, \operatorname{mod} \omega_{n+3}^\beta.$$

But this follows from (2.10) and our conditions.

This result was obtained by Pinkall [12, p. 433] who accomplished the proof by invoking known results for submanifolds of Euclidean space.

3. Tautness

A. Taut Submanifolds of Real Space Forms

We begin by recalling some basic facts about taut submanifolds of real space forms. See Chap. 2 of [7] for more detail.

Let $f: M \to \mathbb{R}^n$ be an immersion of a smooth compact connected manifold M into Euclidean space. For $p \in \mathbb{R}^n$, the Euclidean distance function $L_p: M \to \mathbb{R}$ is defined by $L_p(x) = |p - f(x)|^2$. It is well known that for almost all p, i.e., p not a focal point of M, the function L_p is nondegenerate. By the Morse inequalities, the number $\mu(L_p)$ of critical points of a nondegenerate distance function satisfies $\mu(L_p) \geqq \beta(M; F)$, the sum of the F-Betti numbers of M, where F is any field. The immersion f is said to be *taut* if there exists some field F such that $\mu(L_p) = \beta(M; F)$ for all nondegenerate L_p functions. So far, the field $F = Z_2$ has been satisfactory for all considerations. A taut immersion must, in fact, be an embedding [3], since if $p \in \mathbb{R}^n$ were a double point, then the function L_p would have two absolute minima.

Tautness is preserved by conformal transformations of $\mathbb{R}^n \cup \{\infty\} = S^n$, i.e., by Moebius transformations. Thus, by stereographic projection, one can consider a taut embedding to be an embedding $M \subset S^n \subset \mathbb{R}^{n+1}$ such that every nondegenerate spherical distance function has $\beta(M; F)$ critical points. But a spherical distance function is essentially a Euclidean height function $\ell_p: M \to \mathbb{R}$ defined by

$$\ell_p(x) = p \cdot x, \quad p \in S^n.$$

Thus, $M \subset S^n$ is taut if and only if $\mu(\ell_p) = \beta(M; F)$ for every nondegenerate Euclidean height function ℓ_p. That is, a tight spherical embedding is taut.

Tight and taut submanifolds have also been studied in hyperbolic space H^n. Cecil and Ryan [4] introduced three classes of distance functions L_p, L_π, L_h in H^n whose level sets, respectively, are spheres centered at a point $p \in H^n$, equidistant hypersurfaces to a hyperplane π, and horospheres equidistant from a fixed horosphere h. Suppose that $f: M \to H^n$ is an immersion of a compact, connected manifold and that P is stereographic projection of H^n onto the unit disk D^n in $\mathbb{R}^n$. Then $P \circ f$ is a taut embedding of M into $\mathbb{R}^n$ if and only if every nondegenerate function of each of the three types L_p, L_π, L_h has β critical points [5, p. 563].

As we saw in Sect. 2.B, a submanifold of $\mathbb{R}^n$ of any dimension $k \leqq n-1$ induces a Legendre submanifold of dimension $n-1$ by a consideration of the unit normal bundle B^{n-1}. In the theory of taut embeddings with $F = Z_2$, it also suffices to work with hypersurfaces for the following reason. Let $M \subset S^n$ be a compact submanifold of dimension $k < n-1$, and let M_ε be a tube of sufficiently small radius $\varepsilon > 0$ about M so that M_ε is an embedded hypersurface in S^n. Then, Pinkall [13] has shown that M is taut with $F = Z_2$ if and only if M_ε is taut with $F = Z_2$. One computes directly that if p is a point in S^n and not in M, then ℓ_p is nondegenerate on M_ε if and only if ℓ_p is nondegenerate on M. Furthermore, the number of critical points of ℓ_p on M_ε is twice the number of critical points of ℓ_p on M. Pinkall then used the Gysin sequence for the unit normal bundle of M to show that $\beta(M_\varepsilon; Z_2) = 2\beta(M; Z_2)$.

B. Parallel Transformations

For each of the base spaces $\mathbb{R}^n$, S^n, and H^n there is a Lie transformation P_t which keeps the center of each oriented sphere fixed and adds t to the signed radius. P_t is called *parallel transformation by t*.

Fix a choice of orthonormal basis $B=\{e_1,\ldots,e_{n+3}\}$ with respect to the metric (1.7) on $\mathbb{R}_2^{n+3}$. By (1.8), the sphere in $\mathbb{R}^n$ with center p and oriented radius ϱ is represented by the point

$$[k]=\left[\left(\frac{1+p\cdot p-\varrho^2}{2},\frac{1-p\cdot p+\varrho^2}{2},p,\varrho\right)\right].$$

One can check that the *Euclidean parallel transformation* P_t defined as follows has the desired effect on $[k]$.

$$\begin{aligned}
P_te_1&=\frac{2-t^2}{2}e_1+\frac{t^2}{2}e_2+te_{n+3}\\
P_te_2&=\frac{-t^2}{2}e_1+\frac{2+t^2}{2}e_2+te_{n+3}\\
P_te_{n+3}&=-te_1+te_2+e_{n+3}\\
P_te_i&=e_i,\quad 3\leq i\leq n+2.
\end{aligned}\tag{3.1}$$

In S^n, the sphere with center p and radius ϱ is represented as in (1.12) by $[(\cos\varrho, p, \sin\varrho)]$. The *spherical parallel transformation* P_t is defined by

$$\begin{aligned}
P_te_1&=\cos t e_1+\sin t e_{n+3}\\
P_te_{n+3}&=-\sin t e_1+\cos t e_{n+3}\\
P_te_i&=e_i,\quad 2\leq i\leq n+2.
\end{aligned}\tag{3.2}$$

Finally, in H^n, the sphere with center p and radius ϱ is represented as in (1.14) by $[(p,\cosh\varrho,\sinh\varrho)]$, and the *hyperbolic parallel transformation* P_t is defined by

$$\begin{aligned}
P_te_{n+2}&=\cosh t e_{n+2}+\sinh t e_{n+3}\\
P_te_{n+3}&=\sinh t e_{n+2}+\cosh t e_{n+3}\\
P_te_i&=e_i,\quad 1\leq i\leq n+1.
\end{aligned}\tag{3.3}$$

The important role of parallel transformations in Lie sphere geometry is demonstrated by the following result. Here a "Moebius transformation" is a Lie transformation Φ which takes point spheres to point spheres, i.e., $\Phi[e_{n+3}]=[e_{n+3}]$.

Theorem 3.1. *Any Lie transformation α can be expressed in the form $\alpha=\Phi P_t\Psi$, where Φ and Ψ are Moebius transformations and P_t is some Euclidean, spherical or hyperbolic parallel transformation.*

Proof. Represent α by a transformation $A\in O(n+1,2)$ and suppose that $Ae_{n+3}=v$ for some unit timelike vector v linearly independent from e_{n+3}. The plane $[e_{n+3},v]$ in $\mathbb{R}_2^{n+3}$ can have signature $(-,-)$, $(-,0)$, or $(-,+)$. In the case where $[e_{n+3},v]$ has signature $(-,-)$, we can write

$$v=-\sin t u_1+\cos t e_{n+3},$$

where u_1 is a unit timelike vector orthogonal to e_{n+3}, and $t\in(0,\pi)$. Let Φ be a Moebius transformation such that $\Phi^{-1}u_1=e_1$. Then from (3.2), we see that

$P_{-t}\Phi^{-1}v=e_{n+3}$. Hence, $P_{-t}\Phi^{-1}\alpha(e_{n+3})=e_{n+3}$, i.e., $P_{-t}\Phi^{-1}\alpha$ is a Moebius transformation Ψ. Thus, $\alpha=\Phi P_t\Psi$, as desired.

The two other cases are very similar. If $[e_{n+3},v]$ has signature $(-,0)$, then we can write

$$v=-tu_1+tu_2+e_{n+3},$$

where u_1, u_2 are unit timelike, spacelike, respectively, vectors orthogonal to e_{n+3} and to each other. If Φ is a Moebius transformation such that $\Phi^{-1}u_1=e_1$, $\Phi^{-1}u_2=e_2$, then $P_{-t}\Phi\alpha$ is a Moebius transformation Ψ, where P_t is the Euclidean parallel transformation (3.1). As before $\alpha=\Phi P_t\Psi$. Finally, if $[e_{n+3},v]$ has signature $(-,+)$, then

$$v=\sinh t u_{n+2}+\cosh t e_{n+3},$$

for a unit spacelike vector u_{n+2} orthogonal to e_{n+3}. Define the Moebius transformation Φ so that $\Phi^{-1}u_{n+2}=e_{n+2}$, and conclude that $\alpha=\Phi P_t\Psi$ for the hyperbolic parallel transformation P_t in (3.3).

C. A Distance Function on a Legendre Submanifold

We use the notations of Sect. 2. Let

$$\lambda: B^{n-1}(=B)\rightarrow \Lambda^{2n-1} \tag{3.4}$$

be a Legendre submanifold. Consider the sequence

$$H\in\ell\subset E, \tag{3.5}$$

where E is a timelike plane and ℓ is a timelike line in it, so that H is a timelike point. The intersection $C=E\cap Q$ is a conic (see Fig. 1). Let A_1 and A_2 be the points of intersection of C and $H^\perp$, so that HA_1 and HA_2 are the tangents to C issued from H. Let $D_1=\ell\cap H^\perp$, and let D_2 be the harmonic conjugate of D_1 with respect to A_1, A_2.

We can represent the points A_1 and A_2 by vectors in $\mathbb{R}^{n+3}_2$ with the same names satisfying $\langle A_1,A_1\rangle=\langle A_2,A_2\rangle=0$, $\langle A_1,A_2\rangle=1$ in such a way that D_1 is repre-

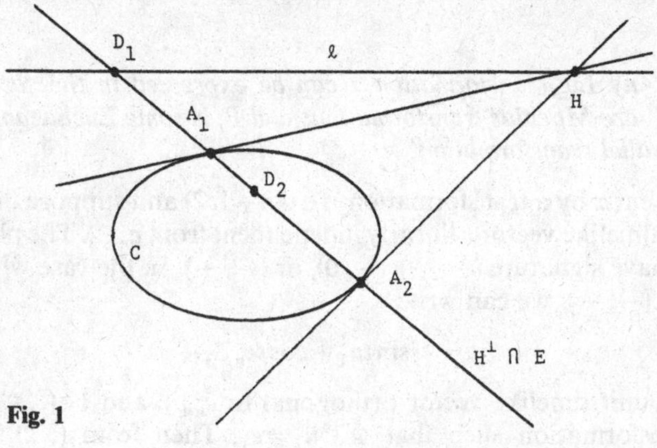

Fig. 1

sented by the unit timelike vector $u_1 = A_1 - A_2/\sqrt{2}$ and D_2 by the unit spacelike vector $u_2 = A_1 + A_2/\sqrt{2}$. For $b \in B$, let $Y_1(b) = \lambda(b) \cap H^\perp$. We will call Y_1 the *Moebius projection of λ determined by H*. We define the *Lie distance function*

$$L(b) = -\frac{\langle Y_1, u_2 \rangle}{\langle Y_1, u_1 \rangle}. \tag{3.6}$$

The conditions $\langle Y_1, H \rangle = 0$ and $\langle Y_1, Y_1 \rangle = 0$ imply that the denominator $\langle Y_1, u_1 \rangle$ is never zero.

We can represent Y_1 by the vector

$$Y_1(b) = u_1 + f(b), \tag{3.7}$$

where $\langle f(b), H \rangle = \langle f(b), u_1 \rangle = 0$ and $\langle f(b), f(b) \rangle = 1$. The set of unit vectors in $\mathbb{R}_2^{n+3}$ orthogonal to both H and u_1 form a Euclidean metric sphere S^n. The map $f: B \to S^n$ defined in (3.7) will be called the *spherical projection of λ determined by the ordered pair $\{H, u_1\}$*.

The Moebius projection Y_1 is said to be *regular* at a point $b \in B$ if $Y_1(b)$ is not a curvature sphere of λ at b. This is equivalent to the condition that the spherical projection f (3.7) be an immersion at b. Note that if Y_1 is represented as in (3.7), then the function L in (3.6) is exactly equal to the Euclidean height function $\ell_p(b) = p \cdot f(b)$, where $p = u_2$. Furthermore, using the correspondence (1.12) with $u_1 = (1, 0, \ldots)$, $H = (0, \ldots, 0, 1)$, the points on the conic $C = E \cap Q$ correspond to a family of concentric spheres in S^n with centers $\pm p$.

The function $L(b)$ in (3.6) depends on the timelike sequence (3.5). However, the location of its critical points is independent of the choice of timelike line ℓ through the point H in the plane E. To see this, suppose that $\ell' \subset E$ is another timelike line through H. Let D'_1 be the point of intersection of ℓ' with the line A_1A_2, and D'_2 its harmonic conjugate with respect to A_1, A_2. We can represent the points in question by the vectors

$$D_1 = A_1 - A_2, \qquad D_2 = A_1 + A_2,$$

$$D'_1 = A_1 - \sigma A_2, \quad D'_2 = A_1 + \sigma A_2.$$

From this, we find

$$2D'_1 = (1+\sigma)D_1 + (1-\sigma)D_2$$

$$2D'_2 = (1-\sigma)D_1 + (1+\sigma)D_2.$$

It follows that under a change from ℓ to ℓ', the function $L(b)$ undergoes a linear fractional transformation with a non-zero determinant, and so the two functions have the same critical points.

The following theorem gives a geometric criterion for a point b to be a critical point of the function L in (3.6). It can be proven by a direct calculation using the differential forms of Sect. 2.A. On the other hand, the theorem follows immediately from well known results in Euclidean geometry once it is noted (as above) that L is equal to the height function ℓ_p, with $p = u_2$, applied to the spherical projection f. If Y_1 is regular at b, then f is an immersion at b. Then, it is known that ℓ_p has a critical point at b if and only if p lies along the normal geodesic in S^n to $f(B)$ at $f(b)$. This happens precisely when one of the two oriented spheres centered at p and containing b is in oriented contact with $f(B)$ at $f(b)$, i.e., $\lambda(b)$ intersects the timelike plane E corresponding to the pencil of concentric spheres centered at p. Furthermore, the critical point is degenerate precisely when p is the center of a curvature sphere to $f(B)$ at $f(b)$. Finally, the index of ℓ_p at a nondegenerate critical point b is determined by the number of focal points on the geodesic segment from p to $f(b)$ in S^n. Hence, we will simply state the following result and omit the calculation.

Theorem 3.2. *Let $\lambda: B^{n-1} \to \Lambda$ be a Legendre submanifold and L a Lie distance function determined by the timelike sequence $H \in \ell \subset E$. Assume that the Moebius projection Y_1 determined by H is regular at the point $b \in B^{n-1}$. Then*

(a) *L has a critical point at b if and only if $\lambda(b)$ intersects E in a point k.*

(b) *b is a degenerate critical point of L if and only if k is a curvature sphere of λ at b.*

(c) *Suppose that b is a nondegenerate critical point. Then the index of L at b is equal to the number of curvature spheres (counting multiplicities) on one of the segments from $Y_1(b)$ to k.*

Motivated by Theorem 3.2, we say that a timelike plane E is *nondegenerate* if E does not contain any curvature spheres of λ and *degenerate* otherwise. If E is nondegenerate and contains a timelike point H whose Moebius projection Y_1 is regular on B, then the Morse inequalities applied to the function L imply that the number $\mu(E)$ of points b such that $\lambda(b)$ intersects E satisfies

$$\beta(B;F) \leqq \mu(E) < \infty,$$

where $\beta(B;F)$ is the sum of the F-Betti numbers of the compact manifold B.

Thus, we are led to a natural generalization of Moebius tautness.

Definition. Let $\lambda: B^{n-1} \to \Lambda$ be a compact Legendre submanifold and H a timelike point in P^{n+2}. Then λ is said to be *H-taut* if $\mu(E) = \beta(B^{n-1}; F)$ for all nondegenerate timelike planes E containing H and for some field F.

In the case where the Moebius projection Y_1 is regular on B^{n-1}, H-tautness is equivalent to the classical Moebius tautness of Y_1 in the following way. A choice of unit timelike vector u_1 orthogonal to H determines a spherical projection $f: B^{n-1} \to S^n$ as in (3.7). If λ is H-taut, then every spherical projection determined by such a pair $\{H, u_1\}$ will be taut in the classical sense, i.e., every nondegenerate ℓ_p, $p \in S^n$, has $\beta(B;F)$ critical points. On the other hand, suppose that one such spherical projection f determined by $\{H, u_1\}$ is taut in the classical sense. Then any other spherical projection determined by a pair $\{H, v_1\}$ is also taut because of the invariance of classical tautness under Moebius transformations. This then implies that λ is H-taut as defined above.

Suppose that $\lambda: B^{n-1} \to \Lambda$ is a Legendre submanifold and α is a Lie transformation. Then the map $\alpha\lambda: B^{n-1} \to \Lambda$, is also a Legendre submanifold which we will denote by αB. If k is a curvature sphere of λ at a point $b \in B$, then $\alpha(k)$ is easily seen to be a curvature sphere of $\alpha\lambda$ at b having the same principal space at b as k.

Proposition 3.3. *Let $\lambda: B^{n-1} \to \Lambda$ be a compact Legendre submanifold. Suppose that α is a Lie transformation and H, H′ are timelike points such that $\alpha H' = H$. Then λ is H′-taut if and only if $\alpha\lambda$ is H-taut.*

Proof. A timelike plane E' containing H' is nondegenerate for λ if and only if the timelike plane $E = \alpha E'$ is nondegenerate for $\alpha\lambda$. Furthermore, $\lambda(b)$ intersects E' if and only if $\alpha(\lambda(b))$ intersects E.

We can now prove the main result of this section. It shows that tautness is invariant under Lie transformations within the class of immersions.

Theorem 3.4 *Let $\lambda: B^{n-1} \to \Lambda$ be a compact Legendre submanifold whose Moebius projection with respect to a certain timelike point H is regular and H-taut. Then λ is also H'-taut for every timelike point H' whose corresponding Moebius projection is regular.*

Proof. The projective line $[H, H']$ can have signature $(-,-)$, $(-,0)$, or $(-,+)$. In these three respective cases, we can write

$$H' = -\sin t e_1 + \cos t H, \tag{3.8}$$

$$H' = -te_1 + te_2 + H, \tag{3.9}$$

$$H' = \sinh t e_{n+2} + \cosh t H, \tag{3.10}$$

where e_1, e_2, e_{n+2} are vectors of lengths $-1, 1, 1$ respectively, which are orthogonal to H and to each other, as in (3.1)–(3.3) with H playing the role of e_{n+3} in those equations.

We will handle the spherical case (3.8) in detail. The others can be done in a very similar manner. In the spherical case, the Moebius projection Y_1 of λ determined by H can be written in the form (3.7),

$$Y_1 = e_1 + f,$$

where $f: B^{n-1} \to S^n$ is the spherical projection determined by $\{H, e_1\}$. Since Y_1 is regular, f is an immersion. Let $Y_{n+3} = \lambda(b) \cap e_1^{\perp}$. Then we can write

$$Y_{n+3} = \xi + H,$$

where $\xi: B^{n-1} \to S^n$ satisfies $\xi \cdot f = 0$, $\xi \cdot df = 0$, because of the Legendre conditions $\langle Y_1, Y_{n+3} \rangle = 0$, $\langle dY_1, Y_{n+3} \rangle = 0$. Thus, ξ is a field of unit normals to the immersion f. Note that $H' = P_t H$, where P_t is the spherical parallel transformation (3.2) with $H = e_{n+3}$. Thus, $P_{-t}H' = H$, and by Proposition 3.3, we know that λ is H'-taut if and only if $P_{-t}\lambda$ is H-taut. We have

$$P_{-t}Y_1 = \cos t e_1 - \sin t H + f,$$

$$P_{-t}Y_{n+3} = \sin t e_1 + \cos t H + \xi.$$

The Moebius projection Z_1 of $P_{-t}\lambda$ with respect to H is then

$$Z_1 = \cos t P_{-t} Y_1 + \sin t P_{-t} Y_{n+3} = e_1 + f_t,$$

where

$$f_t = \cos t f + \sin t \xi.$$

We recognize f_t as the parallel hypersurface to f in S^n at oriented distance t. Z_1 is regular by the assumption that the Moebius projection of λ determined by H' is regular. Thus, f_t is an immersion. Then, the tautness of f implies that f_t is also taut (see, for example, [7, p. 185]), and hence that $P_{-t}\lambda$ is H-taut, as desired.

The other two cases (3.9) and (3.10) are handled in a similar manner using Euclidean, respectively, hyperbolic projections of λ obtained from spherical projections via stereographic projection. In these cases also, the key fact from the classical theory is that within the class of immersions, tautness is preserved by taking parallel hypersurfaces. In the Euclidean case, this is well known. In the hyperbolic case, Theorem C of [4] shows that under the process of taking parallel hypersurfaces, the property that every nondegenerate distance function of each of the three types L_p, L_π, L_h has β critical points is preserved. As we noted in Sect. 3.A, this property is equivalent to the Euclidean tautness of a related Euclidean projection and thus to H-tautness.

D. The Relationship Between Taut and Dupin Submanifolds

From the results of Thorbergsson [15] and Pinkall [13], it is known that the notions of taut and Dupin are equivalent for compact embedded hypersurfaces in S^n whose principal curvatures have constant multiplicities. In that case, each principal distribution is a foliation whose leaves are metric spheres in S^n. However, the multiplicites of the principal curvatures of a taut hypersurface are not necessarily constant. For example, let $T^2 \subset \mathbb{R}^3$ be a torus of revolution and embed $\mathbb{R}^3$ in $\mathbb{R}^4 = \mathbb{R}^3 \times \mathbb{R}$. Let M^3 be a tube of sufficiently small radius ε about T^2 in $\mathbb{R}^4$ so that M^3 is an embedded hypersurface in $\mathbb{R}^4$. Then M^3 is taut, but its principal curvatures do not have constant multiplicities. Specifically, on the two tori $T^2 \times \{\pm\varepsilon\}$ in M^3, one principal curvature is identically equal to zero, and it has multiplicity two. At all other points of M^3, there are three distinct principal curvatures, each of which is constant along its own lines of curvature, which are circles (see [7, pp. 188–190] for more discussion of this and related examples). The two tori and all of the circles are curvature submanifolds of M^3. Along each curvature submanifold, the corresponding principal curvature is constant, i.e., M^3 is Dupin. In fact, Pinkall [13] showed that every taut submanifold of S^n is Dupin.

Using the recent work of Ozawa [10], one can actually make an even stronger statement. Pinkall showed that if M is taut, then along any curvature submanifold N, the corresponding principal curvature is constant. However, he did not show that given any principal space T_i at any point $x \in M$, there exists a curvature submanifold N through x whose tangent space at x is T_i. This is true, however, because of Ozawa's work. He showed that each connected component of a critical set of a height function ℓ_p on a taut submanifold $M \subset S^n$ is a taut smooth submanifold N whose dimension is equal to the nullity of the Hessian of ℓ_p at each point of N. It is then rather easy to show that such an N is a curvature submanifold along which the corresponding principal curvature is constant. Hence, a taut hypersurface $M \subset S^n$ actually satisfies the condition called "semi-Dupin" in [7, p. 189], i.e., each principal space at each point of M is the tangent space to a curvature submanifold and along each curvature submanifold, the corresponding principal curvature is constant. Of course, semi-Dupin is actually stronger than "Dupin" as we have defined it in Sect. 2 but weaker than Thorbergsson's "Dupin" which assumes constant multiplicities. We expect that one can prove that our Dupin implies semi-Dupin. Further, we conjecture that semi-Dupin implies taut, although we do not see how to generalize Thorbergsson's [15] proof at this time.

A major remaining problem in this area is to classify Dupin hypersurfaces up to reducibility as defined by Pinkall [12, p. 437]. It seems possible at this time that the only irreducible ones are Legendre submanifolds generated by isoparametric submanifolds [14], some of which are also reducible. It may even be possible to do the classification using local Lie geometric invariants and thus by restricting one's attention to open subsets on which the curvature spheres have constant multiplicities. This was done successfully for Dupin hypersurfaces in S^4 by Pinkall [11].

References

1. Arnold, V.: Mathematical methods of classical mechanics. Berlin, Heidelberg, New York: Springer 1978
2. Blaschke, W.: Vorlesungen über Differentialgeometrie, Vol. 3. Berlin: Springer 1929
3. Carter, S., West, A.: Tight and taut immersions. Proc. London Math. Soc. **25**, 701–720 (1972)
4. Cecil, T., Ryan, P.: Distance functions and umbilic submanifolds in hyperbolic space. Nagoya Math. J. **74**, 67–75 (1979)
5. Cecil, T., Ryan, P.: Tight and taut immersions into hyperbolic space. J. London Math. Soc. **19**, 561–572 (1979)
6. Cecil, T., Ryan, P.: Tight spherical embeddings. Proc. 1979 Berlin Symposium in Global Differential Geometry. Lect. Notes Math. **838**, 94–104. Berlin, Heidelberg, New York: Springer 1981
7. Cecil, T., Ryan, P.: Tight and taut immersions of manifolds. Res. Notes Math. **107**. London: Pitman 1985
8. Hsiang, W.-Y., Palais, R., Terng, C.-L.: The topology of isoparametric submanifolds. Preprint, University of California, Berkeley 1986
9. Lie, S., Scheffers, G.: Geometrie der Berührungstransformationen. Leipzig: Teubner 1896
10. Ozawa, T.: On critical sets of distance functions to a taut submanifold. Math. Ann. **276**, 91–96 (1986)
11. Pinkall, U.: Dupin'sche Hyperflächen in E^4. Manuscr. Math. **51**, 89–119 (1985)
12. Pinkall, U.: Dupin hypersurfaces. Math. Ann. **270**, 427–440 (1985)
13. Pinkall, U.: Curvature properties of taut submanifolds. Geom. Dedicata **20**, 79–83 (1986)
14. Terng, C.-L.: Isoparametric submanifolds and their Coxeter groups. J. Differential Geometry **21**, 79–107 (1985)
15. Thorbergsson, G.: Dupin hypersurfaces. Bull. Lond. Math. Soc. **15**, 493–498 (1983)

Received October 1, 1986

VECTOR BUNDLES WITH A CONNECTION

SHIING-SHEN CHERN[1]

1. Introduction.

Riemannian geometry, which was the high-dimensional generalization of Gauss' intrinsic surface theory, gives a geometrical structure which is entirely local. It was later realized that much of its geometrical properties derives from the Levi-Civita parallelism, i.e., the connection in the tangent bundle. Recent developments in mathematics and physics have shown the importance of the notion of a "vector bundle with a connection" over a manifold. An introductory account of this notion and some of its applications will be given in this article.

2. Vector Bundles.

We will be dealing with C^∞-manifolds and their C^∞-mappings. A vector bundle over a manifold M is a mapping

$$\pi : E \to M \tag{1}$$

such that the following conditions are satisfied:

B1) $\pi^{-1}(x)$, $x \in M$, is a (real or complex) vector space Y of dimension q. Either $\pi^{-1}(x)$ or Y will be called a fiber.

B2) E is locally a product, i.e., every point $x \in M$ has a neighborhood U such that there is a diffeomorphism

$$\varphi_U : U \times Y \to \pi^{-1}(U) \tag{2}$$

satisfying the condition

$$\pi \circ \varphi_U(x, y_U) = x, \qquad y_U \in Y; \tag{3}$$

[1] Work done under partial support of NSF Grant No. DMS84-03201.

(x, y_U) are the local coordinates of E relative to U.

B3) For two neighborhoods U, V, with $U \cap V \neq \phi$, the relation

$$\varphi_U(x, y_U) = \varphi_V(x, y_V), \qquad x \in U \cap V;\ y_U, y_V \in Y, \tag{4}$$

holds if and only if

$$y_U = g_{UV}(x) y_V, \tag{5}$$

where $g_{UV}(x)$ is a non-degenerate endomorphism of Y. If y_U, y_V are expressed as one-columned matrices, then $g_{UV}(x)$ is a non-singular $(q \times q)$-matrix. The condition B3) means that the linear structure on the fiber has a meaning.

E is called a <u>vector bundle</u> over the <u>base manifold</u> M. A vector bundle can be viewed as a family of vector spaces parametrized by a manifold such that it is locally trivial and the linear structure on the fiber is defined.

A <u>section</u> is a mapping $s: M \to E$ such that $\pi \circ s =$ identity. Two sections can be added, and a section can be multiplied by a (real or complex-valued) function, remaining a section. All the sections form a vector space which we will denote by $\Gamma(E)$.

The functions $g_{UV}(x)$, $x \in U \cap V \neq \phi$, have values in $GL(q;\mathbf{R})$ or $GL(q;\mathbf{C})$, and are called <u>transition functions</u>. If $\{U, V, W, \ldots\}$ form a covering of M, the transition functions satisfying the conditions

$$\begin{aligned} g_{UU}(x) &= \text{identity} \\ g_{UV}(x) g_{VU}(x) &= \quad '' \qquad , \text{in } U \cap V \neq \phi \\ g_{UV}(x) g_{VW}(x) g_{WU}(x) &= \quad '' \qquad , \text{in } U \cap V \cap W \neq \phi. \end{aligned} \tag{6}$$

It can be proved that a family of transition functions relative to the covering, which satisfy the conditions (6) define a vector bundle.

We give some examples of vector bundles:

EXAMPLE 1: $E = M \times Y$, and π is the projection to the first factor. A section is a graph in $M \times Y$ and is exactly a vector-valued function on M (cf. following figure).

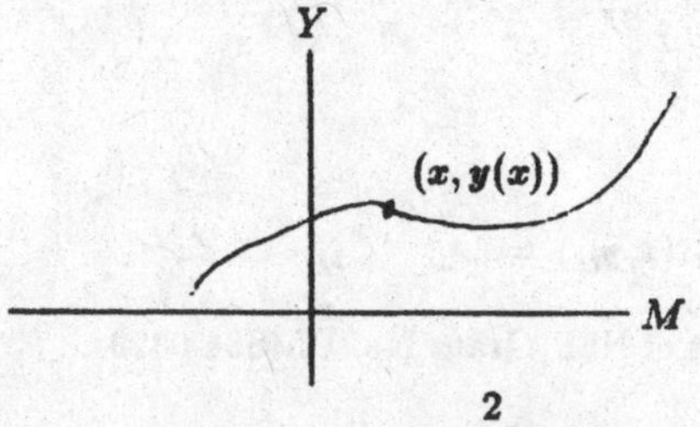

EXAMPLE 2: $E = TM$, the tangent bundle of M. Relative to the local coordinates u^i of M, $1 \leq i \leq n$ (= dim M), a tangent vector can be written as $\sum y_U^i \frac{\partial}{\partial u^i}$. In an open set where the coordinates u^i and v^j are both valid, we have

$$\sum y_U^i \frac{\partial}{\partial u^i} = \sum y_V^j \frac{\partial}{\partial v^j}, \qquad 1 \leq i, j \leq n, \tag{7a}$$

if and only if

$$y_U^i = \sum_j \frac{\partial u^i}{\partial v^j} y_V^j. \tag{7b}$$

Thus the transition functions are given by the Jacobian matrices. A section of $TM \to M$ is a vector field on M.

EXAMPLE 3: If M is immersed in the Euclidean space E^{n+q} of dimension $n+q$, the normal vectors to M form its normal bundle.

3. Connections.

Consider the trivial bundle $E = M \times Y \to M$ in Example 1. A section $s \in \Gamma(E)$ is a Y-valued function on M. If X is a vector field on M, the directional derivative of s along X, which we denote by $\nabla_X s$, is again a section. We wish to generalize this "differentiation" to any vector bundle.

A connection is a map

$$D : \Gamma(E) \to \Gamma(E \otimes T^*M), \tag{8}$$

where T^*M is the cotangent bundle of M, such that the following conditions are satisfied:

D1) $D(s_1 + s_2) = Ds_1 + Ds_2$,

D2) $D(fs) = fDs + s \otimes df, s, s_1, s_2 \in \Gamma(E)$,

where f is a (real or complex-valued) function on M. From Ds we get the directional derivative by the pairing

$$\nabla_X s = \langle X, Ds \rangle. \tag{9}$$

To describe a connection analytically we restrict to a neighborhood and take in it a frame field, i.e., q sections e_i, $1 \leq i \leq q$, which are everywhere linearly independent. Since the e_i's form a basis on each fiber, we can write

$$De_i = \sum \omega_i^j \otimes e_j, \tag{10}$$

when ω_i^j, $1 \leq i, j \leq q$, are linear differential forms. The matrix of linear differential forms:

(11) $$\omega = (\omega_i^j)$$

is called the connection matrix relative to the frame field e_i.

The connection matrix determines the connection completely. For any section can be written

(12) $$s = \sum s^i e_i,$$

and we have, by the Properties D1) and D2),

(13) $$Ds = \sum (ds^i + s^j \omega_j^i) \otimes e_i$$

The section s is said to be parallel, if $Ds = 0$.

Our fundamental formula relates the connection matrices under a change of the frame field. In fact, let

(14) $$e_i' = \sum a_i^j e_j$$

be a new frame field, where $det(a_i^j) \neq 0$. Let

(15) $$De_i' = \sum \omega_i'^j \otimes e_j', \qquad \omega' = (\omega_i'^j).$$

Here it will be most easy to compute with matrix equations. We write

(16) $$e = \begin{pmatrix} e_1 \\ \cdot \\ \cdot \\ \cdot \\ e_q \end{pmatrix}, \quad e' = \begin{pmatrix} e_1' \\ \cdot \\ \cdot \\ \cdot \\ e_q' \end{pmatrix}, \quad A = (a_i^j).$$

Then we have the matrix equations

(17) $$\begin{gathered} De = \omega e, \qquad De' = \omega' e' \\ e' = Ae. \end{gathered}$$

Differentiating the last equation and using the properties D1), D2), we get immediately the fundamental formula

$$\omega' A = dA + A\omega. \tag{18}$$

This relation satisfies the condition of consistency, i.e., if e'' is a third frame field and ω'' is the connection matrix relative to e'', the relation between ω and ω'' follows from the relations between ω, ω', and ω', ω''.

A connection is also called a covariant or absolute differentiation. In physics it is called a gauge potential, for a reason to be given below. It is given relative to a frame field by a matrix of linear differential forms, with the transformation law (18).

REMARK: From (18) we see that connections exist in a vector bundle. The main idea of the proof is as follows: Take an open covering $\{U_\alpha\}$ of M, by coordinate neighborhoods, in each of which choose a frame field. Take any connection matrix in U_1. It suffices to define the connection in $\bigcup_{1\le\alpha\le m} U_\alpha$ by induction on m. Suppose the connection be defined in $V = \bigcup_{1\le\alpha\le m-1} U_\alpha$. Relative to the frame field in U_m the connection matrix is given in $V \cap U_m$. It remains to extend it over U_m, using standard extension theorems in Euclidean space. The complete argument needs a little care, because the extension theorem holds only for functions defined on closed sets. We leave the details to the reader.

Taking the exterior derivative of (18) and using the equation itself, we get

$$\Omega' A = A\Omega, \tag{19}$$

where Ω is a $(q \times q)$-matrix of exterior two-forms, given by

$$\Omega = d\omega - \omega \wedge \omega. \tag{20}$$

In the last term we use the multiplication of matrices, while the multiplication of differential forms is exterior multiplication. The matrix Ω is called the curvature matrix, relative to the frame field e_i.

Exterior differentiation of (20) gives

$$d\Omega = \omega \wedge \Omega - \Omega \wedge \omega. \tag{21}$$

This is called the Bianchi identity.

EXERCISE: Let X, Y be two vector fields on M, and let

$$R(X,Y) = < X \wedge Y, \Omega >, \qquad R'(X,Y) = < X \wedge Y, \Omega' > . \tag{22}$$

Then $R(X,Y)$ and $R'(X,Y)$ are matrices satisfying the equation

$$R'(X,Y) = AR(X,Y)A^{-1} \tag{23}$$

For the section s in (12) let

$$\sigma = (s^1, \ldots, s^q). \tag{24}$$

Then $\sigma Re = \sigma' R' e'$ is independent of the choice of the frame field. The map

$$K(X,Y) : s = \sigma e \rightarrow \sigma R(X,Y) e \tag{25}$$

is called the curvature transformation.

Curvature measures the non-commutativity of covariant differentiation. Prove that, as an operator on $\Gamma(E)$,

$$K(X,Y) = \nabla_X \nabla_Y - \nabla_Y \nabla_X - \nabla_{[X,Y]}. \tag{26}$$

4. Chern Forms and Chern Classes.

We consider complex vector bundles. Let

$$det(I + \frac{\sqrt{-1}}{2\pi} \Omega) = 1 + c_1(\Omega) + \ldots + c_q(\Omega), \tag{27}$$

where I is the unit matrix. Then $c_i(\Omega)$, $1 \leq i \leq q$, is an exterior differential form of degree $2i$. By (19) they are independent of the choice of the frame field, i.e.,

$$c_i(\Omega) = c_i(\Omega'). \tag{28}$$

It follows that $c_i(\Omega)$ are globally defined on M. For we can take an open covering $\{U_\alpha\}$ of M and choose in each U_α a frame field. The resulting c_i's must agree in the intersection of any two members of the covering. We write

$$c_i(E;D) = c_i(\Omega), \tag{29}$$

indicating that the forms now only depend on the connection D (and of course on the bundle E). These c_i are called the Chern forms.

We introduce another set of forms in M as follows:

$$b_i(\Omega) = Tr\left(\frac{\sqrt{-1}}{2\pi}\,\Omega^i\right), \qquad 1 \le i \le q. \tag{30}$$

When Ω is diagonal, both $c_i(\Omega)$ and $b_i(\Omega)$ are symmetrical functions of its diagonal elements. Between them Newton's identities are valid:

$$b_i - c_1 b_{i-1} + c_2 b_{i-2} + \ldots + (-1)^{i-1} c_{i-1} b_1 + (-1)^i i c_i = 0, \qquad 1 \le i \le q \tag{31}$$

Since both $c_i(\Omega)$ and $b_i(\Omega)$ are invariant under the change (19), we can use the latter to put Ω in a normal form, and it is not hard to see that (31) is true in general. It follows that b_i (resp. c_i) is a polynomial in $c_1, \ldots, c_i$ (resp. $b_1, \ldots, b_i$) with integral (resp. rational) coefficients.

THEOREM 1. *The forms b_i and c_i, $1 \le i \le q$, are closed.*

Owing to the above remark, it suffices to show that b_i is closed. In fact, we have, by (21),

$$\begin{aligned} dTr\Omega^i &= iTr(d\Omega \wedge \Omega^{i-1}) \\ &= iTr(\omega \wedge \Omega^i - \Omega \wedge \omega \wedge \Omega^{i-1}) = 0. \end{aligned}$$

This proves the theorem.

The following theorem gives the effect on these forms by changing the connection:

THEOREM 2. *Let D and D_1 be two connections on the bundle E. Then the forms*

$$b_i(E;D) - b_i(E;D_1), \qquad c_i(E;D) - c_i(E;D_1), \qquad 1 \le i \le q,$$

are exact.

Again it is sufficient to prove the theorem for b_i. We observe that

(32) $$D_t = (1 - t)D_0 + tD_1, \qquad 0 \leq t \leq 1, \qquad D_0 = D,$$

is also a connection. Relative to a frame field e_i let ω_t be the connection matrix of D_t. Its curvature matrix is given by

(33) $$\Omega_t = d\omega_t - \omega_t \wedge \omega_t,$$

and, by exterior differentiation, we have the Bianchi identity

(34) $$d\Omega_t = -\Omega_t \wedge \omega_t + \omega_t \wedge \Omega_t.$$

Let

(35) $$\eta = \omega_1 - \omega_0.$$

By (18) we have, under a change of the frame field,

(36) $$\eta' A = A\eta.$$

Hence the form

(37) $$\alpha = Tr(\eta \wedge \Omega_t^{i-1})$$

is globally defined on M. By using (33), we find

$$d\alpha = Tr\{(d\eta - \eta \wedge \omega_t - \omega_t \wedge \eta) \wedge \Omega_t^{i-1}\};$$

(differentiate each of the factors and telescope!) The expression between the parentheses is equal to

$$\beta = d\eta - \eta \wedge \omega_0 - \omega_0 \wedge \eta - 2t\eta \wedge \eta$$

On the other hand, expansion of (33) in t gives

$$\Omega_t = d\omega_0 - \omega_0 \wedge \omega_0 + t(d\eta - \omega_0 \wedge \eta$$

$$- \eta \wedge \omega_0) - t^2 \eta \wedge \eta,$$

so that

$$\frac{d}{dt}\,\Omega_t = \beta.$$

It follows that

$$\frac{1}{i}\,\frac{d}{dt}\,Tr(\Omega_t^i) = d\alpha$$

Integrating with respect to t, we get

$$(38) \qquad Tr(\Omega_1^i) - Tr(\Omega_0^i) = id\int_0^1 \alpha dt.$$

This proves the statement in the theorem for b_i, and the theorem is proved.

The de Rham cohomology on a differentiable manifold can be summarized as follows: Let A^r be the space of all C^∞ differential forms of degree r on M, and C^r its subspace of forms which are closed. The quotient space

$$(39) \qquad H^r(M;\mathbf{C}) = \frac{C^r}{dA^{r-1}}$$

is called the r-dimensional de Rham cohomology group of M. An element of $H^r(M;\mathbf{C})$ is called an r-dimensional de Rham cohomology class. If $\alpha \in C^r$, we shall denote its de Rham class by $\{\alpha\}$.

Theorem 2 says that the de Rham cohomology class $\{c_i(E;D)\}$ is independent of the connection D. It is called the ith Chern class of the bundle E and will be denoted by $c_i(E)$, $1 \le i \le q$.

If M is a compact oriented manifold of even dimension $2m$ and $i_1 + \ldots + i_p = m$, then

$$(40) \qquad \int_M c_{i_1}(E) \ldots c_{i_p}(E) = c_{i_1 \ldots i_p}(E)$$

is called a Chern number of E.

This is a remarkable story. From its very definition it is not clear whether there are non-trivial vector bundles, i.e. whether every vector bundle must be globally a product. The necessity of introducing a covariant differentiation and the possibility that it is non-commutative lead to the curvature. The latter should be considered as a two-form, and some elementary combinations of it, based essentially on the eigenvalues of a matrix and

their elementary symmetric functions, lead to the first and most important invariants of a vector bundle.

The vector bundle E is called hermitian, if there is a C^∞-field of positive definite hermitian scalar products (,), such that

$$\overline{(s,t)} = (t,s), \qquad s,t \in \pi^{-1}(x), \qquad x \in M. \tag{41}$$

Let e_i be a frame field and let

$$(e_i, e_j) = g_{i\bar{j}}, \qquad 1 \leq i,\ j \leq q \tag{42}$$

Then the matrix

$$(g_{i\bar{j}}) \tag{43}$$

is positive-definite hermitian. If

$$s = \sum s^i e_i, \qquad t = \sum t^i e_i, \tag{44}$$

then

$$(s,t) = \sum g_{i\bar{j}} s^i t^{\bar{j}}, \qquad t^{\bar{j}} = \overline{t^j}. \tag{45}$$

The connection D is called admissible, if (s,t) remains constant, when s and t are parallelly displaced, i.e., when $Ds = Dt = 0$. In fact, under these conditions we find

$$d(s,t) = \sum (dg_{i\bar{j}} - \omega_{i\bar{j}} - \omega_{\bar{j}i}) s^i t^{\bar{j}},$$

where

$$\omega_{i\bar{j}} = \sum \omega_i^k g_{k\bar{j}}, \qquad \omega_{\bar{j}i} = \overline{\omega}_{ji}. \tag{46}$$

Hence the conditions for an admissible connection are

$$dg_{i\bar{j}} = \omega_{i\bar{j}} + \omega_{\bar{j}i} \tag{47}$$

It follows by an extension argument (cf. Remark in §3) that on a given hermitian vector bundle admissible connections exist.

The frame e_i is called unitary, if

$$(48) \qquad (e_i, e_j) = \delta_{i\bar{j}} \ (= 1, \text{ if } i = j, \text{ and } = 0 \text{ otherwise})$$

On an hermitian vector bundle with an admissible connection we can restrict ourselves to unitary frames. Then we have, by (47),

$$(49) \qquad \omega_{i\bar{j}} + \omega_{\bar{j}i} = 0,$$

i.e., the connection matrix is skew-hermitian. By (20) it follows that the same is true of the curvature matrix. This implies that $det(I + \frac{i}{2\pi}\Omega_{i\bar{j}})$ in (27) is real, and so are the forms $c_i(\Omega)$, $1 \leq i \leq q$. This differential-geometric argument shows that the Chern classes $c_i(E)$ are real cohomology classes. It can be proved that they are integral cohomology classes, i.e. elements of $H^{2i}(M; \mathbf{Z})$.

Similar notions can be introduced for real vector bundles, so that we have Riemannian vector bundles and their admissible connections. By restricting to orthonormal frames which satisfy the conditions

$$(50) \qquad (e_i, e_j) = \delta_{ij},$$

we find that both the connection matrix and the curvature matrix are skew-symmetric. It follows that $c_i(\Omega) = 0$ if i is odd. The form

$$(51) \qquad p_i(\Omega) = (-1)^i c_{2i}(\Omega)$$

of degree $4i$, is called a Pontrjagin form of a real vector bundle E, and the class $p_i(E) = \{p_i(\Omega)\}$ is a cohomology class of dimension $4i$, to be called a Pontrjagin class.

5. Submanifolds in Euclidean Space.

We consider an immersion

$$(52) \qquad x : M^n \to E^{n+q}$$

of a manifold $M^n = M$ of dimension n into the Euclidean space of dimension $n+q$. The notation x will denote both a point $x \in M$ and its position vector from a fixed point of E^{n+q}. Over M we consider orthonormal frames xe_A, such that $x \in M$, e_α are tangent vectors to M at x, and hence e_i are normal vectors to M. Throughout this section we will use the following ranges of indices:

$$\begin{aligned} 1 &\leq A, B, C \leq n+q, \\ 1 &\leq \alpha, \beta, \gamma \leq n, \\ n+1 &\leq i, j, k \leq n+q. \end{aligned} \tag{53}$$

We can write

$$\begin{aligned} dx &= \sum \omega_\alpha e_\alpha, \\ de_A &= \sum \omega_{AB} e_B, \end{aligned} \tag{54}$$

where

$$\omega_{AB} + \omega_{BA} = 0. \tag{55}$$

Taking the exterior derivative of the first equation of (54) and equating to zero the coefficient of e_i, we get

$$\sum_\alpha \omega_\alpha \wedge \omega_{\alpha i} = 0. \tag{56}$$

Since x is an immersion, the ω_α are linearly independent and it follows by Cartan's lemma that

$$\omega_{\alpha i} = \sum h_{i\alpha\beta}\omega_\beta, \tag{57}$$

where

$$h_{i\alpha\beta} = h_{i\beta\alpha}. \tag{58}$$

The (ordinary) quadratic differential forms

$$\prod_i = \sum \omega_\alpha \omega_{\alpha i} = \sum h_{i\alpha\beta}\omega_\alpha\omega_\beta \tag{59}$$

are the <u>second fundamental forms</u> of M. The second fundamental form, also to be denoted by $\sum \text{II}_i e_i$, is a quadratic differential form with value in the normal bundle.

The exterior differentiation of the second equation of (54) gives

$$d\omega_{AB} = \sum \omega_{AC} \wedge \omega_{CB}. \tag{60}$$

From (54) we get, by projection from the normal bundle,

$$De_\alpha = \sum \omega_{\alpha\beta} e_\beta. \tag{61}$$

This defines a connection in the tangent bundle, as the conditions D1) and D2) in §3 are satisfied. It is a fundamental theorem of local Riemannian geometry that the connection D depends only on the induced metric on M. This is the Levi-Civita connection, as originally defined by him.

Similarly, there is a connection

$$D^\perp e_i = \sum \omega_{ij} e_j, \tag{62}$$

the normal connection, in the normal bundle.

We consider the case $n = 2$, $q = 1$, i.e., surfaces in the ordinary Euclidean space E^3. Suppose M be oriented, so that the unit normal vector e_3 is well-defined, and the orthonormal frame e_1, e_2 in the tangent plane is defined up to the transformation

$$\begin{pmatrix} e_1' \\ e_2' \end{pmatrix} = \begin{pmatrix} \cos\theta & \sin\theta \\ -\sin\theta & \cos\theta \end{pmatrix} \begin{pmatrix} e_1 \\ e_2 \end{pmatrix} \tag{63}$$

The connection matrix relative to e_1, e_2 is

$$\omega = \begin{pmatrix} 0 & \omega_{12} \\ -\omega_{12} & 0 \end{pmatrix}. \tag{64}$$

By (18) we find that under the change of frame (63),

$$\omega_{12}' = \omega_{12} + d\theta \tag{65}$$

The curvature matrix is

$$\Omega = \begin{pmatrix} 0 & \Omega_{12} \\ -\Omega_{12} & 0 \end{pmatrix} \tag{66}$$

and is invariant under a change of frame.

The curvature form Ω_{12} has a simple geometrical interpretation. In fact, by (20) we have

$$\Omega_{12} = d\omega_{12}. \tag{67}$$

The latter is, by (60), equal to

$$d\omega_{12} = -\omega_{13} \wedge \omega_{23} = -(h_{311}h_{322} - h_{312}^2)\omega_1 \wedge \omega_2, \tag{68}$$

where the expression between the parentheses is the determinant of the second fundamental form $\prod_3$ and is equal to the Gaussian curvature K of M. Thus we have

$$d\omega_{12} = \Omega_{12} = -KdA, \tag{69}$$

where $dA = \omega_1 \wedge \omega_2$ is the element of area. This leads immediately to the:

GAUSS-BONNET THEOREM. *Let M be a compact oriented surface in E^3. Then*

$$\frac{1}{2\pi}\int_M KdA = \chi(M), \tag{70}$$

where $\chi(M)$ is the Euler characteristic of M.

To prove this theorem let E_0 be the space of unit tangent vectors xe_1' of M. E_0 can be identified with the space of orthonormal frames of the tangent bundle of M, because e_1' determines e_2' as the unit vector orthogonal to e_1' such that $e_1'e_2'$ agrees with the orientation of M. E_0 is a three-dimensional manifold. As its local coordinates we can take those of M and the angle θ in (63). In E_0 we have the global one-form

$$\omega_{12}' = (de_1', e_2') = \omega_{12} + d\theta, \tag{71}$$

whose exterior derivative is

$$d\omega_{12}' = -KdA. \tag{72}$$

We define a vector field on M with singularities at a finite number of points x_k, $1 \leq k \leq r$; such a vector field always exists. Let Δ_k be a small disk with x_k as center.

The vector field lifts $M - \bigcup_{1 \le k \le r} \Delta_k$ into a surface in E_0 with the boundaries $\partial\Delta_k$. Applying Stokes Theorem to (72) we have

$$\frac{1}{2\pi}\int_{M-\cup\Delta_k} KdA = \frac{1}{2\pi}\sum\int_{-\partial\Delta_k} d\theta + \omega_{12}$$

As the Δ_k's tend to the points x_k, we have at the limit,

$$\frac{1}{2\pi}\int_M KdA = \sum_{1\le k\le r} I_k, \tag{73}$$

where I_k is the index of the vector field at x_k. Thus we have proved that the integral at the left-hand side is equal to the sum of indices, which is therefore independent of the choice of the vector field.

For another discussion of the Gauss-Bonnet theorem we refer the reader to the article "Curves and surfaces in Euclidean space" in this book.

By choosing a special vector field, we show that it is equal to $\chi(M)$. Our proof demonstrates at the same time the theorem that the sum of indices of a continuous vector field on M with a finite number of singularities is equal to $\chi(M)$.

Even the case $n = 1$, $q = 2$, i.e., curves in E^3, leads to interesting conclusions. In this case the normal planes xe_2e_3 form the normal bundle. Its normal connection $D^\perp$ is given by the matrix

$$\omega = \begin{pmatrix} 0 & \omega_{23} \\ -\omega_{23} & 0 \end{pmatrix}. \tag{74}$$

As above, when the normal frames undergo the change

$$\begin{pmatrix} e_2' \\ e_3' \end{pmatrix} = \begin{pmatrix} \cos\theta & \sin\theta \\ -\sin\theta & \cos\theta \end{pmatrix}\begin{pmatrix} e_2 \\ e_3 \end{pmatrix}, \tag{75}$$

the connection form is modified according to:

$$\omega_{23}' = \omega_{23} + d\theta \tag{76}$$

There is no curvature form, because the base manifold M is one-dimensional.

However, an interesting invariant can be introduced as follows: Let M be a closed curve, and let e_2 be a continuous and smooth normal vector field. Then

(77) $$T(e_2) = \frac{1}{2\pi} \int_M \omega_{23}$$

is a real number. From (76) we see that if e_2' is another smooth normal vector field, $T(e_2')$ differs from $T(e_2)$ by an integer. Thus $T(e_2)$ *mod* 1 is an invariant of the closed curve M, to be called the <u>total twist</u> of M. If the curvature of M never vanishes and M is C^3, we can choose e_2 to be the principal normal vector and $T(e_2)$ becomes the total torsion. But the total twist is defined for C^1-curves.

The total twist plays an important rôle in molecular biology. T. Banchoff and J. White proved that it is invariant under conformal transformations [2].

6. Complex Line Bundles.

A complex vector bundle with $q = 1$ is called a <u>complex line bundle</u>. It plays an important rôle in various parts of mathematics.

For a complex line bundle the connection and curvature matrices ω, Ω are one- and two-forms respectively, and (20) becomes

(78) $$\Omega = d\omega.$$

By (27) we have

(79) $$c_1(\Omega) = \frac{i}{2\pi} \Omega.$$

If the bundle is hermitian and the connection is admissible, we will restrict ourselves to unitary frames. Then both ω and Ω are skew-hermitian:

(80) $$\omega + \overline{\omega} = 0, \qquad \Omega + \overline{\Omega} = 0,$$

i.e., both $\sqrt{-1}\omega$ and $\sqrt{-1}\Omega$ are real. We emphasize that ω and Ω in this section are forms, not matrices.

Perhaps the most important complex line bundle is the Hopf bundle, defined as follows: The map

(81) $$\pi : C_{n+1} - \{0\} \to P_n(\mathbf{C})$$

defines the complex projective space $P_n(\mathbf{C})$ of dimension n as the space of all lines through the origin of C_{n+1}. If $(z_0, z_1, \ldots, z_n) = Z \in \mathbf{C}_{n+1} - \{0\}$, $\pi Z = [Z]$ is the point in P_n with Z as the homogeneous coordinate vector.

To study the geometry in $P_n(\mathbf{C})$ we introduce in $\mathbf{C}_{n+1}$ the hermitian scalar product

$$(Z, W) = \sum z_A \overline{w}_A, \tag{82}$$

where

$$Z = (z_0, \ldots, z_n), \qquad W = (w_0, \ldots, w_n), \tag{83}$$

and we use in this section the ranges of indices:

$$0 \leq A, B, C \leq n. \tag{84}$$

Z_A is called a unitary frame, if we have

$$(Z_A, Z_B) = \delta_{A\overline{B}}. \tag{85}$$

The space of all unitary frames can be identified with the unitary group $U(n+1)$. We have the diagram

$$\begin{array}{ccc} \{Z_0, Z_1, \ldots, Z_n\} & \in & U(n+1) \\ \downarrow & & \pi_0 \downarrow \\ Z_0 & \in & S^{2n+1} = \{Z \in C_{n+1} | (Z, Z) = 1\} \\ \downarrow & & \pi \downarrow \\ [Z_0] & \in & P_n(\mathbf{C}) \end{array} \tag{86}$$

The map π defines a circle bundle over $P_n(\mathbf{C})$, called the <u>Hopf bundle</u>. For $n = 1$, $P_1(\mathbf{C}) = S^2$ and the resulting map

$$\pi : S^3 \to S^2, \tag{87}$$

the <u>Hopf map</u>, was the first map discovered from a manifold to one of lower dimension, which is not homotopic to a constant map (which collapses the manifold to a point of the image space).

In the space of unitary frames Z_A we can write

$$dZ_A = \sum \omega_{A\overline{B}} Z_B, \tag{88}$$

where

$$\omega_{A\overline{B}} + \omega_{\overline{B}A} = 0, \qquad \omega_{\overline{B}A} = \overline{\omega}_{B\overline{A}}. \tag{89}$$

The $\omega_{A\overline{B}}$ are the Maurer-Cartan forms of $U(n+1)$. They satisfy the following Maurer-Cartan equations obtained by exterior differentiation of (88):

$$d\omega_{A\overline{B}} = \sum_C \omega_{A\overline{C}} \wedge \omega_{C\overline{B}}. \tag{90}$$

The method of moving frames is to use these equations in $U(n+1)$ to study the geometry of $P_n(\mathrm{C})$ or S^{2n+1}.

S^{2n+1} consists of the unit vectors of the complex line bundle (81). By (88),

$$DZ_0 = \omega_{0\overline{0}} Z_0 \tag{91}$$

defines an admissible connection in this bundle. It is a connection because the conditions D1) and D2) in §3 are satisfied, and is admissible because (89) implies that $\omega_{0\overline{0}}$ is skew-hermitian. The connection form is

$$\omega = \omega_{0\overline{0}}. \tag{92}$$

By (90) the curvature form of this connection is

$$\Omega = d\omega = \sum \omega_{0\overline{\alpha}} \wedge \omega_{\alpha\overline{0}} = -\sum \omega_{0\overline{\alpha}} \wedge \omega_{\overline{0}\alpha}, \qquad 1 \le \alpha \le n. \tag{93}$$

To this corresponds the hermitian differential form

$$\sum \omega_{0\overline{\alpha}} \omega_{\overline{0}\alpha} = (dZ_0, dZ_0) - (dZ_0, Z_0)(Z_0, dZ_0). \tag{94}$$

The last expression remains invariant, when Z_0 undergoes the change $Z_0 \rightarrow \lambda Z_0$, $|\lambda| = 1$. Hence it is a positive-definite hermitian differential from in $P_n(\mathrm{C})$, and defines an hermitian metric in $P_n(\mathrm{C})$. Its Kähler form is

$$K = \frac{i}{2} \sum \omega_{0\overline{\alpha}} \wedge \omega_{\overline{0}\alpha} = -\frac{i}{2}\Omega, \tag{95}$$

which is clearly closed. Thus this metric is kählerian. It is called the Study-Fubini metric on $P_n(\mathbf{C})$.

A fundamental fact is that, because of the complex structures involved, the connection form ω has a "potential", that is, it can be "integrated". In fact, let $Z \in C_{n+1} - \{0\}$, and let

$$Z_0 = Z/|Z|, \qquad |Z|^2 = (Z, Z). \tag{96}$$

Then we find

$$\begin{aligned} \pi_0^* \, \omega_{0\overline{0}} &= (DZ_0, Z_0) \ = \ (dZ_0, Z_0) \\ &= \frac{1}{2|Z|^2}\{(dZ, Z) \ - \ (Z, dZ)\} \ = \ (\partial \ - \ \overline{\partial}) \log |Z|, \end{aligned} \tag{97}$$

where ∂ and $\overline{\partial}$ are differentiations in C_{n+1} with respect to the holomorphic coordinates z_A and the anti-holomorphic coordinates $\overline{z}_A$ respectively. But the last expression is an expression in $C_{n+1} - \{0\}$. To get one in $P_n(\mathbf{C})$ we take a fixed vector $A \in C_{n+1} - \{0\}$. Then the equation

$$(Z, A) \ = \ 0 \tag{98}$$

defines a hyperplane L in $P_n(\mathbf{C})$. The function

$$\frac{|Z, A|^2}{|Z|^2}, \qquad |Z, A|^2 \ = \ |(Z, A)|^2 \tag{99}$$

is well-defined in $P_n(\mathbf{C}) - L$, because it remains unchanged when Z is multplied by a factor. Clearly

$$\partial\overline{\partial} \log |Z, A|^2 \ = \ 0. \tag{100}$$

(This can be described by saying that the logarithm of the absolute value of a holomorphic function is harmonic.) We have therefore the relation

$$\Omega \ = \ \partial\overline{\partial} \log \ \frac{|Z, A|^2}{|Z|^2} \tag{101}$$

in $P_n(\mathbf{C}) - L$. This contains the analytic content of the following theorem:

THEOREM (WINTINGER). *Let M be a compact Riemann surface and $f : M \to P_n(\mathbf{C})$ be a holomorphic mapping. The image $f(M)$ meets all hyperplanes of $P_n(\mathbf{C})$ the same number of times, which is equal to the area of $f(M)$, properly normalized.*

We normalize the element of area of $P_n(\mathbf{C})$ to be K/π, K being the Kähler form defined in (95). Then

$$\frac{K}{\pi} = -\frac{i}{2\pi}\,\Omega = -\frac{i}{\pi}\,\partial\bar{\partial}\log\frac{|Z,A|}{|Z|} = \frac{i}{2\pi}\,d(\partial - \bar{\partial})\log\frac{|Z,A|}{|Z|} \tag{102}$$

We leave it to the reader to show that the total area of $P_1(\mathbf{C})$ is 1.

By hypothesis $f(M)$ is a compact holomorphic curve. We suppose that it does not lie in a hyperplane L. Then it meets L in a finite number of points, say $x_k \in M$, $1 \le k \le r$. About each x_k we take a disc Δ_k. Formula (102) allows us to apply Stokes Theorem to $M - \cup\Delta_k$. It is easy to see that the boundary term will tend, as Δ_k shrinks to x_k, to the number of zeros of the holomorphic function (Z, A), and the theorem is proved.

Wirtinger's Theorem could also be proved by a topological argument. The great advantage of our approach is that it relates the local and global aspects of the problem in the clearest way and extends to the case that M is non-compact. The ideas introduced in our differential-geometric treatment is at the basis of the value distribution theory of non-compact holomorphic curves. Needless to say, the non-compact case is much more delicate. For its development cf. [6], [9].

The tangent bundle of a surface in E^3 and the normal bundle of a curve in E^3, as discussed in §5, can also be considered as complex line bundles. For the fibers are oriented planes with an inner product and a complex structure is defined in which multiplication by i is rotation by 90°. The results at the end of §5 can be put in the notation of this section. We will leave it as an exercise.

7. Maxwell's and Yang-Mills' Equations.

Maxwell's equations in the space-time $(x^1, x^2, x^3, t = x^0)$, as commonly given, can be written as

$$\begin{aligned} div\,\vec{E} &= 4\pi\rho, \\ curl\,\vec{B} - \tfrac{\partial}{\partial t}\,\vec{E} &= 4\pi\,\vec{j}, \end{aligned} \tag{103a}$$

(103b)
$$div\ \vec{B} = 0,\qquad curl\ \vec{E} + \frac{\partial}{\partial t}\vec{B} = 0,$$

where

(104)
$$\begin{aligned}\vec{E} &= (E_1, E_2, E_3) = \text{electric field},\\ \vec{B} &= (B_1, B_2, B_3) = \text{magnetic field},\\ \rho &= \text{charge density},\\ \vec{j} &= (j_1, j_2, j_3) = \text{current vector}.\end{aligned}$$

These can be written in a different form, which will be susceptible to natural and important generalizations.

In fact, introduce the anti-symmetric matrix

(105)
$$(F_{\alpha\beta}) = \begin{pmatrix} 0 & E_1 & E_2 & E_3 \\ -E_1 & 0 & -B_3 & B_2 \\ -E_2 & B_3 & 0 & -B_1 \\ -E_3 & -B_2 & B_1 & 0 \end{pmatrix},$$

and let

(106)
$$J = \rho dx^0 + j_1 dx^1 + j_2 dx^2 + j_3 dx^3$$
$$F = \sum F_{\alpha\beta} dx^\beta \wedge dx^\alpha$$

In the space-time we make use of the Lorentz-metric

(107)
$$ds^2 = -(dx^0)^2 + (dx^1)^2 + (dx^2)^2 + (dx^3)^2$$

and the corresponding $*$-operator in the sense of Hodge. With these notations it is immediately verified that (103a), (103b) can be written respectively as

(108a)
$$d^* F = 4\pi J$$

(108b)
$$dF = 0,$$

where d^* is the codifferential defined by

(109)
$$d^* = *d*.$$

This is exactly the situation discussed in the last section, when M is a four-dimensional Lorentzian manifold and E is an hermitian line bundle over M. There is a discrepancy between our notation and terminology and those of the physicists, and we make the following table:

mathematics	physics
connection form ω	gauge potential A
curvature form Ω	Faraday F or strength

Equation (103b) or (108b) is the Bianchi identity. It is a consequence of the equation

$$dA = F, \tag{110}$$

which rewrites (78). It is for this fact that A is called a gauge potential.

It has recently been realized that in order to describe all phenomena in electricity and magnetism one should use (110) instead of the classical equation (108b), which is a consequence of (110) and has been adequate for most applications. This was the consequence of an important experiment proposed by Y. Aharanov and D. Böhm in 1959 and carried out by R.G. Chambers in 1960; cf. [10]. In other words, the unknown in Maxwell's equations should be the connection form or the gauge potential A and not the curvature form or the strength F, and the equations should be

$$d^*F = 4\pi J, \qquad dA = F. \tag{111}$$

This becomes manifest when one generalizes to the Yang-Mills equations; cf. [1]. The object here is an $SU(2)$-bundle over M, again a four-dimensional Lorentzian manifold. As the gauge group is non-abelian, one has to use covariant differentiation D_A. The Yang-Mills equations are

$$D_A^*F = 4\pi J, \qquad D_AA = F, \tag{112}$$

where

$$D_AA = dA - A \wedge A. \tag{113}$$

Thus we see that the Yang-Mills equations are a straight-forward generalization of the Maxwell's equations, being the cases when the fiber dimensions are $q = 1$ and $q = 2$ respectively. The Yang-Mills equations are playing a far-reaching rôle in the study of four-dimensional manifolds (work of S. Donaldson), the main reason being that their solutions give manifolds which have an important geometrical meaning.

Literature

(Standard books on differential geometry are not included.)

1. M.F. Atiyah, Geometry of Yang-Mills Fields, Pisa 1979.

2. T. Banchoff and J. White, The behaviour of the total twist and the self-linking number of a closed space curve under inversions, Math. Scandinavica 36, 254-262 (1975).

3. S. Chern, Complex Manifolds Without Potential Theory, 2nd edition, Springer 1979.

4. S. Chern, Circle Bundles, Geometry and Topology, III Latin Amer. School of Math., Lecture Notes in Math., No. 597, 114-131, Springer 1977.

5. Johan L. Dupont, Curvature and Characteristic Classes, Lecture Notes in Math. 640, Springer 1978.

6. P. Griffiths, Entire Holomorphic Mappings in One and Several Complex Variables, Annals of Math. Studies, No. 85, Princeton Univ. Press 1976.

7. J.W. Milnor and J. Stasheff, Characteristic Classes, Annals of Math. Studies 76, Princeton 1974.

8. H.V. Pittie, Characteristic Classes of Foliations, Pitman, San Francisco, 1976.

9. H. Wu, The Equidistribution Theory of Holomorphic Curves, Annals of Math. Studies, No. 64, Princeton Univ. Press 1970.

10. T.T. Wu and C.N. Yang, Concept of non-integrable phase-factors and global formulation of gauge fields, Physical Review D, 12,3845-57 (1975).

Reprinted from
Math. Sci. Research Inst. (1987) 1–23.

DUPIN SUBMANIFOLDS IN LIE SPHERE GEOMETRY

Thomas E. Cecil and Shiing-Shen Chern

1. Introduction.

Consider a piece of surface immersed in three-dimensional Euclidean space E^3. Its normal lines are the common tangent lines of two surfaces, the focal surfaces. These focal surfaces may have singularities, and a classical theorem says that if the focal surfaces both degenerate to curves, then the curves are conics, and the surface is a cyclide of Dupin. (See, for example, [CR, pp. 151-166].) Equivalently, the cyclides can be characterized as those surfaces in E^3 whose two distinct principal curvatures are both constant along their corresponding lines of curvature.

The cyclides have been generalized to an interesting class of hypersurfaces in E^n, the Dupin hypersurfaces. Initially, a hypersurface M in E^n was said to be Dupin if the number of distinct principal curvatures (or focal points) is constant on M and if each principal curvature is constant along the leaves of its corresponding principal foliation. (See [CR], [Th], [GH].) More recently, this has been generalized to include cases where the number of distinct principal curvatures is not constant. (See [P3], [CC].)

The study of Dupin hypersurfaces in E^n is naturally situated in the context of Lie sphere geometry, developed by Lie [LS] as part of his work on contact transformations. The projectivized cotangent bundle PT^*E^n of E^n has a contact structure. In fact, if $x^1,\ldots,x^n$ are the coordinates in E^n, the

---

The first author was supported by NSF Grant No. DMS 87-06015, the second author by NSF Grant No. DMS 87-01609.

contact structure is defined by the linear differential form $dx^n - p_1 dx^1 - \ldots - p_{n-1} dx^{n-1}$. Lie proved that the pseudo-group of all contact transformations carrying (oriented) hyperspheres in the generalized sense (i.e., including points and oriented hyperplanes) into hyperspheres is a Lie group, called the Lie sphere group, isomorphic to $O(n+1,2)/\pm I$, where $O(n+1,2)$ is the orthogonal group for an indefinite inner product on $\mathbb{R}^{n+3}$ with signature $(n+1,2)$. The Lie sphere group contains as a subgroup the Moebius group of conformal transformations of E^n and, of course, the Euclidean group. Lie exhibited a bijective correspondence between the set of oriented hyperspheres in E^n and the points on the quadric hypersurface Q^{n+1} in real projective space P^{n+2} given by the equation $\langle x,x \rangle = 0$, where $\langle , \rangle$ is the inner product on $\mathbb{R}^{n+3}$ mentioned above. The manifold Q^{n+1} contains projective lines but no linear subspaces of P^{n+2} of higher dimension. The 1-parameter family of oriented spheres corresponding to the points of a projective line lying on Q^{n+1} consists of all oriented hyperspheres which are in oriented contact at a certain contact element on E^n. Thus, Lie constructed a local diffeomorphism between PT^*E^n and the manifold Λ^{2n-1} of projective lines which lie on Q^{n+1}.

An immersed submanifold $f: M^k \to E^n$ naturally induces a Legendre submanifold $\lambda: B^{n-1} \to \Lambda^{2n-1}$, where B^{n-1} is the bundle of unit normal vectors to f (take $B^{n-1} = M^{n-1}$ in the case $k = n-1$). This Legendre map λ has similarities with the familiar Gauss map, and like the Gauss map, it can be a powerful tool in the study of submanifolds of Euclidean space. In particular, the Dupin property for hypersurfaces in E^n is easily formulated in terms of the Legendre map, and it is immediately seen to be invariant under Lie sphere transformations.

The study of Dupin submanifolds has both local and global aspects. Thorbergsson [Th] showed that a Dupin hypersurface M with g distinct principal curvatures at each point must be taut, i.e., every nondegenerate Euclidean distance function $L_p(x) = |p-x|^2$, $p \in E^n$, must have the minimum number of critical points on M. Tautness was shown to be invariant under Lie transformations in our earlier paper [CC]. Using tautness and the work of Münzner [Mu], Thorbergsson was then able to conclude that the number g must be 1,2,3,4 or 6, as with an isoparametric hypersurface in the sphere S^n. The case g = 1 is, of course, handled by the well-known classification of umbilic hypersurfaces. Compact Dupin hypersurfaces with g=2 and g=3 were classified by Cecil and Ryan (see [CR, p. 168]) and Miyaoka [M1] respectively. In two recent preprints, Miyaoka [M2], [M3] has made further progress on the classification of compact Dupin hypersurfaces in the cases g=4 and g=6. Meanwhile, Grove and Halperin [GH] have completely determined the topology of all compact Dupin hypersurfaces in the cases g=4 and g=6.

In this paper, we study Dupin hypersurfaces in the setting of Lie sphere geometry using local techniques. In Section 2, we give a brief introduction to Lie sphere geometry. In Section 3, we introduce the basic differential geometric notions: the Legendre map and the Dupin property. The case of E^3 is handled in Section 4, where we handle the case of g=2 distinct focal points for E^n. This was first done for n > 3 by Pinkall [P3]. Our main contribution lies in Section 5, where we treat the case E^4 by the method of moving frames. This case was also studied by Pinkall [P2], but our treatment seems to be more direct and differs from his in several essential points. It is our hope that this method will provide a framework and give some direction for the study of Dupin hypersurfaces in E^n for n > 4.

2. Lie Sphere Geometry.

We first present a brief outline of the main ideas in Lie's geometry of spheres in $\mathbb{R}^n$. This is given in more detail in Lie's original treatment [LS], in the book of Blaschke [B], and in our paper [CC].

The basic construction in Lie sphere geometry associates each oriented sphere, oriented plane and point sphere in $\mathbb{R}^n \cup \{\infty\} = S^n$ with a point on the quadric Q^{n+1} in projective space P^{n+2} given in homogeneous coordinates $(x_1,\ldots,x_{n+3})$ by the equation

$$\langle x,x\rangle = -x_1^2 + x_2^2 + \ldots + x_{n+2}^2 - x_{n+3}^2 = 0. \tag{2.1}$$

We will denote real (n+3)-space endowed with the metric (2.1) of signature (n+1,2) by $\mathbb{R}_2^{n+3}$.

We can designate the orientation of a sphere in $\mathbb{R}^n$ by assigning a plus or minus sign to its radius. Positive radius corresponds to the orientation determined by the field of outward normals to the sphere, while a negative radius corresponds to the orientation determined by the inward normal. A plane in $\mathbb{R}^n$ is a sphere which goes through the point ∞. The orientation of the plane can be associated with a choice of unit normal N. The specific correspondence between the points of Q^{n+1} and the set of oriented spheres, oriented planes and points in $\mathbb{R}^n \cup \{\infty\}$ is then given as follows:

(2.2)

<u>Euclidean</u>	<u>Lie</u>
Points: $u \in \mathbb{R}^n$	$\left[\left(\frac{1+u\cdot u}{2}, \frac{1-u\cdot u}{2}, u, 0\right)\right]$
∞	$[(1,-1,0,0)]$
<u>Spheres</u>: Center p, signed radius r	$\left[\left(\frac{1+p\cdot p-r^2}{2}, \frac{1-p\cdot p+r^2}{2}, p, r\right)\right]$
<u>Planes</u>: $u\cdot N = h$, unit normal N	$[(h,-h,N,1)]$.

Here the square brackets denote the point in projective space P^{n+2} given by the homogeneous coordinates in the round brackets, and $u \cdot u$ is the standard Euclidean dot product in $\mathbb{R}^n$.

From (2.2), we see that the point spheres correspond to the points in the intersection of Q^{n+1} with the hyperplane in P^{n+2} given by the equation $x_{n+3} = 0$. The manifold of point spheres is called <u>Moebius</u> <u>space</u>.

A fundamental notion in Lie sphere geometry is that of oriented contact of spheres. Two oriented spheres S_1 and S_2 are in <u>oriented</u> <u>contact</u> if they are tangent and their orientations agree at the point of tangency. If p_1 and p_2 are the respective centers of S_1 and S_2, and r_1 and r_2 are the respective signed radii, then the condition of oriented contact can be expressed analytically by

$$(2.3) \qquad |p_1 - p_2| = |r_1 - r_2|.$$

If S_1 and S_2 are represented by $[k_1]$ and $[k_2]$ as in (2.2), then (2.3) is equivalent to the condition

$$(2.4) \qquad \langle k_1, k_2 \rangle = 0.$$

In the case where S_1 and/or S_2 is a plane or a point in $\mathbb{R}^n$, oriented contact has the logical meaning. That is, a sphere S and plane π are in oriented contact if π is tangent to S and their orientations agree at the point of contact. Two oriented planes are in oriented contact if their unit normals are the same. They are in oriented contact at the point ∞. A point sphere is in oriented contact with a sphere or plane S if it lies on S, and ∞ is in oriented contact with each plane. In each case, the analytic condition for oriented contact is equivalent to (2.4) when the two "spheres" in question are represented in Lie coordinates as in (2.2).

Because of the signature of the metric (2.1), the quadric Q^{n+1} contains lines in P^{n+2} but no linear subspaces of higher dimension. A line on Q^{n+1} is determined by two points [x], [y] in Q^{n+1} satisfying $\langle x,y\rangle = 0$. The lines on Q^{n+1} form a manifold of dimension 2n-1, to be denoted by Λ^{2n-1}. In $\mathbb{R}^n$, a line on Q^{n+1} corresponds to a 1-parameter family of oriented spheres such that any two of the spheres are in oriented contact, i.e., all the oriented spheres tangent to an oriented plane at a given point,i.e., an oriented contact element. Of course, a contact element can also be represented by an element of T_1S^n, the bundle of unit tangent vectors to the Euclidean sphere S^n in E^{n+1} with its usual metric. This is the starting point for Pinkall's [P3] considerations of Lie geometry.

A <u>Lie sphere transformation</u> is a projective transformation of P^{n+2} which takes Q^{n+1} to itself. Since a projective transformation takes lines to lines, a Lie sphere transformation preserves oriented contact of spheres. The group G of Lie sphere transformations is isomorphic to $O(n+1,2)/\{\pm I\}$, where $O(n+1,2)$ is the group of orthogonal transformations for the inner product (2.1). Moebius transformations are those Lie transformations which take point spheres to point spheres. The group of Moebius transformations is isomorphic to $O(n+1,1)/\{\pm I\}$.

3. <u>Legendre Submanifolds</u>.

Here we recall the concept of a Legendre submanifold of the contact manifold $\Lambda^{2n-1}(= \Lambda)$ using the notation of [CC]. In this section, the ranges of the indices are as follows:

(3.1) $$1 \le A,B,C \le n + 3,$$
$$3 \le i,j,k \le n + 1.$$

Instead of using an orthonormal frame for the metric $\langle , \rangle$ defined by (2.1), it is useful to consider a <u>Lie</u> <u>frame</u>, that is, an ordered set of vectors Y_A in $\mathbb{R}^{n+3}_2$ satisfying

(3.2) $$\langle Y_A, Y_B \rangle = g_{AB},$$

with

(3.3) $$(g_{AB}) = \begin{bmatrix} J & 0 & 0 \\ 0 & I_{n-1} & 0 \\ 0 & 0 & J \end{bmatrix},$$

where I_{n-1} is the identity $(n-1) \times (n-1)$ matrix and

(3.4) $$J = \begin{bmatrix} 0 & 1 \\ 1 & 0 \end{bmatrix}.$$

The space of all Lie frames can be identified with the orthogonal group $O(n+1,2)$, of which the Lie sphere group, being isomorphic to $O(n+1,2)/\{\pm I\}$, is a quotient group. In this space, we introduce the Maurer-Cartan forms

(3.5) $$dY_A = \Sigma\, \omega_A^B\, Y_B.$$

Through differentiation of (3.2), we show that the following matrix of 1-forms is skew-symmetric

$$(3.6) \qquad (\omega_{AB}) = \begin{bmatrix} \omega_1^2 & \omega_1^1 & \omega_1^i & \omega_1^{n+3} & \omega_1^{n+2} \\ \omega_2^2 & \omega_2^1 & \omega_2^i & \omega_2^{n+3} & \omega_2^{n+2} \\ \omega_j^2 & \omega_j^1 & \omega_j^i & \omega_j^{n+3} & \omega_j^{n+2} \\ \omega_{n+2}^2 & \omega_{n+2}^1 & \omega_{n+2}^i & \omega_{n+2}^{n+3} & \omega_{n+2}^{n+2} \\ \omega_{n+3}^2 & \omega_{n+3}^1 & \omega_{n+3}^i & \omega_{n+3}^{n+3} & \omega_{n+3}^{n+2} \end{bmatrix}.$$

Next, by taking the exterior derivative of (3.5), we get the Maurer-Cartan equations

$$(3.7) \qquad d\omega_A^B = \sum_C \omega_A^C \wedge \omega_C^B .$$

In [CC], we then show that the form

$$\omega_1^{n+2} = \langle dY_1, Y_{n+3} \rangle$$

gives a contact structure on the manifold Λ.

Let B^{n-1} (= B) be an (n-1)-dimensional smooth manifold. A <u>Legendre map</u> is a smooth map $\lambda : B \to \Lambda$ which annihilates the contact form on Λ, i.e., $\lambda^* \omega_1^{n+2} = 0$ on B. All of our calculations are local in nature. We use the method of moving frames and consider a smooth family of Lie frames Y_A on an open subset U of B, with the line $\lambda(b)$ given by $[Y_1(b), Y_{n+3}(b)]$ for each $b \in U$. The Legendre map λ is called a <u>Legendre submanifold</u> if for a generic choice of Y_1 the forms ω_1^i, $3 \le i \le n+1$, are linearly independent, i.e.,

$$(3.8) \qquad \wedge\, \omega_1^i \neq 0 \text{ on } U .$$

Here and later we pull back the structure forms to B^{n-1} and omit the symbols of such pull-backs for simplicity. Note that the Legendre condition is just

(3.9) $$\omega_1^{n+2} = 0 .$$

We now assume that our choice of Y_1 satisfies (3.8). By exterior differentiation of (3.9) and using (3.6), we get

(3.10) $$\Sigma\, \omega_1^i \wedge \omega_{n+3}^i = 0 .$$

Hence by Cartan's Lemma and (3.8), we have

(3.11) $$\omega_{n+3}^i = \Sigma\, h_{ij}\omega_1^j \text{ , with } h_{ij} = h_{ji} .$$

The quadratic differential form

$$II(Y_1) = \sum_{i,j} h_{ij}\omega_1^i\omega_1^j ,$$

defined up to a non-zero factor and depending on the choice of Y_1, is called the <u>second fundamental form</u>.

This form can be related to the well-known Euclidean second fundamental form in the following way. Let e_{n+3} be any unit timelike vector in $\mathbb{R}_2^{n+3}$. For each $b \in U$, let $Y_1(b)$ be the point of intersection of the line $\lambda(b)$ with the hyperplane $e_{n+3}^{\perp}$. Y_1 represents the locus of point spheres in the Moebius space $Q^{n+1} \cap e_{n+3}^{\perp}$, and we call Y_1 the <u>Moebius projection of λ</u> determined by e_{n+3}. Let e_1 and e_2 be unit timelike, respectively spacelike, vectors orthogonal to e_{n+3} and to each other, chosen so that Y_1 is not the point at infinity $[e_1 - e_2]$ for any $b \in U$. We can represent Y_1 by the vector

$$(3.13) \qquad Y_1 = \frac{1 + f\cdot f}{2} e_1 + \frac{1 - f\cdot f}{2} e_2 + f,$$

as in (2.2), where $f(b)$ lies in the space $\mathbb{R}^n$ of vectors orthogonal to e_1, e_2 and e_{n+3}. We will call the map $f: B \to \mathbb{R}^n$ the Euclidean projection of λ determined by the ordered triple e_1, e_2, e_{n+3}. The regularity condition (3.8) is equivalent to the condition that f be an immersion on U into $\mathbb{R}^n$. For each $b \in U$, let $Y_{n+3}(b)$ be the intersection of $\lambda(b)$ with the orthogonal complement of the lightlike vector $e_1 - e_2$. Y_{n+3} is distinct from Y_1 and thus $\langle Y_{n+3}, e_{n+3} \rangle \neq 0$. So we can represent Y_{n+3} by a vector of the form

$$(3.14) \qquad Y_{n+3} = h(e_1 - e_2) + \xi + e_{n+3},$$

where $\xi: U \to \mathbb{R}^n$ has unit length and h is a smooth function on U. Thus, according to (2.2), $Y_{n+3}(b)$ represents the plane in the pencil of oriented spheres in $\mathbb{R}^n$ corresponding to the line $\lambda(b)$ on Q^{n+1}. Note that the condition $\langle Y_1, Y_{n+3} \rangle = 0$ is equivalent to $h = f\cdot\xi$, while the Legendre condition $\langle dY_1, Y_{n+3} \rangle = 0$ is the same as the Euclidean condition

$$(3.15) \qquad \xi \cdot df = 0 .$$

Thus, ξ is a field of unit normals to the immersion f on U. Since f is an immersion, we can choose the Lie frame vectors $Y_3, \ldots, Y_{n+1}$ to satisfy

$$(3.16) \qquad Y_i = dY_1(X_i) = (f\cdot df(X_i))(e_1 - e_2) + df(X_i), \quad 3 \leq i \leq n+1 ,$$

for tangent vector fields $X_3,\ldots,X_{n+1}$ on U. Then, we have

$$\omega_1^i(X_j) = \langle dY_1(X_j), Y_i\rangle = \langle Y_j, Y_i\rangle = \delta_{ij} \;. \tag{3.17}$$

Now using (3.14) and (3.16), we compute

$$\begin{aligned}\omega^i_{n+3}(X_j) &= \langle dY_{n+3}(X_j), Y_i\rangle = d\xi(X_j)\cdot df(X_i)\\ &= -\,df(AX_j)\cdot df(X_i) = -A_{ij}\;,\end{aligned} \tag{3.18}$$

where $A = [A_{ij}]$ is the Euclidean shape operator (second fundamental form) of the immersion f. But by (3.11) and (3.17), we have

$$\omega^i_{n+3}(X_j) = \Sigma\, h_{ik}\omega^k_1(X_j) = h_{ij}\;.$$

Hence $h_{ij} = -A_{ij}$, and $[h_{ij}]$ is just the negative of the Euclidean shape operator A of f.

<u>Remark 3.1</u>: The discussion above demonstrates how an immersion $f:B^{n-1} \to \mathbb{R}^n$ with field of unit normals ξ induces a Legendre submanifold $\lambda:B^{n-1} \to \Lambda$ defined by $\lambda(b) = [Y_1(b), Y_{n+3}(b)]$, for Y_1, Y_{n+3} as in (3.13), (3.14). Further, an immersed submanifold $f:M^k \to \mathbb{R}^n$ of codimension greater than one also gives rise to a Legendre submanifold $\lambda:B^{n-1} \to \Lambda$, where B^{n-1} is the bundle of unit normals to f in $\mathbb{R}^n$. As in the case of codimension one, $\lambda(b)$ is defined to be the line on Q^{n+1} corresponding to the oriented contact element determined by the unit vector b normal to f at the point $x = \pi(b)$, where π is the bundle projection from B^{n-1} to M^k.

As one would expect, the eigenvalues of the second fundamental form have geometric significance. Consider a curve $\gamma(t)$ on B. The set of points in Q^{n+1} lying on the lines $\lambda(\gamma(t))$ forms a ruled surface in Q^{n+1}. We look for the conditions that this ruled surface be developable, i.e., consist of tangent lines to a curve in Q^{n+1}. Let $rY_1 + Y_{n+3}$ be the point of contact. We have by (3.5) and (3.6)

$$(3.19) \qquad d(rY_1 + Y_{n+3}) \equiv \sum_i (r\omega_1^i + \omega_{n+3}^i)Y_i, \bmod Y_1, Y_{n+3} .$$

Thus, the lines $\lambda(\gamma(t))$ form a developable if and only if the tangent direction of $\gamma(t)$ is a common solution to the equations

$$(3.20) \qquad \sum_j (r\delta_{ij} + h_{ij})\omega_1^j = 0, \quad 3 \leq i \leq n+1 .$$

In particular, r must be a root of the equation

$$(3.21) \qquad \det(r\delta_{ij} + h_{ij}) = 0 .$$

By (3.11) the roots of (3.21) are all real. Denote them by $r_3, \ldots, r_{n+1}$. The points $r_iY_1 + Y_{n+3}$, $3 \leq i \leq n+1$ are called the <u>focal points</u> or <u>curvature spheres</u> (Pinkall [P3]) on $\lambda(b)$. If Y_1 and Y_{n+3} correspond to an immersion $f: U \to \mathbb{R}^n$ as in (3.13) and (3.14), then these focal points on $\lambda(b)$ correspond by (2.2) to oriented spheres in $\mathbb{R}^n$ tangent to f at f(b) and centered at the Euclidean focal points of f. These spheres are called curvature spheres of f and the r_i are just the principal curvatures of f, i.e., eigenvalues of the shape operator A.

If r is a root of (3.21) of multiplicity m, then the equations (3.20) define an m-dimensional subspace T_r of T_bB, the tangent space to B at the point b. The space T_r is called a <u>principal</u> <u>space</u> of T_bB, the latter being decomposed into a direct sum of its principal spaces. Vectors in T_r are called <u>principal</u> <u>vectors</u> corresponding to the focal point $rY_1 + Y_{n+3}$. Of course, if Y_1 and Y_{n+3} correspond to an immersion $f:U \to \mathbb{R}^n$ as in (3.13) and (3.14), then these principal vectors are the same as the Euclidean principal vectors for f corresponding to the principal curvature r.

With a change of frame of the form

(3.22) $$Y_i^* = \sum_i c_i^j Y_j, \quad 3 \leq i \leq n+1 ,$$

where $[c_i^j]$ is an (n-1) x (n-1) orthogonal matrix, we can diagonalize $[h_{ij}]$ so that in the new frame, equation (3.11) has the form

(3.23) $$\omega_{n+3}^i = - r_i\omega_1^i , \quad 3 \leq i \leq n+1 .$$

Note that none of the functions r_i is ever infinity on U because of the assumption that (3.8) holds, i.e, Y_1 is not a focal point. By associating Y_1 to a Euclidean immersion f as in (3.13), we can apply results from Euclidean geometry to our situation. In particular, it follows from a result of Singley [S] on Euclidean shape operators that there is a dense open subset of B on which the number g(b) of distinct focal points on $\lambda(b)$ is locally constant. We will work exclusively on open subsets U of B on which g is constant. In that case, each distinct eigenvalue function $r:U \to \mathbb{R}$ is smooth (see Nomizu [N]), and its corresponding principal distribution is a smooth m-dimensional foliation, where m is the multiplicity of r (see [CR, p. 139]). Thus, on U we can find smooth vector fields $X_3,\dots X_{n+1}$ dual to smooth 1-forms $\omega_1^3,\dots,\omega_1^{n+1}$, respectively, such that each X_i is principal for the smooth focal point map $r_iY_1 + Y_{n+3}$ on U. If $rY_1 + Y_{n+3}$ is a smooth focal point map of multiplicity m on U, then we can assume that

$$r_3 = \ldots = r_{m+2} = r \,. \tag{3.24}$$

By a different choice of the point at infinity, i.e., e_1 and e_2, if necessary, we can also assume that the function r is never zero on U, i.e., Y_{n+3} is not a focal point on U.

We now want to consider a Lie frame Y^*_A for which Y^*_1 is a smooth focal point map of multiplicity m on U. Specifically, we make the change of frame

$$\begin{aligned} Y^*_1 &= r\,Y_1 + Y_{n+3} \\ Y^*_2 &= (1/r)Y_2 \\ Y^*_{n+2} &= Y_{n+2} - (1/r)Y_2 \\ Y^*_{n+3} &= Y_{n+3} \\ Y^*_i &= Y_i \quad , \quad 3 \leq i \leq n+1 \,. \end{aligned} \tag{3.25}$$

We denote the Maurer-Cartan forms in this frame by θ^B_A. Note that

$$dY^*_1 = d(rY_1 + Y_{n+3}) = (dr)Y_1 + rdY_1 + dY_{n+3} = \Sigma\, \theta^A_1 Y^*_A \,. \tag{3.26}$$

By examining the coefficient of $Y^*_i = Y_i$ in (3.26), we see from (3.23) that

$$\theta^i_1 = r\omega^i_1 + \omega^i_{n+3} = (r-r_i)\omega^i_1, \; 3 \leq i \leq n+1 \,. \tag{3.27}$$

From (3.24) and (3.27), we see that

$$(3.28) \qquad \theta_1^a = 0 \;,\; 3 \leq a \leq m+2 \;.$$

This equation characterizes the condition that a focal point map Y_1^* have constant multiplicity m on U.

We now introduce the concept of a Dupin submanifold and then see what further restrictions it allows us to place on the structure forms.

A connected submanifold $N \subset B$ is called a <u>curvature submanifold</u> if its tangent space is everywhere a principal space. The Legendre submanifold is called <u>Dupin</u> if for every curvature submanifold $N \subset B$, the lines $\lambda(b)$, $b \in N$, all pass through a fixed point, i.e., each focal point map is constant along its curvature submanifolds. This definition of "Dupin" is the same as that of Pinkall [P3]. It is weaker than the definition of Dupin for Euclidean hypersurfaces used by Thorbergsson [Th], Miyaoka [M1], Grove-Halperin [GH] and Cecil-Ryan [CR], all of whom assumed that the number g of distinct curvature spheres is constant on B. As we noted above, g is locally constant on a dense open subset of any Legendre submanifold, but g is not necessarily constant on the whole of a Dupin submanifold, as the example of the Legendre submanifold induced from a tube B^3 over a torus $T^2 \subset \mathbb{R}^3 \subset \mathbb{R}^4$ demonstrates (see Pinkall [P3], Cecil-Ryan [CR, p. 188]).

It is easy to see that the Dupin property is invariant under Lie sphere transformations as follows. Suppose that $\lambda : B \to \Lambda^{2n-1}$ is a Legendre submanifold and that α is a Lie transformation. The map $\alpha\lambda : B \to \Lambda^{2n-1}$ is also a Legendre submanifold with $\alpha\lambda(b) = [\alpha Y_1(b), \alpha Y_{n+3}(b)]$. Furthermore, if $k = rY_1 + Y_{n+3}$ is a curvature sphere of λ at a point $b \in B$, then since α is a linear transformation, αk is a curvature sphere of $\alpha\lambda$ at b with the same principal space as k. Thus λ and $\alpha\lambda$ have the same curvature submanifolds on

B, and the Dupin property clearly holds for λ if and only if it holds for $\alpha\lambda$.

We now return to the calculation that led to equation (3.28). We have that $Y_1^* = rY_1 + Y_{n+3}$ is a smooth focal point map of multiplicity m on the open set U and its corresponding principal space is spanned by the vector fields $X_3, \ldots, X_{m+2}$. The Dupin condition that Y_1^* be constant along the leaves of T_r is simply

(3.29) $$dY_1^*(X_a) \equiv 0, \bmod Y_1^*, \quad 3 \le a \le m+2 .$$

From (3.17), (3.27) and (3.28), we have that

(3.30) $$dY_1^*(X_a) = \theta_1^1(X_a)Y_1 + \theta_1^{n+3}(X_a)Y_{n+3}, \quad 3 \le a \le m+2.$$

Comparing (3.29) and (3.30), we see that

(3.31) $$\theta_1^{n+3}(X_a) = 0 , \quad 3 \le a \le m+2 .$$

We now show that we can make one more change of frame and make $\theta_1^{n+3} = 0$. We can write the form θ_1^{n+3} in terms of the basis $\omega_1^3, \ldots, \omega_1^{n+3}$ as

$$\theta_1^{n+3} = \Sigma\, a_i \omega_1^i .$$

From (3.31), we see that we actually have

(3.32) $$\theta_1^{n+3} = \sum_{b=m+3}^{n+1} a_b \omega_1^b .$$

Using (3.17), (3.27) and (3.32), we compute for $m+3 \le b \le n+1$,

$$(3.33)\qquad dY_1^*(X_b) = \theta_1^1(X_b)Y_1^* + \theta_1^b(X_b)Y_b + \theta_1^{n+3}(X_b)Y_{n+3}$$
$$= \theta_1^1(X_b)Y_1^* + (r-r_b)Y_b + a_b\,Y_{n+3}$$
$$= \theta_1^1(X_b)Y_1^* + (r-r_b)(Y_b + (a_b/(r-r_b))Y_{n+3}).$$

We now make the change of frame,

$$\bar{Y}_1 = Y_1^*\ ,\quad \bar{Y}_2 = Y_2^*\ ,$$
$$\bar{Y}_a = Y_a\ ,\quad 3 \le a \le m+2,$$
$$(3.34)\qquad \bar{Y}_b = Y_b + (a_b/r-r_b)Y_{n+3}\ ,\quad m+3 \le b \le n+1,$$
$$\bar{Y}_{n+2} = -\sum_{b=m+3}^{n+1} (a_b/r-r_b)Y_b + Y_{n+2} - 1/2 \sum_{b=m+3}^{n+1} (a_b/r-r_b)^2\, Y_{n+3}\ ,$$
$$\bar{Y}_{n+3} = Y_{n+3}.$$

Let α_A^B be the Maurer-Cartan forms for this new frame. We still have

$$(3.35)\qquad \alpha_1^a = \langle d\bar{Y}_1,\bar{Y}_a\rangle = \langle dY_1^*,Y_a\rangle = \theta_1^a = 0,\quad 3 \le a \le m+2\ .$$

Furthermore, since $\bar{Y}_1 = Y_1^*$, the Dupin condition (3.29) still yields

$$(3.36)\qquad \alpha_1^{n+3}\ (X_a) = 0\quad ,\ 3 \le a \le m+2\ .$$

Finally, for $m+3 \le b \le n+1$, we have from (3.33) and (3.34)

$$(3.37)\qquad \alpha_1^{n+3}\ (X_b) = \langle d\bar{Y}_1(X_b),\ \bar{Y}_{n+2}\rangle$$
$$= \langle \theta_1^1(X_b)\bar{Y}_1 + (r-r_b)\bar{Y}_b,\bar{Y}_{n+2}\rangle = 0\ .$$

From (3.36) and (3.37), we conclude that

$$(3.38) \qquad \alpha_1^{n+3} = 0 .$$

Thus, our main result of this section is that the assumption that the focal point map $\bar{Y}_1 = rY_1' + Y_{n+3}$ has constant multiplicity m and is constant along the leaves of its principal foliation T_r allows to produce a Lie frame $\bar{Y}_A$ whose structure forms satisfy

$$(3.39) \qquad \begin{aligned} &\alpha_1^a = 0 \quad , \ 3 \le a \le m+2 , \\ &\alpha_1^{n+3} = 0 . \end{aligned}$$

4. <u>Cyclides of Dupin</u>.

Dupin initiated the study of this subject in 1822 when he defined a cyclide to be a surface M^2 in E^3 which is the envelope of the family of spheres tangent to three fixed spheres. This was shown to be equivalent to requiring that both sheets of the focal set of M^2 in E^3 degenerate into curves. Then M^2 is the envelope of each of the two families of curvature spheres. The key step in the classical Euclidean proof (see, for example, Eisenhart [E, pp. 312-314] or Cecil-Ryan [CR, pp. 151-166]) is to show that the two focal curves are a pair of so-called "focal conics" in E^3, i.e., an ellipse and hyperbola in mutually orthogonal planes such that the vertices of the ellipse are the foci of the hyperbola and vice-versa, or a pair of parabolas in orthogonal planes such the vertex of each is the focus of the other. This classical proof is local, i.e., one needs only a small piece of the surface to determine the focal conics and reconstruct the whole cyclide. Of course, envelopes of families of spheres can have singularities and some of the cyclides have one or two singular points in E^3. It turns out, however, that all of the different forms of cyclides in Euclidean space induce Legendre submanifolds which are locally Lie equivalent. In other words, they are all various Euclidean projections of one Legendre submanifold. Pinkall [P3] generalized this result to higher dimensional Dupin submanifolds. He defined a cyclide of characteristic (p,q) to be a Dupin submanifold with the property that at each point it has exactly two distinct focal points with respective multiplicites p and q. He then proved the following.

Theorem 4.1: (Pinkall [P3]): (a) Every connected cyclide of Dupin is contained in a unique compact and connected cyclide of Dupin.
(b) Any two cyclides of the same characteristic are locally Lie equivalent, each being Lie equivalent to an open subset of a standard product of spheres in S^n.

In this section, we give a proof of Pinkall's result using the method of Lie frames. Let $\lambda : B \to \Lambda$ be the Dupin cyclide. The main step in the proof of the Theorem 4.1 is to show that the two focal point maps k_1 and k_2 from B to Q^{n+1} are such that the image $k_1(B)$ lies in the intersection of Q^{n+1} with a (p+1)-dimensional subspace E of P^{n+2} while $k_2(B)$ lies in the intersection of the (q+1)-dimensional subspace $E^{\perp}$ with Q^{n+1}. This generalizes the key step in the classical Euclidean proof that the two focal curves are focal conics. Once this fact has been established for k_1 and k_2, it is relatively easy to complete the proof of the Theorem.

We begin the proof by taking advantage of the results of the previous section. As we showed in (3.39), on any neighborhood U in B, we can find a local Lie frame, which we now denote by Y_A, whose Maurer-Cartan forms, now denoted ω_A^B, satisfy

$$\omega_1^a = 0 \quad , \; 3 \leq a \leq p+2, \qquad (4.1)$$
$$\omega_1^{n+3} = 0 \; .$$

In this frame, Y_1 is a focal point map of multiplicity p from U to Q^{n+1}. By the hypotheses of Theorem 4.1, there is one other focal point of multiplicity q = n-1-p at each point of B. By repeating the procedure used in constructing the frame Y_A, we can construct a new frame $\overline{Y}_A$ which has as $\overline{Y}_{n+3}$ the other focal point map $sY_1 + Y_{n+3}$, where s is a root of (3.21) of multiplicity q. The principal space T_s is the span of the vectors $X_{p+3},\ldots,X_{n+1}$ in the notation of the previous section. The fact that $\overline{Y}_{n+3}$ is a focal point yields

$$\overline{\omega}_{n+3}^{\,b} = 0 \quad , \quad p+3 \leq b \leq n+1 \; , \qquad (4.2)$$

in analogy to (3.28). The Dupin condition analogous to (3.29) is

$$d\overline{Y}_{n+3}(X_b) \equiv 0, \text{ mod } \overline{Y}_{n+3} \; , \; p+3 \leq b \leq n+1 \; . \qquad (4.3)$$

This eventually leads to

$$\overline{\omega}_{n+3}^{\,1} = 0 \; . \qquad (4.4)$$

One can check that this change of frame does not affect condition (4.1). We now drop the bars and call this last frame Y_A with Maurer-Cartan forms ω_A^B satisfying,

$$
(4.5)\qquad
\begin{aligned}
&\omega_1^a = 0 \quad , \quad 3 \le a \le p+2 , \\
&\omega_{n+3}^b = 0 \quad , \quad p+3 \le b \le n+1 , \\
&\omega_1^{n+3} = 0 \quad , \quad \omega_{n+3}^1 = 0 .
\end{aligned}
$$

Furthermore, the following forms are easily shown to be a basis for the cotangent space at each point of U,

$$
(4.6)\qquad \{\omega_{n+3}^3,\dots,\omega_{n+3}^{p+2},\omega_1^{p+3},\dots,\omega_1^{n+1}\} .
$$

We begin by taking the exterior derivative of the equations $\omega_1^a = 0$ and $\omega_{n+3}^b = 0$ in (4.5). Using (3.6), (3.7) and (4.5), we obtain

$$
(4.7)\qquad 0 = \omega_1^{p+3} \wedge \omega_{p+3}^a + \dots + \omega_1^{n+1} \wedge \omega_{n+1}^a , \quad 3 \le a \le p+2 ,
$$

$$
(4.8)\qquad 0 = \omega_{n+3}^3 \wedge \omega_3^b + \dots + \omega_{n+3}^{p+2} \wedge \omega_{p+2}^b , \quad p+3 \le b \le n+1 .
$$

We now show that (4.7) and (4.8) imply that

$$
(4.9)\qquad \omega_b^a = 0 \quad , \quad 3 \le a \le p+2 , \ p+3 \le b \le n+1 .
$$

To see this, note that since $\omega_b^a = -\omega_a^b$, each of the terms ω_b^a occurs in exactly one of the equations (4.7) and in exactly one of the equations (4.8). Equation (4.7) involves the basis forms $\omega_1^{p+3},\dots,\omega_1^{n+1}$, while equation (4.8) involves the basis forms $\omega_{n+3}^3,\dots,\omega_{n+3}^{p+2}$. We now show how to handle the form

ω^3_{p+3}; the others are treated in similar fashion. The equations from (4.7) and (4.8), respectively, involving $\omega^3_{p+3} = -\omega^{p+3}_3$ are

$$(4.10) \qquad 0 = \omega^{p+3}_1 \wedge \omega^3_{p+3} + \omega^{p+4}_1 \wedge \omega^3_{p+4} + \ldots + \omega^{n+1}_1 \wedge \omega^3_{n+1} ,$$

$$(4.11) \qquad 0 = \omega^3_{n+3} \wedge \omega^{p+3}_3 + \omega^4_{n+3} \wedge \omega^{p+3}_4 + \ldots + \omega^{p+2}_{n+3} \wedge \omega^{p+3}_{p+2} .$$

We take the wedge product of (4.10) with $\omega^{p+4}_1 \wedge \ldots \wedge \omega^{n+1}_1$ and get

$$0 = \omega^3_{p+3} \wedge (\omega^{p+3}_1 \wedge \omega^{p+4}_1 \wedge \ldots \wedge \omega^{n+1}_1) ,$$

which implies that ω^3_{p+3} is in the span of $\omega^{p+3}_1, \ldots, \omega^{n+1}_1$. On the other hand, taking the wedge product of (4.11) with $\omega^4_{n+3} \wedge \ldots \wedge \omega^{p+2}_{n+3}$ yields

$$0 = \omega^3_{p+3} \wedge (\omega^3_{n+3} \wedge \ldots \wedge \omega^{p+2}_{n+3}) ,$$

and thus that ω^3_{p+3} is in the span of $\omega^3_{n+3}, \ldots, \omega^{p+2}_{n+3}$. We conclude that $\omega^3_{p+3} = 0$, as desired.

We next differentiate $\omega^{n+3}_1 = 0$ and use (3.6), (3.9) and (4.5) to obtain

$$(4.12) \qquad 0 = d\omega^{n+3}_1 = \omega^{p+3}_1 \wedge \omega^{n+3}_{p+3} + \ldots + \omega^{n+1}_1 \wedge \omega^{n+3}_{n+1} .$$

This implies that

$$(4.13) \qquad \omega^{n+3}_b \in \operatorname{Span}\{\omega^{p+3}_1, \ldots, \omega^{n+1}_1\} , \; p+3 \le b \le n+1.$$

Similarly, differentiation of $\omega^1_{n+3} = 0$ yields

(4.14) $$0 = \omega^3_{n+3} \wedge \omega^1_3 + \ldots + \omega^{p+2}_{n+3} \wedge \omega^1_{p+2} \, ,$$

which implies that

(4.15) $$\omega^a_1 \in \mathrm{Span}\{\omega^3_{n+3}, \ldots, \omega^{p+2}_{n+3}\} \, , \; 3 \le a \le p+2 \, .$$

We next differentiate (4.9). Using the skew-symmetry relations (3.6) and equations (4.5) and (4.9), we see that all terms drop out except the following,

$$0 = d\omega^a_b = \omega^2_b \wedge \omega^a_2 + \omega^{n+3}_b \wedge \omega^a_{n+3}$$
$$= -(\omega^1_a \wedge \omega^b_1) + \omega^{n+3}_b \wedge \omega^a_{n+3} \, .$$

Thus,

(4.16) $$\omega^1_a \wedge \omega^b_1 = \omega^{n+3}_b \wedge \omega^a_{n+3} \, , \; 3 \le a \le p+2, \; p+3 \le b \le n+1 \, .$$

We now show that (4.16) implies that there is some function α on U such that

(4.17) $$\omega^1_a = \alpha \, \omega^a_{n+3} \, , \quad 3 \le a \le p+2 \, ,$$
$$\omega^{n+3}_b = -\alpha \, \omega^b_1 \, , \quad p+3 \le b \le n+1 \, .$$

To see this, note that for any a, $3 \le a \le p+2$, (4.15) gives

(4.18) $$\omega^1_a = c_3\omega^3_{n+3} + \ldots + c_{p+2}\omega^{p+2}_{n+3} \quad \text{for some } c_3, \ldots, c_{p+2} \, ,$$

while for any b, $p+3 \leq b \leq n+1$, (4.13) gives

(4.19) $$\omega_b^{n+3} = d_{p+3}\omega_1^{p+3} + \ldots + d_{n+1}\omega_1^{n+1} \text{ for some } d_{p+3}, \ldots, d_{n+1} .$$

Thus,

(4.20) $$\omega_a^1 \wedge \omega_1^b = c_3\omega_{n+3}^3 \wedge \omega_1^b + \ldots + c_a\omega_{n+3}^a \wedge \omega_1^b + \ldots + c_{p+2}\omega_{n+3}^{p+2} \wedge \omega_1^b ,$$

(4.21) $$\omega_b^{n+3} \wedge \omega_{n+3}^a = d_{p+3}\omega_1^{p+3} \wedge \omega_{n+3}^a + \ldots + d_b\omega_1^b \wedge \omega_{n+3}^a + \ldots + d_{n+1}\omega_1^{n+1} \wedge \omega_{n+3}^a .$$

From (4.16), we know that the right-hand sides of (4.20) and (4.21) are equal. But these expressions contain no common terms from the basis of 2-forms except those involving $\omega_{n+3}^a \wedge \omega_1^b$. Thence, all of the coefficients except c_a and d_b are zero, and we have

$$c_a\omega_{n+3}^a \wedge \omega_1^b = d_b\omega_1^b \wedge \omega_{n+3}^a = (-d_b)\, \omega_{n+3}^a \wedge \omega_1^b .$$

Thus $c_a = -d_b$. If we set $\alpha_a = c_a$ and $\mu_b = d_b$, we have shown that (4.18) and (4.19) reduce to

$$\omega_a^1 = \alpha_a\omega_{n+3}^a , \quad \omega_b^{n+3} = \mu_b\omega_1^b \text{ with } \mu_b = -\alpha_a .$$

This procedure works for any choice of a and b in the appropriate ranges. By holding a fixed and varying b, we see that all of the quantities μ_b are equal to each other and to $-\alpha_a$. Similarly, all of the quantities α_a are the same, and (4.17) holds. We now consider the expression (3.5) for dY_a, $3 \leq a \leq p+2$. We omit the terms which vanish because of (3.6), (4.5) or (4.9),

$$dY_a = \omega_a^1 Y_1 + \omega_a^3 Y_3 + \ldots + \omega_a^{p+2} Y_{p+2} + \omega_a^{n+2} Y_{n+2} + \omega_a^{n+3} Y_{n+3} .$$

Using (4.17) and the fact from (3.6) that $\omega_a^{n+2} = -\omega_{n+3}^a$, this becomes

$$(4.22) \qquad dY_a = \omega_{n+3}^a(\alpha Y_a - Y_{n+2}) + \omega_a^3 Y_3 + \ldots + \omega_a^{p+2} Y_{p+2} + \omega_a^{n+3} Y_{n+3} .$$

Similarly, for $p+3 \leq b \leq n+1$, we find

$$(4.23) \qquad dY_b = \omega_b^1 Y_1 + \omega_b^2(Y_2 + \alpha Y_{n+3}) + \omega_b^{p+3} Y_{p+3} + \ldots + \omega_b^{n+1} Y_{n+1} .$$

We make the change of frame

$$(4.24) \qquad \begin{aligned} &Y_2^* = Y_2 + \alpha Y_{n+3} \quad , \quad Y_{n+2}^* = Y_{n+2} - \alpha Y_1 , \\ &Y_\beta^* = Y_\beta \quad , \quad \beta \neq 2 \text{ or } n+2 . \end{aligned}$$

We now drop the asterisks but use the new frame. From (4.22) and (4.23), we see that in this new frame, we have

$$(4.25) \qquad dY_a = \omega_{n+3}^a(-Y_2) + \omega_a^3 Y_3 + \ldots + \omega_a^{p+2} Y_{p+2} + \omega_a^{n+3} Y_{n+3} ,$$

$$(4.26) \qquad dY_b = \omega_b^1 Y_1 + \omega_b^2 Y_2 + \omega_b^{p+3} Y_{p+3} + \ldots + \omega_b^{n+1} Y_{n+1} .$$

That is, in the new frame, we have

$$(4.27) \qquad \omega_a^1 = 0 , \quad 3 \leq a \leq p+2 ,$$

(4.28) $$\omega_b^{n+3} = 0 \quad , \quad p+3 \leq b \leq n+1 \ .$$

Our goal now is to show that the two spaces

(4.29) $$E = \mathrm{Span}\{Y_1, Y_2, Y_{p+3}, \ldots, Y_{n+1}\} \ .$$

and its orthogonal complement

(4.30) $$E^{\perp} = \mathrm{Span}\{Y_3, \ldots, Y_{p+2}, Y_{n+2}, Y_{n+3}\}$$

are invariant under d.

Concerning E, we have that $dY_b \in E$ for $p+3 \leq b \leq n+1$ by (4.26). Furthermore, (3.6), (3.9) and (4.5) imply that

$$dY_1 = \omega_1^1 Y_1 + \omega_1^{p+3} Y_{p+3} + \ldots + \omega_1^{n+1} Y_{n+1} \ ,$$

which is in E. Thus, it only remains to show that dY_2 is in E. To do this, we differentiate (4.27). As before, we omit terms which are zero because of (3.6), (4.5), (4.9) and (4.27). We see that formula (3.7) for $d\omega_a^1$ reduces to

(4.31) $$0 = d\omega_a^1 = \omega_a^{n+2} \wedge \omega_{n+2}^1 = -\omega_{n+3}^a \wedge \omega_{n+2}^1 = \omega_{n+3}^a \wedge \omega_2^{n+3} \ , \ 3 \leq a \leq p+2 \ .$$

Similarly, by differentiating (4.28), we find that

$$0 = d\omega_b^{n+3} = \omega_b^2 \wedge \omega_2^{n+3} = -\omega_1^b \wedge \omega_2^{n+3} \ , \ p+3 \leq b \leq n+1 \ .$$

From this and (4.31), we see that the wedge product of ω_2^{n+3} with every form in the basis (4.16) is zero, and hence $\omega_2^{n+3} = 0$. Using this and the fact that $\omega_2^{n+2} = -\omega_{n+3}^1 = 0$, and that by (3.6) and (4.27),

$$\omega_2^a = -\omega_a^1 = 0 \quad , \quad 3 \leq a \leq p+2 \ ,$$

we have

$$dY_2 = \omega_2^2 Y_2 + \omega_2^{p+3} Y_{p+3} + \ldots + \omega_2^{n+1} Y_{n+1} \ ,$$

which is in E. So E is invariant under d and is thus a fixed subspace of P^{n+2}, independent of the choice of point of U. Obviously then, the space $E^\perp$ in (4.30) is also a fixed subspace of P^{n+2}.

Note that E has signature (1,q+1) as a vector subspace of $\mathbb{R}_2^{n+3}$, and $E^\perp$ has signature (1,p+1). Take an orthonormal basis $e_1, \ldots, e_{n+3}$ of $\mathbb{R}_2^{n+3}$ with e_1 and e_{n+3} timelike and

$$E = \mathrm{Span}\{e_1, \ldots, e_{q+2}\} \ , \ E^\perp = \mathrm{Span}\{e_{q+3}, \ldots, e_{n+3}\} \ .$$

Then $E \cap Q^{n+1}$ is given in homogeneous coordinates $(x_1, \ldots, x_{n+3})$ with respect to this basis by

$$(4.32) \qquad x_1^2 = x_2^2 + \ldots + x_{q+2}^2 \ , \ x_{q+3} = \ldots = x_{n+3} = 0 \ .$$

This quadric is diffeomorphic to the unit sphere S^q in the span E^{q+1} of the spacelike vectors $e_2, \ldots, e_{q+2}$ with the diffeomorphism $\phi: S^q \to E \cap Q^{n+1}$ being given by

(4.33) $$\phi(u) = [e_1 + u] \ , \ u \in S^q \ .$$

Similarly, $E^\perp \cap Q^{n+1}$ is the quadric given in homogeneous coordinates by

(4.34) $$x_{n+3}^2 = x_{q+3}^2 + \ldots + x_{n+2}^2 \ , \ x_1 = \ldots = x_{q+2} = 0 \ .$$

$E^\perp \cap Q^{n+1}$ is diffeomorphic to the unit sphere S^p in the span E^{p+1} of $e_{q+3}, \ldots, e_{n+2}$ with the diffeomorphism $\psi : S^p \to E \cap Q^{n+1}$ being given by

(4.35) $$\psi(v) = [v + e_{n+3}] \ , \ v \in S^p \ .$$

The focal point map Y_1 of our Dupin submanifold is constant on the leaves of the principal foliation T_r, and so Y_1 factors through an immersion of the q-dimensional space of leaves U/T_r into the q-sphere given by the quadric (4.32). Hence, the image of Y_1 is an open subset of this quadric. Similarly, Y_{n+3} factors through an immersion of the p-dimensional space of leaves U/T_s onto an open subset of the p-sphere given by the quadric (4.34). From this it is clear that the unique compact cyclide containing $\lambda : U \to \Lambda$ as an open submanifold is given by the Dupin submanifold $\bar{\lambda} : S^q \times S^p \to \Lambda$ with

$$\bar{\lambda}(u,v) = [k_1(u,v), k_2(u,v)] \ , \ (u,v) \in S^q \times S^p \ ,$$

where

$$k_1(u,v) = \phi(u) \quad \text{and} \quad k_2(u,v) = \psi(v) \ .$$

Geometrically, the image of $\bar{\lambda}$ consists of all lines joining a point on the quadric (4.32) to a point on the quadric (4.34).

Thus, any choice of (q+1)-plane E in P^{n+2} with signature (1,q+1) and corresponding orthogonal complement $E^{\perp}$ determines a unique compact cyclide of characteristic (p,q) and vice-versa. The local Lie equivalence of any two cyclides of the same characteristic is then clear.

From the standpoint of Euclidean geometry, if we consider the point spheres to be given by $Q^{n+1} \cap e_{n+3}^{\perp}$, as in Section 2, then the Legendre submanifold $\bar{\lambda}$ above is induced in the usual way from the unit normal bundle $B^{n-1} = S^q \times S^p$ of the standard embedding of S^q as a great q-sphere $E^{q+1} \cap S^n$, where E^{q+1} is the span of $e_2,\ldots,e_{q+2}$ and S^n is the unit sphere in E^{n+1}, the span of $e_2,\ldots,e_{n+2}$. The spheres S^q and $S^p = S^n \cap E^{p+1}$, where E^{p+1} is the span of $e_{q+3},\ldots,e_{n+2}$, are the two focal submanifolds in S^n of a standard product of spheres $S^p \times S^q$ in S^n (see [CR, p. 295]).

5. Dupin submanifolds for n = 4.

The classification of Dupin submanifolds induced from surfaces in $\mathbb{R}^3$ follows from the results of the last section. In his doctoral dissertation [P1] (later published as [P2]), U. Pinkall obtained a local classification up to Lie equivalence of all Dupin submanifolds induced from hypersurfaces in $\mathbb{R}^4$. As we shall see, this is a far more complicated calculation than that of the previous section, and as yet, no one has obtained a similar classification of Dupin hypersurfaces in $\mathbb{R}^n$ for $n \geq 5$. In this section, we will prove Pinkall's theorem using the method of moving frames.

We follow the notation used in Sections 3 and 4. We consider a Dupin submanifold

(5.1) $$\lambda : B \rightarrow \Lambda$$

where

(5.2) $$\dim B = 3\ ,\ \dim \Lambda = 7\ ,$$

and the image $\lambda(b)$, $b \in B$, is the line $[Y_1, Y_7]$ of the Lie frame $Y_1, \ldots Y_7$. We assume that there are three distinct focal points on each line $\lambda(b)$.

By (3.39), we can choose the frame so that

(5.3)
$$\omega_1^3 = \omega_1^7 = 0\ ,$$
$$\omega_7^4 = \omega_7^1 = 0\ .$$

By making a change of frame of the form

(5.4)
$$Y_1^* = \alpha Y_1\ \ ,\ Y_2^* = (1/\alpha)Y_2$$
$$Y_7^* = \beta Y_7\ \ ,\ \ Y_6^* = (1/\beta)Y_6\ ,$$

for suitable smooth functions α and β on B, we can arrange that Y_1+Y_7 represents the third focal point at each point of B. Then, using the fact that B is Dupin, we can use the method employed at the end of Section 3 to make a change of frame leading to the following equations similar to (3.39) (and to (5.3)),

(5.5)
$$\omega_1^5 + \omega_7^5 = 0,$$
$$\omega_1^1 - \omega_7^7 = 0.$$

This completely fixes the Y_i, $i = 3,4,5$, and Y_1, Y_7 are determined up to a transformation of the form

(5.6) $$Y_1^* = \tau\, Y_1 \ , \ Y_7^* = \tau\, Y_7 \ .$$

Each of the three focal point maps Y_1, Y_7, $Y_1 + Y_7$ is constant along the leaves of its corresponding principal foliation. Thus, each focal point map factors through an immersion of the corresponding 2-dimensional space of leaves of its principal foliation into Q^5. (See Section 4 of Chapter 2 of the book [CR] for more detail on this point.) In terms of moving frames, this implies that the forms ω_1^4, ω_1^5, ω_7^3 are linearly independent on B, i.e.,

(5.7) $$\omega_1^4 \wedge \omega_1^5 \wedge \omega_7^3 \neq 0 \ .$$

This can also be seen by expressing the forms above in terms of a Lie frame $\overline{Y}_1,\ldots,\overline{Y}_7$, where $\overline{Y}_1$ satisfies the regularity condition (3.8), and using the fact that each focal point has multiplicity one. For simplicity, we will also use the notation

(5.8) $$\theta_1 = \omega_1^4 \ , \ \theta_2 = \omega_1^5 \ , \ \theta_3 = \omega_7^3 \ .$$

Analytically, the Dupin conditions are three partial differential equations, and we are treating an over-determined system. The method of moving frames reduces the handling of its integrability conditions to a straightforward algebraic problem, viz. that of repeated exterior differentiations.

We begin by taking the exterior derivatives of the three equations $\omega_1^3 = 0$, $\omega_7^4 = 0$, $\omega_1^5 + \omega_7^5 = 0$. Using the skew-symmetry relations (3.6), as well as (5.3) and (5.5), the exterior derivatives of these three equations yield the system

$$0 = \omega_1^4 \wedge \omega_3^4 + \omega_1^5 \wedge \omega_3^5 ,$$
$$0 = \qquad \omega_1^5 \wedge \omega_4^5 + \omega_7^3 \wedge \omega_3^4 ,$$
$$0 = \omega_1^4 \wedge \omega_4^5 \qquad + \omega_7^3 \wedge \omega_3^5 .$$

If we take the wedge product of the first of these with ω_1^4, we conclude that ω_3^5 is in the span of ω_1^4 and ω_1^5. On the other hand, taking the wedge product of the third equation with ω_1^4 yields that ω_3^5 is in the span of ω_1^4 and ω_3^7. Consequently, $\omega_3^5 = \rho\omega_1^4$ for some smooth function ρ on B. Similarly, one can show that there exist functions σ and τ such that $\omega_3^4 = \sigma\,\omega_1^5$ and $\omega_4^5 = \tau\,\omega_7^3$. Then, if we substitute these into the three equations above, we get that $\rho = \sigma = \tau$, and hence we have

(5.9) $$\omega_3^5 = \rho\,\omega_1^4 ,\ \omega_3^4 = \rho\,\omega_1^5 ,\ \omega_4^5 = \rho\,\omega_7^3 .$$

Next we differentiate the equations $\omega_1^7 = 0$, $\omega_7^1 = 0$, $\omega_1^1 - \omega_7^7 = 0$. As above, use of the skew-symmetry relations (3.6) and the equations (5.3), (5.5) yields the existence of smooth functions a,b,c,p,q,r,s,t,u on B such that the following relations hold:

(5.10)
$$\omega_4^7 = -\,\omega_6^4 = a\,\omega_1^4 + b\,\omega_1^5 ,$$
$$\omega_5^7 = -\,\omega_6^5 = b\,\omega_1^4 + c\,\omega_1^5 ;$$

(5.11)
$$\omega_3^1 = -\,\omega_2^3 = p\,\omega_7^3 - q\,\omega_1^5 ,$$
$$\omega_5^1 = -\,\omega_2^5 = q\,\omega_7^3 - r\,\omega_1^5 ;$$

(5.12)
$$\omega_4^1 = -\,\omega_2^4 = b\,\omega_1^5 + s\,\omega_1^4 + t\,\omega_7^3 ,$$
$$\omega_6^3 = -\,\omega_3^7 = q\,\omega_1^5 + t\,\omega_1^4 + u\,\omega_7^3 .$$

We next see what can be deduced from taking the exterior derivatives of the equations (5.9)-(5.12). First, we take the exterior derivatives of the three basis forms ω_1^4, ω_1^5, ω_7^3. For example, using the relations that we have derived so far, we have from the Maurer-Cartan equation (3.7),

$$d\omega_1^4 = \omega_1^1 \wedge \omega_1^4 + \omega_1^5 \wedge \omega_5^4 = \omega_1^1 \wedge \omega_1^4 - \rho\omega_1^5 \wedge \omega_7^3 .$$

We obtain similar expressions for $d\omega_1^5$ and $d\omega_7^3$. When we use the forms $\theta_1, \theta_2, \theta_3$ defined in (5.8) for ω_1^4, ω_1^5, ω_7^3, we have

(5.13)
$$\begin{aligned} d\theta_1 &= \omega_1^1 \wedge \theta_1 - \rho\, \theta_2 \wedge \theta_3 , \\ d\theta_2 &= \omega_1^1 \wedge \theta_2 - \rho\, \theta_3 \wedge \theta_1 , \\ d\theta_3 &= \omega_1^1 \wedge \theta_3 - \rho\, \theta_1 \wedge \theta_2 . \end{aligned}$$

We next differentiate (5.9). We have $\omega_3^4 = \rho\omega_1^5$. On the one hand,

$$d\omega_3^4 = \rho\, d\omega_1^5 + d\rho \wedge \omega_1^5 .$$

Using the second equation in (5.13) with $\omega_1^5 = \theta_2$, this becomes

$$d\omega_3^4 = \rho\, \omega_1^1 \wedge \omega_1^5 - \rho^2\omega_7^3 \wedge \omega_1^4 + d\rho \wedge \omega_1^5 .$$

On the other hand, we can compute $d\omega_3^4$ from the Maurer-Cartan equation (3.7) and use the relationships that we have derived to find

$$d\omega_3^4 = (-p-\rho^2-a)(\omega_1^4 \wedge \omega_7^3) - q\, \omega_1^5 \wedge \omega_1^4 + b\, \omega_7^3 \wedge \omega_1^5 .$$

If we equate these two expressions for $d\omega^4_3$, we get

$$(5.14)\qquad (-p-a-2\rho^2)\ \omega^4_1 \wedge \omega^3_7 = (d\rho + \rho\ \omega^1_1 - q\ \omega^4_1 - b\ \omega^3_7) \wedge \omega^5_1 .$$

Because of the independence of ω^4_1, ω^5_1 and ω^3_7, both sides of the equation above must vanish. Thus, we conclude that

$$(5.15)\qquad 2\rho^2 = -a-p\ ,$$

and that $d\rho + \rho\omega^1_1 - q\omega^4_1 - b\omega^3_7$ is a multiple of ω^5_1. Similarly, differentiation of $\omega^5_4 = \rho\omega^3_7$, yields the following analogue of (5.14),

$$(5.16)\qquad (s-a-r + 2\rho^2)\ \omega^4_1 \wedge \omega^5_1 = (d\rho + \rho\ \omega^1_1 + t\ \omega^5_1 - q\ \omega^4_1) \wedge \omega^3_7\ ,$$

and differentiation of $\omega^5_3 = \rho\omega^4_1$ yields

$$(5.17)\qquad (c+p+u - 2\rho^2)\ \omega^5_1 \wedge \omega^3_7 = (-d\rho - \rho\ \omega^1_1 - t\ \omega^5_1 + b\ \omega^3_7) \wedge \omega^4_1 .$$

In each of the equations (5.14), (5.16), (5.17) both sides of the equation must vanish. From the vanishing of the left-hand sides of the equations, we get the fundamental relationship,

$$(5.18)\qquad 2\rho^2 = -a-p = a+r-s = c+p+u\ .$$

Furthermore, from the vanishing of the right-hand sides of the three equations (5.14), (5.15) and (5.17), we can determine after some algebra that

(5.19) $$d\rho + \rho\,\omega^1_1 = q\,\omega^4_1 - t\,\omega^5_1 + b\,\omega^3_7 .$$

The last equation shows the importance of the function ρ. Following the notation introduced in (5.8), we write (5.19) as

(5.20) $$d\rho + \rho\omega^1_1 = \rho_1\theta_1 + \rho_2\theta_2 + \rho_3\theta_3,$$

where

(5.21) $$\rho_1 = q,\ \rho_2 = -t,\ \rho_3 = b\ ,$$

are the "covariant derivatives" of ρ.

Using the Maurer-Cartan equations, we can compute

$$\begin{aligned} d\omega^1_1 &= \omega^4_1 \wedge \omega^1_4 + \omega^5_1 \wedge \omega^1_5 \\ &= \omega^4_1 \wedge (b\,\omega^5_1 + t\,\omega^3_7) + \omega^5_1 \wedge (q\,\omega^3_7 - r\,\omega^5_1) \\ &= b\,\omega^4_1 \wedge \omega^5_1 + q\,\omega^5_1 \wedge \omega^3_7 - t\,\omega^3_7 \wedge \omega^4_1 . \end{aligned}$$

Using (5.8) and (5.21), this can be rewritten as

(5.22) $$d\omega^1_1 = \rho_3\,\theta_1 \wedge \theta_2 + \rho_1\,\theta_2 \wedge \theta_3 + \rho_2\,\theta_3 \wedge \theta_1 .$$

The trick now is to express everything in terms of ρ and its successive covariant derivatives.

We first derive a general form for these covariant derivatives. Suppose that σ is a smooth function which satisfies a relation of the form

$$(5.23)\qquad d\sigma + m\,\sigma\omega^1_1 = \sigma_1\theta_1 + \sigma_2\theta_2 + \sigma_3\theta_3$$

for some integer m. (Note that (5.19) is such a relationship for ρ with m=1.) By taking the exterior derivative of (5.23) and using (5.13) and (5.22) to express both sides in terms of the standard basis of two forms $\theta_1 \wedge \theta_2$, $\theta_2 \wedge \theta_3$ and $\theta_3 \wedge \theta_1$, one finds that the functions $\sigma_1, \sigma_2, \sigma_3$ satisfy equations of the form

$$(5.24)\qquad d\sigma_\alpha + (m+1)\sigma_\alpha\omega^1_1 = \sigma_{\alpha 1}\theta_1 + \sigma_{\alpha 2}\theta_2 + \sigma_{\alpha 3}\theta_3,\ \alpha=1,2,3\ ,$$

where the coefficient functions $\sigma_{\alpha\beta}$ satisfy the commutation relations

$$(5.25)\qquad \begin{aligned} \sigma_{12} - \sigma_{21} &= -m\sigma\rho_3 - \rho\sigma_3\ , \\ \sigma_{23} - \sigma_{32} &= -m\sigma\rho_1 - \rho\sigma_1\ , \\ \sigma_{31} - \sigma_{13} &= -m\sigma\rho_2 - \rho\sigma_2\ . \end{aligned}$$

In particular, from equation (5.20), we have the following commutation relations on ρ_1, ρ_2, ρ_3:

$$(5.26)\qquad \begin{aligned} \rho_{12} - \rho_{21} &= -2\rho\rho_3\ , \\ \rho_{23} - \rho_{32} &= -2\rho\rho_1\ , \\ \rho_{31} - \rho_{13} &= -2\rho\rho_2\ . \end{aligned}$$

We next take the exterior derivatives of the equations (5.10)-(5.12). We first differentiate the equation

(5.27) $$\omega_4^7 = a\,\omega_1^4 + b\,\omega_1^5 .$$

On the one hand, from the Maurer-Cartan equation (3.7) for $d\omega_4^7$, we have (by not writing those terms which have already been shown to vanish),

(5.28) $$\begin{aligned} d\omega_4^7 &= \omega_4^2 \wedge \omega_2^7 + \omega_4^3 \wedge \omega_3^7 + \omega_4^5 \wedge \omega_5^7 + \omega_4^7 \wedge \omega_7^7 \\ &= -\,\omega_1^4 \wedge \omega_2^7 + (-\rho\,\omega_1^5) \wedge (-q\,\omega_1^5 - t\,\omega_1^4 - u\,\omega_7^3) \\ &\quad + \rho\,\omega_7^3 \wedge (b\,\omega_1^4 + c\,\omega_1^5) + (a\,\omega_1^4 + b\,\omega_1^5) \wedge \omega_1^1 . \end{aligned}$$

On the other hand, differentiation of the right-hand side of (5.27) yields

(5.29) $$\begin{aligned} d\omega_4^7 &= da \wedge \omega_1^4 + a\,d\omega_1^4 + db \wedge \omega_1^5 + b\,d\omega_1^5 \\ &= da \wedge \omega_1^4 + a(\omega_1^1 \wedge \omega_1^4 - \rho\,\omega_1^5 \wedge \omega_7^3) \\ &\quad + db \wedge \omega_1^5 + b(\omega_1^1 \wedge \omega_1^5 - \rho\,\omega_1^4 \wedge \omega_7^3) . \end{aligned}$$

Equating (5.28) and (5.29) yields

(5.30) $$\begin{aligned} &(da + 2a\,\omega_1^1 - 2b\rho\,\omega_7^3 - \omega_2^7) \wedge \omega_1^4 \\ &+ (db + 2b\,\omega_1^1 + (a+u-c)\rho\,\omega_7^3) \wedge \omega_1^5 \\ &+ \rho t\,\omega_1^4 \wedge \omega_1^5 = 0 . \end{aligned}$$

Since $b = \rho_3$, it follows from (5.19) and (5.24) that

(5.31) $$db + 2b\,\omega_1^1 = d\rho_3 + 2\rho_3\,\omega_1^1 = \rho_{31}\theta_1 + \rho_{32}\theta_2 + \rho_{33}\theta_3 .$$

By examining the coefficient of $\omega_1^5 \wedge \omega_7^3 = \theta_2 \wedge \theta_3$ in equation (5.30) and using (5.31), we get that

(5.32) $$\rho_{33} = \rho(c-a-u) \ .$$

Furthermore, the remaining terms in (5.30) are

(5.33) $$(da + 2a\,\omega_1^1 - \omega_2^7 - 2\rho b\,\omega_7^3 - (\rho t + \rho_{31})\omega_1^5) \wedge \omega_1^4$$
$$+ \text{ terms involving } \omega_1^5 \text{ and } \omega_7^3 \text{ only.}$$

Thus, the coefficient in parentheses must be a multiple of ω_1^4, call it $\bar{a}\omega_1^4$. We can write this using (5.8) and (5.21) as

(5.34) $$da + 2a\,\omega_1^1 = \omega_2^7 + \bar{a}\theta_1 + (\rho_{31} - \rho\rho_2)\theta_2 + 2\rho\rho_3\theta_3 \ .$$

In a similar manner, if we differentiate

$$\omega_5^7 = b\,\omega_1^4 + c\,\omega_1^5 \ ,$$

we obtain,

(5.35) $$dc + 2c\,\omega_1^1 = \omega_2^7 + (\rho_{32} + \rho\rho_1)\theta_1 + \bar{c}\theta_2 - 2\rho\rho_3\theta_3 .$$

Thus, from the two equations in (5.10), we have obtained (5.32), (5.34) and (5.35). In completely analogous fashion, we can differentiate the two equations in (5.11) to obtain

(5.36) $$\rho_{11} = \rho(s+r-p) \ ,$$

(5.37) $$dp + 2p\,\omega^1_1 = -\omega^7_2 + 2\rho\rho_1\theta_1 + (-\rho_{13}-\rho\rho_2)\theta_2 + \bar{p}\theta_3\ ,$$

(5.38) $$dr + 2r\,\omega^1_1 = -\omega^7_2 - 2\rho\rho_1\theta_1 + \bar{r}\theta_2 + (-\rho_{12} + \rho\rho_3)\theta_3,$$

while differentiation of (5.12) yields

(5.39) $$\rho_{22} + \rho_{33} = \rho(p-r-s),$$

(5.40) $$ds + 2s\,\omega^1_1 = \bar{s}\theta_1 + (\rho_{31} + \rho\rho_2)\theta_2 + (-\rho_{21} + \rho\rho_3)\theta_3\ ,$$

(5.41) $$du + 2u\,\omega^1_1 = (-\rho_{23}-\rho\rho_1)\theta_1 + (\rho_{13}-\rho\rho_2)\theta_2 + \bar{u}\theta_3\ .$$

In these equations, the coefficients $\bar{a},\bar{c},\bar{p},\bar{r},\bar{s},\bar{u}$ remain undetermined. However, by differentiating (5.18) and using the appropriate equations among those involving these quantities above, one can show that

(5.42) $$\begin{aligned} \bar{a} &= -6\rho\rho_1\ , & \bar{c} &= 6\rho\rho_2\ , \\ \bar{p} &= -6\rho\rho_3\ , & \bar{r} &= 6\rho\rho_2\ , \\ \bar{s} &= -12\rho\rho_1\ , & \bar{u} &= 12\rho\rho_3\ . \end{aligned}$$

From equations (5.32), (5.36), (5.39) and (5.18), we easily compute that

(5.43) $$\rho_{11} + \rho_{22} + \rho_{33} = 0\ .$$

Using (5.42), equations (5.40) and (5.41) can be rewritten as

(5.44) $$ds + 2s\,\omega^1_1 = -12\rho\rho_1\theta_1 + (\rho_{31} + \rho\rho_2)\theta_2 + (-\rho_{21} + \rho\rho_3)\theta_3\ ,$$

(5.45) $$du + 2u\,\omega^1_1 = (-\rho_{23}-\rho\rho_1)\theta_1 + (\rho_{13}-\rho\rho_2)\theta_2 + 12\rho\rho_3\theta_3\ .$$

By taking the exterior derivatives of these two equations and making use of (5.43) and of the commutation relations (5.25) for ρ and its various derivatives, one ultimately can show after a lengthy calculation that the following fundamental equations hold:

$$
\begin{aligned}
&\rho\rho_{12} + \rho_1\rho_2 + \rho^2\rho_3 = 0 ,\\
&\rho\rho_{21} + \rho_1\rho_2 - \rho^2\rho_3 = 0 ,\\
&\rho\rho_{23} + \rho_2\rho_3 + \rho^2\rho_1 = 0 ,\\
&\rho\rho_{32} + \rho_2\rho_3 - \rho^2\rho_1 = 0 ,\\
&\rho\rho_{31} + \rho_3\rho_1 + \rho^2\rho_2 = 0 ,\\
&\rho\rho_{13} + \rho_3\rho_1 - \rho^2\rho_2 = 0 .
\end{aligned}
\tag{5.46}
$$

We now briefly outline the details of this calculation. By (5.44), we have

$$s_1 = -12\rho\rho_1 \ , \quad s_2 = \rho_{31} + \rho\rho_2 \ , \quad s_3 = \rho\rho_3 - \rho_{21} . \tag{5.47}$$

The commutation relation (5.25) for s with m=2 gives

$$s_{12} - s_{21} = -2s\rho_3 - \rho s_3 = -2s\rho_3 - \rho(\rho\rho_3 - \rho_{21}) . \tag{5.48}$$

On the other hand, we can directly compute by taking covariant derivatives of (5.47) that

$$s_{12} - s_{21} = -12\rho\rho_{12} - 12\rho_2\rho_1 - (\rho_{311} + \rho_1\rho_2 + \rho\rho_{21}) . \tag{5.49}$$

The main problem now is to get ρ_{311} into a usable form. By taking the covariant derivative of the third equation in (5.26), we find

$$\rho_{311} - \rho_{131} = -2\rho_1\rho_2 - 2\rho\rho_{21} . \tag{5.50}$$

Then using the commutation relation

$$\rho_{131} = \rho_{113} - 2\rho_1\rho_2 - \rho\rho_{12} ,$$

we get from (5.50)

$$\rho_{311} = \rho_{113} - 4\rho_1\rho_2 - \rho\rho_{12} - 2\rho\rho_{21} . \tag{5.51}$$

Taking the covariant derivative of $\rho_{11} = \rho(s+r-p)$ and substituting the expression obtained for ρ_{113} into (5.51), we get

$$\rho_{311} = \rho_3(s+r-p)-3\rho\rho_{21}-2\rho\rho_{12} + 8\rho^2\rho_3-4\rho_1\rho_2 . \tag{5.52}$$

If we substitute (5.52) for ρ_{311} in (5.49) and then equate the right-hand sides of (5.48) and (5.49), we obtain the first equation in (5.46). The cyclic permutations are obtained in a similar way from $s_{23} - s_{32}$, etc.

Our frame attached to the line $[Y_1, Y_7]$ is still not completely determined, viz., the following change is allowable:

$$Y_2^* = \alpha^{-1}Y_2 + \mu Y_7 , \quad Y_6^* = \alpha^{-1}Y_6 - \mu Y_1 . \tag{5.53}$$

The Y_i's, $i = 3,4,5$ being completely determined, we have under this change,

$$\omega_1^{4*} = \alpha\omega_1^4,\ \omega_1^{5*} = \alpha\omega_1^5,\ \omega_7^{3*} = \alpha\omega_7^3,$$
$$\omega_4^{7*} = \alpha^{-1}\omega_4^7 + \mu\omega_1^4,$$
$$\omega_3^{1*} = \alpha^{-1}\omega_3^1 - \mu\omega_7^3,$$

which implies that

$$a^* = \alpha^{-2}a + \alpha^{-1}\mu,$$
$$p^* = \alpha^{-2}p - \alpha^{-1}\mu.$$

We choose μ to make $a^* = p^*$. After dropping the asterisks, we have from (5.18) that

$$(5.54) \qquad a = p = -\rho^2,\ r = 3\rho^2 + s,\ c = 3\rho^2 - u.$$

Now using the fact that a = p, we can subtract (5.37) from (5.34) and get that

$$(5.55) \qquad \omega_2^7 = 4\rho\rho_1\theta_1 - ((\rho_{31}+\rho_{13})/2)\theta_2 - 4\rho\rho_3\theta_3.$$

We are finally in position to proceed toward the main results. Ultimately, we show that the frame can be chosen so that the function ρ is constant, and the classification naturally splits into the two cases $\rho = 0$ and $\rho \neq 0$.

<u>The case $\rho\neq 0$</u>.

We now assume that the function ρ is never zero on B. The following lemma is the key in this case. This is Pinkall's Lemma [P2, p. 108], where his function c is the negative of our function ρ. Since $\rho \neq 0$, the fundamental equations (5.46) allow one to express all of the second covariant derivatives $\rho_{\alpha\beta}$ in terms of ρ and its first derivatives ρ_α. This enables us to give a somewhat simpler proof than Pinkall gave for the lemma.

Lemma 5.1: Suppose that ρ never vanishes on B. Then $\rho_1 = \rho_2 = \rho_3 = 0$ at every point of B.

Proof: First note that if the function ρ_3 vanishes identically, then (5.46) and the assumption that $\rho \neq 0$ imply that ρ_1 and ρ_2 also vanish identically. We now complete the proof of the lemma by showing that ρ_3 must vanish everywhere. This is accomplished by considering the expression $s_{12} - s_{21}$. By the commutation relations (5.25), we have

$$s_{12} - s_{21} = -2s\rho_3 - \rho s_3 .$$

By (5.46) and (5.47), we see that

$$\rho s_3 = \rho^2\rho_3 - \rho\rho_{21} = \rho_1\rho_2 ,$$

and so

$$s_{12} - s_{21} = -2s\rho_3 - \rho_1\rho_2 . \tag{5.56}$$

On the other hand, we can compute s_{12} directly from the equation $s_1 = -12\rho\rho_1$. Then using the expression for ρ_{12} obtained from (5.46), we get

$$s_{12} = -12\rho_2\rho_1 - 12\rho\rho_{12} = -12(\rho_2\rho_1 + \rho\rho_{12})$$

(5.57)

$$= -12(\rho_2\rho_1 + (-\rho_2\rho_1 - \rho^2\rho_3)) = 12\rho^2\rho_3 .$$

Next we have from (5.47),

$$s_2 = \rho_{31} + \rho\rho_2 .$$

Using (5.46), we can write

$$\rho_{31} = -\rho_3\rho_1\rho^{-1} - \rho\rho_2 ,$$

and thus

(5.58)
$$s_2 = -\rho_3\rho_1/\rho .$$

Then, we compute

$$s_{21} = -(\rho(\rho_3\rho_{11} + \rho_{31}\rho_1) - \rho_3\rho_1^2)/\rho^2 .$$

Using (5.36) for ρ_{11} and (5.46) to get ρ_{31}, this becomes

(5.59)
$$s_{21} = -\rho_3(s+r-p) + 2\rho_3\rho_1^2\rho^{-2} + \rho_1\rho_2 .$$

Now equate the expression (5.56) for $s_{12} - s_{21}$ with that obtained by subtracting (5.59) from (5.57) to get

$$-2s\rho_3 - \rho_1\rho_2 = 12\rho^2\rho_3 + \rho_3(s+r-p)-2\rho_3\rho_1^2\rho^{-2} - \rho_1\rho_2 .$$

This can be rewritten as

$$0 = \rho_3(12\rho^2 + 3s+r-p - 2\rho_1^2\rho^{-2}) . \tag{5.60}$$

Using the expressions in (5.54) for r and p, we see that $3s+r-p = 4s + 4\rho^2$, and so (5.60) can be written as

$$0 = \rho_3(16\rho^2 + 4s - 2\rho_1^2\rho^{-2}) . \tag{5.61}$$

Suppose that $\rho_3 \neq 0$ at some point $b \in B$. Then ρ_3 does not vanish on some neighborhood U of b. By (5.61), we have

$$16\rho^2 + 4s - 2\rho_1^2\rho^{-2} = 0 \tag{5.62}$$

on U. We now take the θ_2-covariant derivative of (5.62) and obtain

$$32\rho\rho_2 + 4s_2 - 4\rho_1\rho_{12}\rho^{-2} + 4\rho_1^2\rho_2\rho^{-3} = 0 . \tag{5.63}$$

We now substitute the expression (5.58) for s_2 and the formula

$$\rho_{12} = -\rho_1\rho_2\rho^{-1} - \rho\rho_3$$

obtained from (5.46) into (5.63). After some algebra, (5.63) reduces to

$$\rho_2(32\rho^4 + 8\rho_1^2) = 0 .$$

Since $\rho \neq 0$, this implies that $\rho_2 = 0$ on U. But then the left side of the equation (5.46)

$$\rho\rho_{21} + \rho_1\rho_2 = \rho^2\rho_3$$

must vanish on U. Since $\rho \neq 0$, we conclude that $\rho_3 = 0$ on U, a contradiction to our assumption. Hence, ρ_3 must vanish identically on B and the lemma is proven.

We now continue with the case $\rho \neq 0$. According to Lemma 5.1, all the covariant derivatives of ρ are zero, and our formulas simplify greatly. Equations (5.32) and (5.36) give

$$c-a-u = 0 \ , \ s+r-p = 0 \ .$$

These combined with (5.54) give

$$c = r = \rho^2 \ , \ u = -s = 2\rho^2 \ . \tag{5.64}$$

By (5.55) we have $\omega_2^7 = 0$. So the differentials of the frame vectors can now be written

$$\begin{aligned}
&dY_1 - \omega_1^1 Y_1 = \omega_1^4 Y_4 + \omega_1^5 Y_5 \ , \\
&dY_7 - \omega_1^1 Y_7 = \omega_7^3 Y_3 - \omega_1^5 Y_5 \ , \\
&dY_2 + \omega_1^1 Y_2 = \rho^2(\omega_7^3 Y_3 + 2\omega_1^4 Y_4 + \omega_1^5 Y_5) \ , \\
&dY_6 + \omega_1^1 Y_6 = \rho^2(2\omega_7^3 Y_3 + \omega_1^4 Y_4 - \omega_1^5 Y_5) \ , \\
&dY_3 = \omega_7^3 Z_3 + \rho(\omega_1^5 Y_4 + \omega_1^4 Y_5) \ , \\
&dY_4 = - \omega_1^4 Z_4 + \rho(- \omega_1^5 Y_3 + \omega_7^3 Y_5) \ , \\
&dY_5 = \omega_1^5 Z_5 + \rho(- \omega_1^4 Y_3 - \omega_7^3 Y_4) \ ,
\end{aligned} \tag{5.65}$$

where

$$Z_3 = -Y_6 + \rho^2(-Y_1 - 2Y_7) ,$$
$$(5.66) \qquad Z_4 = Y_2 + \rho^2(2Y_1 + Y_7) ,$$
$$Z_5 = -Y_2 + Y_6 + \rho^2(-Y_1 + Y_7) .$$

From this, we notice that

$$(5.67) \qquad Z_3 + Z_4 + Z_5 = 0 ,$$

so that the points Z_3, Z_4, Z_5 lie on a line.

From (5.20) and (5.22) and the lemma we see that

$$(5.68) \qquad d\rho + \rho\,\omega_1^1 = 0 \quad , \quad d\omega_1^1 = 0 .$$

We now make a change of frame of the form

$$(5.69) \qquad Y_1^* = \rho Y_1,\ Y_7^* = \rho Y_7,\ Y_2^* = (1/\rho)Y_2,\ Y_6^* = (1/\rho)Y_6,$$
$$Y_i^* = Y_i,\ i = 3,4,5.$$

Then set

$$(5.70) \qquad Z_i^* = (1/\rho)Z_i \quad , \quad i = 3,4,5 .$$
$$\omega_1^{4*} = \rho\,\omega_1^4,\ \omega_1^{5*} = \rho\,\omega_1^5,\ \omega_7^{3*} = \rho\,\omega_7^3 .$$

The effect of this change is to make $\rho^* = 1$ and $\omega_1^{1*} = 0$, for we can compute the following:

$$
(5.71) \qquad
\begin{aligned}
dY_1^* &= \omega_1^{4*} Y_4 + \omega_1^{5*} Y_5 , \\
dY_7^* &= \omega_7^{3*} Y_3 - \omega_1^{5*} Y_5 , \\
dY_2^* &= \omega_7^{3*} Y_3 + 2\omega_1^{4*} Y_4 + \omega_1^{5*} Y_5 , \\
dY_6^* &= 2\omega_7^{3*} Y_3 + \omega_1^{4*} Y_4 - \omega_1^{5*} Y_5 , \\
dY_3 &= \omega_7^{3*} Z_3^* + \omega_1^{5*} Y_4 + \omega_1^{4*} Y_5 , \\
dY_4 &= - \omega_1^{4*} Z_4^* - \omega_1^{5*} Y_3 + \omega_7^{3*} Y_5 , \\
dY_5 &= \omega_1^{5*} Z_5 - \omega_1^{4*} Y_3 - \omega_7^{3*} Y_4 ,
\end{aligned}
$$

with

$$
(5.72) \qquad
\begin{aligned}
dZ_3^* &= 2(-2\omega_7^{3*} Y_3 - \omega_1^{4*} Y_4 + \omega_1^{5*} Y_5) , \\
dZ_4^* &= 2(\omega_7^{3*} Y_3 + 2\omega_1^{4*} Y_4 + \omega_1^{5*} Y_5) , \\
dZ_5^* &= 2(\omega_7^{3*} Y_3 - \omega_1^{4*} Y_4 - 2\omega_1^{5*} Y_5) ,
\end{aligned}
$$

and

$$
\begin{aligned}
d\omega_1^{4*} &= - \omega_1^{5*} \wedge \omega_7^{3*} , \text{ i.e., } d\theta_1^* = -\theta_2^* \wedge \theta_3^* , \\
d\omega_1^{5*} &= - \omega_7^{3*} \wedge \omega_1^{4*} , \text{ i.e., } d\theta_2^* = -\theta_3^* \wedge \theta_1^* , \\
d\omega_7^{3*} &= - \omega_1^{4*} \wedge \omega_1^{5*} , \text{ i.e., } d\theta_3^* = -\theta_1^* \wedge \theta_2^* .
\end{aligned}
$$

Comparing the last equation with (5.13), we see that $\omega_1^{1*} = 0$ and $\rho^* = 1$.

This is the final frame which we will need in this case $\rho \neq 0$. So, we again drop the asterisks.

We are now ready to prove Pinkall's classification result for the case $\rho \neq 0$ [P2, p. 117]. As with the cyclides, there is only one compact model, up to Lie equivalence. This is Cartan's isoparametric hypersurface M^3 in S^4. It is a tube of constant radius over each of its two focal submanifolds, which are standard Veronese surfaces in S^4. (See [CR, pp. 296-299] for more detail.) We will describe the Veronese surface after stating the theorem.

__Theorem 5.2__: (Pinkall [P2]): (a) Every connected Dupin submanifold with $\rho \neq 0$ is contained in a unique compact connected Dupin submanifold with $\rho \neq 0$.
(b) Any two Dupin submanifolds with $\rho \neq 0$ are locally Lie equivalent, each being Lie equivalent to an open subset of Cartan's isoparametric hypersurface in S^4.

Our method of proof differs from that of Pinkall in that we will prove directly that each of the focal submanifolds can naturally be considered to be an open subset of a Veronese surface in a hyperplane $P^5 \subset P^6$. The Dupin submanifold can then be constructed from these focal submanifolds.

We now recall the definition of a Veronese surface. First consider the map from the unit sphere $y_1^2 + y_2^2 + y_3^2 = 1$ in $\mathbb{R}^3$ into $\mathbb{R}^5$ given by

$$(x_1,\ldots,x_5) = (2y_2y_3, 2y_3y_1, 2y_1y_2, y_1^2, y_2^2) \; . \tag{5.73}$$

This map takes the same value on antipodal points of the 2-sphere, so it induces a map $\phi: P^2 \to \mathbb{R}^5$. One can show by an elementary direct calculation that ϕ is an embedding of P^2 and that ϕ is substantial in $\mathbb{R}^5$, i.e., does not lie in any hyperplane. Any embedding of P^2 into P^5 which is projectively equivalent to ϕ is called a __Veronese surface__. (See Lane [L, pp. 424-430] for more detail.)

Let $k_1 = Y_1$, $k_2 = Y_7$, $k_3 = Y_1 + Y_7$ be the focal point maps of the Dupin submanifold $\lambda: B \to \Lambda$ with $\rho \neq 0$. Each k_i is constant along the leaves of its corresponding principal foliation T_i, so each k_i factors through an immersion ϕ_i of the 2-dimensional space of leaves B/T_i into P^6. We will show that each of these ϕ_i is an open subset of a Veronese surface in some $P^5 \subset P^6$.

We wish to integrate the differential system (5.71), which is completely integrable. For this purpose we drop the asterisks and write the system as follows:

$$\begin{aligned} dY_1 &= \theta_1 Y_4 + \theta_2 Y_5 , \\ dY_7 &= \theta_3 Y_3 - \theta_2 Y_5 , \\ dY_2 &= \theta_3 Y_3 + 2\theta_1 Y_4 + \theta_2 Y_5 , \\ dY_6 &= 2\theta_3 Y_3 + \theta_1 Y_4 - \theta_2 Y_5 , \\ dY_3 &= \theta_3 Z_3 + \theta_2 Y_4 + \theta_1 Y_5 , \\ dY_4 &= - \theta_1 Z_4 - \theta_2 Y_3 + \theta_3 Y_5 , \\ dY_5 &= \theta_2 Z_5 - \theta_1 Y_3 - \theta_3 Y_4 ; \end{aligned} \tag{5.74}$$

with

$$\begin{aligned} dZ_3 &= 2(-2\theta_3 Y_3 - \theta_1 Y_4 + \theta_2 Y_5) , \\ dZ_4 &= 2(\theta_3 Y_3 + 2\theta_1 Y_4 + \theta_2 Y_5) , \\ dZ_5 &= 2(\theta_3 Y_3 - \theta_1 Y_4 - 2\theta_2 Y_5) , \end{aligned} \tag{5.75}$$

where

$$\begin{aligned} d\theta_1 &= - \theta_2 \wedge \theta_3 , \\ d\theta_2 &= - \theta_3 \wedge \theta_1 , \\ d\theta_3 &= - \theta_1 \wedge \theta_2 , \end{aligned} \tag{5.76}$$

and

$$\begin{aligned} Z_3 &= - Y_1 - Y_6 - 2Y_7 , \\ Z_4 &= 2Y_1 + Y_2 + Y_7 , \\ Z_5 &= -Y_1 - Y_2 + Y_6 + Y_7 , \end{aligned} \tag{5.77}$$

so that

$$Z_3 + Z_4 + Z_5 = 0 . \tag{5.78}$$

Put

$$W_1 = -Y_1 + Y_6 - 2Y_7, \quad W_2 = -2Y_1 + Y_2 - Y_7 . \tag{5.79}$$

We find from (5.74) that

$$dW_1 = dW_2 = 0 , \tag{5.80}$$

so that the points W_1, W_2 are fixed. Their inner products are

$$\langle W_1, W_1\rangle = \langle W_2, W_2\rangle = -4, \quad \langle W_1, W_2\rangle = -2 , \tag{5.81}$$

and the line $[W_1, W_2]$ consists entirely of timelike points. Its orthogonal complement in $\mathbb{R}^7_2$ is spanned by Y_3, Y_4, Y_5, Z_4, Z_5. It consists entirely of spacelike points and has no point in common with Q^5. We will denote it as $\mathbb{R}^5$.

It suffices to solve the system (5.74) in $\mathbb{R}^5$ for Y_3, Y_4, Y_5, Z_4, Z_5. For we have

$$\begin{aligned} &d(Z_4 - Z_5 - 6Y_1) = 0 , \\ &d(Z_4 + 2Z_5 - 6Y_7) = 0 , \end{aligned} \tag{5.82}$$

so that there exist constant vectors C_1, C_2 such that

$$\begin{aligned} Z_4 - Z_5 - 6Y_1 &= C_1 , \\ Z_4 + 2Z_5 - 6Y_7 &= C_2 . \end{aligned} \tag{5.83}$$

Thus, Y_1 and Y_7 are determined by these equations, and then Y_2 and Y_6 are determined from (5.79). Note that C_1 and C_2 are timelike points and the line $[C_1, C_2]$ consists entirely of timelike points.

Equations (5.76) are the structure equations of SO(3). It is thus natural to take the latter as the parameter space, whose points are the 3x3 matrices

$$A = [a_{ik}] \ , \ 1 \le i,j,k \le 3 \ ,$$

satisfying

(5.84) $$^tAA = A^tA = I \ , \ \det A = 1 \ .$$

The first equations above, when expanded, are

(5.85) $$\Sigma\, a_{ij}a_{ik} = \Sigma\, a_{ji}a_{ki} = \delta_{jk} \ .$$

The Maurer-Cartan forms of SO(3) are

(5.86) $$\alpha_{ik} = \Sigma\, a_{kj}da_{ij} = -\alpha_{ki} \ .$$

They satisfy the Maurer-Cartan equations

(5.87) $$d\alpha_{ik} = \Sigma\, \alpha_{ij} \wedge \alpha_{jk} \ .$$

If we set

(5.88) $$\theta_1 = \alpha_{23} \ , \ \theta_2 = \alpha_{31} \ , \ \theta_3 = \alpha_{12} \ ,$$

these equations reduce to (5.76). With the θ_i given by (5.88), we shall write down an explicit solution of (5.74).

Let E_A, $1 \le A \le 5$, be a fixed linear frame in $\mathbb{R}^5$. Let

$$(5.89)\quad F_i = 2a_{i2}a_{i3}E_1+2a_{i3}a_{i1}E_2+2a_{i1}a_{i2}E_3+a_{i1}^2E_4+a_{i2}^2E_5,\ 1 \le i,j,k \le 3 .$$

Since $$a_{i1}^2 + a_{i2}^2 + a_{i3}^2 = 1 ,$$

we see from (5.73), with $y_j = a_{ij}$, that F_i is a Veronese surface for $1 \le i \le 3$. Using (5.85), we compute that

$$(5.90)\quad F_1 + F_2 + F_3 = E_4 + E_5 = \text{constant}.$$

Since the coefficients in F_i are quadratic, the partial derivatives $\partial^2F_i/\partial a_{ij}\partial a_{ik}$ are independent of i. Moreover, the quantities G_{ik} defined below satisfy

$$(5.91)\quad G_{ik} = \sum_j a_{ij} \frac{\partial F_k}{\partial a_{kj}} = G_{ki} .$$

We use these facts in the following computation:

$$dG_{ik} = \Sigma \frac{\partial F_k}{\partial a_{kj}} da_{ij} + \Sigma a_{ij} \frac{\partial^2 F_k}{\partial a_{kj}\partial a_{k\ell}} da_{k\ell}$$

(5.92)

$$= \Sigma \frac{\partial F_k}{\partial a_{kj}} da_{ij} + \Sigma a_{ij} \frac{\partial^2 F_i}{\partial a_{ij}\partial a_{i\ell}} da_{k\ell} = \Sigma \frac{\partial F_k}{\partial a_{kj}} da_{ij} + \Sigma \frac{\partial F_i}{\partial a_{ij}} da_{kj} ,$$

where the last step follows from the linear homogeneity of $\partial F_i/\partial a_{i\ell}$. In terms of α_{ij}, we have

(5.93) $$dG_{ik} = \Sigma \frac{\partial F_k}{\partial a_{kj}} a_{\ell j}\alpha_{i\ell} + \Sigma \frac{\partial F_i}{\partial a_{ij}} a_{\ell j}\alpha_{k\ell} ,$$

which gives, when expanded,

(5.94) $$dG_{23} = 2(F_3-F_2)\theta_1 + G_{12}\theta_2 - G_{13}\theta_3 ,$$

and its cyclic permutations.

On the other hand, by the same manipulation, we have

(5.95) $$dF_i = \Sigma \frac{\partial F_i}{\partial a_{ij}} da_{ij} = \Sigma \frac{\partial F_i}{\partial a_{ij}} a_{kj}\alpha_{ik} ,$$

giving

$$dF_1 = - G_{31}\theta_2 + G_{12}\theta_3 ,$$

and its cyclic permutations.

One can now immediately verify that a solution of (5.74) is given by

(5.96) $$\begin{aligned} &Y_3 = G_{12} ,\ Y_4 = - G_{23} ,\ Y_5 = G_{31} , \\ &Z_3 = 2(F_2-F_1),\ Z_4 = 2(F_3-F_2),\ Z_5 = 2(F_1-F_3) . \end{aligned}$$

This is also the most general solution of (5.74), for the solution is determined up to a linear transformation, and our choice of frame E_A is arbitrary.

By (5.96), the functions Z_1, Z_2, Z_3 are expressible in terms of F_1, F_2, F_3, and then by (5.83), so also are Y_1, Y_7, Y_1+Y_7. Specifically, by (5.83), (5.90), and (5.96) we have

$$
\begin{aligned}
6Y_1 &= Z_4 - Z_5 - C_1 = 2(-F_1 - F_2 + 2F_3) - C_1 \\
&= 6F_3 - 2(E_4 + E_5) - C_1 ,
\end{aligned}
$$

so that the focal map Y_1, up to an additive constant vector, is the Veronese surface F_3. Similarly, the focal maps Y_7 and Y_1+Y_7 are the Veronese surfaces F_1 and $-F_2$, respectively, up to additive constants.

We see from (5.79) that

$$\langle Y_1, W_1\rangle = 0,\ \langle Y_7, W_2\rangle = 0,\ \langle Y_1+Y_7, W_1-W_2\rangle = 0 .$$

Thus Y_1 is contained in the Moebius space $\Sigma^4 = Q^5 \cap W_1^{\perp}$ .Let $e_7 = W_1/2$ and $e_1 = (2W_2 - W_1)/\sqrt{12}$. Then e_1 is the unique unit vector on the timelike line $[W_1, W_2]$ which is orthogonal to W_1. In a manner similar to that of Section 3, we can write

$$Y_1 = e_1 + f ,$$

where f maps B into the unit sphere S^4 in the Euclidean space $\mathbb{R}^5 = [e_1, e_7]^{\perp}$ in $\mathbb{R}^7_2$. We call f the <u>spherical projection</u> of the Legendre map λ determined by the ordered pair (e_7, e_1) (see [CC] for more detail). We know that f is constant along the leaves of the principal foliation T_1 corresponding to Y_1.

and f induces a map $\tilde{f}:B/T_1 \to S^4$. By what we have shown above, $\tilde{f}$ must be an open subset a spherical Veronese surface.

Note that the unit timelike vector $W_2/2$ satisfies

$$W_2/2 = (\sqrt{3}/2)e_1 + (1/2)e_7 = \sin(\pi/3)e_1 + \cos(\pi/3)e_7 .$$

If we consider the points in the Moebius space Σ to represent point spheres in S^4, then as we show in [CC], points in $Q^5 \cap W_2^{\perp}$ represent oriented spheres in S^4 with oriented radius $-\pi/3$. In a way similar to that above, the second focal submanifold $Y_7 \subset Q^5 \cap W_2^{\perp}$ induces a spherical Veronese surface. When considered from the point of view of the Moebius space Σ, the points in Y_7 represent oriented spheres of radius $-\pi/3$ centered at points of this Veronese surface. These spheres must be in oriented contact with the point spheres of the first Veronese surface determined by Y_1 in $Q^5 \cap P^5$. Thus, the points in the second Veronese surface must lie at a distance $\pi/3$ along normal geodesics in S^4 to the first Veronese surface $\tilde{f}$. In fact (see, for example, [CR,pp. 296-299]), the set of all points in S^4 at distance $\pi/3$ from a spherical Veronese surface is another spherical Veronese surface.

Thus, with this choice of coordinates, the Dupin submanifold in question is simply an open subset of the Legendre submanifold induced in the standard way by considering B to be the unit normal bundle to the spherical Veronese embedding $\tilde{f}$ induced by Y_1. For values of $t = k\pi/3$, $k \in Z$, the parallel hypersurface at oriented distance t to $\tilde{f}$ in S^4 is a Veronese surface. For other values of t, the parallel hypersurface is an isoparametric hypersurface in S^4 with three distinct principal curvatures (Cartan's isoparametric hypersurface). All of these parallel hypersurfaces are Lie equivalent to each other and to the Legendre submanifolds induced by the Veronese surfaces.

<u>The case $\rho = 0$</u>.

We now consider the case where ρ is identically zero on B. It turns out that no new examples occur here, in that these Dupin submanifolds can all be constructed from Dupin cyclides by certain standard constructions. To make this precise, we recall Pinkall's [P3, p. 437] notion of reducibility. Our Dupin submanifold can be considered, as in Section 3, to have been induced from a Dupin hypersurface $M^3 \subset E^4$. The Dupin submanifold is <u>reducible</u> if M^3 is obtained from a Dupin surface $S \subset E^3 \subset E^4$ by one of the four following standard constructions.

(5.97)

i. M is a cylinder $S \times \mathbb{R}$ in E^4.

ii. M is the hypersurface of revolution obtained by revolving S about a plane π disjoint from S in E^3.

iii. Project S stereographically onto a surface $N \subset S^3 \subset E^4$. M is the cone $\mathbb{R}\cdot N$ over N.

iv. M is a tube of constant radius around S in E^4.

Pinkall proved [P3, p. 438] that the Dupin submanifold $\lambda: B \rightarrow \Lambda^7$ is reducible if and only if some focal point map is contained in a 4-dimensional subspace $P^4 \subset P^6$.

If ρ is identically zero on B, then by (5.20), all of the covariant derivatives of ρ are also equal to zero. From (5.21) and (5.54), we see that the functions in equations (5.10)-(5.12) satisfy

$$q = t = b = 0 \ ,$$
$$a = p = 0 \ , \ r = s, \ c = -u.$$

Then from (5.55), we have that $\omega_2^7 = 0$. From these and the other relations among the Maurer-Cartan forms which we have derived, we see that the differentials of the frame vectors can be written

$$
\begin{aligned}
dY_1 - \omega_1^1 Y_1 &= \omega_1^4 Y_4 + \omega_1^5 Y_5 , \\
dY_7 - \omega_1^1 Y_7 &= \omega_7^3 Y_3 - \omega_1^5 Y_5 , \\
dY_2 + \omega_1^1 Y_2 &= s(- \omega_1^4 Y_4 + \omega_1^5 Y_5) , \\
dY_6 + \omega_1^1 Y_2 &= u(\omega_7^3 Y_3 + \omega_1^5 Y_5) , \\
dY_3 &= \omega_7^3(-Y_6 + uY_7) , \\
dY_4 &= \omega_1^4(sY_1 - Y_2) , \\
dY_5 &= \omega_1^5(-sY_1 - Y_2 + Y_6 - uY_7) .
\end{aligned} \tag{5.98}
$$

Note that from (5.44), (5.45), we have

$$ds + 2s\,\omega_1^1 = 0 \quad , \quad du + 2u\,\omega_1^1 = 0 , \tag{5.99}$$

and from (5.13) that

$$d\theta_i = \omega_1^1 \wedge \theta_i \quad , \quad i = 1,2,3 . \tag{5.100}$$

From (5.22), we have that $d\omega_1^1 = 0$. Hence on any local disk neighborhood U in B, we have that

$$\omega_1^1 = d\sigma , \tag{5.101}$$

for some smooth function σ on U. We next consider a change of frame of the form

$$(5.102)\qquad \begin{aligned} &Y_1^* = e^{-\sigma}Y_1,\ Y_7^* = e^{-\sigma}Y_7,\ Y_2^* = e^{\sigma}Y_2,\ Y_6^* = e^{\sigma}Y_6,\\ &Y_i^* = Y_i,\quad i = 3,4,5\,. \end{aligned}$$

The effect of this change is to make $\omega_1^{1*} = 0$ while keeping $\rho^* = 0$. If we set

$$\omega_1^{4*} = e^{-\sigma}\omega_1^4,\ \omega_1^{5*} = e^{-\sigma}\omega_1^5,\ \omega_7^{3*} = e^{-\sigma}\omega_7^3,$$

then we can then compute that from (5.98) that

$$(5.103)\qquad \begin{aligned} dY_1^* &= \omega_1^{4*}Y_4 + \omega_1^{5*}Y_5,\\ dY_7^* &= \omega_7^{3*}Y_3 - \omega_1^{5*}Y_5,\\ dY_2^* &= s^*(-\omega_1^{4*}Y_4 + \omega_1^{5*}Y_5),\\ dY_6^* &= u^*(\omega_7^{3*}Y_3 + \omega_1^{5*}Y_5),\\ dY_3^* &= \omega_7^{3*}Z_3^*,\ \text{where } Z_3^* = -Y_6^* - u^*Y_7^*,\\ dY_4^* &= \omega_1^{4*}Z_4^*,\ \text{where } Z_4^* = s^*Y_1^* - Y_2^*,\\ dY_5^* &= \omega_1^{5*}Z_5^*,\ \text{where } Z_5^* = -s^*Y_1^* - Y_2^* + Y_6^* - u^*Y_7^*, \end{aligned}$$

where

$$(5.104)\qquad s^* = se^{2\sigma},\ u^* = ue^{2\sigma}.$$

Using (5.99) and (5.104), we can then compute that

$$ds^* = 0 \quad , \quad du^* = 0 \; , \tag{5.105}$$

i.e., s^* and u^* are constant functions on the local neighborhood U.

The frame (5.102) is our final frame, and we will now drop the asterisks in further references to (5.102)-(5.105). Since the functions s and u are now constant, we can compute from (5.103) that

$$\begin{aligned} dZ_3 &= -\,2u\,\omega_7^3\,Y_3 \;, \\ dZ_4 &= 2s\,\omega_1^4\,Y_4 \;, \\ dZ_5 &= 2(u-s)\omega_1^5\,Y_5 \;. \end{aligned} \tag{5.106}$$

From this we see that the following 4-dimensional subspaces,

$$\begin{aligned} &\mathrm{Span}\{Y_1, Y_4, Y_5, Z_4, Z_5\} \;, \\ &\mathrm{Span}\{Y_7, Y_3, Y_5, Z_3, Z_5\} \;, \\ &\mathrm{Span}\{Y_1+Y_7, Y_3, Y_4, Z_3, Z_4\} \;, \end{aligned} \tag{5.107}$$

are invariant under exterior differentiation and are thus constant. Thus, each of the three focal point maps Y_1, Y_7 and Y_1+Y_7 is contained in a 4-dimensional subspace of P^6, and our Dupin submanifold is reducible in three different ways. Each of the three focal point maps is thus an immersion of the space of leaves of its principal foliation onto an open subset of a cyclide of Dupin in a space $\Sigma^3 = P^4 \cap Q^5$.

We state this result due to Pinkall [P2] as follows:

<u>Theorem 5.3:</u> Every Dupin submanifold with $\rho = 0$ is reducible. Thus, it is obtained from a cyclide in $\mathbb{R}^3$ by one of the four standard constructions (5.97).

Pinkall [P2, p. 111] then proceeds to classify Dupin submanifolds with $\rho = 0$ up to Lie equivalence. We will not prove his result here. The reader can follow his proof using the fact that his constants α and β are our constants s and -u, respectively.

REFERENCES

[B] W. Blaschke, Vorlesungen über Differentialgeometrie, Vol. 3, Springer, Berlin, 1929.

[CC] T. Cecil and S.S. Chern, Tautness and Lie sphere geometry, Math. Ann. 278 (1987), 381-399.

[CR] T. Cecil and P. Ryan, Tight and taut immersions of manifolds, Res. Notes Math. 107, Pitman, London, 1985.

[E] L. Eisenhart, A treatise on the differential geometry of curves and surfaces, Ginn, Boston, 1909.

[GH] K. Grove and S. Halperin, Dupin hypersurfaces, group actions and the double mappings cylinder, J. Differential Geometry 26 (1987), 429-459.

[L] E.P. Lane, A treatise on projective differential geometry, U. Chicago Press, Chicago, 1942.

[LS] S. Lie and G. Scheffers, Geometrie der Berührungstransformationen, Teubner, Leipzig, 1896.

[M1] R. Miyaoka, Compact Dupin hypersurfaces with three principal curvatures, Math. Z. 187 (1984), 433-452.

[M2] _____, Dupin hypersurfaces with four principal curvatures, Preprint, Tokyo Institute of Technology.

[M3] _____, Dupin hypersurfaces with six principal curvatures, Preprint, Tokyo Institute of Technology.

[Mu] H.F. Münzner, Isoparametrische Hyperflächen in Sphären, I and II, Math. Ann. 251 (1980), 57-71 and 256 (1981), 215-232.

[N] K. Nomizu, Characteristic roots and vectors of a differentiable family of symmetric matrices, Lin. and Multilin. Alg. 2 (1973), 159-162.

[P1] U. Pinkall, Dupin'sche Hyperflächen, Dissertation, Univ. Freiburg, 1981.

[P2] _____, Dupin'sche Hyperflächen in E^4, Manuscr. Math 51 (1985), 89-119.

[P3] _____, Dupin hypersurfaces, Math. Ann. 270 (1985), 427-440.

[S] D. Singley, Smoothness theorems for the principal curvatures and principal vectors of a hypersurface, Rocky Mountain J. Math., 5 (1975), 135-144.

[Th] G. Thorbergsson, Dupin hypersurfaces, Bull. Lond. Math. Soc. 15 (1983), 493-498.

Thomas E. Cecil
Department of Mathematics
College of the Holy Cross
Worcester, MA 01610

Shiing-Shen Chern
Department of Mathematics
University of California
Berkeley, CA 94720
and
Mathematical Sciences Research Institute
1000 Centennial Drive
Berkeley, CA 94720

Topics in Differential Geometry

SHIING-SHEN CHERN

The Institute for Advanced Study
1951

Introduction

These notes are based on lectures which I gave in Professor Veblen's Seminar during the spring of 1949. In these two years, some of the materials have been further clarified and some problems solved. An attempt is made to include a few of the latest results. As a consequence the presentation given here is hardly the original form, particularly in Chapters III and IV.

It is my pleasure to express here my thanks to Professor Veblen for his interest in this work. I wish also to acknowledge my privilege of having frequent conversations with André Weil. An unpublished manuscript of his has greatly influenced the presentation in Chapter III.

S. S. C., April 24, 1951

Chapter I
General Notions on Differentiable Manifolds

This chapter gives a summary, for later applications, of some notions and results on the topology of differentiable manifolds and the algebra of exterior differential forms. Proofs are only indicated in the simple cases. For lengthy proofs we content ourselves by giving a reference.

1. Homology and cohomology groups of an abstract complex

An *abstract complex* K is a collection of *cells* $\{\sigma_i^r\}$, with the following properties:

1) To each cell there is associated a non-negative integer, its *dimension* (which will be denoted by the superscript), and to two cells of consecutive dimensions $\sigma_i^r, \sigma_j^{r-1}$, there is associated an integer $[\sigma_i^r : \sigma_j^{r-1}]$, their *incidence number*.

2) To a cell σ_i^r there exists only a finite number of cells σ_j^{r-1}, such that $[\sigma_i^r : \sigma_j^{r-1}] \neq 0$.

3) To two cells $\sigma_i^{r+1}, \sigma_j^{r-1}$ whose dimensions differ by two, the incidence numbers satisfy the relation

$$\sum_k [\sigma_i^{r+1} : \sigma_k^r][\sigma_k^r : \sigma_j^{r-1}] = 0. \tag{1}$$

It is of such an abstract complex that we shall define the homology and cohomology groups.

Let G be an abelian group. A finite sum

$$c_r = \sum \lambda_i \sigma_i^r, \qquad \lambda_i \in G$$

is called a *chain*, r being its dimension. If

$$d_r = \sum \mu_i \sigma_i^r, \quad \mu_i \in G$$

is another r-dimensional chain, we define addition by

$$c_r + d_r = \sum (\lambda_i + \mu_i)\sigma_i^r. \tag{2}$$

With this addition all r-dimensional chains form a group $C_r(K, G)$.

To the chains a *boundary operation* is defined, by

$$\partial c_r = \sum_i \lambda_i \partial \sigma_i^r = \sum_{i,j} \lambda_i [\sigma_i^r : \sigma_j^{r-1}]\sigma_j^{r-1}. \tag{3}$$

By definition the boundary operation commutes with the addition of chains:

$$\partial(c_r + d_r) = \partial c_r + \partial d_r, \tag{4}$$

so that it defines a homomorphism

(5) $$\partial\colon C_r(K, G) \to C_{r-1}(K, G).$$

Elements of the kernel of this homomorphism, that is, chains whose boundaries are zero, are called *cycles*. The r-dimensional cycles form a subgroup $Z_r(K, G) \subset C_r(K, G)$.

It follows from Property 3) of K that the boundary operation has the property:

(6) $$\partial\partial c_r = 0,$$

so that the boundary of a chain is a cycle, called a *bounding cycle*. All r-dimensional bounding cycles form a subgroup $B_r(K, G) \subset Z_r(K, G)$.

The difference group

$$H_r(K, G) = Z_r(K, G) - B_r(K, G)$$

is called the *r-dimensional homology group of* K, *with coefficient group* G.

Let $C_r(K)$ be the group of r-dimensional chains of K, with integer coefficients, and let G be a topological group. A linear function over $C_r(K)$, with values in G, is called an *r-dimensional cochain*. An r-dimensional cochain γ^r satisfies therefore the conditions:

$$1)\quad \gamma^r(c_r + d_r) = \gamma^r(c_r) + \gamma^r(d_r),$$
$$2)\quad \gamma^r(-c_r) = -\gamma^r(c_r).$$

If β^r and γ^r are two r-cochains, we define their *sum* $\beta^r + \gamma^r$ to be the cochain given by

(7) $$(\beta^r + \gamma^r)(c_r) = \beta^r(c_r) + \gamma^r(c_r),$$

With this addition all the r-cochains form a group $C^r(K, G)$.

Of an r-cochain γ^r an $(r + 1)$-cochain can be defined, called its *coboundary*, by means of the relation

(8) $$\delta\gamma^r(c_{r+1}) = \gamma^r(\partial c_{r+1}).$$

The coboundary operation commutes with the addition of cochains:

(9) $$\delta(\beta^r + \gamma^r) = \delta\beta^r + \delta\gamma^r,$$

and therefore defines a homomorphism

(10) $$\delta\colon C^r(K, G) \to c^{r+1}(K, G).$$

A cochain whose coboundary is zero is called a *cocycle*. The r-cocycles form the kernel of the homomorphism δ and hence a subgroup $Z^r(K, G) \subset C^r(K, G)$.

It follows from (6) and (8) that

(11) $$\delta\delta\gamma^r = 0,$$

so that the coboundary of a cochain is a cocycle. All r-dimensional coboundaries form a subgroup $B^r(K, G) \subset Z^r(K, G)$. Their difference group

$$H^r(K, G) = Z^r(K, G) - B^r(K, G)$$

is called the *r-dimensional cohomology group of* K, *with coefficient group* G.

An important part of algebraic topology consists in the identification of the homology and cohomology groups of different complexes constructed from a space. For instance, the main theorem in the topology of polyhedra asserts that the complex of its singular cells and the complex of its simplicial decomposition have isomorphic homology and cohomology groups.

Let K and K' be two complexes. A mapping f of the cells of K into the cells of K' is called a *chain mapping*, if it commutes with the boundary operation:

$$f\partial = \partial f. \tag{12}$$

It follows that a chain mapping induces the homomorphisms

$$f\colon Z_r(K, G) \to Z_r(K', G),$$

$$f\colon B_r(K, G) \to B_r(K', G),$$

and hence the homomorphism

$$f\colon H_r(K, G) \to H_r(K', G). \tag{13}$$

To the chain mapping f we can define a dual mapping

$$f^*\colon C^r(K', G) \to C^r(K, G) \tag{14}$$

as follows: Let $\gamma'^r \in C^r(K', G)$. Then

$$(f^*\gamma'^r)(c_r) = \gamma'^r(f(c_r)). \tag{15}$$

It is easily verified that the dual cochain mapping commutes with the coboundary operation

$$f^*\delta = \delta f^*. \tag{16}$$

Hence there results the *dual homomorphism*

$$f^*\colon H^r(K', G) \to H^r(K, G). \tag{17}$$

This fact will play an important role in differential geometry. Reference: S. Eilenberg, Singular homology theory, Annals of Math, Vol. 45, 407–447 (1944).

2. Product theory

In order to develop a satisfactory product theory for a complex some additional notions and assumptions are necessary.

A cell σ_j^{r-1} is called a *face* of σ_i^r, if the incidence number $[\sigma_i^r : \sigma_j^{r-1}] \neq 0$. In general, σ^{r-p} is called a *face* of σ^r, if either $p = 0$ and the two cells are identical or $p > 0$ and there exists a sequence of cells $\sigma^{r-p}, \sigma^{r-p+1}, \ldots, \sigma^r$ such that each is a face of the next one. The closure $\bar{\sigma}_i^r$ of a cell σ_i^r is the subcomplex formed by all its faces. The *star* st σ_i^r is the subcomplex of all cells having σ_i^r as a face. A cycle is called *boundary-like*, if it is either of dimension > 0 or is of dimension 0 and has the sum of its coefficients equal to zero.

We shall first use the ring of integers as the coefficient ring for the product theory. Denote by I the 0-dimensional cochain which has the value one for every 0-cell.

Two further conditions will now be imposed on the complex in the establishment of a product theory:

(I) Every boundary-like cycle in $\bar{\sigma}_i^r$ bounds a chain in $\bar{\sigma}_i^r$.

(II) I is a cocycle.

For simplicity the notation σ_i^r will be used to denote at the same time the cell σ_i^r, the chain $1 \cdot \sigma_i^r$, and the cochain which has the value 1 for σ_i^r and 0 for other cells of dimension r. More generally, the notation $\sum \lambda_i \sigma_i^r$ will occasionally be used to denote the cochain having the value λ_i for σ_i^r and the value zero for other cells.

The *cup* product of two cochains of dimensions r and s is a cochain of dimension $r + s$ which satisfies the following conditions:

$$(\cup 1) \qquad (\beta_1^r + \beta_2^r) \cup \gamma^s = \beta_1^r \cup \gamma^s + \beta_2^r \cup \gamma^s,$$

$$\beta^r \cup (\gamma_1^s + \gamma_2^s) = \beta^r \cup \gamma_1^s + \beta^r \cup \gamma_2^s,$$

$$(\lambda\beta^r) \cup \gamma^s = \beta^r \cup (\lambda\gamma^s) = \lambda(\beta^r \cup \gamma^s), \qquad \lambda = \text{integer}$$

$$(\cup 2) \qquad \delta(\beta^r \cup \gamma^s) = \delta\beta^r \cup \gamma^s + (-1)^r \beta^r \cup \delta\gamma^s$$

$(\cup 3)$ $\sigma_i^r \cup \sigma_j^s$ is a cochain which has the value zero for any cell $\bar{\varepsilon}(\text{st } \sigma_i^r)(\text{st } \sigma_j^s)$.

$$(\cup 4) \qquad I \cup \sigma_i^r = \sigma_i^r \cup I = \sigma_i^r.$$

Theorem 1 (Fundamental Existence and Uniqueness Theorem). *The cup product of cochains induces a multiplication of cohomology classes. For a complex satisfying the conditions* (I), (II) *there exists a multiplication of cochains which fulfills the conditions* $(\cup 1)$–$(\cup 4)$. *Any two kinds of multiplications with these properties lead to the same multiplication of the cohomology classes.*

Let R be a commutative ring and consider the chains and cochains with R as coefficient group. The *cup product* of the cochains $\beta \in C^r(K, R), \gamma \in C^s(K, R)$ is defined by the conditions:

1. $\beta \cup \gamma$ is bilinear in both variables;
2. If $\lambda\sigma^r \in C^r(K, R), \mu\sigma^s \in C^s(K, R)$, then

$$(\lambda\sigma^r) \cup (\mu\sigma^s) = \lambda\mu(\sigma^r \cup \sigma^s).$$

Let K, K' be simplicial complexes, and $f: K \to K'$ a simplicial mapping. If $f^*: H^r(K', R) \to H^r(K, R)$ is the dual homomorphism of the cohomology groups, then

$$(18) \qquad f^*(\beta') \cup f^*(\gamma') = f^*(\beta' \cup \gamma').$$

This is called *Hopf's inverse homomorphism*, which can be described by simply saying that the dual homomorphism preserves the cup product.

Theorem 2 (Topological invariance). *Let P be a polyhedra and K its simplicial decomposition. There exists between the cohomology groups of K and the cohomology groups of the singular complex of P an isomorphism which preserves the cup product.*

For later applications we shall only be interested in the case that P is either the ring of integers or the real field or the finite field mod 2. We shall therefore assume that R is the ring of integers or a field. Then, if $\beta^r \cdot \gamma^s$ are cochains and c^{r+s} a chain (all with coefficients in R), the relation

(19) $$\beta^r \cdot (\gamma^s \cap c^{r+s}) = (\beta^r \cup \gamma^s)(c^{r+s}),$$

for β^r arbitrary, defines a chain $\gamma^s \cap c^{r+s}$ of dimension r, called the *cap product* of γ^s and c^{r+s}. Under our assumption for R a cup product determines a cap product, and vice versa.

Let M be a manifold, oriented if R is the ring of integers and otherwise if R is the field mod 2. In both cases there is a fundamental cycle which we also denote by M. Define

(20) $$\Theta\gamma^r = \gamma^r \cap M.$$

Then Θ establishes an isomorphism between $H^r(M, R)$ and $H_{n-r}(M, R)$. For $u_r \in H_r(M, R)$, $u_s \in H_s(M, R)$, define

(21) $$u_r \circ u_s = \Theta(\Theta^{-1}u_r \cup \Theta^{-1}u_s).$$

Theorem 3. *The product $u_r \circ u_s$ of homology classes on a manifold defined by* (21) *is identical with the intersection class of u_r and u_s.*

This theorem gives the connection between product theory and intersection theory. Reference: H. Whitney, On products in a complex, Annals of Math. Vol. 39, 397–432 (1938).
For the uniqueness in Theorem 1 we have to assume that the complex K is also star-finite and that the cochains under consideration are finite.

3. An example

As an illustration we consider the n-dimensional real projective space P^n and take as coefficient field the field mod 2. P^n contains a sequence of projective spaces of lower dimensions

$$P^n \supset P^{n-1} \supset P^{n-2} \supset \cdots \supset P' \supset P^0,$$

and has a cellular subdivision consisting of the cells

$$P^n - P^{n-1}, P^{n-1} - P^{n-2}, \ldots, P' - P^0, P^0.$$

It is easy to verify that each of these cells is a cycle and $H_r(P^n, I_2)$, $H^r(P^n, I_2)$, $n \geqq r \geqq 0$, are cyclic groups of order two. Without danger of confusion we can denote the generator of $H_r(P^n, I_2)$ by P^r and the generator of $H^r(P^n, I_2)$ by ζ^r.

We shall prove that the cohomology ring of P^n is

$$H(P^n) = \{1, \zeta(=\zeta^1), (\zeta)^2, \ldots, (\zeta)^n\},$$

when the superscripts outside the parentheses denote powers in the sense of the cup product. We notice that the isomorphism Θ maps ζ^r into P^{n-r}, so that $\zeta^r(P^r) = KI(P^{n-r}, P^r) = 1$. By applying induction on r we suppose $(\zeta)^r(P^r) = 1$ and then find

$$\Theta((\zeta)^{r+1}) = \Theta(\zeta)^r \cup \zeta) = \Theta(\Theta^{-1}(P^{n-r}) \cup \Theta^{-1}(P^{n-1})) = P^{n-r}{}_0P^{n-1} = \\ = P^{n-r-1}.$$

This proves that $(\zeta)^{r+1}$ is the generator of $H^{r+1}(P^n, I_2)$ and hence the above form of the cohomology ring of P^n.

A) Let $g: P^{r-1} \to P^{n-1}$ be a continuous mapping such that the induced homomorphism g carries a projective line (that is, the homology class of it) into a projective line. Then $n \geqq r$.

Proof. Let ξ and ζ be the generators of the cohomology rings of P^{r-1} and P^{n-1} respectively. Then

$$g^*(\zeta)(P^1) = \zeta(g(P^1)) = \zeta(P^1) = 1.$$

It follows that $g^*(\zeta) = \xi$. Since $\zeta^n = 0$, we have

$$\xi^n = (g^*(\zeta))^n = g^*(\zeta^n) = 0.$$

Hence $n \geqq r$.

B) Let $g_1, \dots, g_n$ be continuous odd functions in $x_1, \dots, x_r$ defined on the sphere

$$x_1^2 + \cdots + x_r^2 = 1.$$

If $n < r$, the functions $g_1, \dots, g_n$ have a common zero on the sphere.

Proof. Suppose there be no common zero. We can then assume that

$$g_1^2 + \cdots + g_n^2 = 1.$$

The functions $g_i = g_i(x_1, \dots, x_r)$, $i = 1, \dots, n$, then define a mapping of a sphere S^{r-1} into a sphere S^{n-1} and, after identifying the antipodal pairs of both spheres, a mapping $g: P^{r-1} \to P^{n-1}$. Moreover, since the functions g_i are odd, the mapping g has the property that it carries the homology class of a projective line into the homology class of a projective line. But this contradicts A).

C) (Borsuk-Ulam) Let an n-sphere S^n be mapped continuously into the n- dimensional Euclidean space E^n. There exists in S^n at least one pair of antipodal points which are mapped into the same point of E^n.

Proof. Let $x_1, \dots, x_n$ be the coordinates of E^n. Suppose the mapping be

$$x_i = f_i(p), \qquad p \in S^n, \qquad i = 1, \dots, n.$$

Denote by p^* the antipodal point of p. Put

$$g_i(p) = f_i(p) - f_i(p^*).$$

From B) it follows that $g_i(p)$ have a common zero p_0. At this p_0 we have

$$f_i(p_0) = f_i(p_0^*), \qquad i = 1, \dots, n.$$

4. Algebra of a vector space

The differentiable manifolds which will be our later concern have the property that there is associated at each point a finite dimensional vector space. The study of the algebraic properties of the vector space and of various associated vector spaces will therefore constitute a necessary prerequisite for later developments.

We denote by V^n or V an n-dimensional vector space over the real field. To V there is associated its *dual space* V^*, the space of all linear functions over V, and the relation between V and V^* is reciprocal. We shall denote elements of V by small Gothic letters and elements of V^* by small Greek letters. Then $\alpha(\mathfrak{m})$ or $\alpha \cdot \mathfrak{m}$, $\mathfrak{m} \in V, \alpha \in V^*$, is a real number.

Consider the direct product

$$V(k, l) = \underbrace{Vx \ldots xV}_{k} \times \underbrace{V^*x \ldots xV^*}_{l}.$$

A *tensor of type* (k, l) is a multilinear function in $V(k, l)$, with real values, that is, a real-valued function linear in each argument when the other $k + l - 1$ arguments are kept fixed. k is called the covariant order and l the contravariant order. The tensor is called *covariant* or *contravariant*, when $l = 0$ or $k = 0$ and is in general called *mixed*. Covariant (contravariant) tensors of order one are called covariant (contravariant) vectors. Given a tensor f of type (k, l) and a tensor y of type (k', l'), we define $f \times g$ to be the tensor of type $(k + k', l + l')$ by the relation

$$(22) \quad (f \times g)(\mathscr{C}_1, \ldots, \mathscr{C}_k, \mathscr{C}_{k+1}, \ldots, \mathscr{C}_{k+k'}; \alpha_1, \ldots, \alpha_l, \alpha_{l+1}, \ldots, \alpha_{l+l'}) \\ = f(\mathscr{C}_1, \ldots, \mathscr{C}_k; \alpha_1, \ldots, \alpha_l) g(\mathscr{C}_{k+1}, \ldots, \mathscr{C}_{k+k'}; \alpha_{l+1}, \ldots, \alpha_{l+l'}).$$

With a natural addition all tensors of type (k, l) form a vector space of dimension n^{k+l}. Because of the duality between V and V^* the space of all covariant vectors can be identified with V^* and the space of all contravariant vectors with V itself.

A tensor may have the property of symmetry or anti-symmetry. For simplicity take a covariant tensor of order two, given by $f(\mathscr{C}, \mathscr{Y})$. The tensor is called *symmetric* or *anti-symmetric*, according as $f(\mathscr{C}, \mathscr{Y}) = f(\mathscr{Y}, \mathscr{C})$ of $f(\mathscr{C}, \mathscr{Y}) = -f(\mathscr{Y}, \mathscr{C})$ holds. From a given covariant (contravariant) tensor f of order k we can construct its *symmetrized* or *alternated* tensor respectively by the equations

$$(23) \qquad S(f)(\mathscr{C}_1, \ldots, \mathscr{C}_k) = \frac{1}{k!} \sum f(\mathscr{C}_{i_1}, \ldots, \mathscr{C}_{i_k}),$$

$$(24) \qquad T(f)(\mathscr{C}_1, \ldots, \mathscr{C}_k) = \frac{1}{k!} \sum \varepsilon_{i_1 \ldots i_k} f(\mathscr{C}_{i_1}, \ldots, \mathscr{C}_{i_k}),$$

where the summations are extended over all permutations $i_1, \ldots, i_k$ of $1, \ldots, k$ and $\varepsilon_{i_1 \cdots i_k} = +1$ or -1 according as $i_1, \ldots, i_k$ is an even or odd permutation of $1, \ldots, k$.

Of particular importance will be the vector spaces $A^r, r = 1, \ldots, n$, of anti-symmetric or alternating tensors of order $(r, 0)$. Let A^0 be the (one-dimensional) vector space isomorphic to the real field, and let

$$(25) \qquad A = A^0 \dotplus A^1 \dotplus \cdots \dotplus A^n.$$

Then A is a vector space of dimension 2^n.

We shall convert A into a ring by defining a multiplication which has the properties:

1) It is distributive:

$$(26) \qquad f \wedge (g_1 + g_2) = f \wedge g_1 + f \wedge g_2,$$
$$(f_1 + f_2) \wedge g = f_1 \wedge g + f_2 \wedge g.$$

2) If $f \in A^r$, $g \in A^s$, then

$$(27) \qquad f \wedge g = T(f \times g).$$

With this multiplication the vector space A becomes a ring (of dimension 2^n), called the *Grassmann ring* associated to V. An element of the Grassmann ring which belongs to one of the A^r's, that is, whose other components in the direct summand are zero, is called *homogeneous dimensional* or an *alternating form*. It follows from definition that if f A^r, g A^s, then

$$(28) \qquad f \wedge g = (-1)^{rs} g \wedge f.$$

If α_i, $i = 1, \ldots, n$, form a base in V^* (which is then identified to A^1), a base in the associated Grassmann ring will be formed by the elements

$$(29) \quad 1, \alpha_i, \alpha_i \wedge \alpha_j \ (i < j), \alpha_i \wedge \alpha_j \wedge \alpha_k \ (i < j < k), \ldots, \alpha_1 \wedge \alpha_2 \wedge \cdots \wedge \alpha_n,$$

which are 2^n in number.

We shall give two theorems in the Grassmann ring, which are particularly useful later on.

A) Let $\omega_1, \ldots, \omega_r \in V^*$. Then $\omega_1, \ldots, \omega_r$ are linearly dependent, if and only if

$$\omega_1 \wedge \cdots \wedge \omega_r = 0.$$

This can either be proved by induction on r, or by choosing a base α_i, $i = 1, \ldots, n$ in V^*, writing

$$\omega_s = \sum_{i=1}^{n} l_{si} \alpha_i, \qquad s = 1, \ldots, r,$$

and observing that

$$(30) \qquad \omega_1 \wedge \cdots \wedge \omega_r = \sum_{i_1 < \cdots < i_r} \begin{vmatrix} l_{1i_1} \cdots l_{1i_r} \\ \cdots\cdots\cdots \\ l_{ri_1} \cdots l_{ri_r} \end{vmatrix} \alpha_{i_1} \ldots \alpha_{i_r}.$$

B) If $\omega_1, \ldots, \omega_r$ are linearly independent and $\pi_s \in V^*$, $s = 1, \ldots, r$, are such that

$$(31) \qquad \sum_{s=1}^{r} \pi_s \wedge \omega_s = 0,$$

then

$$(32) \qquad \pi_s = \sum_{t=1}^{r} a_{st} \omega_t,$$

$$(33) \qquad a_{st} = a_{ts}.$$

A *tensorial* form of order (k, l) and degree r is a multilinear function of $V(k, l)$, with values in A^r. Clearly, all tensorial forms of given order and degree form a vector space.

There is an operation, called the *tensor product*, which pairs two abelian groups (and hence two vector spaces) into an abelian group. Let A and B be two abelian groups, with the elements a_i and b_i respectively. Take the finite sums of the formal products $\sum a_i b_i$. Two such sums are called *equivalent*, if one can be transformed into the other by a finite number of the following elementary transformations: 1) $(a_i + a_i')b_i = a_i b_i + a_i' b_i$; 2) $a_i(b_i + b_i') = a_i b_i + a_i b_i'$; 3) addition or deletion of $a.0.$ or $0.b.$ Among the equivalence classes so obtained we can define an addition which, for the representatives, is defined just by adding the terms formally. The equivalence classes with such an addition form a group, called the tensor product of A and B and to be denoted by $A \otimes B$.

5. Differentiable manifolds

A (topological) manifold M is said to have a *differentiable structure*, if the following conditions are satisfied:

1) There is an open covering $\{U_i\}$ of M such that for each i there exists a homeomorphism Θ_i of an n-cell E into U_i.

2) For any two open sets U_i, U_j of the covering the mapping $\Theta_j^{-1}\Theta_i(S)$, $S = \Theta_i^{-1}(U_i \cap U_j)$, of S into E is differentiable of class $r > 0$.

A manifold with a differentiable structure is called a *differentiable manifold*, r (>0) being its *class*. U_i are called *coordinate neighborhoods*; the coordinates in $E = \Theta_i^{-1}(U_i)$ are called *local coordinates*.

If M is of dimension n, the tangent vectors at p form a vector space T_p of dimension n, called the *tangent space* at p. To this space the considerations of the last section will apply, so that we can consider its dual space, the spaces of tensors of different types, and the Grassmann ring, etc. Let F_p be the space of tensors of a definite type associated to T_p, and let

$$X = \bigcup_{p \in M} F_p.$$

A natural topology can be defined in X, so that X becomes a topological space. X is then called a tensor bundle over M. There is a natural mapping, called *projection*,

$$\psi : X \to M,$$

defined by

$$\psi(F_p) = p.$$

We shall give the relation of the tensors defined here with those of classical differential geometry. For definiteness consider a tensor Ξ of type (1, 1). Let $x^1, \ldots, x^n$ be a system of local coordinates at p. For $f \in D(p)$ define

$$\mathfrak{n}_i(f) = \left(\frac{\partial f}{\partial x^i}\right)_p, \qquad i = 1, \ldots, n.$$

Then $\mathfrak{n}_i$ are vectors and span the tangent space T_p. In the dual space T_p^* of T_p we choose the base α^i defined by

$$\alpha^i(\mathfrak{n}_j) = \delta_j^i.$$

We put

$$\xi_i^j = \Xi(m_i, \alpha^j),$$

which are called the *components* of Ξ *relative to the local coordinate system* x^i. Suppose $\bar{x}^1$ be another system of local coordinates at p. Put

$$a_j^i = \left(\frac{\partial \bar{x}^i}{\partial x^j}\right)_p, \quad b_j^i = \left(\frac{\partial x^i}{\partial \bar{x}^j}\right)_p,$$

so that

$$\sum_j a_j^i b_k^j = \sum_j b_j^i a_k^j = \delta_k^i.$$

Then we have by definition, for the vectors $\bar{m}_i$, $\bar{\alpha}^i$ relative to the coordinate system $\bar{x}^i$.

$$\bar{m}_i = \sum_j b_i^j m_j, \qquad \bar{\alpha}^i = \sum_j a_j^i \alpha^j.$$

It follows that the components of Ξ relative to the local coordinate system $\bar{x}^i$ are

(34) $$\bar{\xi}_i^j = \sum_{k,l} b_i^l a_k^j \xi_l^k.$$

These are the well-known equations of transformation in classical differential geometry.

6. Multiple integrals

For the theory of multiple integrals on a differentiable manifold M we have to consider the *bundle of Grassmann rings over* M, which is the union of the Grassmann rings associated to the tangent spaces T_p, $p \in M$, with a natural topology. We denote by A_p the Grassmann ring associated to p, and let $\mathscr{U} = \bigcup_{p \in M} A_p$. A differential polynomial is a mapping $\omega: M \to \mathscr{U}$ such that the projection of $\omega(p)$, $p \in M$, is p itself. The mapping is assumed to be locally differentiable of class $\geqq 2$. If $\omega(p)$, $p \in M$, is a form of degree r, ω is called a differential form of degree r.

We shall define an operation d, called *exterior differentiation*, which carries differential polynomials into differential polynomials, by the following properties:

1) $d(\omega + \Theta) = d\omega + d\Theta$,

2) $d(\omega \wedge \Theta) = d\omega \wedge \Theta + (-1)^r \omega \wedge d\Theta$,

where ω is a differential form of degree r.

3) If f is a scalar (that is, a differential form of degree zero), df is the covariant vector such that

$$df(m) = m(f)$$

holds for every contravariant vector.

4) For every scalar f,

$$d(df) = 0.$$

In terms of a local coordinate system x^i we can take as a base for A^1 the differentials dx^i, $i = 1, \ldots, n$, which are covariant vectors defined by

$$dx^i(\mathfrak{m}) = \mathfrak{m}(x^i), \tag{35}$$

$\mathfrak{m}$ being any contravariant vector. Then a base for A^r is formed by

$$dx^{i_1} \wedge dx^{i_2} \wedge \cdots \wedge dx^{i_r}, \qquad i_1 < i_2 < \cdots < i_r,$$

so that a differential form of degree r can be written

$$\omega = \sum_{i_1 < \cdots < i_r} a_{i_1 \ldots i_r} dx^{i_1} \wedge dx^{i_2} \wedge \cdots \wedge dx^{i_r}, \tag{36}$$

where the coefficients may be assumed to be anti-symmetric.

It follows from 2) and 4), by induction on the degree r, that

$$d(dx^{i_1} \wedge dx^{i_2} \wedge \cdots \wedge dx^{i_r}) = 0.$$

Therefore we have the formula

$$d\omega = \sum_{i_1 < \cdots < i_r} da_{i_1 \ldots i_r} \wedge dx^{i_1} \wedge \cdots \wedge dx^{i_r}. \tag{37}$$

We also have, by 3),

$$df = \sum_i \frac{\partial f}{\partial x^i} dx^i. \tag{38}$$

Concerning the exterior differentiation of differential forms two facts are of importance:

A) For any differential form ω,

$$d(d\omega) = 0. \tag{39}$$

B) Let $f: M \to M'$ be a differentiable mapping of a manifold M into a manifold M'. f induces a differential mapping df of the tangent space T_p of M at p into the tangent space $T_{p'}$, $p' = f(p)$. The dual mapping f^* of df is a linear mapping of the Grassmann ring $A_{p'}$ at p' into A_p. Then

$$f^*(d\omega) = d(f^*\omega). \tag{40}$$

In other words, exterior differentiation commutes with the induced dual mapping of a mapping of one differentiable manifold into another.

A differential form ω is called *exact* or *closed*, if $d\omega = 0$, and is called *derived*, if there exists a differential form Θ such that $\omega = d\Theta$. It follows from (39) that every derived form is exact.

Differential forms can be taken integrands of multiple integrals in the manifold. The details of a satisfactory integration theory will be too long to be reproduced here. For simplicity we take a cellular decomposition K of M which is so fine that each cell belongs to a coordinate neighborhood. Let ω be a differential form of degree r and C_r an r-chain of K. By the equation

$$\omega(c_r) = \int_{c_r} \omega, \tag{41}$$

ω defines a cochain of dimension r, with real coefficients. An important link of exterior differentiation with cohomology theory is now the generalized Stokes' formula:

$$\int_{c_r} \omega = \int_{\partial c_r} d\omega. \tag{42}$$

In other words, if ω is (or defines) a cochain, $d\omega$ is its coboundary, and is a cocycle, if $d\omega = 0$.

The study of the ring of differential forms and its exterior differentiation is justified by the following fundamental theorem due to de Rham:

Theorem. *Let M be a compact differentiable manifold. To every cohomology class of M, with real coefficients, there exists an exact differential form which defines a cocycle belonging to this class. The cohomology class containing the product (in the sense of the Grassmann ring) of two exact differential forms is the cup product of the classes which contain the factors.*

Example. The de Rham theorem asserts the existence of a differential form, while in concrete cases it is important to construct the forms explicitly, and the ones with simple properties. Consider the unit n-sphere in an $(n + 1)$-dimensional Euclidean space E^{n+1}. $H^n(S^n, R) \approx R$, so that we wish to construct the exact differential form which defines a generator of $H^n(S^n, R)$.

Let O be the center of S^n, and consider the frames $O\mathfrak{n}_1 \ldots \mathfrak{n}_{n+1}$ formed by O and $n + 1$ mutually perpendicular unit vectors in a definite orientation. Identify a point of S^n with the end-point of $\mathfrak{n}_{n+1}$, and put

$$\omega_{i,n+1} = (d\mathfrak{n}_{n+1}\mathfrak{n}_i), \qquad i = 1, \ldots, n, \tag{43}$$

where the product in the right-hand side is the scalar product in E^{n+1}. The differential form

$$\omega = \frac{1}{O_n}\omega_{1,n+1} \wedge \cdots \wedge \omega_{n,n+1}, \tag{44}$$

where O_n is the area of S^n, has the property that the value of its integral over a fundamental cycle of S^n is ± 1. Hence ω defines a generator of $H^n(S^n, R)$.

Let $\sum$ be a hypersurface in E^{n+1} and let $\mathfrak{m}(p)$, $p \in \sum$, be a continuous vector field on $\sum$. Choose $\mathfrak{n}_{n+1}$ to be parallel to $\mathfrak{m}(p)$. Then $f(p) = \mathfrak{n}_{n+1}$ defines a mapping $f: \sum \to S^n$. Its induced dual mapping f^* maps ω into a differential form $f^*\omega$ in $\sum$. The integral

$$\int_{\Sigma} f^*\omega = \int_{f(\Sigma)} \omega \tag{45}$$

is equal to the *index of the vector field.* It is the *Kronecker integral.* Reference: Hodge, Theory and Applications of Harmonic Integrals.

Chapter II
Riemannian Manifolds

This chapter is devoted to the theory of Riemannian manifolds, and in particular to the Gauss-Bonnet formula which, for a compact orientable Riemannian manifold, expresses its Euler-Poincaré characteristic as an integral of a scalar invariant over the manifold. We begin with the study of Riemannian manifolds imbedded in an Euclidean space, because this case appeals more to geometrical intuition and usually furnishes a first test for properties in general Riemannian manifolds. To simplify the formulas repeated indices always denote summation.

1. Riemannian manifolds in Euclidean space

Let E^{n+N} be an oriented Euclidean space of dimension $n + N$. E^{n+N} is transformed transitively by the group of motions. We call a *frame* the figure $P\imath_1 \ldots \imath_{n+N}$ formed by a point p and an ordered set of $n + N$ mutually perpendicular unit vectors through p. The set of frames has the property that there exists one and only one motion which carries one frame to another. It can therefore be made a differentiable manifold isomorphic to the group of motions.

Since the vectors are in Euclidean space, we can write*

$$(1)\qquad \begin{aligned} dp &= \theta_A \imath_A, \\ d\imath_A &= \theta_{AB} \imath_B. \end{aligned}$$

The differential forms θ_A, θ_{AB} are in the manifold of frames and satisfy

$$(2)\qquad \theta_{AB} + \theta_{BA} = 0.$$

Since

$$d(dp) = d(d\imath_A) = 0,$$

we derive from (1) that

$$(3)\qquad \begin{aligned} d\theta_A &= \theta_B \wedge \theta_{BA}, \\ d\theta_{AB} &= \theta_{AC} \wedge \theta_{CB}. \end{aligned}$$

Formulas (3) are called the *equations of structure* of the Euclidean space.

* We agree in this section to use the following ranges of indices: $A, B, C = 1, \ldots, n + N$; $\alpha, \beta, \gamma = 1, \ldots, n$; $r, s = n + 1, \ldots, n + N$.

Let M be an n-dimensional manifold, differentially imbedded (of class $\geqq 3$) in E^{n+N}. M has a Riemannian metric, induced by E^{n+N}. To study M we consider the submanifold of the frames $p\mathfrak{n}_1 \ldots \mathfrak{n}_{n+N}$, such that $p \in M$, and $\mathfrak{n}_1, \ldots, \mathfrak{n}_n$ are the tangent vectors of M at p. This submanifold is mapped into the manifold of frames by the inclusion mapping ι. Let ι^* be its dual mapping of differential forms, and let

$$(4) \qquad \omega_A = \iota^* \Theta_A, \qquad \omega_{AB} = \iota^* \Theta_{AB}.$$

Since ι^* commutes with both exterior differentiation and multiplication of the Grassmann ring, we have

$$(5) \qquad d\omega_r = \omega_\alpha \wedge \omega_{\alpha r} = 0.$$

It follows that

$$(6) \qquad \omega_{\alpha r} = \lambda_{r\alpha\beta}\omega_\beta, \qquad \lambda_{r\alpha\beta} = \lambda_{r\beta\alpha}.$$

The second equation of (3) gives

$$(7) \qquad d\omega_{\alpha\beta} = \omega_{\alpha\gamma} \wedge \omega_{\gamma\beta} + \Omega_{\alpha\beta},$$

where

$$(8) \qquad \Omega_{\alpha\beta} = -\omega_{\alpha r}\omega_{\beta r} = -\lambda_{r\alpha\rho}\lambda_{r\beta\sigma}\omega_\rho \wedge \omega_\sigma.$$

We shall prove that $\Omega_{\alpha\beta}$ depends only on the Riemannian metric of M. The proof depends on the following

Lemma. *There exists only one set of $\omega_{\alpha\beta}$, which are anti-symmetric in its indices and which satisfy the equations*

$$d\omega_\alpha = \omega_\beta \wedge \omega_{\beta\alpha}.$$

Proof. Suppose a second set $\bar{\omega}_{\beta\alpha}$ also possesses these properties. Then

$$\omega_\beta \wedge (\bar{\omega}_{\beta\alpha} - \omega_{\beta\alpha}) = 0,$$

and we have

$$\bar{\omega}_{\beta\alpha} - \omega_{\beta\alpha} = P_{\beta\alpha\gamma}\omega_\gamma,$$

where $P_{\beta\alpha\gamma}$ is symmetric in β, γ. Since $\bar{\omega}_{\beta\alpha} - \omega_{\beta\alpha}$ is anti-symmetric in β, α and ω_γ are linearly independent, $P_{\beta\alpha\gamma}$ is anti-symmetric in β, α. But a three-indexed symbol symmetric in one pair and anti-symmetric in another is zero. Hence $\bar{\omega}_{\beta\alpha} = \omega_{\beta\alpha}$.

Our statement that $\Omega_{\alpha\beta}$ depends only on the Riemannian metric of M then follows from the lemma. $\Omega_{\alpha\beta}$ will be called the *curvature forms*. Equations (8) are the Gauss equations. The quadratic differential forms

$$(9) \qquad \Phi_r = \omega_\alpha\omega_{\alpha r} = \lambda_{r\alpha\beta}\omega_\alpha\omega_\beta$$

are called the *second fundamental forms* of M.

We shall make two applications of the above formulas.

A) Let $N = 1$, so that M is a closed hypersurface. The normals $\mathfrak{n}_{n+1}$ define a vector field over M, whose index is, according to Kronecker's integral formula,

$$I = \frac{1}{O_n} \int_M \omega_{1n+1} \ldots \omega_{nn+1} = \frac{1}{n!\,O_n} \int_M \varepsilon_{i_1 \ldots i_n} \omega_{i_1 n+1} \ldots \omega_{i_n n+1},$$

where O_n is the area of the n-dimensional unit hypersphere. If *n is even*, the integrand can be written

$$\varepsilon_{i_1 \ldots i_n} \omega_{i_1 n+1} \ldots \omega_{i_n+1} = (-1)^{n/2} \varepsilon_{i_1 \ldots i_n} \Omega_{i_1 i_2} \ldots \Omega_{i_{n-1} i_n},$$

and depends only on the metric in M. On the other hand, it can be proved that $I = \frac{1}{2}\chi$, where χ is the Euler-Poincare characteristic of M. We get therefore the Gauss-Bonnet formula for a hypersurface:

$$\text{(10)} \qquad \frac{2(-1)^{n/2}}{n!\,O_n} \int \varepsilon_{i_1 \ldots i_n} \Omega_{i_1 i_2} \ldots \Omega_{i_{n-1} i_n} = \chi.$$

B) Let $m(p)$ be the minimum number of linear differential forms in which the second fundamental forms Φ_r can be expressed. $m(p)$ is obviously equal to the number of linearly independent equations in the system

$$\frac{\partial \Phi_r}{\partial \omega_\alpha} = 0.$$

We put $m = \operatorname{Max}_{p \in M} m(p)$. Then we have the theorem:

Let M be a closed manifold of dimension n, differentially imbedded in E^{n+N}. Then $m \geqq n$.

Proof. We take a fixed point O and a fixed system of coordinate axes n_A^0 through O. For $p \in M$ let $s = \overrightarrow{Op}^2$ (square of the distance Op). Then s attains a maximum at a point p_0. Writing

$$\overrightarrow{Op} = x_A n_A^0,$$

we have

$$\tfrac{1}{2}\, ds = x_A\, dx_A = \overrightarrow{Op}\, dp,$$

$$\tfrac{1}{2}\, d^2 s = dp\, dp + \overrightarrow{Op}\, d^2 p = \omega_\alpha \omega_\alpha + \overrightarrow{Op}\, d^2 p.$$

At p_0 we have

$$ds = 0, \qquad d^2 s \leqq 0.$$

Since ω_α are linearly independent, the first equation implies

$$\overrightarrow{op}\, n_\alpha = 0.$$

Now

$$\overrightarrow{op}\, d^2 p = (\ldots)(\overrightarrow{op}\, n_\alpha) + \Phi_r(\overrightarrow{op}\, n_r),$$

so that the inequality implies

$$\omega_\alpha \omega_\alpha + \Phi_r(\overrightarrow{op}\, n_r) \leqq 0.$$

If $m \leqq n - 1$, there exists at least a direction in M for which $\Phi_r = 0$ and along this direction we have

$$\omega_\alpha \omega_\alpha \leqq 0,$$

which contradicts the positive definiteness of the quadratic differential form in the left-hand side.

From this theorem we shall deduce the following theorem which was first proved by Tompkins:

A closed flat Riemannian manifold of dimension n cannot be isometrically imbedded in an Euclidean space of dimension $2n-1$.

Proof. We suppose that such an imbeeding exists, so that $N = n - 1$. By the last theorem it suffices to prove that $m \leqq n - 1$.

Suppose that $m = n$. Consider the vectors

$$\mathscr{L}_{ij} = (\lambda_{1ij}, \ldots, \lambda_{Nij})$$

in an N-dimensional Euclidean space. By (8) the flatness of the induced metric implies that

$$(*) \qquad (\mathscr{L}_{ik}\mathscr{L}_{jl}) = (\mathscr{L}_{il}\mathscr{L}_{jk}),$$

where the products in the parentheses are scalar products of vectors (in the auxiliary N-space). Since $m = n$, the matrix of vectors

$$V = (\mathscr{L}_{ik})$$

has the property that no column is a linear continuation of the other columns. We also observe that the conditions (*) remain invariant, if we add to a column a linear combination of the other columns and if we permute the columns.

Since $N = n - 1$, the vectors $\mathscr{L}_{1k}$ are linearly dependent. By applying the above elementary transformations on the columns, we can assume $\mathscr{L}_{1n} = 0$. The vectors $\mathscr{L}_{in}$ span a linear space of dimension $\geqq 1$, and condition (*) implies that $\mathscr{L}_{11}, \ldots, \mathscr{L}_{1n-1}$ are perpendicular to this linear space, and hence belong to a linear space of dimension $\leqq n-2$. It follows that $\mathscr{L}_{11}, \ldots, \mathscr{L}_{1n-1}$ are linearly dependent and we can assume $\mathscr{L}_{1n-1} = 0$. Since there is no linear combination of the columns $\mathscr{L}_{in-1}$, $\mathscr{L}_{in}$ which is zero, the vectors $\mathscr{L}_{in-1}$, $\mathscr{L}_{in}$ span a linear space of dimension $\geqq 2$. The above process can again be applied, and finally we prove that $\mathscr{L}_{1k} = 0$ and $\mathscr{L}_{ik} = 0$. But this is a contradiction, and the theorem follows.

Remark. The above argument can be applied to establish the following slightly more general theorem:

Let $k(p)$ be the minimum number of linear differential forms in terms of which the curvature forms Ω_{ij} at p can be expressed, and let $k = \max_{p \in M} k(p)$. Then a closed Riemannian manifold of dimension n can not be isometrically imbedded in an Euclidean space of dimension $2n - k - 1$.

2. Imbedding and rigidity problems in Euclidean space

Let M be an *abstract Riemannian manifold*, that is, a differentiable manifold with a positive definite symmetric covariant tensor field of the second order or, what is the same, a positive definite (ordinary) quadratic differential form. Two abstract Riemannian manifolds are *isometric*, if they are differentially homeomorphic and if under the homeomorphism the differential forms are mapped to each other.

Concerning the relations between abstract Riemannian manifolds and submanifolds of an Euclidean space, two problems naturally arise:

1) Imbedding problem: Is an abstract Riemannian manifold isometric to a submanifold in an Euclidean space?

2) Rigidity problem: Are two isometric submanifolds in an Euclidean space necessarily congruent or symmetric?

Our knowledge of the first problem is extremely meagre. It is not known whether the hyperbolic plane (that is, the two-dimensional R. M. with Gaussian curvature $= -1$) can be imbedded in an Euclidean space of sufficiently high dimension.

The purpose of this section is to study the rigidity problem for a hypersurface in Euclidean space. The most interesting case is that of surfaces in three-dimensional Euclidean space, about which the rigidity theorem was first proved by Cohn-Vossen under the further assumption that the surfaces are convex. We shall present a proof due to G. Herglotz and apply this idea to prove a generalization of a theorem of Christoffel.

We use the notation of the last section, with $N = 1$. Two hypersurfaces are now given in E^{n+1} and we denote the quantities for the second hypersurface by the same symbols with dashes. From the isometry of the hypersurfaces it follows that the homeomorphism h between them can be extended into a homeomorphism h' between the manifolds of their tangent frames such that

$$\omega_\alpha = h'^*\bar{\omega}_\alpha, \tag{11}$$

h'^* being again the dual homomorphism on the differential forms under h'.

Taking the exterior derivative of this equation and applying the lemma of the last section, we get

$$\omega_{\alpha\beta} = h'^*\bar{\omega}_{\alpha\beta} \tag{12}$$

Another exterior differentiation gives

$$\Omega_{\alpha\beta} = h'^*\bar{\Omega}_{\alpha\beta} \tag{13}$$

Put

$$\lambda_{\alpha\beta} = \lambda_{n+1\alpha\beta}, \qquad \lambda'_{\alpha\beta} = h'^*\bar{\lambda}_{n+1\alpha\beta}, \tag{14}$$

and consider the matrices

$$\bigwedge = (\lambda_{\alpha\beta}), \qquad \bigwedge' = (\lambda'_{\alpha\beta}). \tag{15}$$

Condition (13) signifies in matrix language that corresponding two-rowed determinants of the matrices $\bigwedge$ and $\bigwedge'$ are equal. If $n \geqq 3$ and the ranks of $\bigwedge$ and $\bigwedge'$ are $\geqq 2$, we conclude that $\bigwedge' = \pm\bigwedge$, which implies that the hypersurfaces are congruent or symmetric.

More interesting is therefore the case $n = 2$, about which we shall prove the following theorem:

Theorem 1 (Cohn-Vossen). *Two closed convex surfaces in E^3 which are isometric are congruent or symmetric.*

Proof. We put

$$y_A = (pn_A),$$
$$\omega'_{\alpha,n+1} = h'^* \bar{\omega}_{\alpha n+1}.$$

Then we have

$$dy_1 = \omega_1 + y_2\omega_{12} + y_3\omega_{13},$$
$$dy_2 = \omega_2 + y_1\omega_{21} + y_3\omega_{23},$$
$$d(y_1\omega'_{23} - y_2\omega'_{13}) = \omega_1 \wedge \omega'_{23} - \omega_2 \wedge \omega'_{13} + y_3(\omega_{13} \wedge \omega'_{23} - \omega_{23} \wedge \omega'_{13})$$
$$= \{(\lambda'_{11} + \lambda'_{22}) + y_3(\lambda_{11}\lambda'_{22} + \lambda'_{11}\lambda_{22} - 2\lambda_{12}\lambda'_{12})\}\omega_1 \wedge \omega_2.$$

Introduce the mean and Gaussian curvatures

$$H = \tfrac{1}{2}(\lambda_{11} + \lambda_{22}), \qquad K = \lambda_{11}\lambda_{22} - \lambda_{12}^2, \quad \text{etc.}$$

and define

$$J = \lambda_{11}\lambda'_{22} + \lambda_{22}\lambda'_{11} - 2\lambda_{12}\lambda'_{12} = 2K - \begin{vmatrix} \lambda'_{11} - \lambda_{11} & \lambda'_{12} - \lambda_{12} \\ \lambda'_{12} - \lambda_{12} & \lambda'_{22} - \lambda_{22} \end{vmatrix}.$$

From the above formula we have, on integrating over the surface M,

$$2\int H'\,dS + \int y_3\,JdS = 0,$$

where $dS = \omega_1 \wedge \omega_2$ is the element of area of M. In particular, if we identify the surfaces M and $\overline{M}$,

$$2\int HdS + \int 2y_3\,KdS = 0,$$

Substracting, we get

$$2\int HdS - 2\int H'dS = \int y_3(J - 2K)\,dS = -\int y_3 \begin{vmatrix} \lambda'_{11} - \lambda_{11} & \lambda'_{12} - \lambda_{12} \\ \lambda'_{12} - \lambda_{12} & \lambda'_{22} - \lambda_{22} \end{vmatrix} dS.$$

We can choose the origin inside M, so that $y_3 < 0$. Since $K \geqq 0$, the integrand in the right-hand side of the equation is $\geqq 0$. It follows that

$$\int HdS - \int H'dS \leqq 0.$$

By symmetry we must also have

$$\int H'dS - \int HdS \leqq 0.$$

Therefore

$$\int H'dS - \int HdS = 0,$$

and

$$\int y_3 \begin{vmatrix} \lambda'_{11} - \lambda_{11} & \lambda'_{12} - \lambda_{12} \\ \lambda'_{12} - \lambda_{12} & \lambda'_{22} - \lambda_{22} \end{vmatrix} dS = 0.$$

But this is possible, only when

$$\lambda'_{11} = \lambda_{11}, \qquad \lambda'_{12} = \lambda_{12}, \qquad \lambda'_{22} = \lambda_{22}.$$

Hence the two surfaces are congruent.

Theorem 2. *Let two convex closed hypersurfaces M and $\overline{M}$ in E^{n+1} be differentiably homeomorphic such that at corresponding points the third fundamental forms and the sums of the principal radii of curvature are equal. Then M and $\overline{M}$ are congruent or symmetric.*

For $n = 2$ this theorem is due to Christoffel. Its generalization to arbitrary n was made by Kubota. The classical proof, due to A. Hurwitz, makes use of spherical harmonics.

Proof. By definition the third fundamental form of M is

$$\mathrm{III} = \omega_{n+1,1}^2 + \cdots + \omega_{n+1,n}^2.$$

Since the total curvature of M is >0, we have $(\lambda_{\alpha\beta}) \neq 0$ and we can write

$$\omega_\alpha = l_{\alpha\beta}\omega_{\beta,n+1},$$

where $(l_{\alpha\beta})$ is the inverse matrix of $(\lambda_{\alpha\beta})$. By considering the normal forms of these matrices under orthogonal transformations, it is easy to see that $L = l_{\alpha\alpha}$ is the sum of the principal radii of curvature.

Suppose that M and $\overline{M}$ are two convex hypersurfaces satisfying the hypotheses. Using the notations of the proof of Theorem 1, we find

$$\begin{aligned} &d(\varepsilon_{\alpha_1\ldots\alpha_n} y_{\alpha_1}\omega'_{\alpha_2} \wedge \omega_{\alpha_3 n+1} \wedge \cdots \wedge \omega_{\alpha_n n+1}) \\ &= \varepsilon_{\alpha_1\ldots\alpha_n}(\omega'_{\alpha_1} \wedge \omega_{\alpha_2} \wedge \omega_{\alpha_3 n+1} \wedge \cdots \wedge \omega_{\alpha_n n+1} + y_{n+1}\varepsilon_{\alpha_1\ldots\alpha_n}\omega'_{\alpha_1} \wedge \omega_{\alpha_2 n+1} \wedge \cdots \wedge \omega_{\alpha_n n+1} \\ &= \{\varepsilon_{\alpha_1\ldots\alpha_n}(l'_{\alpha_1\alpha_1} l_{\alpha_2\alpha_2} - l'_{\alpha_1\alpha_2} l_{\alpha_1\alpha_2}) + y_{n+1}\varepsilon_{\alpha_1\ldots\alpha_n} l'_{\alpha_1\alpha_1}\}\omega_{\alpha_1 n+1} \wedge \cdots \wedge \omega_{\alpha_n n+1}, \end{aligned}$$

where $\varepsilon_{\alpha_1\ldots\alpha_n}$ is $+1$ or -1 according as $\alpha_1, \ldots, \alpha_n$ form an even or odd permutation of $1, \ldots, n$, and is otherwise zero. We denote by

$$dV = \omega_{1n+1} \wedge \cdots \wedge \omega_{nn+1}$$

the element of volume of the spherical image of M. It follows from the last formula by integration that

$$\int \Big\{ \sum_{\substack{\alpha,\beta \\ \alpha\neq\beta}} (l'_{\alpha\alpha} l_{\beta\beta} - l'_{\alpha\beta} l_{\alpha\beta}) + y_{n+1} \sum_\alpha l'_{\alpha\alpha} \Big\} dV = 0,$$

the integration being taken over M. Taking $\overline{M} = M$, we get

$$\int \Big\{ \sum_{\substack{\alpha,\beta \\ \alpha\neq\beta}} (l_{\alpha\alpha} l_{\beta\beta} - l^2_{\alpha\beta}) + y_{n+1} \sum_\alpha l_{\alpha\alpha} \Big\} dV = 0,$$

Using the condition $L' = L$, we get

$$\int \Big\{ \sum_{\substack{\alpha,\beta \\ \alpha\neq\beta}} (l'_{\alpha\alpha} l_{\beta\beta} - l'_{\alpha\beta} l_{\alpha\beta}) - \sum_{\substack{\alpha,\beta \\ \alpha\neq\beta}} (l_{\alpha\alpha} l_{\beta\beta} - l^2_{\alpha\beta}) \Big\} dV = 0.$$

By symmetry between M and $\overline{M}$ we have a similar relation with $l_{\alpha\beta}$ and $l'_{\alpha\beta}$ interchanged. Combining these two relations we can write

$$\int \sum_{(\alpha,\beta)} \begin{vmatrix} l'_{\alpha\alpha} - l_{\alpha\alpha} & l'_{\alpha\beta} - l_{\alpha\beta} \\ l'_{\alpha\beta} - l_{\alpha\beta} & l'_{\beta\beta} - l_{\beta\beta} \end{vmatrix} dV = 0,$$

where the summation under the integral sign is taken over all combinations of α, β, with $\alpha \neq \beta$. But

$$\begin{aligned} (l'_{\alpha\alpha} - l_{\alpha\alpha})(l'_{\beta\beta} - l_{\beta\beta}) &= \tfrac{1}{2} \sum_{\alpha,\beta} (l'_{\alpha\alpha} - l_{\alpha\alpha})(l'_{\beta\beta} - l_{\beta\beta}) - \tfrac{1}{2} \sum_{\alpha} (l'_{\alpha\alpha} - l_{\alpha\alpha})^2 \\ &= -\tfrac{1}{2} \sum_{\alpha} (l'_{\alpha\alpha} - l_{\alpha\alpha})^2. \end{aligned}$$

It follows that

$$\int \Big\{ -\sum_{\substack{(\alpha,\beta) \\ \alpha\neq\beta}} (l'_{\alpha\beta} - l_{\alpha\beta})^2 - \tfrac{1}{2} \sum_{\alpha} (l'_{\alpha\alpha} - l_{\alpha\alpha})^2 \Big\} dV = 0.$$

Since the integrand is $\leqq 0$, this is possible only when it is identically zero, which gives

$$l'_{\alpha\beta} = l_{\alpha\beta}.$$

From this it follows that M and $\overline{M}$ are congruent.

3. Affine connection and absolute differentiation

In order that differentiation of tensors be defined on a differentiable manifold M, which will be intrinsic, that is, independent of the choice of local coordinates, we shall need an additional structure, an *affine connection.*

To define the affine connection we consider the space X of all ordered sets of n linearly independent tangent vectors with the same origin (cf. §5, Chapter I). Again denote by $\psi: X \to M$ the projection. Relative to a base every such set of vectors can be identified with an element of the group G of all $n \times n$ regular matrices. It follows that, corresponding to a coordinate neighborhood U, there is a homeomorphism

(16) $$\varphi_U: U \times G \to \psi^{-1}(U),$$

called a *coordinate function*, such that

(17) $$\psi\varphi_U(p, g) = p, \qquad p \in U, g \in G.$$

If $p \in U \cap V$, there are matrices $g_{UV}(p)$ defined by

$$\varphi_U(p, g_{UV}(p)g) = \varphi_V(p, g), \qquad g \in G.$$

We suppose the elements of $g_{UV}(p)$ to be differentiable functions, and put

(18) $$\theta_{UV} = g_{UV}^{-1}\, dg_{UV},$$

so that θ_{UV} is an $n \times n$ matrix of linear differential forms. In $U \cap V \cap W$ we have

(19) $$\theta_{UW} = g_{VW}^{-1}\theta_{UV}g_{VW} + \theta_{VW}.$$

An affine connection is defined, by giving in every coordinate neighborhood U, a matrix ω of linear differential forms, such that, in $U \cap V$,

(20) $$\theta_{UV} = \omega_V - g_{UV}^{-1}\omega_U g_{UV}.$$

It is easily verified that this definition is compatible.

We put

(21) $$\Omega_U = d\omega_U + \omega_U \wedge \omega_U,$$

so that Ω_U is an $n \times n$ matrix of quadratic differential forms. A little calculation will give

(22) $$\Omega_V = g_{UV}^{-1}\Omega_U g_{UV}.$$

We shall call Ω_U the *curvature matrix* of the affine connection.

To define absolute differentiation let T be a finite-dimensional vector space and let R be a linear representation of G in T, that is, a homomorphism of G into the group of linear endomorphisms of T. We consider entities of the form (p, U, t), $p \in U$, $t \in T$, where U is a coordinate neighborhood. Two such entities (p, U, t), (p', V, t') are called *equivalent*, if $p = p'$, $p \in U \cap V$, $t = R(g_{UV})t'$. A natural topology can be defined in the set of such equivalence classes, and the space so obtained is called the *tensor bundle* of type $R(G)$. A tensor of type $R(G)$ is a *cross section* of this bundle, and is thus defined in each coordinate neighborhood U by $t = f_U(p) \in T$, and is such that $f_U(p) = R(g_{UV})f_V(p)$ for $p \in U \cap V$. More generally, we can consider the tensor product $T \otimes A^r$ of T and the vector space of exterior differential forms of degree r, operated on by a linear representation of G. A cross section in this bundle is called a tensorial differential form of degree r and type $R(G)$. It is thus defined in each coordinate neighborhood U by $t = f_U(p) \in T \otimes A^r$ and is such that $f_U(p) = R(g_{UV})f_V(p)$, $p \in U \cap V$.

We take a base in $T \otimes A^r$ and consider its linear endomorphisms as matrices. Put

$$\bar{R} = R(g)^{-1}\, dR(g),$$

which is then a matrix of linear differential forms. It is also left invariant, so that its elements are linear combinations with constant coefficients of the Maurer-Cartan forms of G. We denote the matrix by $\bar{R}(\omega_U)$, when the Maurer-Cartan forms are replaced by the corresponding forms of the connection. (In terms of a base the Maurer-Cartan forms of a linear group can be considered as the elements of an $n \times n$ matrix.) Put

(23) $$Df_U = df_U + \bar{R}(\omega_U)f_U.$$

Then we have, in $U \cap V$,

$$dg_{UV} = g_{UV}\omega_V - \omega_U g_{UV},$$
$$dR(g_{UV}) = R(g_{UV})\bar{R}(\omega_V) - \bar{R}(\omega_U)R(g_{UV}),$$

and it follows that

$$Df_U = R(g_{UV})Df_{V*}$$

Hence Df_U is a tensorial form of type $R(G)$ and degree $r + 1$. We call Df_U the *absolute differential* of f_U.

To every differentiable manifold there is naturally defined a tensor of type (1, 1), which defines the identity mapping of the tangent vectors. For geometrical reasons we denote it by dp. The affine connection is called *without torsion*, if

$$D(dp) = 0.$$

In a coordinate neighborhood U put

$$\omega_U = (\omega_i^j) = (\Gamma_{ik}^j \, dx^k).$$

Then we have

$$\bar{R}(\omega_U) = \omega_U,$$

and dp has the components dx^i with respect to the coordinate system whose coordinate vectors are tangent vectors to the parametrized coordinate curves. The condition for the affine connection to be without torsion then becomes

$$\Gamma_{ik}^j \, dx^i \wedge dx^k = 0 \qquad \text{or} \qquad \Gamma_{ik}^j = \Gamma_{ki}^j.$$

4. Riemannian metric

A differentiable manifold M is called *Riemannian*, if there is given a quadratic differential form

$$ds^2 = g_{ij} \, dx^i \, dx^j.$$

We shall always assume this form to be *positive definite*. It is clear that the quadratic differential form defines a scalar product in each tangent space, so that we can talk about the length of a vector, the angle between two vectors with the same origin, etc.

The group G in the definition of an affine connection plays an important role. Sometimes the matrices g_{UV} can be so chosen that they belong to a subgroup of G. In particular, if the subgroup is the orthogonal group, then θ_{UV} is anti-symmetric. If ω_U is also anti-symmetric, the affine connection is called a *metric connection*.

The fundamental theorem on local Riemannian geometry is the following:

Theorem. *On a Riemannian manifold there is exactly one metrical connection without torsion.*

To prove this theorem we first notice that in a coordinate neighborhood U, ds^2 can be written as a sum of squares:

$$ds^2 = \omega_1^2 + \cdots + \omega_n^2.$$

It follows that we can choose G to be the orthogonal group. ω_i defines a tensorial form of degree 1. Let (ω_{ij}) be the matrix in U, which defines the metrical connection. The condition that the connection is without torsion gives

(24) $$D\omega_i = d\omega_i + \omega_{ij} \wedge \omega_j = 0.$$

By the Lemma of §1 a set of $\omega_{ij} = -\omega_{ji}$ satisfying these conditions is unique.

To prove the existence write

(25) $$d\omega_i = A_{ijk}\omega_j \wedge \omega_k, \qquad A_{ijk} + A_{ikj} = 0.$$

It is sufficient to put

(26) $$\omega_{ik} = -(A_{kij} + A_{ijk} + A_{jik})\omega_j.$$

This proves the theorem.

We can derive from the above considerations the theory of curvatures of a Riemannian manifold. In fact, exterior differentiation of (24) gives

$$(d\omega_{ik} + \omega_{ij} \wedge \omega_{jk}) \wedge \omega_k = 0.$$

We therefore put

$$d\omega_{ik} + \omega_{ij} \wedge \omega_{jk} = \Omega_{ik},$$

with

$$\Omega_{ik} \wedge \omega_k = 0, \qquad \Omega_{ik} + \Omega_{ki} = 0.$$

From the last equation it follows that Ω_{ik} is of the form

$$\Omega_{ik} = \theta_{ikj} \wedge \omega_j + R_{ikjl}\omega_j \wedge \omega_l,$$

when θ_{ikj} does not contain ω_l. But then we see that θ_{ikj} is symmetric in k, j and is skew-symmetric in i, k. Therefore $\theta_{ikj} = 0$ and we have

$$\Omega_{ik} = R_{ikjl}\omega_j \wedge \omega_l.$$

Let us summarize the fundamental equations for local Riemannian geometry in the following form

(27) $$\begin{cases} d\omega_i = -\omega_j \wedge \omega_{ji}, \\ d\omega_{ik} = -\omega_{ij} \wedge \omega_{jk} + \Omega_{ik}, \\ \omega_{ik} + \omega_{ki} = 0, \qquad \Omega_{ik} + \Omega_{ki} = 0, \\ \Omega_{ik} = R_{iklj}\omega_l \wedge \omega_j, \\ R_{iklj} + R_{ikjl} = 0, R_{iklj} + R_{kilj} = 0. \end{cases}$$

These equations will be called the *equations of structure* of the Riemann space. Ω_{ik} are called the *curvature forms*. R_{iklj} gives essentially what is known as the *Riemann-Christoffel curvature tensor.*

Applying exterior differentiation to the first two equations of (27), we get

(28) $$\begin{aligned} \omega_j \wedge \Omega_{ji} &= 0, \\ -\Omega_{ij} \wedge \omega_{jk} + \omega_{ij} \wedge \Omega_{jk} + d\Omega_{ik} &= 0. \end{aligned}$$

These are called the *Bianchi identities.*

Over a Riemannian manifold M there are associated different tensor bundles. The more important ones are: 1) the *principal bundle*, the bundle of all frames $p\mathfrak{n}_1 \ldots \mathfrak{n}_n$, where $\mathfrak{n}_i$ are mutually perpendicular unit vectors at p; 2) the bundle of unit tangent vectors. We denote their total spaces by $X^{(n)}$ and $X^{(1)}$ respectively. There are natural projections

$$X^{(n)} \xrightarrow[\psi_{n,1}]{} X^{(1)} \xrightarrow[\psi_{1,0}]{} M, \tag{29}$$

defined by

$$\begin{aligned} \psi_{n,1}(p\mathfrak{n}_1 \cdots \mathfrak{n}_n) &= p\mathfrak{n}_n, \\ \psi_{1,0}(p\mathfrak{n}_n) &= p, \end{aligned} \tag{30}$$

and we put

$$\psi = \psi_{1,0}\psi_{n,1}. \tag{31}$$

These projections induce dual mappings of the differential forms in the inverse direction. Our problem is to determine the differential forms in $X^{(n)}$ and $X^{(1)}$, which are dual images of differential forms in $X^{(1)}$ or M, onto which they have been projected. Perhaps the simplest way to do this is to examine the effect of a change of frame. We put

$$\mathfrak{n}_i = u_{ik}\mathfrak{n}_k^*, \tag{32}$$

where (u_{ik}) is an orthogonal matrix, so that $\mathfrak{n}_k^*$ are new frames. Denote the differential forms relative to $\mathfrak{n}_k^*$ by the same symbols but preceded with asterisks. Then we find

$$\omega_i = u_{ik}\omega_k^*, \qquad \omega_k^* = u_{ik}\omega_i,$$

$$d\omega_i = \omega_j \wedge (-du_{ik}u_{jk} + u_{ik}u_{jl}\omega_{lk}^*),$$

so that

$$\omega_{ji} = du_{ik}u_{jk} - u_{ik}u_{jl}\omega_{lk}^*.$$

It follows that

$$d\omega_{ik} + \omega_{ij} \wedge \omega_{jk} = u_{ij}u_{kl}\Omega_{jl}^*,$$

and hence that

$$\Omega_{ik} = u_{ij}u_{kl}\Omega_{jl}^*. \tag{33}$$

From (33) we see that the following are forms in $X^{(n)}$ which are dual images of forms in M:

$$\begin{aligned} \Delta_4 &= \Omega_{ij}\Omega_{ji}, \\ \Delta_8 &= \Omega_{ij}\Omega_{jk}\Omega_{kl}\Omega_{li}, \end{aligned} \tag{34}$$

and more generally, Δ_{4m}, $4m \leqq n$. We shall denote by the same symbols the originals of these forms in M, and we say simply that these forms are *in* M.

If M is orientable, we can restrict ourselves to the frames $p\mathfrak{n}_1 \cdots \mathfrak{n}_n$ whose orientations are coherent with an orientation of the manifold. Then two frames at a point p

are related by a proper orthogonal transformation, and we can assume (u_{ij}) to be properly orthogonal. In this case the form

$$\Delta_0 = \varepsilon_{i_1 \ldots i_n} \Omega_{i_1 i_2} \ldots \Omega_{i_{n-1} i_n}, \tag{35}$$

which exists when the manifold is of even dimensions, is also in M.

By the Bianchi identities it can be verified that Δ_0, Δ_{4m} are closed. A main theorem in differential geometry in the large asserts that their cohomology classes depend only on the differentiable structure of the manifold.

We shall add the following remarks:

A) If M is the Euclidean space with the Euclidean metric, then $\Omega_{ij} = 0$ and ω_i, ω_{ij} are the Maurer-Cartan forms of the group of motions. This can be seen by comparing the discussion of this section to §1.

B) The relation of our presentation to that of classical Riemannian geometry can be given as follows: Let S_{ikjl} be the Riemann-Christoffel tensor in the classical sense, and let n_i have the components u_i^k (relative to the coordinate system x^j). Then

$$\Omega_{ik} = u_i^m u_k^q S_{mqjl}\, dx^j \wedge dx^l. \tag{36}$$

5. The Gauss-Bonnet formula

In this section we assume our Riemannian manifold M to be compact and orientable. With Δ_0 defined by (35), put

$$\Omega = \begin{cases} (-1)^p \dfrac{1}{2^{2p}\pi^p p!} \Delta_0, & \text{if } n = 2p \text{ is even} \\ 0, & \text{if } n \text{ is odd.} \end{cases} \tag{37}$$

The Gauss-Bonnet formula states that

$$\int_M \Omega = \chi(M), \tag{38}$$

where $\chi(M)$ is the Euler-Poincare characteristic of M.

We shall make use of the projections as defined in (29) and shall deduce differential forms which are in $X^{(1)}$. Perform a change of frames which leaves pn_n unaltered:*

$$n_\alpha = u_{\alpha\beta} n_\beta^*, \tag{39}$$

where $(u_{\alpha\beta})$ is properly orthogonal. Denoting the quantities formed from the new frames by the same notations preceded with asterisks, we get

$$\begin{aligned} \omega_{\alpha n} &= u_{\alpha\beta} \omega_{\beta n}^*, \\ \Omega_{\alpha\beta} &= u_{\alpha\gamma} u_{\beta\delta} \Omega_{\gamma\delta}. \end{aligned} \tag{40}$$

Define

* Greek indices in this section run from 1 to $n - 1$.

$$\Phi_k = \varepsilon_{\alpha_1\dots\alpha_{n-1}}\Omega_{\alpha_1\alpha_2}\cdots\Omega_{\alpha_{2k-1}\alpha_{2k}}\omega_{\alpha_{2k+1}n}\cdots\omega_{\alpha_{n-1}n},$$
(41)
$$\Psi_k = 2(k+1)\varepsilon_{\alpha_1\dots\alpha_{n-1}}\Omega_{\alpha_1\alpha_2}\cdots\Omega_{\alpha_{2k-1}\alpha_{2k}}\Omega_{\alpha_{2k+1}n}\omega_{\alpha_{2k+2}n}\cdots\omega_{\alpha_{n-1}n},$$

where $k = 0, 1, \dots, [\frac{n}{2}] - 1$, $[\frac{n}{2}]$ = largest integer $\leqq \frac{n}{2}$. If n is odd, $\Phi_{[n/2]}$ is also defined. By convention we set

(42) $$\Psi_{-1} = \Psi_{[n/2]} = 0.$$

Using (40) it is easily verified that Φ_k and Ψ_k, of degrees $n-1$ and n respectively, are differential forms in $X^{(1)}$.

By exterior differentiation we have

$$d\Phi_k = k\varepsilon_{\alpha_1\dots\alpha_{n-1}}d\Omega_{\alpha_1\alpha_2}\Omega_{\alpha_3\alpha_4}\cdots\Omega_{\alpha_{2k-1}\alpha_{2k}}\omega_{\alpha_{2k+1}n}\cdots\omega_{\alpha_{n-1}n}$$
$$+(n-2k-1)\varepsilon_{\alpha_1\dots\alpha_{n-1}}\Omega_{\alpha_1\alpha_2}\cdots\Omega_{\alpha_{2k-1}\alpha_{2k}}d\omega_{\alpha_{2k+1}n}\omega_{\alpha_{2k+2}n}\cdots\omega_{\alpha_{n-1}n}.$$

This exterior differentiation is carried out in $X^{(n)}$. But as the resulting forms are in $X^{(1)}$, the terms involving $\omega_{\alpha\beta}$ must cancel with each other. Hence we immediately get

(43) $$d\Phi_k = \Psi_{k-1} + \frac{n-2k-1}{2(k+1)}\Psi_k.$$

Solving for Ψ_k, we get

$$\Psi_k = d\Theta_k, \qquad k = 0, 1, \dots, [\tfrac{n}{2}] - 1,$$

where

$$\Theta_k = \sum_{\lambda=0}^{k}(-1)^{k-\lambda}\frac{(2k+2)\dots(2\lambda+2)}{(n-2\lambda-1)\dots(n-2k-1)}\Phi_\lambda, \qquad k = 0, 1, \dots, [\tfrac{n}{2}] - 1.$$

If n is even, say $= 2p$, then we have

$$d\Theta_{p-1} = \Psi_{p-1},$$

where

$$\Psi_{p-1} = n\varepsilon_{\alpha_1\dots\alpha_{n-1}}\Omega_{\alpha_1\alpha_2}\cdots\Omega_{\alpha_{n-1}n} = \varepsilon_{i_1\dots i_n}\Omega_{i_1i_2}\cdots\Omega_{i_{n-1}i_n}.$$

If n is odd, say $= 2q+1$, then

$$d\Theta_{q-1} = \Psi_{q-1},$$

But in this case we have also

$$d\Phi_q = \Psi_{q-1},$$

so that

$$d(\Theta_{q-1} - \Phi_q) = 0.$$

By defining

(44) $$\Pi = \begin{cases} \dfrac{1}{\pi^p}\displaystyle\sum_{\lambda=0}^{p-1}(-1)^\lambda\frac{1}{1.3\dots(2p-2\lambda-1)2^{p+\lambda}\lambda!}\Phi_\lambda, & n = 2p, \\ \dfrac{1}{2^{2q+1}\pi^q q!}\displaystyle\sum_{\lambda=0}^{q}(-1)^{\lambda+1}\binom{q}{\lambda}\Phi_\lambda, & n = 2q+1, \end{cases}$$

we therefore get

$$-d\Pi = \Omega. \tag{45}$$

Π is a differential form in $X^{(1)}$ whose exterior derivative is in M.

Let K^{n-1} be the $(n-1)$-dimensional skeleton of a cellular decomposition of M. It is well-known that a continuous non-zero vector field can be defined in K^{n-1}. We extend this field over M, with the possibility of introducing a number of isolated singularities where the vector is zero. Draw about each singularity a small sphere with radius ε, and call M_ε the domain of these spheres. The vector field defines a mapping $f: M - M_\varepsilon \to X^{(1)}$, with $\psi_{1,0} f =$ identity. Applying the Theorem of Stokes, we get

$$\int_{M-M_\varepsilon} \Omega = \int_{f(M-M_\varepsilon)} \Omega = -\int_{f(M-M_\varepsilon)} d\Pi = -\int_{\partial f(M-M_\varepsilon)} \Pi = I + \eta,$$

where I is the sum of indices of the vector field (Cf. §6, Chapter I), and $\eta \to 0$ as $\varepsilon \to 0$. It follows that, as $\varepsilon \to 0$,

$$\int_M \Omega = I.$$

This shows that I is independent of the choice of the vector field. By considering a particular field, we verify that it is equal to the Euler-Poincare characteristic $\chi(M)$ of M. Hence the formula is proved.

Chapter III
Theory of Connections

We shall develop in this chapter the theory of connections in a fiber bundle in the sense of Elie Cartan. Its modern treatment was first carried out by Ehresmann and Weil. Our main theorem consists in giving a relationship between a characteristic homomorphism defined by topological properties of the fiber bundle and a homomorphism defined by the local properties of the connection. As we shall see, the Gauss-Bonnet formula for a compact orientable Riemann manifold is a corollary of this theorem.

1. Resume on fiber bundles

There are now available several accounts of the general theory of fiber bundles, in particular, N. E. Steenrod's forthcoming book. We can therefore restrict ourselves to a resume of the notions and results which are necessary for our purpose. To conform with some notations currently in use (at least in Princeton), we change our previous notation by interchanging B and X, so that B will be the bundle and X the base space.

Let F be a space acted on by a topological group G of homeomorphisms. A *fiber bundle* with the *director space F* and *structural group G* consists of topological spaces B, X and a mapping ψ of B onto X, together with the following:

1) X is covered by a family of neighborhoods $\{U_\alpha\}$, called the *coordinate neighborhoods*, and to each U_α there is a homeomorphism (a coordinate function), $\varphi_\alpha: U_\alpha \times F \to \varphi^{-1}(U_\alpha)$, with $\psi\varphi_\alpha(x, y) = x$, $x \in U_\alpha$, $y \in F$.

2) As a consequence of 1) a point of $\psi^{-1}(U_\alpha)$ has the coordinates (x, y), and a point of $\psi^{-1}(U_\alpha \cap U_\beta)$ has two sets of coordinates (x, y) and (x, y'), satisfying $\varphi_\alpha(x, y) = \varphi_\beta(x, y')$. It is required that $g_{\alpha\beta}(x)$: $y' \to y$ is a continuous mapping $g_{\alpha\beta}$ of $U_\alpha \cap U_\beta$ into G.

The spaces X and B are called the *base space* and the *bundle* respectively. Each subset $\psi^{-1}(x) \subset B$ is called a *fiber*.

Since we wish to allow changes of coordinate neighborhoods and coordinate functions in a bundle, an equivalence relation is introduced. Two bundles (B, X), (B', X) with the same base space and the same F, G are called *equivalent*, if, $\{U_\alpha, \varphi_\alpha\}$, $\{V'_\alpha, \theta_{\alpha'}\}$ being respectively their coordinate neighborhoods and coordinate functions, there is a fiber-preserving homeomorphism $T: B \to B'$ such that the mapping $h_{\alpha\alpha'}(x)$: $y \to y'$ defined by $\theta_{\alpha'}(x, y) = T\phi_\alpha(x, y')$ is a continuous mapping of $U_\alpha \cap V_{\alpha'}$ into G.

An important operation on fiber bundles is the construction from a given bundle of other bundles with the same structural group, in particular, the principal fiber bundle which has G as director space acted upon by G itself as the group of left translations. It can be defined as follows: For $x \in X$ let G_x be the totality of all maps $\varphi_{\alpha,g}(x)$: $F \to \psi^{-1}(x)$ defined by $y \to \varphi_\alpha(x, g(y))$, $y \in F$, $g \in G$, relative to a coordinate neighborhood U_α containing x. G_x depends only on x. Let $B^* = \bigcup_{x \in X} G_x$ and define the mapping $\psi^*\colon B^* \to X$ by $\psi^*(G_x) = x$ and the coordinate functions $\varphi_\alpha^*(x, g) = \varphi_{\alpha,g}(x)$. Topologize B^* such that the φ_α^*'s define homeomorphisms of $U_\alpha \times G$ into B^*. The bundle (B^*, X) so obtained is called a *principal fiber bundle.* This construction is an operation on the equivalence classes of bundles in the sense that two fiber bundles are equivalent if and only if their principal fiber bundles are equivalent. Similarly, an inverse operation can be defined, which will permit us to construct bundles with a given principal bundle and having as director space a given space acted upon by the structural group G. Such bundles are called *associated bundles.* An important property of the principal fiber bundle is that B^* is acted upon by G as right translations.

Let G' be a subgroup of G. If the mappings $g_{\alpha\beta}\colon U_\alpha \cap U_\beta \to G$ have their images in G' for every pair of coordinate neighborhoods U_α and U_β with a non-empty intersection, we say that the bundle has the structural group G'. A bundle is called *trivial,* if it is equivalent to a bundle with a structural group which consists of the unit element only.

A *cross section* of a bundle is a mapping $f\colon X \to B$ such that φf is the identity. Using this notion, it is easy to establish the following statements, of which the second is a consequence of the first:

1) A bundle with the group G is equivalent to a bundle with the group $G' \subset G$, if and only if the associate bundle having G/G' as director space has a cross section.

2) A bundle is trivial if and only if its principal bundle has a cross section.

Suppose a bundle be given, with the above notations. Let f be a mapping of a space Y into X. The neighborhoods $\{f^{-1}(U_\alpha)\}$ then form a covering of Y and coordinate functions $\varphi_\alpha'\colon f^{-1}(U_\alpha) \times F \to f^{-1}(U_\alpha) \times \psi^{-1}(U_\alpha)$ can be defined by $\varphi_\alpha'(\eta, y) = \eta \times \varphi_\alpha(f(\eta), y)$. This defines a fiber bundle $Y \times \psi^{-1}(f(Y))$ over Y, with the same director space F and the same group G. The new bundle is said to be *induced* by the mapping f.

This method of generating new fiber bundles from a given bundle is very useful, particularly in the case when a so-called universal bundle exists. Let the director space and the group G be given and fixed for our present considerations. A bundle with the base space X_0 is called *universal* relative to a space X, if every bundle over X is equivalent to a bundle induced by a mapping $X \to X_0$ and if two such induced bundles are equivalent when and only when the mappings are homotopic. If, for a space X, there exists a universal bundle with the base space X_0, then the classes of bundles over X are in one-one correspondence with the homotopy classes of mappings $X \to X_0$, so that the enumeration of the bundles over X reduces to a homotopy classification problem.

For our purpose the existence of a universal bundle has another consequence. Let $H(X_0, R)$, $H(X, R)$ denote the cohomology rings of the spaces X_0, X respectively, with the coefficient ring R. It follows from the above that the induced dual homomorphism

$$h'\colon H(X_0, R) \to H(X, R)$$

is completely determined by the bundle. h' will be called the *characteristic homomorphism,* its image $h'(H(X_0, R)) \subset H(X, R)$ the *characteristic ring,* and an element of the characteristic ring a *characteristic* (cohomology) *class.*

A necessary and sufficient condition that a bundle is universal relative to all polyhedra of dimension n is that its principal bundle B satisfies the conditions; $\pi_i(B) = 0$, $0 \leqq i \leqq n$, where $\pi_0(B) = 0$ means that B is connected. The proof of this theorem can be found in Steenrod's book or in: S. S. Chern and Y. Sun, The imbedding theorem for fiber bundles, Trans. Amer. Math. Soc. 67, 286–303 (1949).

When a universal bundle exists, it may not be unique. However, we shall show that the characteristic homomorphism is independent of the choice of the universal bundle by proving the theorem: *Let $\psi_0: B_0 \to X_0$, $\psi_0': B_0' \to X_0'$ be two universal principal bundles relative to complexes of dimension n. There are one-one isomorphisms*

$$H^{\gamma}(X_0, G) \cong H^{\gamma}(X_0', G), \qquad r \leqq n,$$

For simplicity we assume X_0 and X_0' to be cellular complexes. Denote by X_0^n and $X_0'^n$ their n-dimensional skeletons. Since every continuous mapping is homotopic to a cellular mapping, the sub-bundles over X_0^n, $X_0'^n$ can be induced respectively by mappings

$$f: X_0^n \to X_0'^n, \qquad g: X_0'^n \to X_0^n.$$

It follows that these sub-bundles are equivalent to bundles induced by the mappings gf and fg. Since the given bundles are universal, we conclude that gf and fg are homotopic to the identity mapping. This proves the theorem. The theorem is valid under more general assumptions of the base space of the universal bundle. The proof will then make use of the singular complex and is more complicated.

We shall show that a universal bundle exists whenever the base space X is compact and the structural group is a connected Lie group. According to a theorem due to E. Cartan, Malcev, Iwasawa, and Mostow, all the maximal compact subgroups of a connected Lie group G are conjugate to each other and the homogeneous space G/G_1 is homeomorphic to an Euclidean space, where G_1 is a maximal compact subgroup. (Cf. in particular, K. Iwasawa, On some types of topological groups, Annals of Math. 50, p. 530 (1949)). Using this theorem, it follows that every bundle with the group G is equivalent to a bundle with the group G_1 and that two bundles with the group G are equivalent when and only when their equivalent bundles with the group G_1 are equivalent relative to G_1. In other words, so long as the equivalence classes of bundles are concerned, we can replace G by its maximal compact subgroup G_1.

G_1 being a compact Lie group, it can be considered as a subgroup of the rotation group $R(m)$ in m variables. Imbed $R(m)$ as a subgroup of $R(m + n + 1)$ which operates on the first m variables, while $R(n + 1)$ operates on the last $n + 1$ variables. Then we have

$$R(m + n + 1) \supset G \times R(n + 1) \supset I_m \times R(n + 1),$$

where I_m denotes the group of the identity. By the natural projection

$$\psi: R(m + n + 1)/I_m \times R(n + 1) \to R(m + n + 1)/G \times R(n + 1),$$

we get a principal fiber bundle with the group G. It is universal relative to complexes of dimension n, since, by the covering homotopy theorem, all homotopy groups of the bundle up to the dimention n inclusive are zero.

2. Connections

We consider a fiber bundle and adopt the notation of the last section. For the purpose of differential geometry the following assumptions will be made: 1) B, X, F are differentiable manifolds; 2) G is a Lie group which acts differentiably on F; 3) the projection of B onto X is differentiable.

Let $L(G)$ be the Lie algebra of G. $L(G)$ is invariant under the left translations of G, while the right translations and the inner automorphisms of G induce on $L(G)$ a group of linear endomorphisms $ad(G)$, called the *adjoint group* of G. Relative to a base of $L(G)$ there are the left-invariant linear differential forms ω^i and the right-invariant linear differential forms π^i, each set consisting of linearly independent forms whose number is equal to the dimension of G. A fundamental theorem on Lie groups asserts that their exterior derivatives are given by

$$(1)\qquad \begin{aligned} d\omega^i &= \tfrac{1}{2}\sum_{j,k} c^i_{jk}\omega^j \wedge \omega^k, \\ d\pi^i &= -\tfrac{1}{2}\sum_{j,k} c^i_{jk}\pi^j \wedge \pi^k, \qquad 1 \leqq i, j, k \leqq \dim G. \end{aligned}$$

where c^i_{jk} are the so-called constants of structure. They are anti-symmetric in the lower indices and satisfy the Jacobi relations obtained by expressing that the exterior derivative of the right-hand side of (1) is equal to zero:

$$(2)\qquad \sum_m (c^i_{mj}c^m_{kl} + c^i_{mk}c^m_{lj} + c^i_{ml}c^m_{jk}) = 0.$$

This being said about the structural group G, the dual mapping of the mapping $g_{\alpha\beta}\colon U_\alpha \cap U_\beta \to G$ carries ω^i and π^i into linear differential forms in $U_\alpha \cap U_\beta$, which we shall denote by $\omega^i_{\alpha\beta}$ and $\pi^i_{\alpha\beta}$ respectively. Since $g_{\alpha\gamma} = g_{\alpha\beta}g_{\beta\gamma}$ in $U_\alpha \cap U_\beta \cap U_\gamma$, we have

$$(3)\qquad \begin{aligned} \omega^i_{\alpha\gamma} &= \sum_j ad(g_{\beta\gamma})^i_j\omega^j_{\alpha\beta} + \omega^i_{\beta\gamma}, \\ \pi^i_{\alpha\gamma} &= \pi^i_{\alpha\beta} + \sum ad(g^{-1}_{\alpha\beta})^i_j\pi^j_{\beta\gamma}. \end{aligned}$$

We can also interpret $\omega^i_{\alpha\beta}$ as a vector-valued linear differential form in $U_\alpha \cap U_\beta$, with values in $L(G)$, and shall denote it simply by $\omega_{\alpha\beta}$, when so interpreted.

The generalization of the notion of a tensor field in classical differential geometry leads to the following situation: Let E be a vector space acted on by a representation $M(G)$ of G. A *tensorial differential form of degree r and type* $M(G)$ is an exterior differential form u_α of degree r in each coordinate neighborhood U_α, with values in E, such that, in $U_\alpha \cap U_\beta$, $u_\alpha = M(g_{\alpha\beta})u_\beta$. The exterior derivative du_α of u_α is in general not a tensorial differential form. To recover its tensorial character, a connection is introduced into the fiber bundle.

A *connection* in the fiber bundle is a set of linear differential forms θ_α in U_α, with values in $L(G)$, such that

$$(4)\qquad \omega_{\alpha\beta} = -ad(g_{\alpha\beta})\theta_\alpha + \theta_\beta.$$

It follows from (3) that such relations are consistent, in $U_\alpha \cap U_\beta \cap U_\gamma$. With the help of a connection *absolute* or *covariant differentiation* can be defined as follows: Let $\overline{M}(X)$,

$X \in L(G)$, be the representation of the Lie algebra $L(G)$ induced by the representation $M(G)$ of G. $M(G)$ being a linear group, the elements of $\overline{M}(X)$ can be identified with linear endomorphisms in E. If a base is chosen in the vector space E, both $M(G)$ and $\overline{M}(X)$ can be represented by matrices. Moreover, we can take as their elements differential forms with values in $L(G)$. With this understanding equation (4) goes under the homomorphic mapping $G \to M(G)$ into the equation

$$dM(g_{\alpha\beta}) = M(g_{\alpha\beta})\overline{M}(\theta_\beta) - \overline{M}(\theta_\alpha)M(g_{\alpha\beta}). \tag{4'}$$

This, together with the equation obtained by exterior differentiation of $u_\alpha = M(g_{\alpha\beta})u_\beta$, shows that

$$Du_\alpha = du_\alpha + \overline{M}(\theta_\alpha) \wedge u_\alpha \tag{5}$$

is a tensorial differential form of degree $r + 1$ and the same type $M(G)$.

It is easy to prove that a connection can always be defined in a fiber bundle. In fact, let $\{U_\alpha\}$ be a finite or countable covering of X by coordinate neighborhoods. There exists an open covering $\{V_\alpha\}$ of X, such that $\bar{V}_\alpha \subset U_\alpha$. We define the connection in $V_1 + \cdots + V_\alpha$ by induction on α. If the connection has been defined in $V_1 + \ldots + V_{\alpha-1}$, θ_α is given in $V_\alpha \cap (V_1 + \ldots + V_{\alpha-1})$ and is consistent there because of (4). By an elementary extension theorem, θ_α exists in V_α such that it becomes the pre-assigned form in $V_\alpha \cap (V_1 + \ldots + V_{\alpha-1})$ and is zero in $X - U_\alpha$. Thus we define a connection relative to the covering $\{V_\alpha\}$. We shall give a second proof of the existence of a connection at the end of this section.

The above definition of a connection makes use of the coordinate neighborhoods and is entirely analytic. We shall give equivalent but intrinsic definitions and at the same time interpret the definition geometrically.

For this purpose we consider the principal bundle, whose bundle space we again denote by B. If (x, s), (x, t), $s, t \in G$, are the coordinates of a point of B relative to the coordinate neighborhoods U_α and U_β respectively, we have $g_{\alpha\beta}t = s$. Since the left-invariant differential forms of G can be regarded as the components of such a form ω with values in $L(G)$, the form $sd(s)\theta_\alpha + \omega$ is a linear differential form in $U_\alpha \times G$, with values in $L(G)$. From (4) we can verify that in $U_\alpha \times U_\beta$ this differential form is equal to the same form constructed from U_β. The set of these differential forms defines therefore an $L(G)$-valued linear differential form in B, which we shall denote by $\varphi(G)$, $b \in B$. Since B is acted on by G as right translations, we can study the effect of such a translation on $\varphi(b)$, and we find $\varphi(bg) = ad(g^{-1})\varphi(b)$. This leads to the following definition of a connection:

A connection in a fiber bundle is defined by an $L(G)$-valued linear differential form in the principal fiber bundle having the properties: 1) Under right translations it is transformed according to the adjoint group, $\varphi(bg) = ad(g^{-1})\varphi(b)$; 2) It is transformed by the dual mapping of the identity mapping of a fiber into the bundle into the left-invariant form of the fiber. The second property has a sense, because a fiber in the principal bundle has a group structure defined up to a left-translation.

Since an $L(G)$-valued linear differential form can be interpreted as a linear mapping of the tangent space of B into $L(G)$, it follows from the above definition that a connection is a linear mapping at every point of B such that: 1) every tangent vector is mapped into a tangent vector to the fiber; 2) every tangent vector to the fiber remains invariant; 3) the mapping is invariant under right translations. It is an elementary theorem of linear algebra that under such a mapping the set of tangent vectors which are mapped into zero form a linear space complementary to the tangent space of the fiber. A connection therefore gives rise to a family of tangent subspaces in the principal fiber bundle which are transversal to the fibers (that is, which span with the tangent space of the fiber the tangent space of the bundle) and are invariant under right translations.

From these geometrical considerations it follows easily that we can define a connection in a homogeneous space whose group is a semi-simple Lie group. In fact, let R be a semi-simple Lie group and H a closed subgroup of R. Then R is a principal fiber bundle over R/H with the director space H. It is known that a positive definite Riemannian metric can be defined in R, which is both left and right invariant. Using this metric, we can define a mapping which maps a tangent vector of R to its orthogonal projection in the fiber through the origin of the vector. This mapping satisfies the three conditions above and therefore defines a connection in the bundle.

In particular, it follows that we can define a connection in the particular universal bundle chosen above. Let $f: X' \to X$ be a mapping which induces a bundle over X'. If the original bundle has a connection given by the differential form θ_α in U_α, the dual mapping f^* of f carries θ_α into $f^* \theta_\alpha$ in $f^{-1}(U_\alpha)$ for which the relation corresponding to (4) is valid. The forms $f^*\theta_\alpha$ therefore define an induced connection in the induced bundle. Since every bundle is equivalent to one induced by mapping its base space into the base space of a universal bundle and since the latter has a connection, it follows that a connection can be defined in any bundle. This gives a second proof of the statement that a connection can always be defined in a bundle.

3. Local theory of connections; the curvature tensor

To study the local properties of the connection we again make use of a base of the Lie algebra, relative to which the form θ_α has the components θ_α^i. We put

$$\Theta_\alpha^i = d\theta_\alpha^i - \tfrac{1}{2}\sum_{j,k} c_{jk}^i \theta_\alpha^j \wedge \theta_\alpha^k \quad 1 \leqq i, j, k \leqq \dim G. \tag{6}$$

The form Θ_α, whose components relative to the base are Θ_α^i, is then an exterior quadratic differential form of degree 2, with values in $L(G)$. It is easy to verify that $\Theta_\alpha = ad(g_{\alpha\beta})\Theta_\beta$ in $U_\alpha \cap U_\beta$. The Θ's therefore define a tensorial differential form of degree 2 and type $ad(G)$, called the *curvature tensor* of the connection.

The following formulas for absolute differentiation can easily be verified:

$$\begin{aligned} \overline{M}(\Theta_\alpha) &= d\overline{M}(\theta_\alpha) + \overline{M}(\theta_\alpha)^2, \\ D\Theta_\alpha &= 0, \\ D^2 u_\alpha &= \overline{M}(\Theta_\alpha) \wedge u_\alpha. \end{aligned} \tag{7}$$

The second relation is known as the *Bianchi identity*. It shows that absolute differentiation of the curvature tensor does not give further invariants.

It follows from our intrinsic definition of connection that a connection in a bundle gives rise to a connection in every bundle of its equivalence class, so that we can speak of a connection in an equivalence class of bundles. The connections in two equivalent bundles are called *equivalent*, if they define the same connection in the class of bundles.

More interesting is the notion of *local equivalence of two connections.* Given two bundles with the same structural group and with a connection defined in each bundle. The structures pertaining to the second bundle we denote by the same notation with dashes. We shall define the notion that the connections are equivalent at a point $x \in X$ and a point $x' \in X'$. In fact, let U_α and $U'_{\alpha'}$ be coordinate neighborhoods containing x and x' respectively. The two connections are said to be equivalent at x and x' if: 1) there exist open sets V, V' satisfying $x \in V \subset U_\alpha$, $x' \in V' \subset U'_{\alpha'}$; 2) there exist a differentiable homeomorphism $f\colon V \to V'$ and a differentiable mapping $g\colon V \to G$ such that

$$\theta_\alpha = g^*\omega + ad(g)(f^*\theta'_{\alpha'}).$$

We can verify that this condition is independent of the choice of U_α and $U'_{\alpha'}$. Instead of making this verification we can also formulate the definition in an intrinsic form. We shall say that a mapping $1\colon \psi^{-1}(V) \to \psi'^{-1}(V')$ is *admissible* if there is a mapping $1_*\colon V \to V'$ such that $1\,\psi^{-1}(x) = \psi'^{-1}(1_*(x))$ and that the mapping $g \to g'$ defined by $1\varphi_\alpha(x, g) = \varphi'_{\alpha'}(1_*(x), g')$ is a left translation of G. Clearly the last condition is independent of the choice of the coordinate neighborhoods U_α and $U'_{\alpha'}$ in the definition. Remembering that a connection gives rise to an $L(G)$-valued linear differential form $\varphi(b)$, $b \in B$ in the principal bundle B, we can formulate the definition of local equivalence of two connections as follows: The two connections are equivalent at x and x' if there are neighborhoods V and V' of x and x' respectively such that there is an admissible homeomorphism between $\psi^{-1}(V)$ and $\psi'^{-1}(V')$ under which $\varphi(b)$ and $\varphi'(b')$ are equal. The equivalence of the two definitions can be easily verified by making use of (3).

A connection is called *locally flat* at $x \in X$ if $\theta_\alpha = 0$ in a neighborhood of x (relative to a coordinate neighborhood U_α containing x) or is equivalent to one with this property. It follows that a necessary and sufficient condition for a connection to be locally flat at x is that there exist an open set V containing x and contained in a coordinate neighborhood U_α and a mapping f of V into G such that $\theta_\alpha = f^*\omega$ in V. When a connection is locally flat at x, the curvature tensor vanishes in a neighborhood of x. Conversely, when the curvature tensor is zero, it follows from the theorem of Frobenius that the system of differential equations

$$\theta_\alpha - f^*\omega = 0,$$

where the mapping f is the unknown function, is completely integrable. Hence f exists in a neighborhood of x and the connection is flat at x.

As a first instance of the problem concerning the relationship between properties of a bundle with those of the connections which can be defined on it, we consider the case that the connection is everywhere flat. Then, by Frobenius's Theorem, there exists to every coordinate neighborhood U_α a mapping $f_\alpha\colon U_\alpha \to G$ such that $f_\alpha^*\omega = \theta_\alpha$. In $U_\alpha \cap U_\beta$ we define $g'_{\alpha\beta} = f_\alpha g_{\alpha\beta} f_\beta^{-1}$. Then we have $g'^*_{\alpha\beta}\omega = 0$, which in turn implies that the mapping $g'_{\alpha\beta}$ is constant in every connected component of $U_\alpha \cap U_\beta$. These mappings $g'_{\alpha\beta}$ define a bundle over X equivalent to B. It follows that the bundle will not be affected if we replace the topology in G by the discrete topology. Such a fiber bundle is a covering space. If X is simply connected, B is a topological product of G and X. In general, B is a product of the connected component of G and a covering of X. Thus the flatness of a connection implies topological properties of the bundle.

4. The homomorphism h and its independence of connection

We consider a fiber bundle (B, X), with a structural group G which is a Lie group, and we assume that a connection is given in the bundle.

A real-valued multilinear function $P(Y_1, \ldots, Y_k)$, with arguments $Y_1, \ldots, Y_k \in L(G)$, is called *invariant*, if

$$P(ad(a)\,Y_1, \ldots, ad(a)\,Y_k) = P(Y_1, \ldots, Y_k) \qquad \text{for all } a \in G.$$

Replacing a by an infinitesimal transformation, the condition of invariance implies

$$\sum_{i=1}^{k} P(Y_1, \ldots, Y_{i-1}, [Z, Y_i], Y_{i+1}, \ldots, Y_k) = 0, \tag{8}$$

where Z is any element of $L(G)$. If G is connected, condition (8) is equivalent to the definition of invariance. Instead of taking Y_i to be in $L(G)$, we can take them to be $L(G)$-valued exterior differential forms of degree p_i. If, therefore, Y_i are tensorial differential forms of degree p_i and type $ad(G)$, P is a differential form of degree $\sum_{i=1}^{k} p_i$ defined in the whole manifold X.

We can define the bracket operation $[Z, W]$, where Z, W are $L(G)$-valued exterior differential forms. In fact, let Z^i, W^i be the components of Z, W relative to a base of $L(G)$. The components $\sum_{j,k} c^i_{jk} Z^j \wedge W^k$ define an $L(G)$-valued differential form which is independent of the choice of the base and will be denoted by $[Z, W]$. The degree of $[Z, W]$ is the sum of the degrees of Z and W.

The formula (8) can be generalized to the case that Z, Y_i are differential forms. In particular, when Z is a linear $L(G)$-valued differential form and $p_i = \dim Y_i$, we have

$$\sum_{i=1}^{k} (-1)^{p_1 + \cdots + p_{i-1}} P(Y_1, \ldots, Y_{i-1}\, [Z, Y_i], Y_{i+1}, \ldots, Y_k) = 0. \tag{9}$$

We now suppose our invariant functions P to be symmetric in their arguments and call them for simplicity *invariant polynomials*. We shall make them into a ring. By the definition of addition

$$(P + Q)(Y_1, \ldots, Y_k) = P(Y_1, \ldots, Y_k) + Q(Y_1, \ldots, Y_k), \tag{10}$$

all invariant polynomials of *degree* k form an abelian group. Let $I(G)$ be the direct sum of these abelian groups for all $k \geqq 0$. If P and Q are invariant polynomials of degrees k and l respectively, we define their product PQ to be an invariant polynomial of degree $k + l$ given by

$$(PQ)(Y_1, \ldots, Y_{k+l}) = \frac{1}{N} \sum P(Y_{i_1}, \ldots, Y_{i_k}) Q(Y_{i_{k+1}}, \ldots, Y_{i_{k+l}}), \tag{11}$$

where the summation is extended over all permutations of the vectors Y_i and N is the number of such permutations. This definition of multiplication, together with the distributive law, makes $I(G)$ into a commutative ring, the ring of invariant polynomials of G.

Let P be an invariant polynomial of degree k and let us substitute for its arguments the curvature tensor Θ of the connection. Then $P(\Theta) = P(\Theta, \ldots, \Theta)$ is a differential form in X, of degree $2k$. Its exterior derivative is, by the Bianchi identity (7_2) and by the observation that the absolute derivative of a product follows the same rule as the exterior derivative,

$$dP(\Theta) = DP(\Theta) = 0.$$

Hence $P(\Theta)$ is a closed differential form and defines, according to the de Rham theory, an element of the cohomology ring $H(X)$ of X having as coefficient ring the field of real numbers. We shall denote the resulting mapping by

$$h\colon I(G) \to H(X). \tag{12}$$

It is clear that the ring operations in $I(G)$ are so defined that h is a ring homomorphism. Notice that $I(G)$ depends only on G and that h is defined with the help of a connection in the bundle.

Concerning the homomorphism h the following *theorem of Weil* is fundamental: h is independent of the choice of the connection. In other words, two different connections in the bundle give rise to the same homomorphism h.

We proceed to give Weil's proof of this theorem. Let two connections be given in the same bundle, defined by the differential forms θ_α and θ'_α respectively. Then $u_\alpha = \theta_\alpha - \theta'_\alpha$ is a linear differential form of the type $ad(G)$, and their curvature tensors are related by the formula

$$\Theta' = \Theta - Du - \tfrac{1}{2}[u, u]. \tag{13}$$

The theorem will be proved, if we express $P(\Theta') - P(\Theta)$ as an exterior derivative. For this purpose we introduce the notations

$$P(Y) = P(Y, \ldots, Y), \qquad Q(Z, Y) = P(Z, Y, \ldots, Y), \tag{14}$$

where Y, Z are $L(G)$-valued differential forms. With an auxiliary variable t we put

$$F(t) = P(W - tY - t^2Z) - P(W),$$

Then

$$F'(t) = -kQ(Y + 2tZ, W - tY - t^2Z),$$

and we have

$$P(W - Y - Z) - P(W) = -k \int_0^1 Q(Y + 2tZ, W - tY - t^2Z)\, dt. \tag{15}$$

Making use of (9) and the relations

$$D^2u = -[\Theta, u], \qquad [[u, u], u] = 0, \tag{16}$$

we get the formula

$$dQ\left(u, \Theta - tDu - \frac{t^2}{2}[u, u]\right) = Q\left(Du + t[u, u], \Theta - tDu - \frac{t^2}{2}[u, u]\right). \tag{17}$$

Integrating with respect to t from 0 to 1, we find

$$dR(u, \Theta, Du, \tfrac{1}{2}[u, u]) = P(\Theta') - P(\Theta), \tag{18}$$

where R is defined by

$$R(V, W, Y, Z) = -k \int_0^1 Q(V, W - tY - t^2 Z)\, dt. \tag{19}$$

This proves the theorem.

Another consequence can be derived from the above considerations. We consider the principal bundle B and the $L(G)$-valued linear differential form φ which defines the connection. Recalling the definition of φ in terms of a coordinate neighborhood, we immediately get

$$d\varphi - \tfrac{1}{2}[\varphi, \varphi] = \Phi, \tag{20}$$

where Φ is an $L(G)$-valued quadratic differential form in B, which, in a coordinate neighborhood U_α, has the representation $ad(s)\Theta_\alpha$. Manipulations similar to those in the above proof then give the formula:

$$dR(\varphi, \Phi, d\varphi, -\tfrac{1}{2}[\varphi, \varphi]) = -P(\Phi). \tag{21}$$

Since $P(\Phi)$ is clearly the dual image of the form $P(\Theta)$ in X, it follows that $\psi^*P(\Theta)$ is cohomologous to zero in B. We identify G with a fiber of B and denote the inclusion mapping by $i: G \to B$. Then $i^*R(\varphi, \Phi, d\varphi, -\frac{1}{2}[\varphi, \varphi])$ is closed and defines an element of $H(G)$, which is defined up to an element of $i^*H(B)$. The result is a group homomorphism

$$t: I(G) \to H(G)/i^*H(B) \tag{22}$$

which maps the polynomial P into an element of the quotient group on the right-hand side having as representative the differential form $c_k Q(\omega, d\omega)$, where

$$c_k = -k \int_0^1 (-t + t^2)^{k-1}\, dt,$$

We notice that $i^*\varphi = \omega$.

5. The homomorphism h for the universal bundle

The results of the last section can in particular be applied to the universal bundle

$$R(m + n + 1)/I_m \times R(n + 1) \to R(m + n + 1)/G \times R(n + 1)$$

considered in §1. In this case we denote the base space by X_0 and the homomorphism h by

$$h_0: I(G) \to H(X_0).$$

The purpose of this section is to prove the theorem: In the dimensions $\leqq n$, h_0 is a one-one isomorphism.

X_0 being a homogeneous space, the structure of the ring $H(X_0)$ can be studied by the method of integral invariants of Elie Cartan (Cf. E. Cartan, Sur les invariants intégraux de certains espaces homogènes clos et les propriétés topologiques de ces espaces, Annales de la Soc. Pol. de Math. t. 8, pp. 181–225, 1929). The main ideas and results are as follows: Let G/K be the homogeneous space, G a compact Lie group of dimension r and K a closed subgroup of dimension s of G. The Lie algebra $L(K)$ is then a subalgebra of $L(G)$. In the dual vector space $L^*(G)$ of $L(G)$ there is determined a subspace $M^*(K)$ consisting of all the elements of $L^*(G)$ which are perpendicular to $L(K)$. The adjoint group $ad(G)$ acts on both vector spaces $L(G)$ and $L^*(G)$. Its subgroup $ad(K)$, consisting of the linear endomorphisms $ad(a)$, $a \in K$, leaves $L(K)$ and $M^*(K)$ invariant. If we identify the space $L^*(G)$ with the space of left-invariant linear differential forms (Maurer-Cartan forms), a base $\omega_1, \ldots, \omega_r$ in $L^*(G)$ can be so chosen that $\omega_{s+1}, \ldots, \omega_r$ span $M^*(K)$ and that the system of differential equations

$$\omega_{s+1} = \ldots = \omega_r = 0$$

is completely integrable. Since $ad(K)$ leaves $M^*(K)$ invariant, it induces a group of linear endomorphisms on $M^*(K)$ and on the exterior algebra $\Lambda(M^*(K))$ of $M^*(K)$. Denote by $R(G/K)$ the subring of $\Lambda(M^*(K))$ consisting of all elements invariant under the action of $ad(K)$. Cartan proved that $R(G/K)$ is stable under exterior differentiation and that its cohomology ring (that is, the quotient ring of the subring of closed forms over the ideal of derived forms) is isomorphic to the cohomology ring of the space G/K.

Returning to our problem, consider the rotation group $R(m)$ in which G is imbedded as a subgroup. A positive definite Riemann metric can be defined in $R(m)$, which is invariant under both left and right translations. If we choose the Maurer-Cartan forms of $R(m)$ such that the ds^2 of the Riemann metric is equal to the sum of their squares, the constants of structure will be anti-symmetric in all three indices.

To the base space of our universal bundle we now apply the method of Cartan. We agree on the following ranges of indices:

$$1 \leqq \alpha, \beta, \gamma \leqq m, \qquad m+1 \leqq r, s, t \leqq m+n+1, \qquad 1 \leqq A, B, C \leqq m+n+1.$$
$$1 \leqq \lambda, \mu, \nu \leqq \dim R(m) - \dim G, \qquad 1 \leqq i, j, k \leqq \dim G.$$

Setting $K = G \times R(n+1)$ and remembering that our underlying group is here $R(m+n+1)$, we see that the vector space $M^*(K)$ is apanned by τ_λ, $\omega_{\alpha r}$, where τ_λ span the space $M^*(G)$ in $L^*(R(m))$. For $a \in I_m \times R(n+1)$ the induced endomorphism of $ad(a)$ on $\omega_{\alpha r}$ is given by

$$\omega'_{\alpha r} = \sum_s a_{rs}\omega_{\alpha s},$$

where (a_{rs}) is a proper orthogonal matrix. Now the so-called first main theorem on vector invariants asserts that any integral rational invariant of a system of vectors under the rotation group is an integral rational function of their scalar products and their determinants. It follows that in order that a form of degree $\leqq n$ generated by τ_λ, $\omega_{\alpha r}$ be invariant under $ad(K)$, it must contain $\omega_{\alpha r}$ in the combinations

$$\Omega_{\alpha\beta} = -\sum_r \omega_{\alpha r} \wedge \omega_{\beta r}. \tag{23}$$

By the equations of Maurer-Cartan for $R(m+n+1)$ these forms satisfy the relations

$$d\omega_{\alpha\beta} = \sum_\gamma \omega_{\alpha\gamma} \wedge \omega_{\gamma\beta} + \Omega_{\alpha\beta}.$$

We take in $L^*(R(m))$ a base (θ_i, τ_λ) such that: 1) τ_λ span $M^*(G)$; the invariant Riemann metric ds^2 is equal to the sum of the squares of θ_i, τ_λ, so that the constants of structure are anti-symmetric in all three indices. We then have

$$d\theta_i = \tfrac{1}{2}\sum_{j,k} c_{ijk}\theta_j \wedge \theta_k + \tfrac{1}{2}\sum_{\mu,\nu} c_{i\mu\nu}\tau_\mu \wedge \tau_\nu + \Theta'_i,$$
$$d\tau_\lambda = \tfrac{1}{2}\sum_{\mu,i} c_{\lambda\mu i}\tau_\mu \wedge \theta_i + \tfrac{1}{2}\sum_{\mu,\nu} c_{\lambda\mu\nu}\tau_\mu \wedge \tau_\nu + T'_\lambda,$$

where (θ_i, τ_λ) are related to the set $(\omega_{\alpha\beta})$ by a non-singular linear transformation. By introducing the forms

(24) $$\Theta_i = \tfrac{1}{2}\sum_{\mu,\nu} c_{i\mu\nu}\tau_\mu \wedge \tau_\nu + \Theta'_i, \qquad T_\lambda = \tfrac{1}{2}\sum_{\mu,\nu} c_{\lambda\mu\nu}\tau_\mu \wedge \tau_\nu + T'_\lambda,$$

we can write

(25) $$d\theta_i = \tfrac{1}{2}\sum_{j,k} c_{ijk}\theta_j \wedge \theta_k + \Theta_i,$$
$$d\tau_\lambda = \tfrac{1}{2}\sum_{\mu,i} c_{\lambda\mu i}\tau_\mu \wedge \theta_i + T_\lambda.$$

The ring of invariant forms is generated by 1, τ_λ, Θ_i, T_λ.

The forms θ_i are in $R(m+n+1)$, but can be regarded to be in the bundle $R(m+n+1)/I_m \times R(n+1)$, because there are uniquely determined forms in the latter of which they are the dual images under the natural projection. Regarded as in the bundle, these forms θ_i define a connection, of which Θ_i is the curvature tensor. Relative to this connection an absolute differentiation is defined in $M^*(G)$ in which $ad(G)$ acts. Since τ_λ can be regarded as an $M^*(G)$-valued linear differential form in X_0, it has an absolute derivative which is a quadratic differential form of the same type. We find

(26) $$D\tau_\lambda = T_\lambda.$$

To describe the ring of invariant forms generated by 1, τ_λ, Θ_i, T_λ, we consider real-valued multilinear functions

$$P(Y_1, \ldots, Y_k; Z_1, \ldots, Z_p; W_1, \ldots, W_q),\ Y_a, Z_b \in M^*(K),\ W_c \in L^*(G),$$

which are anti-symmetric in the Y's, and symmetric in the Z's and in the W's, and are invariant under the action of $ad(G)$:

(27) $$P(ad(a)Y_1, \ldots, ad(a)Y_k; ad(a)Z_1, \ldots, ad(a)Z_p; ad(a)W_1, \ldots, ad(a)W_q)$$
$$= P(Y_1, \ldots, Y_k; Z_1, \ldots, Z_p; W_1, \ldots, W_q).$$

An invariant form $P(\tau, T, \Theta)$ is obtained, by substituting τ for each Y, T for each Z, and Θ for each W. The degree of the form is $k + 2(p+q)$ and the form itself is said to be of type (k, p, q). As in the case of the ring of invariant polynomials these functions P can be made into a ring by the definition of an addition and a multiplication in such a way that one gets a ring homomorphism under the substitution described above. Denote by R the ring of differential forms $P(\tau, T, \Theta)$. If D denotes the absolute differentiation relative to the connection defined above, we have, for $P \in R$,

(28) $$dP(\tau, T, \Theta) = DP(\tau, T, \Theta).$$

It follows that R is stable under D and that the derived ring relative to D is isomorphic in the dimensions $\leqq n$ to the cohomology ring $H(X_0)$.

Under D we have

$$D\Theta = 0, \quad DT = 0, \tag{29}$$

the first being the Bianchi identity and the second following from a similar calculation. We shall prove: For $P \in R$ with $DP = 0$ there exist Q, $P_1 \in R$ such that

$$P(\tau, T, \Theta) = DQ + P_1(\Theta),$$

where P_1 is of type $(0, 0, q')$.

The proof follows an idea of algebraic topology in the construction of a homotopy operator. We define an operator f in R which is an anti-derivation (that is, a differential operator with $f(ab) = f(a)b + (-1)^r af(b)$, $r = \dim(a)$ and is such that

$$f\tau = 0, \qquad fT = \tau, \qquad f\Theta = 0.$$

Then we have

$$(Df + fD)T = \frac{1}{k+p}T, \qquad (Df + fD)\tau = \frac{\tau}{k+p}, \qquad (Df + fD)\Theta = 0.$$

Now P is a sum of terms of types (k, p, q). We can assume P to be homogeneous in the sense that $k + 2(p + q) = \text{const}$. Among the terms of P let m be the largest value for $k + p$. Then the value of $k + p$ in each term of

$$P - \frac{1}{k+p}(fD + Df)P$$

is smaller than m. Since $DP = 0$, we prove the above statement by induction on m.

It follows that every class of $H(X_0)$ contains as a representative a differential form $P(\Theta)$ of the type $(0, 0, q')$. Since the latter is clearly never a derived form in R relative to D, the mapping so established is a one-one isomorphism. Thus the theorem stated in the beginning of this section is completely proved.

6. The fundamental theorem

We consider a fiber bundle with a Lie group as structural group and having as base space a differentiable manifold. A connection is supposed to be defined in the bundle. The problem on the relationship between properties of the connection and topological properties of the bundle can be described as follows:

Let G_1 be a maximal compact subgroup of the structural group G. Since an invariant polynomial under G is an invariant polynomial under G_1, there is a natural homomorphism

$$\sigma\colon I(G) \to I(G_1). \tag{30}$$

We have also defined a homomorphism

$$h\colon I(G) \to H(X),$$

which, according to the theorem of Weil, is independent of the choice of the connection. On the other hand, the bundle defines a characteristic homomorphism

$$h': H(X_0) \to H(X).$$

By the theorem of the last section $H(X_0)$ is isomorphic to $I(G_1)$ in the dimensions $\leq n$. The characteristic homomorphism can therefore be written

$$h': I(G_1) \to H(X).$$

Our fundamental theorem asserts that

$$h = h'\sigma. \tag{31}$$

Let us notice that h is defined by the connection, h' by the topological properties of the bundle, and σ by the relation between the groups G and G_1.

To prove this theorem we observe that the bundle is equivalent to one with the structural group G_1. Define a connection in the bundle with the group G_1. Since $L(G_1)$ is a subalgebra of $L(G)$, the connection can be regarded as relative to the group G. Using this connection, we see that the two sides of (31) are identical.

An important particular case of the theorem is one for which G is itself compact. Then h is identical with the characteristic homomorphism. In particular, it follows that if a connection can be defined in the bundle which is locally flat, then the characteristic ring is zero.

Another consequence of the theorem is that the kernel of σ in $I(G)$ is mapped into zero by h. On the other hand, as we shall see later from examples, σ is not necessarily onto.

The description of the homomorphism σ depends on the study of the relation between the Lie algebras L(G) and $L(G_1)$, and the cohomology structure of these Lie algebras. Their study has recently been successfully carried out by H. Cartan, Chevalley, Koszul, and Weil. (Cf. Koszul, J. L., Homologie et cohomologie des algebres de Lie, Bull. Soc. Math. de France 78, 65–127 (1950))

Chapter IV
Bundles with the Classical Groups as Structural Groups

Fiber bundles which have as structural groups the classical groups, namely the orthogonal group, the rotation group, the unitary group, and the symplectic group, play an important role in problems of geometry. In fact, such bundles include those naturally associated to differentiable manifolds and complex manifolds. When a Riemann metric is given on a differentiable manifold or an Hermitian metric on a complex manifold, the metrics will define intrinsically connections in the tangent bundles, and the determination of the characteristic homomorphism by the connection gives rise to a relationship between the tangent bundle and the metric. Moreover, at least at the present stage, we have a better knowledge of the characteristic homomorphisms of such bundles. This chapter will be devoted to the study of these particular cases.

1. Homology groups of Grassmann manifolds

Let $R(n)$, $O(n)$, $U(n)$ denote respectively the rotation group, the orthogonal group, and the unitary group in n variables. For bundles with these groups as structural groups and with base spaces of dimension $\leqq k$, universal bundles are respectively given by

$$\begin{aligned} &R(n+N)/R(N) \to R(n+N)/R(n) \times R(N), && k \leqq N-1, \\ (1)\qquad &R(n+N)/R(N) \to R(n+N)/R(n+N) \cap (R(n) \times R(N)), && k \leqq N-1, \\ &U(n+N)/U(N) \to U(n+N)/U(n) \times U(N), && k \leqq 2N. \end{aligned}$$

That these are universal bundles follows simply from the vanishing of homotopy groups of dimensions r, $0 \leqq r \leqq k$, of the bundle.

In order to describe the characteristic homomorphism, it is therefore necessary to study the homology properties of the base spaces of these bundles. While this gives rise to a new treatment in the case of real coefficients, it is wider in scope in the sense that more general coefficients can be taken into consideration. Geometrically these base spaces are the so-called Grassmann manifolds and are respectively: 1) the manifold $\tilde{G}(n, N, R)$ of oriented n-dimensional linear spaces through a point in a real Euclidean space $E^{n+N}(R)$ of dimension $n + N$; 2) the manifold $G(n, N, R)$ of non-oriented n-dimensional linear spaces through a point in a real Euclidean space $E^{n+N}(R)$ of dimension $n + N$; 3) the manifold $G(n, N, C)$ of n-dimensional linear spaces through a point in a complex Euclidean space $E^{n+N}(C)$ of dimension $n + N$. The homology groups of the Grassmann manifolds have been studied by Ehresmann who defined cellular decompositions which we proceed to describe.

As the three cases admit a common treatment, we shall adopt the convention to denote the Grassmann manifold by $G(n, N)$ and the Euclidean space by E^{n+N}, when the results in question are valid for all three cases.

Let O be a point of E^{n+N}, the n-dimensional linear spaces through which constitute the Grassmann manifold $G(n, N)$ under consideration. Take through O a sequence of linear spaces

$$L^{a_1+1} \subset L^{a_2+2} \subset \cdots \subset L^{a_n+n}, \qquad 0 \leqq a_1 \leqq \cdots \leqq a_n \leqq N, \tag{2}$$

whose superscripts are the dimensions. The set of all elements X of $G(n, N)$, i.e., n-dimensional linear spaces through O, satisfying the conditions

$$\dim(X \cap L_i^{a_i+i}) \geqq i, \qquad i = 1, \ldots, n, \tag{C_i}$$

is called a *Schubert variety*, to be denoted by $(a_1 \ldots a_n)$. The Schubert varieties have the following properties:

1) $\dim(a_1 \ldots a_n) = \sum_{i=1}^n a_i$.

2) $(a_1 \ldots a_n)$ depends on the choice of the sequence (2). However, relative to the same set of integers $a_1, \ldots, a_n$, with $0 \leqq a_1 \leqq a_2 \leqq \ldots \leqq a_n \leqq N$, the Schubert varieties defined by different sequences (2) are equivalent under the group of motions about O in E^{n+N}.

3) If $a_i = a_{i+1}$, condition (C_i) is a consequence of (C_{i+1}).

4) Define the *open Schubert varieties*:

$$(a_1 \ldots a_n)^* = (a_1 \ldots a_n) - \sum_{\substack{i=1,\ldots,n \\ a_{i-1} < a_i}} (a_1 \ldots a_{i-1} a_i - 1 \ldots a_n). \tag{3}$$

If the sequences used in the definition of the Schubert varieties are chosen from a fixed sequence of linear spaces

$$0 \subset L^1 \subset L^2 \subset \cdots \subset L^{n+N-1} \subset L^{n+N} (= E^{n+N}), \tag{4}$$

the open Schubert varieties form a family of disjoint subsets of $G(n, N)$, which cover $G(n, N)$.

5) An open Schubert variety is an open cell in the case of $G(n, N, R)$ and $G(n, N, C)$ and is the union of two open cells in the case of $\tilde{G}(n, N, R)$.

The properties 1), 2), 3), 4) are easily verified. We shall give a proof of 5). Let $(a_1 \ldots a_n)^*$ be the open Schubert variety, defined by means of the sequence (4). The statement being clear for $n = 1$, we assume its truth for $n - 1$ and give the proof by induction on n. If $a_1 = 0$, we take a hyperplane K through 0 and not containing L^1. For $X \in (a_1 \ldots a_n)^*$ we have $\dim(X \cap K) = n - 1$. Moreover, $X \in (a_1 \ldots a_n)^*$ if and only if $X \cap K \in (a_2 \ldots a_n)'^*$, the latter being defined by means of the sequence in which K intersects the sequence (4). Hence the theorem follows by our induction hypothesis.

Suppose now that $a_1 > 0$. We take a hyperplane M satisfying the conditions: 1) $M \cap L^{a_1+1} = L^{a_1}$; 2) M intersects $L^{a_1+2}, \ldots, L^{n+N-1}$ respectively in the linear spaces

$$L'^{a_1+1}, \ldots, L'^{n+N-2}. \tag{5}$$

Put $X' = X \cap M$, $Y = X \cap L^{a_1+1}$. If $X \in (a_1 \dots a_n)^*$, then

$$dim\ X' = n - 1, \qquad \dim Y = 1,$$

and we have

$$Y \subset L^{a_1+1} - L^{a_1}, \qquad X' \in (a_2 \dots a_n)'^*,$$

the latter being defined by the sequence (5). Conversely, the Y and X' satisfying the last conditions span an $X \in (a_1 \dots a_n)^*$. Since the locus of Y is an a_1-cell, the theorem follows by induction.

Since a Grassmann manifold is an algebraic variety, it follows from general theorems on the covering of algebraic varieties by complexes that it has a simplicial decomposition of which our cellular decomposition in 4) is a consolidation. The additive homology structure can therefore be determined by the boundary relations of our cellular decomposition. To determine the boundary relations we must first orient the open cells. For $G(n, N, C)$ the open cells have a complex structure which determines a natural orientation. Let $(a_1 \dots a_n)$ denote the chain carried by the oriented open cell and also the cochain taking the value one for this chain and the value zero for other chains of the same dimension. Then we have:

6) The boundary relations of the cellular decomposition of $G(n, N, C)$ defined in 4) are

(6) $$\partial(a_1 \dots a_n) = 0, \qquad \delta(a_1 \dots a_n) = 0.$$

It follows that the $(a_1 \dots a_n)$, $0 \leqq a_1 \leqq a_2 \leqq \cdots \leqq a_n \leqq N$, form a homology or cohomology base of $G(n, N, C)$. In particular, all Betti numbers of odd dimension are zero.

Concerning the real Grassmann manifolds we first observe that $\tilde{G}(n, N, R)$ is a covering space of $G(n, N, R)$. An element $X \in (a_1 \dots a_n)^*$ is spanned by the vectors

$$\xi_i = (x_{iA}) \in L^{a_i+i} - L^{a_i+i-1}, \qquad i = 1, \dots, n; \qquad A = 1, \dots, n + N,$$

where x_{iA} are the components of ξ_i in a coordinate system having the linear spaces in the sequence (4) as the coordinate linear spaces. Moreover, we have

$$\Delta = \prod_{i=1}^{n} x_{ia_i+i} \neq 0.$$

In the case of $\tilde{G}(n, N, R)$, $(a_1 \dots a_n)^*$ consists of two open cells, to be denoted by $(a_1 \dots a_n)^+$ and $(a_1 \dots a_n)^-$ and defined respectively by $\Delta > 0$ and $\Delta > 0$. In order to determine the vectors ξ_i uniquely, we also assume

(7)
$$x_{ia_i+i} = 1, \qquad x_{ia_j+j} = 0, \qquad j < i, \qquad i = 1, \dots, n, \qquad j = 1, \dots, n-1 \qquad \text{for } (a_1 \dots a_n)^+,$$

$$x_{ka_k+k} = 1, \qquad x_{na_n+n} = -1, \qquad x_{ia_j+j} = 0, \qquad j < i, \qquad i = 1, \dots, n, \qquad j, k = 1, \dots, n-1 \qquad \text{for } (a_1 \dots a_n)^-.$$

The remaining x's, namely,

$$x_{11}, \dots, x_{1a_1}, x_{21}, , x_{2_1a_1}, x_{2_1a_1+2}, \dots, x_{2_1a_2+1}, x_{31_1}, \dots, x_{3_1a_3+2}, \dots, x_{n1}, \dots, x_{n,a_n+n-1},$$

then form a coordinate system on each of the open cells. We orient the open cell by the above lexicographic order of the coordinates. In this way the cells of $\tilde{G}(n, N, R)$ are oriented. By projecting into $G(n, N, R)$ by the covering mapping, we orient the cells of $G(n, N, R)$. This orientation of the cells has the property that under a covering transformation in $\tilde{G}(n, N, R)$ which transforms an $X \in \tilde{G}(n, N, R)$ into the same linear space with the opposite orientation, the oriented cell $(a_1 \ldots a_n)^+$ goes into the oriented cell $(a_1 \ldots a_n)^-$, and vice versa.

It is clear that the incidence number of two cells is zero, unless they are of the forms $(a_1 \ldots a_i + 1 \ldots a_n)^\pm$ and $(a_1 \ldots a_n)^\pm$. Denoting these cells by E^{m+1} and E^m respectively, we shall determine their incidence number by the following process: Let an element X of E^m be determined by the vectors ξ_i normalized by the conditions (7). Put $\xi_i' = \xi_i + te_{a_i+i}$, where e_{a_i+i} is the $(a_i + i)$th coordinate vector. For $t > 0$, t and the coordinates in E^m provide a system of coordinates in E^{m+1}. When E^{m+1} is oriented by this ordered system of coordinates, it has $-E^m$ as its boundary. Since the relation between the two systems of coordinates is easily determined, we find the incidence number between E^m and E^{n+1}. The result may be summarized by the following theorem:

7) The boundary relations of $\tilde{G}(n, N, R)$ are

$$\begin{aligned}
\partial(a_1 \ldots a_n)^+ &= \sum (-1)^{a_1+\cdots+a_i}[(-1)^{n-i+1}(a_1 \ldots a_i - 1 \ldots a_n)^+ \\
&\qquad + (-1)^{a_i+1}(a_1 \ldots a_i - 1 \ldots a_n)^-], \\
\partial(a_1 \ldots a_n)^- &= \sum (-1)^{a_1+\cdots+a_i}[(-1)^{a_i+1}(a_1 \ldots a_i - 1 \ldots a_n)^+ \\
&\qquad + (-1)^{n-i+1}(a_1 \ldots a_i - 1 \ldots a_n)^-], \\
\delta(a_1 \ldots a_n)^+ &= \sum (-1)^{a_1+\cdots+a_i}[(-1)^{n-i}(a_1 \ldots a_i + 1 \ldots a_n)^+ \\
&\qquad + (-1)^{a_i+1}(a_1 \ldots a_i + 1 \ldots a_n)^-], \\
\delta(a_1 \ldots a_n)^- &= \sum (-1)^{a_1+\cdots+a_i}[(-1)^{a_i+1}(a_1 \ldots a_i + 1 \ldots a_n)^+ \\
&\qquad + (-1)^{n-i}(a_1 \ldots a_i + 1 \ldots a_n)^-],
\end{aligned} \tag{8}$$

The boundary relations of $G(n, N, R)$ are

$$\begin{aligned}
\partial(a_1 \ldots a_n) &= \sum (-1)^{a_1+\cdots+a_i}[(-1)^{n-i+1} + (-1)^{a_i+1}](a_1 \ldots a_i - 1 \ldots a_n), \\
\delta(a_1 \ldots a_n) &= \sum (-1)^{a_1+\cdots+a_i}[(-1)^{n-i} + (-1)^{a_i+1}](a_1 \ldots a_i + 1 \ldots a_n),
\end{aligned} \tag{9}$$

In all these formulas the sums in the right-hand sides are taken for $i = 1, \ldots, n$ such that the symbols have a sense, that is, that the inequalities in (2) are satisfied.

Having the boundary relations, the determination of the additive homology structure is a purely combinatorial problem. For definiteness we shall carry this out for the cohomology groups of dimensions $< N$ of $G(n, N, R)$. We shall call cochains of the type $(a_1 \ldots a_n)$ *elementary cochains*. For such an elementary cochain define two sets of integers b_k, i_k, by the conditions

$$\begin{gathered}
a_1 = \cdots = a_{i_1} < a_{i_1+1} = \cdots = a_{i_1+i_2} < \cdots < a_{i_1+\cdots+i_{s-1}+1} = \cdots = a_{i_1\cdots+i_s}, \\
i_1 + \cdots + i_s = n, \\
b_1 = a_{i_1}, \ldots, b_s = a_{i_1+\cdots+i_s},
\end{gathered} \tag{10}$$

so that

$$(11)\qquad \begin{gathered} 0 \leqq b_1 < b_2 < \cdots < b_s \leqq N, \\ 1 \leqq i_1, \ldots, i_s, \qquad i_1 + \cdots + i_s = n. \end{gathered}$$

Every elementary cochain determines therefore two sequences of non-negative integers $(b_1, \ldots, b_s)$, $(i_1, \ldots, i_s)$, satisfying the relations (11), and is determined by them. According to the nature of these sequences the elementary cochains are classified into three kinds: 1) It is of the first kind if all the b's and i's are even or if $b_2, \ldots, b_s, i_2, \ldots, i_s$ are even and $b_1 = 0$; 2) It is of the second kind if $b_{k+1}, \ldots, b_s, i_{k+1}, \ldots, i_s$ are even and b_k is odd; 3) It is of the third kind if $b_k, \ldots, b_s, i_{k+1}, \ldots, i_s$ are even, $b_k \neq 0$, and i_k is odd. It follows from (9) that an elementary cochain of the first kinds is an integral cocycle. Moreover, if y is an elementary cochain of the second kind, with $b_k < N$, we have

$$\tfrac{1}{2}\delta y = \pm z + \sum \text{elem. cochains of second kind},$$

where z is an elementary cochain of the third kind obtained from y by replacing b_k by $b_k + 1$. This correspondence between y and z is one-one.

As stated above, we restrict ourselves to the study of cochains of dimension $r < N$. From the above remark we have the following consequences: 1) A cochain of dimension r is a linear combination of cochains x^r, y^r, $\frac{1}{2}\delta y^{r-1}$, where x^r is a linear combination of cochains of dimension r of the first kind and y^r, y^{r-1} those of cochains of dimensions r, $r - 1$ respectively of the second kind. A sum $\lambda x^r + \mu y^r + \nu \frac{1}{2}\delta y^{r-1} = 0$ only if $\lambda = \mu = \nu = 0$; 2) y^r is a cocycle only if it is zero.

It follows that an integral cocycle of dimension r is of the form $x + \frac{1}{2}\delta y^{r-1}$ and that the latter is a coboundary only when $x^r = 0$, $y^{r-1} = 0$. We have therefore the theorem:

8) An integral cohomology base of dimension $r < N$ is formed by x^r, $\frac{1}{2}\delta y^{r-1}$, where x^r runs over the elementary cochains of dimension r of the first kind and y^{r-1} those of dimension $r - 1$ of the second kind.

Weak cohomology and cohomology with coefficients in a finite field have an even simpler structure. They are given by the theorem:

9) All elementary cochains $(a_1 \ldots a_n)$ of dimension r are cocycles and form a cohomology base mod 2. All elementary cocycles of the first kind of dimension $r < N$ from a cohomology base with rational coefficients and with coefficients mod $p \geqq 3$.

Since the dimension of an elementary cocycle of the first kind is a multiple of 4, a cocycle of dimension r such that $r \not\equiv 0$ (4), $r < N$, is cohomologous to 0 in rational coefficients or in coefficients mod $p \geqq 3$.

Before concluding this section, it may be of interest to show how a formula of Whitney can be derived from the boundary relations (8). On $\tilde{G}(n, N, R)$ we consider the cochain

$$w^r = (0 \cdots \underbrace{1 \cdots 1}_{r})^+ - (0 \cdots \underbrace{1 \cdots 1}_{r})^-$$

From (8) we get

$$\delta w^r = \{(-1)^r + 1\} w^{r+1}$$

Hence w^r is a cocycle if r is odd and is a cocycle mod 2 if r is even. Moreover, we have

(12) $$w^{2r+1} = \tfrac{1}{2}\delta w^{2r}.$$

When a bundle over X is induced by a mapping $f: X \to \tilde{G}(n, N, R)$, the cohomology classes of f^*w^r are called the Stiefel-Whitney classes. In that context the above formula was first given by Whitney.

If X is an orientable differentiable manifold of dimension n and B is the tangent bundle over X, it can be verified that

(13) $$(f^*w^r)X = \chi(X)$$

is equal to the Euler-Poincaré characteristic of X.

2. Differential forms in Grassmann manifolds

The Grassmann manifolds are compact homogeneous spaces acted on transitively by a compact Lie group. It follows that the method of integral invariants described in §5, Chapter III, is applicable and that it is possible to describe the cohomology classes with real coefficients by invariant differential forms. We shall set up this relationship in this section. Because of the different features of the results we divide the discussions into cases:

Case A. Complex Grassmann manifolds $G(n, N, C)$.

The Maurer-Cartan forms ω_{AB}, $A, B = 1, \ldots, n + N$, of the unitary group $U(n + N)$ are the elements of an Hermitian matrix. We put

(14) $$\Omega_{ik} = \sum_r \omega_{ir}\omega_{rk}, \qquad i, k = 1, \ldots, n, \qquad r = n + 1, \ldots, n + N$$

From these differential forms we construct the following forms:

(15) $$\begin{aligned} \Psi_m &= \sum \delta(p_1 \ldots p_m; q_1 \ldots q_m)\Omega_{p_1q_1} \ldots \Omega_{p_mq_m}, \\ \Phi_m &= \sum \delta^2(p_1 \ldots p_m; q_1 \ldots q_m)\Omega_{p_1q_1} \ldots \Omega_{p_mq_m}, \\ \Delta_m &= \sum \Omega_{p_1p_2}\Omega_{p_2p_3} \ldots \Omega_{p_mp_1}, \end{aligned}$$

where $p_1, \ldots, p_m$ is a permutation of $q_1, \ldots, q_m$, the summation is over all such permutations and all $p_1, \ldots, p_m = 1, \ldots, n$, and $\delta(p_1, \ldots, p_m; q_1, \ldots, q_m)$ is the Kronecker index, equal to $+1$ or -1 according as the permutation is even or odd. These can be regarded as differential forms in $G(n, N, C)$ in the sense that there are uniquely determined forms in $G(n, N, C)$ of which they are the dual images under the natural projection $U(n + N) \to G(n, N, C)$. It is clear that each of the three sets of forms in (15) can be expressed as polynomials of the forms of another, with numerical coefficients. A simple direct computation shows that the forms Δ_m are closed, so that the same is true of the forms Φ_m and Ψ_m. Moreover, by the first main theorem on vector invariants for the unitary group, it follows that every invariant form in $G(n, N, C)$ is a polynomial in the forms of one of the three sets, with numerical coefficients.

An invariant differential form Ω of degree r in $G(n, N, C)$ defines a cochain γ according to the equation $\gamma . z = \int_z \Omega$, $z =$ any r-dim cycle. γ is a cocycle if Ω is closed. We say that Ω belongs to the cohomology class of γ. Of interest is therefore the question of deciding the cohomology class to which a given closed form belongs. Concerning this we have the following theorem:

Theorem 1. *The form* $\frac{1}{(2\pi\sqrt{-1})^m m!}\Psi_m$ *belongs to the class* $(0\ldots0\ 1\ldots1)$. *The form* $\frac{1}{(2\pi\sqrt{-1})^m m!}\Phi_m$ *belongs to the class* $(0\ldots0m)$.

To prove this theorem we shall integrate the forms over the Schubert varieties. We introduce, for a Schubert variety S, the integers b_k, i_k, defined in (10), (11), so that its dimension is $2m = 2(b_1 i_1 + \cdots + b_s i_s)$. If X belongs to the corresponding open Schubert variety, we define a linear space of dimension $i_1 + i_2 + \cdots + i_k$ by

$$Y^{i_1+\cdots+i_k} = X \cap E^{i_1+\cdots+i_k+b_k}, \qquad k = 1, \ldots, s.$$

Then we have $Y^{i_1+\cdots+i_k} \subset E^{i_1+\cdots+i_k+b_k}$, $k = 1, \ldots, s$, and

$$Y^{i_1} \subset Y^{i_1+i_2} \subset \cdots \subset Y^{i_1+\cdots+i_s} = X.$$

Define the vectors $e_1, \ldots, e_{i_1+\cdots+i_s}, f_1, \ldots, f_{b_1+\cdots+b_s}$, which form an orthonormal system and which are such that: 1) $e_1, \ldots, e_{i_1+\cdots+i_k}$ span $Y^{i_1+\cdots+i_k}$; 2) $e_1, \ldots, e_{i_1+\cdots+i_k}$, $f_{b_1+\cdots+b_{k-1}+1}, \ldots, f_{b_1+\cdots+b_k}$ span $E^{i_1+\cdots+i_k+b_k}$, $k = 1, \ldots, s$.

We now notice that the group $U(n + N)$ can be identified with the space of orthonormal systems of vectors or frames, each element to the frame to which it carries the coordinate frame. If we denote the frame by $e_1, \ldots, e_n, f_1, \ldots, f_N$, we have

$$\omega_{ir} = de_i \cdot \bar{f}_{r-n}, \qquad i = 1, \ldots, n, \qquad r = n + 1, \ldots, n + N \tag{16}$$

where the product in the right-hand side is the scalar product.

To integrate a differential form over S we shall find what it reduces on S, under the above choice of vectors. Every polynomial of degree m in Ω_{ij} reduces on S to a multiple of the form

$$\Pi(\Pi(de_\rho \cdot \bar{f}_\sigma)(d\bar{e}_\rho \cdot f_\sigma))$$

$$k = 1, \ldots, s \qquad \rho = i_1 + \cdots + i_{k-1} + 1, \ldots, i_1 + \cdots + i_k \tag{17}$$

$$\sigma = b_1 + \cdots + b_{k-1} + 1, \ldots, b_1 + \cdots + b_k$$

It follows that the form

$$\Omega_{p_1 q_1} \Omega_{p_2 q_2} \cdots \Omega_{p_m q_m}$$

reduces to a non-zero form on S only when the set $p_1, \ldots, p_m$ contains $b_k i_k$ indices among $i_1 + i_2 + \cdots + i_{k-1} + 1, \ldots, i_1 + i_2 + \cdots + i_k$, $k = 1, \ldots, s$.

We apply this criterion to the forms Ψ_m, Φ_m defined in (15). In the case of Ψ_m, it is clear that the summation can be restricted to the indices $p_1, \ldots, p_m$, which are mutually distinct. Hence its integral over S will be non-zero, only when S has the symbol $(0\ldots0\ 1\ldots1)$.

As for the forms Φ_m, we first observe that $\Phi_m = 0$, if $m > N$. In fact, in this case, Φ_m is a sum of terms as

$$\sum_{\substack{(p_1 \ldots p_m)=(q_1 \ldots q_m) \\ p_1, \ldots, p_m = 1, \ldots, n}} \omega_{p_1 r}\omega_{rq_1}\omega_{p_2 r}\omega_{rq_2} \ldots \omega_{p_m s}\omega_{sq_m},$$

where the index r occurs at least twice and is not summed. The sum changes its sign, when ω_{rq_1} and ω_{rq_2} are interchanged. Since the summation is taken over all $p_1, \ldots, p_m = 1, \ldots, n$, we have $\Phi_m = 0$, for $m > N$.

We shall next prove that

$$\int_{(a_1 \ldots a_n)} \Phi_m = 0, \quad \sum_{i=1}^{n} a_i = m,$$

if $a_{n-1} \neq 0$. This will be done by induction on N. For $N = 1$ it is easily verified. Suppose the statement be true for $N - 1$. If $a_n < N$, S lies in a complex Euclidean space of dimension $n + N - 1$ and the result follows from our induction hypothesis. If $a_n = N$, we have $m > N$ and the result follows from the lemma of the last paragraph. Thus it follows that the integral of Φ_m over S is non-zero only when S has the symbol $(0 \ldots 0\ m)$.

It remains to evaluate these integrals. The Schubert variety $(0 \ldots 0 \underbrace{1 \ldots 1}_{m})$ consists by definition of all n-dimensional planes through a fixed E^{n-m} and belonging to E^{n+1}. Let $e_1, \ldots, e_{n+1}$ be a frame in E^{n+1} such that $e_1, \ldots, e_{n-m}$ span E^{n-m} and that $e_1, \ldots, e_n$ span X, a general element of S. Then Ψ_m reduces on S to the form

$$(-1)^m (m!)^2 \prod_{r=n-m+1, \ldots, n} \omega_{n+1,r}\omega_{r,n+1}.$$

The latter may be considered as the measure of all lines lying in the space E^{m+1} in E^{n+1} orthogonal to E^{n-m}. Similarly, the Schubert variety $(0 \ldots 0\ m)$ consists of all X containing a fixed E^{n-1} and contained in a fixed E^{n+m}. On this the form Φ_m reduces to

$$(m!)^2 \prod_{r=n+1, \ldots, m+n} \omega_{nr}\omega_{rn}$$

In each case the evaluation of the integral in question reduces to the determination of the volume of $G(1, m, C)$.

Let E^{m+1} be the complex Euclidean space such that $G(1, m, C)$ is the manifold of all lines through the origin O of E^{m+1}. A vector in E^{m+1} can be written as $v = v' + \sqrt{-1}\, v''$, where v' and v'' are real vectors. To v we can therefore associate in a real Euclidean space R of dimension $2m + 2$ the vectors

$$f(v) = (v', v''), \qquad g(v) = (-v'', v').$$

The relation between the scalar products in E^{m+1} and R is given by

$$\begin{aligned} v\bar{w} &= f(v)f(w) - \sqrt{-1} f(w)g(v) \\ &= g(v)g(w) + \sqrt{-1} g(w)f(v). \end{aligned}$$

It follows that the vectors $f(e_1)$, $g(e_i)$, $i = 1, \ldots, m + 1$, associated to the vectors e_i of a frame in E^{m+1} form a frame in the real Euclidean space R.

As the volume element in $G(1, m, C)$ we can take

$$\Lambda = \prod_{i=1, \ldots, m} \omega_{m+1,i}\omega_{i,m+1}.$$

Since

$$\omega_{m+1,i}\omega_{i,m+1} = 2\sqrt{-1}(df(e_{m+1})\cdot f(e_i))(df(e_{m+1})\cdot g(e_i)),$$

we find by substitution

$$\Lambda = (2\sqrt{-1})^m \prod_{i=1,\ldots,m} (df(e_{m+1})\cdot f(e_i))(df(e_{m+1})\cdot g(e_i)).$$

To integrate this over $G(1, m, C)$ we consider the unit hypersphere S^{2m+1} described by the end-points of the vectors v satisfying $v\bar{v} = 1$. $G(1, m, C)$ is obtained by identifying all v which differ from each other by a scalar factor of absolute value 1. In other words, S^{2m+1} can be fibered by circles with $G(1, m, C)$ as base space. It was shown in Chapter I that the volume element of S^{2m+1} is

$$\prod_{i=1,\ldots,m} (df(e_{m+1})\cdot f(e_i))(df(e_{m+1})\cdot g(e_i))\cdot(df(e_{m+1})\cdot g(e_{m+1})),$$

if we identify v with e_{m+1}.

To evaluate its integral over S^{2m+1} we fix a point of $G(1, m, C)$ and integrate over the fiber, giving

$$V(S^{2m+1}) = \frac{2\pi}{(2\sqrt{-1})^m}\int_{G(1,m,C)} \Lambda.$$

Since the volume of S^{2m+1} is known to be

$$V(S^{2m+1}) = \frac{2\pi^{m+1}}{m!},$$

we get

$$\int_{G(1,m,C)} \Lambda = \frac{(2\pi\sqrt{-1})^m}{m!}.$$

This completely proves our Theorem 1.

Case B. Real Grassmann manifolds $G(n, N, R)$.

As we shall show in the next section, a multiplicative base of the cohomology ring up to dimensions $\leq N$, will be formed by each of the sets of cohomology classes having the symbols

$$(18)\qquad \begin{aligned} P^{4k} &= (0\ldots0\,2\ldots2), \qquad k = 1, \ldots, [\tfrac{1}{2}n]. \\ \bar{P}^{4k} &= (0\ldots0\,2k\,2k), \qquad k = 1, \ldots, [\tfrac{1}{2}N], \end{aligned}$$

The P^{4k} will be called the Pontrjagin classes and their images in the base space of a sphere bundle the Pontrjagin characteristic classes. As in the complex case we take the Maurer-Cartan forms of the orthogonal group $O(n + N)$, which are elements of a skew-symmetric matrix: $(\tilde{\omega}_{AB})$. We put

$$(19)\qquad \tilde{\Omega}_{ij} = \sum_A \tilde{\omega}_{iA}\tilde{\omega}_{Aj}, \qquad i, j = 1, \ldots, n,$$

and from these we construct the forms

$$\tilde{\Psi}_{2m} = \sum \delta(p_1 \ldots p_m; q_1 \ldots q_m)\tilde{\Omega}_{p_1 q_1} \ldots \tilde{\Omega}_{p_m q_m},$$
(20)
$$\tilde{\Phi}_{2m} = \sum \delta^2(p_1 \ldots p_m; q_1 \ldots q_m)\tilde{\Omega}_{p_1 q_1} \ldots \tilde{\Omega}_{p_m q_m},$$

where the summation convention is as above. These forms can be regarded as differential forms in $G(n, N, R)$ and are invariant under the action of $G(n, N, R)$ by the orthogonal group $O(n + N)$. Because of the skew-symmetry of $\tilde{\Omega}_{ij}$ in its two indices, it can be seen that $\tilde{\Psi}_{2m}$, $\tilde{\Phi}_{2m}$ are zero, when m is odd. The relation between these forms and the cohomology classes of $G(n, N, R)$ is given by the following theorem:

Theorem 2. *The form* $\frac{1}{(2\pi)^{2k}(2k)!}\tilde{\Psi}_{4k}$ *belongs to the class* P^{4k} *and the form* $\frac{1}{(2\pi)^{2k}(2k)!}\tilde{\Phi}_{4k}$ *to the class* $\bar{P}^{4k}$.

It is of course possible to prove this theorem by a process similar to the one used in the complex case. However, a simpler and more conceptual procedure would be to reduce the proof to the complex case.

We assume the real Euclidean space of dimension $n + N$ to be imbedded in the complex Euclidean space of the same (complex) dimension:

$$i: \tilde{E}^{n+N} \to E^{n+N}.$$

This induces a mapping

$$f: G(n, N, R) \to G(n, N, C).$$

Denote by C^{2m} and $\bar{C}^{2m}$ respectively the classes having the symbols $(0, \ldots, 0\underbrace{1 \ldots 1}_{m})$ and $(0 \ldots 0\ m)$. Then we have the lemma:

Lemma. Under the mapping

$$f^*C^{2m} = \begin{cases} 0, & m \text{ odd}, \\ (-1)^m P^{2m}, & m \text{ even}, \end{cases}$$
(21)
$$f^*\bar{C}^{2m} = \begin{cases} 0, & m \text{ odd} \\ (-1)^m \bar{P}^{2m}, & m \text{ even} \end{cases}$$

Before proceeding to the proof, we remark that the Schubert varieties C^{2m} and $\bar{C}^{2m}$, being complex manifolds, have orientations defined by their complex structure, so that the real cohomology classes with the same symbols are well defined. The statements for m odd are trivial, because the non-zero classes of $G(n, N, R)$ have as dimensions multiples of 4. We shall give the proof for $f^*\bar{C}^{2m}$, m even, the proof concerning the expression for $f^*\bar{C}^{2m}$, m even, being similar.

Between two Schubert varieties of complementary dimensions on $G(n, N, C)$ we can define their intersection number or Kronecker index. It can be proved that

$$KI((b_1 \ldots b_n), (N - a_n, \ldots, N - a_1)) = 0 \text{ or } 1,$$

according as $(b_1 \ldots b_n)$ is distinct from $(a_1 \ldots a_n)$ or not. (Cf. next section.) Denote by S the Schubert variety having the symbol

$$(\underbrace{N-1\ldots N-1}_{2k} \quad N\ldots N).$$

It reduces to show that

$$KI(S, f(a_1 \ldots a_n)) = 0, \sum_{i=1}^{n} a_i = 4k,$$

unless $(a_1 \ldots a_n) = P^{4k}$, and that in the latter case the Kronecker index is $(-1)^k$.

The first statement is easy to prove. In fact, we have, by definition,

$$S = \{X | \text{comp. dim.}(X \cap E^{N+2k-1}) \geqq 2k\}.$$

If $(a_1 \ldots a_n) \neq P^{4k}$, the fact that the a's are even implies that $a_{n-2k+1} = 0$. If Y is an element of $G(n, N, R)$ belonging to the corresponding Schubert variety, Y must contain a fixed linear space of dimension $n - 2k + 1$, and $f(Y)$ contains a fixed linear space of complex dimension $n - 2k + 1$. When the latter is chosen to be in general position with E^{N+2k-1}, they will have only the origin in common, by dimension considerations. It follows that S and $f((a_1 \ldots a_n))$ are then set-theoretically disjoint and hence that their intersection number is zero.

To prove the second statement we shall determine the set-theoretical intersection of S and $f(P^{4k})$. The latter consists of all n-dimensional linear spaces Y satisfying

$$\tilde{E}^{n-2k} \subset Y \subset \tilde{E}^{n+2},$$

where $\tilde{E}^{n-2k}$ and $\tilde{E}^{n+2}$ are fixed linear spaces of real dimensions $n - 2k$ and $n + 2$ respectively. Let $\tilde{E}^{2k+2}$ be the linear space orthogonal to $\tilde{E}^{n-2k}$ in $\tilde{E}^{n+2}$. Y is then determined by a linear space ξ of real dimension $2k$ in $\tilde{E}^{2k+2}$. Denote by E^{n-2k}, E^{n+2}, E^{2k+2} the complex linear spaces determined by $\tilde{E}^{n-2k}$, $\tilde{E}^{n+2}$, $\tilde{E}^{2k+2}$ respectively. Since E^{N+2k-1} is in general position with them, we have

$$E^{n-2k} \cap E^{N+2k-1} = 0,$$

$$E^{n+2} \cap E^{N+2k-1} = L^{2k+1}, \text{ say.}$$

In order that $f(Y)$ belongs to S, ξ must belong to $L^{2k+1} \cap \tilde{E}^{2k+2}$. But L^{2k+1} and $\tilde{E}^{2k+2}$, both belonging to $\tilde{E}^{2k+2}$, have exactly one linear space of real dimension $2k$ in common. It follows that S and $f(P^{4k})$ have exactly one element in common.

By using a local coordinate system, whose details we shall not give here, we verify that the intersection number is actually equal to $(-1)^k$. This completes the proof of the lemma.

We now observe that under the mapping f we have

$$f^*\Omega_{ij} = \tilde{\Omega}_{ij}.$$

Hence Theorem 2 is an immediate consequence of Theorem 1 and the above lemma.

Remark. The mapping f induces a unitary bundle over $G(n, N, R)$. It is equivalent to the Whitney product of the universal bundle over $G(n, N, R)$ with itself. For if $e_1, \ldots, e_n$ are n vectors in an element X of $G(n, N, R)$, then $f(X)$ will be spanned by e_i, $\sqrt{-1}\, e_i$, $i = 1, \ldots, n$. Thus $f(X)$ can be considered as the vector space spanned by two copies of X, with the orthogonal group acting coherently.

Since a sphere bundle can always be induced by mapping its base space into $G(n, N, R)$, this relationship is valid for a general sphere bundle. We express it by saying that the Whitney square of a sphere bundle has an almost complex structure. The above lemma then gives the relationship between the real characteristic ring of a sphere bundle with that of the almost complex structure of its Whitney square. This result is useful in differential geometry, where the primary concern is the real characteristic ring.

3. Multiplicative properties of the cohomology ring of a Grassmann manifold

Since $G(n, N, C)$ has no torsion, it is sufficient to determine the multiplication in the real cohomology ring. For both $G(n, N, C)$ and $G(n, N, R)$, when the coefficient ring is the real field, the cohomology classes can be described by differential forms and their multiplication by the exterior multiplication of the latter. According to the general theory described in §5, Chapter III, these differential forms can be supposed to be invariant under the actions of the unitary group and the rotation group respectively. For definiteness of description take the case of $G(n, N, C)$. Applying the so-called first main theorem on vector invariants under the unitary group (Weyl, Classical Groups, p. 45) and using an argument of §5, Chap. III, it follows that an invariant differential form of degree $\leqq 2N$ is a polynomial of Δ_m, $m = 1, \ldots, N$, and hence a polynomial of Ψ_m, $m = 1, \ldots, N$, with constant coefficients. Since all these differential forms are of even degree, it follows from Cartan's theorem that they are closed and are not cohomologous to zero unless identically equal to zero.

The case of $G(n, N, R)$ can be described in a similar manner. Here we restrict ourselves to classes of dimension $< N$. Every invariant differential form of degree $< N$ is a polynomial of $\tilde{\Psi}_{2m}$ or of $\tilde{\Phi}_{2m}$ with constant coefficients. Such a form has as degree a multiple of 4. For $\tilde{G}(n, N, R)$ there exists for even n the further invariant form

$$\sum_i \varepsilon_{i_1 \ldots i_n} \tilde{\Omega}_{i_2 i_2} \ldots \tilde{\Omega}_{i_{n-1} i_n},$$

corresponding to the determinant in vector invariants.

For coefficient rings other than the real field, more topological methods have to be employed to describe the multiplicative structure of the cohomology rings. We shall illustrate this method by determining the cohomology ring of $G(n, N, R)$ mod 2.

In this case every cochain $(a_1 \ldots a_n)$ is a cocycle and all these symbols form an independent cohomology base. We shall use the convention of omitting the zeros of such a symbol, thus defining

$$(a_{i+1} \ldots a_n) = (0 \ldots 0\, a_{i+1} \ldots a_n) \tag{23a}$$

and

$$(0) = 1, (c) = 0, \text{ if } c < 0. \tag{23b}$$

Using the same symbols to denote also the cohomology classes, the main multiplication formula to be proved is

$$(a_1 \ldots a_n) \cup (h) = \sum (b_1 \ldots b_n), \tag{24}$$

where the summation is extended over all combinations $b_1, \ldots, b_n$, such that

$$
\begin{gathered}
0 \leqq b_1 \leqq b_2 \leqq \cdots \leqq b_n \leqq N, \\
a_i \leqq b_i \leqq a_{i+1} \qquad (a_{n+1} = N), \quad i = 1, \ldots, n, \\
\sum_{i=1}^{n} a_i + h = \sum_{i=1}^{n} b_i.
\end{gathered} \tag{25}
$$

Formula (24) implies the following multiplication formula:

$$
(a_1 \ldots a_n) = \begin{vmatrix} (a_1) & (a_1 - 1) & \ldots & (a_1 - \overline{n-1}) \\ (a_2 + 1) & (a_2) & \ldots & (a_2 - \overline{n-2}) \\ & \ldots\ldots & & \\ (a_n + \overline{n-1}) & (a_n + \overline{n-2}) & \ldots & (a_n) \end{vmatrix} \tag{26}
$$

where the right-hand side is to be expanded by the Laplace development with cup product as multiplication. This is easily proved by induction on n. In fact, assuming the truth of the formula for $n - 1$ and expanding the determinant according to the first column, we see that the determinant is equal to

$$
\sum_{i=1}^{n} (a_i + i - 1) \cup (a_1 - 1 \ldots a_{i-1} - 1 \quad a_{i+1} \ldots a_n).
$$

Using (24), it can be seen that the sum is equal to the left-hand side of (26).

Since the multiplication by cup product is associative, formulas (24) and (26) together give the multiplication of any two cohomology classes.

We shall prove the formula (24) by establishing a corresponding formula on intersection numbers. Consider first the cycles

$$
(b_1 \ldots b_n), \qquad (N - a_n, \ldots, N - a_1). \tag{27}
$$

Suppose that they are defined by two sequences of linear spaces in general position:

$$
\begin{gathered}
E_1 \subset E_2 \subset \cdots \subset E_n, \\
F_1 \subset F_2 \subset \cdots \subset F_n,
\end{gathered} \tag{28}
$$

whose dimensions are given by

$$
\begin{gathered}
\dim E_i = b_i + i, \\
\dim F_{n-i+1} = N - a_i + n - i + 1, \qquad i = 1, \ldots, n.
\end{gathered}
$$

Since these two intersections both belong to X, they intersect in a linear space of dimension $\geqq 1$. It follows that the same is true of E_i and F_{n-i+1}. Hence we have

$$
(b_i + i) + (N - a_i + n - i + 1) \geqq N + n + 1,
$$

or $b_i \geqq a_i$.

On the other hand, by making use of a coordinate system, we can arrange that $(a_1 \ldots a_n)$ and $(N - a_n \ldots N - a_1)$ intersect in exactly one element and have there the intersection number 1. It follows that when $(b_1 \ldots b_n)$ and $(N - a_n \ldots N - a_1)$ are of complementary dimensions, their intersection number is 1 or 0 according as $(a_1 \ldots a_n)$ is equal to $(b_1 \ldots b_n)$ or not.

Using this result, it is seen that formula (24) is equivalent to the statement that the intersection number of the three cycles

$$(29) \qquad (b_1 \ldots b_n), \qquad (N - a_n \ldots N - a_1), \qquad (N - h\, N \ldots N)$$

is one if the conditions (25) are satisfied and is otherwise zero. We have shown in the above that in order the intersection number be non-zero it is necessary that $a_i \leqq b_i$.

Suppose that these necessary conditions are satisfied. We put

$$M_i = E_i \cap F_{n-i+1}, \qquad P_i = E_i \cap F_{n-i}, \qquad i = 1, \ldots, n,$$

and let M be the space spanned by $M_1, \ldots, M_n$. Since $\dim M_i = b_i - a_i + 1$, we have $\dim M \leqq h + n$, and this dimension is equal to $h + n$, if any two distinct M_i have only the point 0 in common. An element X belonging to the intersection of the first two cycles of (29) meets each M_i in a linear space of dimension $\geqq 1$. Such an X must therefore belong to M. It follows that in order that the three cycles in (29) have a non-empty intersection we must have $\dim M = h + n$. A necessary condition for this is $M_i \cap M_{i+1} = 0$. Since both contain P_i, this condition implies $P_i = 0$. But $\dim P_i = b_{i+1}$, so that we have $b_i \leqq a_{i+1}$.

It remains to show that when the conditions (25) are satisfied, the three cycles in (29) have the intersection number one. In fact, it can be arranged that they have exactly one element in common and we verify the intersection to be simple. For details, compare S. Chern, On the multiplication in the characteristic ring of a sphere bundle, Annals of Math. 49, 362–372 (1948). We remark that the proof given above is a slight simplication of the one given in this paper.

From the above multiplication formulas we can draw a number of consequences. The coefficient ring is always the field mod 2.

1) We introduce the classes

$$(30) \qquad \begin{aligned} w^i &= (0 \ldots 0 \underbrace{1 \ldots 1}_{i}) \qquad 1 \leqq i \leqq n, \\ \bar{w}^k &= (k), \qquad 1 \leqq k \leqq N, \\ w^0 &= \bar{w}^0 = 1. \end{aligned}$$

Applying (24) we get the formula

$$(31) \qquad \sum_{0, r-N \leqq i \leqq n} w^i \cup \bar{w}^{r-i} = 0, \qquad r > 0.$$

This permits us to express the w's in terms of the $\bar{w}$'s, and vice versa.

2) Every cohomology class of $G(n, N, R)$ is a polynomial in $\bar{w}^k$, $k = 0, \ldots, N$, and is a polynomial in w^i, $i = 0, \ldots, n$.

This follows from (26) and (30).

3) There is a natural homeomorphism

$$f\colon G(n, N, R) \to G(N, n, R),$$

under which an n-dimensional linear space is mapped into its N-dimensional orthogonal space. Denote the classes of $G(n, N, R)$ by the same symbols with dashes. The dual homomorphism induced by f is given by

$$(32)\qquad \begin{aligned} f^*\overline{w}'^i &= w^i, \qquad i = 1, \ldots, n, \\ f^*w'^k &= \overline{w}^k, \qquad \mathrm{k} = 1, \ldots, N. \end{aligned}$$

To prove this, we show that both sides have the same value over any homology class, an argument which has been applied several times before. We shall omit the details here.

4) There is no non-trivial relation between the w^i, $i = 0, \ldots, n$ or the $\overline{w}^k$, $k = 0, \ldots, N$. When the classes $\overline{w}^k$ are concerned, we see this from (26). In fact, let $F(\overline{w})$ be a polynomial in $\overline{w}^k$, which is a sum of terms of the form

$$\overline{w}^{i_1}\overline{w}^{i_2}\ldots\overline{w}^{i_r}, \qquad i_1 + \cdots + i_r = d, \text{ say}.$$

We introduce an ordering of such terms by defining

$$\overline{w}^{i_1}\ldots\overline{w}^{i_r} < \overline{w}^{k_1}\ldots\overline{w}^{k_s},$$

if $r < s$ or $r = s$, $i_1 = k_1, \ldots, i_t = k_t, i_{t+1} < k_{t+1}$. Relative to this ordering let $\overline{w}^{m_1}\ldots\overline{w}^{m_r}$ be the largest term with non-zero coefficient in $F(\overline{w})$. We now carry out the multiplication by writing $F(\overline{w})$ as a sum of Schubert symbols. From the form of (26) it is observed that the expansion contains the term $(m_1 \ldots m_r)$ with non-zero coefficient. Since the Schubert symbols are homologically independent, it follows that the class $F(\overline{w})$ is not zero, unless the polynomial is identically zero.

The statement concerning w^i follows from a consideration of $G(N, n, R)$.

4. Some applications

It goes without saying that our interest in Grassmann manifolds lies in the fact that their study leads to a description of the characteristic homomorphism of a general sphere bundle or of a general differentiable manifold. A sphere bundle (B, X), whose base space is of dimension $< N$, can always be induced by a mapping

$$f: X \to G(n, N, R),$$

which is determined up to a homotopy. For any cohomology class γ of $G(n, N, R)$ the class $f^*\gamma$ is therefore an invariant of the bundle. The geometric interpretation of such invariants was given by Pontrjagin and later by the author (Pontrjagin, C. R. Doklady), 35, 34–37 (1942); Mat. Sbornik N. S., 24(66), 129–162 (1949); Chern, Proc. Nat. Acad. Sci., USA, 33, 78–82 (1947)). It consists in their identification with cohomology classes defined in obstruction theory. We shall carry this out for the classes $W^i = f^*w^i$.

To the bundle (B, X) of $(n-1)$-spheres over X with the orthogonal group as the structure group we consider its associate bundle $(E_{n,p}, X)$, whose fibers are the Stiefel manifolds $V_{n,p}$, being the manifolds of ordered sets of p mutually perpendicular unit vectors in an n-space. It is known that $V_{n,p}$ is connected and that its first non-vanishing homotopy group is of dimension $n - p$, which is infinite cyclic if $n - p$ is even or $p = 1$ and is otherwise cyclic of order 2. Suppose X be a simplicial complex, and X^k its k-dimensional skeleton. According to a well-known procedure due to Stiefel, Whitney, and Steenrod, a cross-section can be defined over X^{n-p}. To such a cross-section h we define an $(n-p+1)$-dimensional cochain c^{n-p+1} as follows: To an $(n-p+1)$-cell σ,

$\psi^{-1}(\sigma)$ is homeomorphic to $\sigma \times V_{n,p}$. By taking its projection into $V_{n,p}$, the cross-section $h/\partial\sigma$ defines a mapping of an $(n-p)$-sphere into $V_{n,p}$, and hence an element of $\pi_{n-p}(V_{n,p})$. This we take to be the value of c^{n-p+1} for σ. The cochain c^{n-p+1} so defined is to be understood with local coefficients in the sense that the homotopy groups $\pi_{n-p}(V_{n,p})$ related to different cells are connected by isomorphisms. This cochain is a cocycle and its cohomology class is independent of the choice of the cross-section h over X^{n-p}. (For details, cf. Steenrod, Fibre Bundles, 155–183.) We call it the primary obstruction class of the bundle.

Each of the groups $\pi_{n-p}(V_{n,p})$, whether infinite cyclic or cyclic of order 2, can be mapped homomorphically into the cyclic group of order 2, I_2, such that a generator of the former goes into a generator of the latter. Applying this homomorphism to the primary obstruction c^{n-p+1}, we get a cohomology class $\bar{c}^{n-p+1}$, with coefficients in I_2. Since the homomorphism commutes with the isomorphisms of the local groups, the class $\bar{c}^{n-p+1}$ is an ordinary cohomology class. It will be called the reduced primary obstruction. A theorem due to Pontrjagin can be stated as follows:

Theorem 1. $W^i = f^*\bar{c}^i$, $i = 1, \ldots, n$.

To prove this theorem, notice that it is sufficient to establish it for the universal bundle. In E^{n+N} we take a system of p linearly independent vectors, say the first p coordinate vectors $e_1, \ldots, e_p$. To an element X of $G(n, N, R)$ let x_i be the orthogonal projection of e_i in X. If R^{n+N-p} is the linear space orthogonal to e_i, $i = 1, \ldots, p$, it is seen that the vectors x_i, $i = 1, \ldots, p$, are linearly dependent if and only if X satisfies the condition $\dim(X \cap R^{n+N-p}) \geqq n - p + 1$. The latter form the Schubert variety $(\underbrace{N-1 \ldots N-1}_{n-p+1} N \ldots N)$ of dimension $nN - (n - p + 1)$, whose dual cohomology class is W^{n-p+1}. It follows that if X does not belong to this Schubert variety, a field of p linearly independent vectors can be defined in X. By a well-known orthogonalization process they can be taken to be mutually perpendicular. The remaining part of the proof can be achieved by taking a simplicial decomposition of $G(n, N, R)$ such that the Schubert variety in question is a sub-complex and calculating the reduced primary obstruction relative to this cross-section.

As a second application we shall discuss the question of relations between the characteristic classes. Since the characteristic homomorphism f^* preserves multiplication, relations between the cohomology classes on $G(n, N, R)$ remain valid for the characteristic classes. As an illustration we shall give a proof of a theorem of Pontrjagin on a relation between the characteristic classes of a four-dimensional manifold (that is, relative to its tangent bundle).

Related to this question is the question whether there are relations between the characteristic classes. With coefficients mod 2 it follows from 4), §3 that there is no nontrivial relation between the W's. However, the question assumes a different aspect if the bundle is not a general one, in particular, if it is the tangent bundle of a differentiable manifold or the normal bundle of an imbedded differentiable manifold in an Euclidean space. The following theorem is due to Whitney, but proved in a different way:

Theorem 2. *For the tangent bundle of a compact orientable n-dimensional differentiable manifold, the characteristic class* $\overline{W}^n = 0$, *the coefficients being mod* 2.

We imbed X differentiably in an Euclidean space of dimension $n + N$: $X \subset E^{n+N}$. Consider the normal bundle of X and denote its characteristic classes by dashes. By dimension considerations we have $W'^r = 0$, $r \geqq n + 1$. It also follows from dimension considerations that an $N - n$ field of normal vectors can be defined over X, because the primary obstruction, being a cohomology class of dimension $n + 1$, is zero. We take such a field and consider the bundle A of its normal spaces of dimension n over X. Let A' be the bundle of $(n - 1)$-spheres over X obtained from A by taking the unit sphere in each fiber. We assert that the bundle (A', X) has a cross-section. In fact, A' can be realized as a small tube of unit vectors about X. Suppose its fiber F be ~ 0 in A'. Let D be a disc of unit vectors with F as its boundary. Then $D + C$, where C is a chain in A' with F as boundary, is a cycle in the Euclidean space having with X the intersection number 1. But this contradicts the fact that such a cycle is ~ 0 in E^{n+N}. It follows that no fiber of A' is ~ 0 in A', and that this is true with any coefficient group. By a theorem of Gysin (Comm. Math. Helv. 14, 61–122 (1942); cf. also Chern and Spanier, Proc. Nat. Acad. Sci. 36, 248–255)), the primary obstruction class of the bundle is zero. Hence the bundle (A', X) has a cross-section.

On the other hand, it is clear that the characteristic classes of (A', X) are the same as those of the normal bundle over X. Hence $W'^n = 0$ and, by 3), §3, we have $\overline{W}^n = 0$.

Let X be now a compact orientable four-dimensional manifold, and P^4 its characteristic class which has the symbol $f^*(22)$. The following theorem was stated by Pontrjagin:

Theorem 3. *With coefficients mod* 2,

$$P^4 + W^4 = 0. \tag{33}$$

Since the manifold is orientable, we have $W^1 = 0$. Equations (31) then give

$$\overline{W}^1 = 0,$$
$$\overline{W}^2 + W^2 = 0,$$
$$\overline{W}^4 + W^4 + (W^2)^2 = 0.$$

On the other hand, (26) gives

$$P^4 + (W^2)^2 = 0.$$

Combining these equations and using Theorem 2, we get (33).

The problem of showing that the characteristic classes of a differentiable manifold are non-trivial is undoubtedly of interest. In particular, the question has been raised whether there exists an orientable differentiable manifold with a non-zero Stiefel-Whitney class (Steenrod, Fibre Bundles, p. 212). We shall give such an example (Wu Wen-tsun, C. R. Acad. Paris, 230, 508–511 (1950)).

Let M^4 be the complex projective plane, and $M^4 \times I$ its topological product with the unit interval, so that its points can be represented as (p, t), $p \in M^4$, $t \in I$. Identify $(p, 0)$ and $(\bar{p}, 1)$, where $\bar{p}$ is the conjugate complex point of p. The resulting space N is a 5-dimensional orientable manifold. To see this, let L be a complex projective line in M^4. The conjugation $\tau: p \to \bar{p}$, which maps a point p of M^4 to its conjugate complex point $\bar{p}$, preserves the orientation of M^4 and reverses the orientation of L. The space M^4 has a cellular decomposition consisting of the 4-cell $\sigma^4 = M^4 - L$, the 2-cell $\sigma^2 = L - \sigma^0$, and the 0-cell σ^0, where σ^0 is a point of L. Denoting by I' the open unit interval, we see that N has a cellular decomposition with the cells $\sigma^0, \sigma^2, \sigma^4, \sigma^0 \times I'$, $\sigma^2 \times I', \sigma^4 \times I'$. The above remark on the conjugation gives the following incidence relations

$$\partial(\sigma^4 \times I') = 0, \qquad (34)$$

$$\partial(\sigma^2 \times I') = \pm 2\sigma^2.$$

In particular, the first relation implies that N is an orientable manifold. If, as usual, we use the same notation to denote both a chain and a cochain, then $C^2 = \sigma^2$ is a cocycle mod 2 and $C^3 = \sigma^2 \times I'$ is an integral cocycle.

We shall show that the class W^3 of N is $\neq 0$. We first consider the space M^4. Its characteristic classes can be determined directly by suitably chosen fields of vectors. Just to show the usefulness of some recent results of Wu (C. R. Paris, loc. cit.), we shall find them as follows: Our coefficients are mod 2. Since the Steenrod squares define homomorphisms and since M^4 is a manifold, there exist uniquely determined cohomology classes U^1, U^2, of dimensions 1 and 2 respectively, satisfying

$$Sq^1 Y^3 = U^1 \cup Y^3, \qquad Sq^2 Y^2 = U^2 \cup Y^2, \qquad (35)$$

where Y^2 or Y^3 is an arbitrary class whose dimension is given by the superscript. Then we have the formula

$$W^i = \sum_{p=0}^{i} Sq^p U^{i-p} \qquad (36)$$

Since M^4 is orientable, we have $W^1 = 0$. Notice that $H^2(M^4)$ and $H^4(M^4)$ are both cyclic of order two. The above formula shows that W^2 and W^4 are the non-zero elements.

We can therefore define a 3-field on M^4 with a single singularity in L. From this a 4-field can be constructed in N by taking as the fourth vector the one in the direction of I. Hence the class W^2 of N is $\neq 0$. Using Whitney's formula

$$W^3 = \tfrac{1}{2}\delta W^2. \qquad (37)$$

we find $W^3 \neq 0$ for N.

The manifold N is 5-dimensional. But $N \times S^1$ is 6-dimensional, and its characteristic class W^3 is also $\neq 0$. The latter cannot therefore have an almost complex structure.

As a third application of our results, we mention the fact that the characteristic classes give rise to necessary conditions that a differentiable manifold can be differentiably imbedded in a certain Euclidean space. Clearly we have the theorem;

Theorem 4. *In order that a compact differentiable manifold of dimension n can be differentiably imbedded in an Euclidean space of dimension $n + N$, it is necessary that*

(38) $$\overline{W}^r = 0, \qquad r \geqq N + 1,$$

(39) $$\bar{P}^{4k} = 0, \qquad 2k \geqq N + 1.$$

By means of the multiplication formulas these conditions can be transformed into a different form. In particular, (38) gives, for $N = 1$, the conditions

(40) $$W^r = (W^1)^r, \qquad r = 1, 2, \ldots, n,$$

and, for $N = 2$, the conditions

(41) $$W^r + W^{r-1}W^1 + W^{r-2}(W^2 + (W^1)^2) = 0, \qquad 3 \leqq r \leqq n.$$

It is of course possible to have similar conditions for larger values of N. Notice also that (40) permits us to express all W^r as polynomials of W^1 and (41) permits us to express all W^r as polynomials of W^1 and W^2. The conditions (41) have some easy consequences which would perhaps be simpler to apply to practical problems. In fact, if in addition $W^1 = 0$, then (41) implies

(41a) $$W^{2r} = (W^2)^r, \qquad 2r \leqq n.$$

If in addition $W^2 = 0$ (without necessarily having $W^1 = 0$), (41) implies

(41b) $$W^r = \begin{cases} (W^1)^r, & r \not\equiv 2\ (3), \\ 0, & r \equiv 2\ (3). \end{cases}$$

We proceed to apply these conditions to the real projective space Π^n of n dimensions. We have shown before that its cohomology groups mod 2 are

$$H^r(\Pi^n) = I_2, \qquad 0 \leqq r \leqq n.$$

Moreover, if ζ is a generator of $H^1(\Pi^n)$, then $(\zeta)^r$ is a generator of $H^r(\Pi^n)$, the power being in the sense of the cup product. The Stiefel-Whitney classes of Π^n have been determined by Stiefel (Comm. Math. Helv. 13, 201–218). They are

$$W^r = \binom{n+1}{r}(\zeta)^r.$$

We now suppose that these classes satisfy the conditions (41). Since $\binom{n+1}{3} = \binom{n+1}{2}\frac{n-1}{3}$, the condition that $\binom{n+1}{2}$ is even implies that $\binom{n+1}{3}$ is also even. This means that $W^2 = 0$ implies $W^3 = 0$, and by (41b), that $W^1 = 0$. Then all the W's are 0. Hence $W^2 = 0$ implies all $W^r = 0$. Suppose next $W^2 \neq 0$. If $W^1 = 0$, we derive from (41a) that $W^r = 0$ or not according as r is odd or even. If $W^1 \neq 0$, then $W^2 + (W^1)^2 = 0$ and from (41) we see that W^r is $= 0$ for $0 \leqq r \leqq n$. It follows that if Π^n can be differentiably imbedded in E^{n+2}, n must be a positive integer satisfying one of the following conditions

$$\begin{aligned}(1 + t)^{n+1} &\equiv 1 + t + t^2 + \cdots + t^{n+1}, \ (2) \\ &\equiv 1 + t^{n+1}, \qquad (2) \\ &\equiv 1 + t^2 + t^4 + \cdots + t^{n+1}, (2).\end{aligned}$$

An elementary argument shows that these conditions are respectively equivalent to the conditions that n be of one of the forms: 1) $n = 2^k - 2$, $k \geqq 2$; 2)$n = 2^k - 1$, $k \geqq 2$; 3) $n = 2^k - 3$, $k \geqq 3$.

We get therefore the theorem:

Theorem 5. A real projective space of dimension n cannot be differentiably imbedded in E^{n+2}, if n is not of one of the forms: 1) $n = 2^k - 2, k \geqq 2$; 2) $n = 2^k - 1, k \geqq 2$; 3) $n = 2^k - 3, k \geqq 3$.

Further applications of our criteria can be made, in particular, the application of conditions (39) to the imbedding of the complex projective space in an Euclidean space. It may be remarked that conditions (39) are expressed in terms of cohomology classes with real coefficients. As characteristic classes with real coefficients they can be represented by differential forms obtained from a Riemann metric. There is no difficulty in finding the actual expressions. Thus we get in this way criteria that a Riemann manifold cannot be imbedded in a certain Euclidean space in terms of curvature properties of the manifold. These criteria are particularly useful, when the Riemann manifold admits a transitive group of transformations.

5. Duality Theorems

An operation on sphere bundles which has various geometrical applications was introduced by Whitney. Let (B_1, X) and (B_2, X) be two principal bundles over the same base space X, whose structural groups are orthogonal groups $0(n_1)$, $0(n_2)$ in n_1 and n_2 variables respectively. Then the group $0(n_1) \times 0(n_2)$ can be imbedded in $0(n_1 + n_2)$, and we get a bundle over X with $0(n_1 + n_2)$ as structural group, called the product of the given bundles. Denote by W^r, P^{4k} the Whitney and Pontrjagin characteristic classes of the product bundle and by W_1^r, P_1^{4k}, W_2^r, P_2^{4k} those of the given bundles. We recall here that the W's are with coefficients mod 2 and the P's with real coefficients. The so-called duality theorems deal with relations between the characteristic classes of the three bundles. More precisely, they express the classes of one of these bundles in terms of those of the other two. To express this relationship we introduce the polynomials

$$(42)\qquad \begin{aligned} W(t) &= \sum_{r=0}^{n} W^r t^r \qquad W^0 = 1, \\ P(t) &= \sum_{0 \leqq 4k \leqq n} (-1)^k P^{4k}, \qquad P^0 = 1, \end{aligned}$$

with the independent variable t. Then we have the following formulas

$$(43)\qquad \begin{aligned} W(t) &= W_1(t) W_2(t), \\ P(t) &= P_1(t) P_2(t), \end{aligned}$$

of which the first one is due to Whitney.

The proof of both formulas can be reduced to the case of universal bundles, following an idea of Wu Wen-tsun. Take an Euclidean space $E^{n_1+n_2+N_1+N_2}$, spanned by two Euclidean spaces $E^{n_1+N_1}$ and $E^{n_2+N_2}$, with N_1 and N_2 sufficiently large. Let $G_1 = G(n_1, N_1)$, $G_2 = G(n_2, N_2)$ be the Grassmann manifolds in $E^{n_1+N_1}$ and $E^{n_2+N_2}$ respectively, and $G = G(n_1 + n_2, N_1 + N_2)$ that in $E^{n_1+n_2+N_1+N_2}$. An element in G_1 and an element in G_2 span an element in G, thus giving rise to a mapping

$$f: G_1 \times G_2 \to G.$$

Denote the projections of $G_1 \times G_2$ into its two factors by

$$p_1: G_1 \times G_2 \to G_1,$$

$$p_2: G_1 \times G_2 \to G_2.$$

Suppose the two given bundles be induced by the mappings

$$h_1: X \to G_1, h_2: X \to G_2.$$

We compose these two mappings into a mapping

$$h: X \to G_1 \times G_2,$$

defined by $h(x) = (h_1(x), h_2(x))$, $x \in X$, so that the two bundles are induced by the mappings $p_1 h$ and $p_2 h$. Then the bundle over X induced by the mapping fh is their product. It is therefore sufficient to prove the formulas (43) for the characteristic classes in $G_1 \times G_2$ relative to the bundles induced by the mappings p_1, p_2, and f.

Let us restrict ourselves to the proof of the first formula of (43). Denote the classes in G, G_1, G_2 by w^r, w_1^r, w_2^r respectively. It will be sufficient to prove the following formula

$$f^* w^r = \sum_{i=0}^{r} p_1^* w_1^i \cup p_2^* w_2^{r-i}. \tag{44}$$

To prove this, we show that both sides of this equation have the same value for any homology class of $G_1 \times G_2$ of dimension r. Such a class is of the form $z_1^k \times z_2^{r-k}$, where z_1^k and z_2^{r-k} are homology classes of G_1 and G_2 respectively. Suppose, for instance, that $z \neq (\underbrace{1 \cdots 1}_{k})$. A representative Schubert variety of this class will consist of linear spaces which contain a fixed linear space of dimension $n_1 - k + 1$. The linear spaces of a Schubert variety of the class z_2^{r-k} contain a fixed linear space of dimension $n_2 - (r - k)$. A linear space spanned by them will then contain a linear space L_0 of dimension $n_1 + n_2 - r + 1$. This condition is therefore satisfied by the linear spaces of a representative cycle of the class $f(z_1^k \times z_2^{r-k})$. To prove that w^r has the value 0 for $f(z_1^k \times z_2^{r-k})$, we notice that the value is also the intersection number of the latter with the cycle $(\underbrace{N_1 + N_2 - 1 \ldots N_1 + N_2 - 1}_{r}\ \underbrace{N_1 + N_2 \ldots N_1 + N_2}_{n_1 + n_2 - r})$, which consists of all linear spaces of dimension $n_1 + n_2$ in a fixed linear space of dimension $N_1 + N_2 + r - 1$. When this is in general position with L_0, they have only the origin 0 in common. Hence the intersection number is actually zero.

It remains to show that w^r has the value 1 when z_1^k and z_2^{r-k} are of the forms $z_1 = (1 \ldots 1)$, $z_2 = (1 \ldots 1)$. This will be reduced to the calculation of an intersection number. We shall omit the details here.

So far we have studied the additive and multiplicative homology structures of Grassmann manifolds in order to get conclusions on the characteristic classes of sphere bundles. Recently it has been found useful to study other topological operations in the Grassmann manifold (Cf., for instance, Wu Wen-tsun, C. R. Paris 230, 918–920 (1950)), in particular, the Steenrod squaring operations. It has been found possible to express the Steenrod squares of the Stiefel-Whitney classes as their quadratic polynomials. The formulas are

$$Sq^r W^s = \sum_{t=0}^{r} \binom{s-r+t-1}{t} W^{r-t} W^{s+t}, \qquad s \geqq r > 0, \tag{45}$$

where $\binom{p}{q}$ is the binomial coefficient reduced mod 2, with the following conventions:

$$\begin{aligned} \binom{p}{q} &= 0, \qquad \text{if } p < q, q > 0, \\ &= 1, \qquad \text{if } q = 0. \end{aligned} \tag{46}$$

Let $n - 1$ be the dimension of the spheres. We prove (45) by induction on n. For $n = 1$ it is trivial. Suppose therefore that (45) is true for bundles of spheres of dimension $< n - 1$. It is sufficient to prove this on the universal bundle over $G = G(n, N) = G(n_1 + n_2, N_1 + N_2)$. We use the above notation and take $n_1 = n - 1, n_2 = 1$, so that there is a mapping

$$f: G_1 \times G_2 \to G.$$

We put

$$F^{rs} = Sq^r W^s + \sum_{t=0}^{r} \binom{s-r+t-1}{t} W^{r-t} W^{s+t},$$

and denote the corresponding expressions for the bundles over G_1, G_2 by F_1^{rs}, F_2^{rs} respectively. Then we find

$$\begin{aligned} f^*F^{rs} &= Sq^r f^*W^s + \sum_{t=0}^{r} \binom{s-r+t-1}{t} f^*W^{r-t} f^*W^{s+t} \\ &= Sq^r(W_1^s \otimes 1 + W_1^{s-1} \otimes W_2^1) + \sum_{t=0}^{r} \binom{s-r+t-1}{t} \\ &\quad \times (W_1^{r-t} \otimes 1 + W_1^{r-t-1} \otimes W_2^1)(W_1^{s+t} \otimes 1 + W_1^{s+t-1} \otimes W_2^1, \\ &= F_1^{rs} \otimes 1 + F_1^{r,s-1} \otimes W_2^1 + F_1^{r-s,s-1} \otimes (W_2^1)^2, \end{aligned}$$

which is zero by our induction hypothesis. But by (44) f^* is an isomorphism in the dimensions $\leqq N_1, N_2$. Hence $F^{rs} = 0$, and (45) is proved.

The interest in the formula (45) lies in the fact that it enables us to formulate necessary conditions that a sphere bundle is a tangent bundle. In fact, we introduce the cohomology classes U^r of Wu by the equations

$$W^i = \sum_{p \geqq 0} Sq^{i-p} U^p, \qquad i \geqq 0 \tag{47}$$

which completely determine U^p. Comparing with the equation (36), we have the following theorem:

A necessary condition for a sphere bundle to be a tangent bundle is the vanishing of the following classes of Wu:

$$U^p = 0, \qquad p > \frac{n}{2}, \tag{48}$$

$n - 1$ being the dimension of the spheres.

6. An application to projective differential geometry

We shall conclude these notes by giving an application of a different kind, namely, to a problem of projection differential geometry.

Let P be the three-dimensional real projective space, and E the four-dimensional space of its lines. We define a ruled surface in P to be a differentiable mapping $f: S^1 \to E$ of a circle into E such that no two lines intersect. In our early notation E was written as $G(2, 2)$. The mapping f therefore induces a bundle of circles over S^1 and the ruled surface is a realization of the bundle space in P. Since $\pi_1(E) \approx I_2$, there are two such bundles according as the mapping f defines the zero or non-zero element of $\pi_1(E)$, the ruled surface being homeomorphic to a torus and a Klein bottle respectively. We shall prove the following theorem due to Wu Wen Tsun:

A ruled surface in P is homeomorphic to a torus.

In other words, the non-orientable bundle cannot be realized as a ruled surface in P.

To prove this theorem let us recall that if x_i, $i = 1, \ldots, 4$, are the homogeneous coordinates in P, the Plücker coordinates p_{ij} of a line joining the points x_i, y_i are defined by

$$p_{ij} = x_i y_j - x_j y_i, \qquad i, j = 1, \ldots, 4. \tag{49}$$

These coordinates p_{ij} are homogeneous and satisfy the identity

$$p_{12}p_{34} + p_{13}p_{42} + p_{14}p_{23} = 0. \tag{50}$$

Instead of these we introduce another set of coordinates by

$$\begin{aligned} p_{12} &= \xi_1 + \eta_1, \qquad & p_{13} &= \xi_2 + \eta_2, \qquad & p_{14} &= \xi_3 + \eta_3, \\ p_{34} &= \xi_1 - \eta_1, \qquad & p_{42} &= \xi_2 - \eta_2, \qquad & p_{23} &= \xi_3 - \eta_3. \end{aligned} \tag{51}$$

Then the identity (50) becomes

$$\xi_1^2 + \xi_2^2 + \xi_3^2 = \eta_1^2 + \eta_2^2 + \eta_3^2 \tag{52}$$

The coordinates ξ_i, η_i being still homogeneous, we can normalize them so that

$$\xi_1^2 + \xi_2^2 + \xi_3^2 = 1, \qquad \eta_1^2 + \eta_2^2 + \eta_3^2 = 1. \tag{53}$$

The normalized coordinates are determined up to a sign. We can therefore take two spheres (two-dimensional) S_1, S_2 and represent the lines of P as pairs of points of these spheres such that the pairs (ξ, η), (ξ^*, η^*), where ξ^*, η^* are the antipodal points of ξ, η, determine the same line. If we take the lines to be oriented, then the oriented lines are in one-one correspondence with the pairs of points of the two spheres.

The mapping f will be represented by a pair of curves $\xi(t)$, $\eta(t)$, $0 \leq t \leq 1$, such that

$$(\xi(0), \eta(0)) = (\xi(1), \eta(1) \quad \text{or} \quad (\xi(0), \eta(0)) = (\xi^*(1), \eta^*(1)).$$

The bundle is orientable or non-orientable, according as the first or second case happens. Suppose now that the second is the case. Denote by d_1, d_2 the spherical distances on the two spheres, and write

$$d_1(t, t') = d_1(\xi(t), \xi(t')),$$

$$d_2(t, t') = d_2(\eta(t), \eta(t')).$$

Then we have

$$d_1(0, t) + d_1(t, 1) = d_1(0, 1),$$

$$d_2(0, t) + d_2(t, 1) = d_2(0, 1),$$

An elementary argument will then give the following lemma: There exist two values $0 \leqq t \leqq 1, 0 \leqq t' \leqq 1$, with $t \neq t'$, $(t, t') \neq (0, 1)$, such that

$$d_1(t, t') = d_2(t, t').$$

In terms of the coordinates this can be written

$$\xi_1(t)\xi_1(t') + \xi_2(t)\xi_2(t') + \xi_3(t)\xi_3(t') = \eta_1(t)\eta_1(t') + \eta_2(t)\eta_2(t') + \eta_3(t)\eta_3(t'),$$

which is the condition that the lines corresponding to the parameters t, t' intersect. But this contradicts our assumption that no two distinct lines intersect.

UNIVERSITY OF KANSAS
DEPARTMENT OF MATHEMATICS

MINIMAL SUBMANIFOLDS IN A RIEMANNIAN MANIFOLD

by

S.S. Chern

Technical Report 19
(New Series)

Research done under NSF Grant GP-3460

Lawrence, Kansas
November, 1968

TABLE OF CONTENTS

§1. Review of Riemannian Geometry 1

§2. The first variation 8

§3. Minimal submanifolds in euclidean space 11

§4. Minimal surfaces in euclidean space 19

§5. Minimal submanifolds on the sphere 29

§6. Laplacian of the second fundamental form 36

§7. Inequality of Simons 38

§8. The second variation 44

§9. Minimal cones in euclidean space 49

NOTATION

The ambient riemannian manifold will denoted by X^N or X, of dimension N. In most cases it will be the euclidean space E^N or the unit sphere S^N in E^{N+1}. The immersed manifold or submanifold will be given by

$$M^n \longrightarrow X,$$

where M^n or M is of dimension n. We will set $N = n + p$, so that p is the codimension of M in X. All manifolds are C^∞ and connected.

The following ranges of indices will be used throughout this paper:

$$1 \leqq A, B, C, \ldots \leqq N = n + p,$$

$$1 \leqq i, j, k, \ell, m, \ldots \leqq n,$$

$$n + 1 \leqq \alpha, \beta, \gamma, \ldots \leqq n + p.$$

ACKNOWLEDGEMENT

These notes are based on lectures delivered at the University of Kansas in October 1968. The author wishes to thank Professors N. Aronszajn and B. Price for their invitation and continued interest.

§1. Review of Riemannian Geometry.

Our aim is to study the minimal submanifolds in a Riemannian manifold. We begin by a review of Riemannian geometry, using the method of moving frames.

Let X be a C^∞ Riemannian manifold of dimension N, i.e., there is given in each tangent space T_x, $x \in X$, a positive definite scalar product which varies in a C^∞ manner with x. The scalar product we will denote by (ξ, η), $\xi, \eta \in T_x$. We will suppose X oriented, and we will agree on the following range of indices:

(1) $$1 \leq A, B, C, \ldots \leq N.$$

By an orthogonal frame (or simply frame) e_A is meant an ordered set of N vectors in the same tangent space T_x, which defines the orientation of X and which satisfies the relations

(2) $$(e_A, e_B) = \delta_{AB}.$$

The frame e_A defines uniquely a dual coframe ω_B in the cotangent space T^*_x, and vice versa. Condition (2) is equivalent to the following expression of the element of arc:

(3) $$ds^2 = \sum_A \omega_A^2.$$

The method of moving frames consists of developing the geometrical properties of X by the use of a frame field $_A(x)$, $x \in U$ (= a neighborhood of X). The properties themselves will be independent of the choice of the frame field.

The fundamental theorem of local Riemannian geometry is the following:

A) In a neighborhood U of X let ω_A be a coframe field. There exists uniquely a set of linear differential forms ω_{AB} satisfying the conditions:

(4)
$$\omega_{AB} + \omega_{BA} = 0;$$

(5)
$$d\omega_A = \sum_B \omega_B \wedge \omega_{BA}.$$

We first prove the uniqueness. Suppose ω'_{AB} be a set of forms satisfying the same conditions. Let

$$\varphi_{AB} = \omega'_{AB} - \omega_{AB}.$$

Condition (5) implies

$$\sum_B \omega_B \wedge \varphi_{BA} = 0.$$

By Cartan's lemma we must have

$$\varphi_{BA} = \sum_C a_{BAC}\omega_C$$

where a_{BAC} is symmetric in B, C. By (4), a_{BAC} is anti-symmetric in B, A. It follows that $a_{BAC} = 0$, which proves the uniqueness.

The existence is proven by solving the equations (4), (5). We leave it to the reader.

The ω_{AB} are called the <u>connection forms</u> and allow the definition of <u>covariant differentiation.</u> In fact, let

(6)
$$\xi = \sum_A \xi_A e_A$$

be a vector field. We define its covariant differential to be

(7)
$$D\xi = \sum_A D\xi_A \otimes e_A,$$

where

$$D\xi_A = d\xi_A + \sum_B \xi_B \omega_{BA}. \tag{8}$$

The vector field ξ is said to be parallel along a curve, if $D\xi = 0$. Under a parallelism the scalar product of two vector fields is preserved, as readily verified.

For the vectors e_A themselves equations (7) gives

$$De_A = \sum_B \omega_{AB} \otimes e_B. \tag{9}$$

The covariant differentiation can be defined for tensor fields of higher order. For definiteness let

$$T = \sum_{A, B, C} T_{ABC}\, e_A \otimes e_B \otimes e_C, \tag{10}$$

be a tensor field of order 3. We define

$$DT = \sum_{A, B, C} DT_{ABC} \otimes e_A \otimes e_B \otimes e_C, \tag{11}$$

where

$$DT_{ABC} = dT_{ABC} + \sum_E T_{EBC}\omega_{EA} + \sum_E T_{AEC}\omega_{EB} + \sum_E T_{ABE}\omega_{EC} = \tag{12}$$

$$\sum_E T_{ABC,E}\,\omega_E \quad \text{(say)}$$

The $T_{ABC,E}$ are called the covariant derivatives of T and define a tensor field of order 4.

The exterior differentiation of equations (5) gives

$$\sum_B \omega_B \wedge \Omega_{BA} = 0, \tag{13}$$

where

(14) $$\Omega_{BA} = d\omega_{BA} - \sum_{C} \omega_{BC} \wedge \omega_{CA} = -\frac{1}{2} R_{BACE} \omega_C \wedge \omega_E$$

are exterior differential forms of degree two. The coefficients R_{ABCE} satisfy the symmetry relations

(15) $$R_{ABCE} = -R_{BACE} = -R_{ABEC},$$

and, as a consequence of (13), also the relations

(16) $$R_{ABCE} + R_{ACEB} + R_{AEBC} = 0.$$

From (15) and (16) one derives

(17) $$R_{ABCE} = R_{CEAB}.$$

The tensor field R_{ABCE} is called the Riemann- Christoffel tensor; it gives all the local properties of the Riemannian metric.

From R_{ABCE} the Ricci tensor and the scalar curvature are defined respectively by

(18) $$R_{AB} = R_{BA} = \sum_{C} R_{ACBC},$$

(19) $$R = \sum_{A} R_{AA}$$

For N = 2 we have

(20) $$R = 2K,$$

when K is the gaussian curvature.

From the second covariant derivatives of a tensor field T we can define its Laplacian which is a tensor field of the same order. For example, if T is given by (10), its Laplacian has the components

(21) $$(\Delta T)_{ABC} = \sum_{E} T_{ABC,E,E}.$$

In particular, if $u: U \longrightarrow R$ is a C^{∞}-function, we set

(22) $$du = \sum u_A \omega_A,$$

(23) $$Du_A = du_A + \sum u_B \omega_{BA} = \sum u_{AB} \omega_B,$$

and the Laplacian $\Delta u : U \longrightarrow R$ is given by

(24) $$\Delta u = \sum_A u_{AA}.$$

We also define

(25) $$|\text{grad } u|^2 = \sum_A u_A^2,$$

which is the square of the length of the gradient vector of u.

If $\varphi(u)$ is a smooth function of u, the following formula is immediately verified:

(26) $$\Delta\varphi(u) = \varphi'(u)\Delta u + \varphi''(u)|\text{grad } u|^2.$$

Let M be a C^{∞}- manifold of dimension n and let

$$f : M \longrightarrow X$$

be a differentiable immersion. By imposing on M the induced metric we can suppose M to be Riemannian and f to be an isometric immersion. We will denote by $p = N-n$ the codimension and agree in addition to (1), the following ranges of indices:

(27) $$1 \leqq i, j, k, \ldots \leqq n; \; n+1 \leqq \alpha, \beta, \gamma, \ldots \leqq n+p.$$

If TX denotes the tangent bundle of X, its induced bundle over M splits into a direct sum:

(28) $$f^*(TX) = TM \oplus (TM)^{\perp}$$

where TM and $(TM)^{\perp}$ are respectively the <u>tangent bundle</u> and <u>normal bundle</u> of M. We restrict to a neighborhood of M and consider a frame field $\ell_A(m), m \in M$ of the bundle $f^*(TX)$ such that $\ell_i(m)$ are tangent vectors and $\ell_a(m)$ are normal vectors at m. Let θ_A, θ_{AB} be the forms previously denoted by $\omega_A \omega_{AB}$ relative to this particular frame field. Then we have

(29) $$\theta_a = 0.$$

Taking its exterior derivative and making one of (5), we get

(30) $$\sum_i \theta_i \wedge \theta_{ia} = 0$$

By Cartan's lemma we have

(31) $$\theta_{ia} = \sum_j h_{iaj} \theta_j,$$

where

(32) $$h_{iaj} = h_{jai}$$

The form

(33) $$\Theta = \sum \Theta_a \otimes \ell_a,$$

where

(34) $$\Theta_a = \sum_{i,j} h_{iaj} \theta_i \theta_j,$$

is called the <u>second fundamental form</u> of M in X. It describes the simplest metrical properties of M as a submanifold of X.

The mean curvature vector is defined by

(35) $$H = \frac{1}{n} \sum_{i,a} h_{iai} \ell_a;$$

It is a normal vector field over M. M is called a minimal submanifold if H = 0. It is called totally geodesic if θ = 0. For n = 1 there notions coincide and minimal submanifolds of dimension 1 are precisely the geodesics.

§2. The first variation.

We follow the notations of §1. If M is compact, possibly with boundary, its total volume is given by the integral

$$V = \int_M \theta_1 \wedge \ldots \wedge \theta_n . \tag{1}$$

We apply a variation of M as follows: Let I be the interval $-\frac{1}{2} < t < \frac{1}{2}$. Let $F: M \times I \to X$ be a differentiable mapping such that its restriction to $M \times t$, $t \in I$, is an immersion and that $F(m, 0) = f(m)$ $m \in M$. We consider a frame field $e_A(m,t)$ over $M \times I$ such that for every $t \in I$, $e_i(m,t)$ are tangent vectors to $F(M \times t)$ at (m,t) and hence $e_\alpha(m,t)$ are normal vectors. The forms ω_A, ω_{AB} can then be written

$$\begin{aligned} \omega_i &= \theta_i + a_i dt \ , \quad \omega_\alpha = a_\alpha dt, \\ \omega_{i\alpha} &= \theta_{i\alpha} + a_{i\alpha} dt, \end{aligned} \tag{2}$$

where θ_i, $\theta_{i\alpha}$ are linear differential forms in M with coefficients which may depend on t. For t = 0 they reduce to the forms with the same notation on M. The vector $\sum a_A e_A$ at t = 0 will be called the <u>deformation vector</u>. We write the operator d on $M \times I$ as

$$d = d_M + dt \frac{\partial}{\partial t} . \tag{3}$$

From (1.5)* we get

$$d(\omega_1 \wedge \ldots \wedge \omega_n) = \sum_\alpha \omega_\alpha \Omega_\alpha , \tag{4}$$

where

$$\Omega_\alpha = -\sum_i \omega_1 \wedge \ldots \wedge \omega_{i-1} \wedge \omega_{i\alpha} \wedge \omega_{i+1} \wedge \ldots \wedge \omega_n . \tag{5}$$

---

* (1.5) means formula (5) in §1; the same notation will be used throughout.

Substituting into (4) the expression in (2), we get, as its two sides,

$$\text{LHS} = d\{\theta_1\wedge\ldots\wedge\theta_n + dt\wedge\sum_i (-1)^{i-1}a_i\theta_1\wedge\ldots\wedge\theta_{i-1}\wedge\theta_{i+1}\wedge\ldots\wedge\theta_n\},$$

$$\text{RHS} = dt\wedge\sum_\alpha a_\alpha\tilde{\Theta}_\alpha,$$

where

(6) $$\tilde{\Theta}_\alpha = -\sum_i \theta_1\wedge\cdots\wedge\theta_{i-1}\wedge\theta_{i\alpha}\wedge\theta_{i+1}\wedge\cdots\wedge\theta_n.$$

Equating the terms in dt, we get

(7) $$\frac{\partial}{\partial t}(\theta_1\wedge\ldots\wedge\theta_n) = d_M\sum_i (-1)^{i-1}a_i\theta_1\wedge\ldots\wedge\theta_{i-1}\wedge\theta_{i+1}\wedge\ldots\wedge\theta_n + \sum_\alpha a_\alpha\Theta_\alpha.$$

Integrating over M and setting $t = 0$, we find the first variation of volume:

$$V'(0) = \frac{\partial}{\partial t}\int_M \theta_1\wedge\ldots\wedge\theta_n\Big|_{t=0}$$

(8)

$$= \int_M\sum_\alpha a_\alpha\tilde{\Theta}_\alpha + \int_{\partial M}\sum_i (-1)^{i-1}a_i\theta_1\wedge\ldots\wedge\theta_{i-1}\wedge\theta_{i+1}\wedge\ldots\wedge\theta_n,$$

where we set $t = 0$ in the integrands of the last two integrals.

The second term at the right-hand side of (8) vanishes if $a_i(m, t) = 0$, $m\in\partial M$, i.e., if the deformation vector is orthogonal to M along the boundary ∂M. This condition is a fortiori satisfied if the boundary ∂M remains fixed. The first integral is zero for arbitrary a_α if and only if

$$\tilde{H}_\alpha = 0,$$

which is the condition for M to be a minimal submanifold. Hence we have the theorem:

A) A minimal submanifold of a riemanniam manifold is locally characterized by the property that a piece of it has a stationary volume under deformations with its boundary fixed.

§3. Minimal submanifolds in euclidean space.

We study first the case when the ambient space X is the euclidean space E^N of dimension N. In this case the space has a global parallelism and all the tangent spaces can be identified with E^N itself. If xe_A denotes an orthonormal frame, x being the origin, the connection forms can be defined by the equations

$$(1) \qquad dx = \sum_A \omega_A \otimes e_A, \qquad de_A = \sum_B \omega_{AB} \otimes e_B$$

Let

$$x: M \to E^N$$

be an immersed submanifold of dimension n, so that, to $m \in M$, x(m) is the position vector of the image point. Locally over M we choose a frame field $x(m)e_A(m)$ so that $e_i(m)$ are tangent vectors to x(M) at x(m). Then we can write

$$(3) \qquad \begin{aligned} dx &= \sum_i \theta_i \otimes e_i, \\ De_i &= de_i - \sum_j \theta_{ij} \otimes e_j = \sum_\alpha \theta_{i\alpha} \otimes e_\alpha, \\ de_\alpha &= \sum_j \theta_{\alpha j} \otimes e_j + \sum_\beta \theta_{\alpha\beta} \otimes e_\beta \end{aligned}$$

Our notations and ranges of indices are consistent with those of the last two sections.

Let a' be a fixed vector in E^N. The function (a, x), when restricted to M, is the height function in the direction of a'. From (3) we have

$$d(a,x) = \sum_i (a, e_i)\theta_i,$$

$$D(a, e_i) = \sum_\alpha (a, e_\alpha)\theta_{i\alpha} = \sum_{\alpha, j} (a, e_\alpha)h_{i\alpha j}\theta_j.$$

It follows from the definition of the Laplacian that

$$\Delta(a, x) = \sum_{\alpha, i} (a, e_\alpha)h_{i\alpha i} = n(a, H), \tag{4}$$

where H is the mean curvature vector. This gives the theorem:

A) An immersed submanifold of dimension > 0 in euclidean space is a minimal submanifold if and only if all the coordinate functions are harmonic functions relative to the induced metric. Hence there is no compact minimal submanifold (of dimension > 0) without boundary in euclidean space.

The last statement follows from the fact that a harmonic function on a compact Riemannian manifold without boundary must be constant. (Our manifolds are always supposed to be connected.)

The curvature of M can be calculated from (1.14), which gives in this case

$$d\theta_{ij} - \sum \theta_{ik}\wedge\theta_{kj} = -\sum_\alpha \theta_{i\alpha}\wedge\theta_{j\alpha} = -\frac{1}{2}\sum_{\alpha, k, \ell} (h_{i\alpha k}h_{j\alpha\ell} - h_{i\alpha\ell}h_{j\alpha k})\theta_k\wedge\theta_\ell \tag{5}$$

If we denote by $S_{ijk\ell}$ the curvature tensor of the metric on M, we have therefore

$$S_{ijk\ell} = \sum_\alpha (h_{i\alpha k}h_{j\alpha\ell} - h_{i\alpha\ell}h_{j\alpha k}). \tag{6}$$

If M is minimal, its Ricci tensor is

$$S_{ik} = \sum_j S_{ijkj} = -\sum h_{i\alpha j}h_{k\alpha j}, \tag{7}$$

which is negative semi-definite, and its scalar curvature is

(8) $$S = -\sum_{\alpha, i, j} h_{i\alpha j}^2 \leq 0.$$

Hence we have:

B) The Ricci tensor of a minimal submanifold M in the euclidean space is negative semi-definite. M is totally geodesic, and is therefore a linear subspace if and only if its scalar curvature is zero.

Remark. A minimal submanifold $M \to E^N$ is, according to the above, an isometric immersion of the Riemannian manifold M by means of its harmonic functions. Necessary conditions for such an immersion to exist are: 1) M be non-compact; 2) the Ricci tensor of M be negative semi-definite. Further necessary conditions are not known.

We now consider the special case of minimal hypersurface in E^N, i. e., the case where the codimension p = 1. We will then write

(9) $$h_{i, n+1, j} = h_{ij}$$

and we have

(10) $$\theta_{i, n+1} = \sum_j h_{ij}\theta_j.$$

Taking its exterior derivative and making use of (1.14), we get

(11) $$\sum_j Dh_{ij} \wedge \theta_j = 0,$$

where

(12) $$Dh_{ij} = dh_{ij} + \sum h_{kj}\theta_{ki} + \sum h_{ik}\theta_{kj} = \sum_k h_{ijk}\theta_k.$$

It follows that the "covariant derivatives" h_{ijk} are symmetric in j, k. Since h_{ij} is symmetric in i, j, we see that h_{ijk} is symmetric in any two of the indices i, j, k. Therefore, if M is minimal, the contraction of h_{ijk} with respect to any two of its indices is zero.

The vector e_{n+1} being one of the two unit normal vectors to M, we wish to calculate $\Delta(a, e_{n+1})$ by (3), a' being a fixed vector in E^{n+1}. In fact, we have,

(13) $$d(a, e_{n+1}) = -\sum h_{ij}(a, e_i)\theta_j,$$

and, by (3) and (12)

$$D(-\sum_i h_{ij}(a, e_i)) = -\sum h_{ijk}(a, e_i)\theta_k - \sum h_{ij}h_{ik}\theta_k(a, e_{n+1}).$$

We have therefore, for a minimal hypersurface, the formula

(14) $$\Delta(a, e_{n+1}) = S(a, e_{n+1}).$$

Let $x_1, \ldots, x_n z$ be coordinates in E^{n+1}. We consider minimal hypersurfaces which can be represented by an equation of the form

(15) $$z = z(x_1, \ldots, x_n), \text{ all } x_i,$$

i.e., which have a one-one projection onto a hyperplane. The Bernstein problem is to ask whether such a minimal hypersurface is always a hyperplane, i.e., whether the function in (14) is necessarily linear. The answer is known to be affirmative in the following cases:

$n = 2$, Bernstein 1914

$n = 3$, de Giorgi 1965

n = 4 , Almgren 1966
n = 5, 6, 7, Simons 1968.

Our developments allow us to give a geometrical proof of the classical Bernstein theorem. Unlike most known proofs, no complex function theory will be used.

C) (Bernstein's theorem). Let

(15a) $$z = z(x_1, x_2)$$

be a minimal surface in E^3, which is defined for all x_1, x_2. Then $z(x_1, x_2)$ is a linear function.

We put

(16) $$p_i = \frac{\partial z}{\partial x_i}, \quad W = (1 + \sum_i p_i^2)^{1/2} \geq 0,$$

so that the unit normal vector e_{n+1} has the components $(\frac{p_1}{W}, \ldots, \frac{p_n}{W}, -\frac{1}{W})$. If a' is the unit vector along the z-axis, we have

(17) $$(a, e_{n+1}) = -\frac{1}{W}.$$

The proof of C) depends on the identity

(18) $$\Delta \log(1 + \frac{1}{W}) = K,$$

where $K = \frac{1}{2} S$ is the Gaussian curvature of M. In fact, (14) gives

$$\Delta(\frac{1}{W}) = \frac{2K}{W}.$$

Now , <u>for n = 2</u>,we have

(19) $$\sum_i h_{ij} h_{ik} = -K \delta_{jk}, \quad n = 2.$$

From (13) and (19) we get

(20) $$|\operatorname{grad} \frac{1}{W}|^2 = \sum_{i,j,k} h_{ij} h_{kj} (a, e_i)(a, e_k) = -K \sum_i (a, e_i)^2 = -K(1 - \frac{1}{W^2}).$$

Formula (18) then follows immediately from (1. 26).

To complete the proof of C) consider on M the new metric

$$d\sigma = \left(1 + \frac{1}{W}\right)ds, \tag{21}$$

where ds is the metric induced by the immersion $M \rightarrow E^3$. Since ds is complete, it is clear that $d\sigma \geqq ds$ is complete. By (18) the Gaussian curvature of $d\sigma$ is zero, i.e., $d\sigma$ is a flat metric. By a well-known theorem in Riemannian geometry, M, with its metric $d\sigma$, is isometric to the (ξ, η)-plane with its standard flat metric, i.e.,

$$d\sigma^2 = d\xi^2 + d\eta^2. \tag{22}$$

Since $K \leqq 0$ and since Δ differs from the operator $(\partial^2/\partial\xi^2) + (\partial^2/\partial\eta^2)$ by a positive factor, we have

$$\left(\frac{\partial^2}{\partial\xi^2} + \frac{\partial^2}{\partial\eta^2}\right) \log\left(1 + \frac{1}{W}\right) \leqq 0.$$

The function $\log(1 + (1/W))$, considered as a function in the (ξ, η)-plane, is non-negative and superharmonic. Hence it must be constant [10, p. 130]. This implies in turn that $K = 0$. By B) M must be a plane. Thus the proof of C) is complete.

<u>Remark</u>. The above argument can also be completed by observing that the function $-(1/W)$ in (17) is a negative solution of the equation (14). Since $S \leqq 0$, (14) implies that $-(1/W)$ is subharmonic. Hence $-(1/W)$ = constant and C) follows.

This suggests the problem: On a simply-connected complete Riemannian manifold M with negative semi-definite Ricci tensor consider the equation

$$\Delta u = Su, \tag{23}$$

when $S \leq 0$ is the scalar curvature. Does (23) have a non-constant positive solution?

The above arguments show that a negative answer to this question will imply the Bernstein conjecture for all dimensions.

To understand the analytical implication of the condition of a minimal hypersurface we will derive the differential equation for the hypersurface (15) to be minimal. By the notation in (16) the induced metric on M is

$$(24) \qquad ds^2 = \sum_i dx_i^2 + (\sum_i p_i dx_i)^2,$$

or

$$(25) \qquad ds^2 = \sum_{i,j} g_{ij} dx_i dx_j,$$

where

$$(26) \qquad g_{ij} = \delta_{ij} + p_i p_j.$$

The elements of its inverse matrix are

$$(27) \qquad g^{ij} = \delta_{ij} - \frac{p_i p_j}{W^2}.$$

By (3) we see that the second fundamental form can be written

$$(28) \qquad \varpi_{n+1} = -(dx, de_{n+1}) = -\sum dx_i d(\frac{p_i}{W}) + dzd\,(\frac{1}{W}) = -\frac{1}{W}\sum dx_i dp_i.$$

The condition for a minimal hypersurface is that ϖ_{n+1} should have trace zero. It can therefore be written

$$(29) \qquad \sum p_{ij} g^{ij} = 0,$$

where

$$(30) \qquad p_{ij} = \frac{\partial^2 z}{\partial x_i \partial x_j}.$$

By (27) equation (29) can be put in the form

(31) $$\sum_i \frac{\partial}{\partial x_i} \left(\frac{p_i}{W} \right) = 0.$$

Equation (29) or (31) is therefore the equation of a non-parametric minimal hypersurface.

§4. Minimal surfaces in euclidean space [3].

A submanifold of dimension two will be called a surface. If it is oriented, a paramount fact is that a Riemannian structure on it has an underlying complex structure and makes it into a Riemann surface. Its study will be facilatated by the use of results of complex function theory.

Let M be an oriented surface with the Riemannian metric

(1) $$ds^2 = \theta_1^2 + \theta_2^2 = \alpha\bar{\alpha},$$

where

(2) $$\alpha = \theta_1 + i\theta_2.$$

Suppose it be oriented that

(3) $$\frac{i}{2}\alpha\wedge\bar{\alpha} = \theta_1\wedge\theta_2 > 0.$$

Locally there is a complex coordinate $z = x + iy$ such that

(4) $$\alpha = \lambda dz, \quad \lambda \neq 0.$$

The coordinate z is defined up to a holomorphic transformation

(5) $$z^* = z^*(z),$$

with $((dz^*)/(dz)) \neq 0$ and makes M into a Riemann surface. The metric (1) can be written

(6) $$ds^2 = |\lambda|^2 dz\, d\bar{z} = |\lambda|^2(dx^2 + dy^2),$$

and x, y, are called the isothermal coordinates.

Let

(7) $$x\colon M \to E^N, \quad N = 2 + p$$

be an immersed oriented surface. Let Gr_p be the Grassmann manifold of all oriented two-dimensional planes through the origin 0 of E^N. The

Gauss mapping

(8) $$g: M \longrightarrow Gr_p$$

is defined by the condition that $g(m)$, $m \in M$, is the oriented plane through 0 parallel to the tangent plane to $x(M)$ at $x(m)$.

Now Gr_p has a natural complex structure defined as follows: Suppose the two-plane be spanned by the vectors ξ, η (in that order, as the plane is oriented), satisfying

(9) $$(\xi, \xi) = (\eta, \eta) = 1, \quad (\xi, \eta) = 0.$$

The vectors ξ, η are defined up to a rotation; if ξ', η' are vectors defining the same oriented plane and satisfying conditions analogous to (9), we have

$$\xi' + i\eta' = (\exp i\varphi)\,(\xi + i\eta), \quad \varphi \text{ real}.$$

Extending the scalar product over complex vectors, we have, using (9),

(10) $$(\xi + i\eta, \xi + i\eta) = 0$$

Regarding $\xi + i\eta$ as the homogenous coordinates of a point in the complex projective space $P_{p+1}(C)$ of complex dimension $p+1$, equation (10) defines a non-singular hyperquadric. It can be verified that our mapping defines a diffeomorphism of Gr_p with the hyperquadric (10). Identification of the two spaces gives rise to a complex structure on Gr_p.

To describe the complex structure analytically we consider frames e_A and put

(11) $$e_1 = \xi, \quad e_2 = \eta,$$

so that the two-plane is spanned by the first two vectors of the frame. Writing

(12) $$de_i = \sum_j \omega_{ij} \otimes e_j + \sum_a \omega_{ia} \otimes e_a,$$

we have

(13) $$d(e_1 + ie_2) = -i\omega_{12}\otimes(e_1 + ie_2) + \sum_a (\omega_{1a}+ i\omega_{2a})\otimes e_a.$$

Thus $\omega_{1a} + i\omega_{2a}$ are the forms of type (1, 0) on Gr_p.

Similarly, using (3.3), we have

$$d(e_1 + ie_2) = -i\theta_{12}\otimes(e_1 + ie_2) + \sum_a (\theta_{1a} + i\theta_{2a})\otimes e_a,$$

which are now equations in M relative to a local frame field defined in §3. Using

(14) $$\theta_{ia} = \sum_j h_{iaj}\theta_j, \quad h_{iaj} = h_{jai},$$

we find

(15) $$\theta_{1a} + i\theta_{2a} = (h_{1a1} + ih_{1a2})(\theta_1 - i\theta_2) + i(h_{1a1} + h_{2a2})\theta_2.$$

This gives the theorem

A) M is a minimal surface in E^N if and only if the Gauss mapping is anti-holomorphic.

Since M and Gr_p are both complex manifolds, a continuous mapping $M \longrightarrow Gr_p$ is holomorphic (resp. anti-holomorphic) if locally the mapping is defined by expressing the local coordinates of the image point as holomorphic functions of the local coordinates (resp. of their conjugate complex coordinates) of the original point. Since M is complex one-dimensional, its image g(M) can be viewed as an anti-holomorphic curve on Gr_p.

We can write

(16) $$dx = \theta_1\otimes e_1 + \theta_2\otimes e_2 = \frac{1}{2}(\theta_1 + i\theta_2)\otimes(e_1 - ie_2) + \frac{1}{2}(\theta_1 - i\theta_2)\otimes(e_1 + ie_2).$$

Using the complex structure on M, we get

(17) $$\partial x = \frac{1}{2}(\theta_1 + i\theta_2) \otimes (e_1 - ie_2).$$

Since the components of x are harmonic functions, we have

(18) $$\bar{\partial}(\partial x) = 0,$$

which means that ∂x are abelian differentials on M. From ∂x we recover x by the Weierstrass formula

(19) $$x = 2\mathrm{Re}\int \partial x.$$

We will write out this formula more explicitly relative to a complex coordinate ζ on M. We put

(20) $$\partial x = y(\zeta)d\zeta,$$

where $y(\zeta)$ is holomorphic in ζ. Since $y(\zeta)$ is a multiple of $e_1 - ie_2$, it satisfies the conditions

(21) $$(y(\zeta), y(\zeta)) = 0,$$
$$(y(\zeta), \bar{y}(\zeta)) \neq 0,$$

where the scalar products in question are extensions of that in E^{2+p} to complex vectors.

An application of the Weierstrass formula (19) is to derive examples of complete minimal surfaces. Suppose, for definiteness, that $p = 2q-1$ be odd. Set

(22) $$\begin{aligned} &y_1 = a_1(1+\zeta^{p+1}),\ y_3 = a_2(\zeta + \zeta^p), \ldots, y_{2q-1} = a_q(\zeta^{q-1} + \zeta^{q+1}),\\ &y_2 = ia_1(1-\zeta^{p+1}),\ y_4 = ia_2(\zeta - \zeta^p), \ldots, y_{2q} = ia_q(\zeta^{q-1} - \zeta^{q+1}),\\ &y_{2q+1} = 2a_{q+1}\zeta^q, \end{aligned}$$

where a_λ are non-zero constants satisfying

$$a_1^2 + \ldots + a_{q+1}^2 = 0, \tag{23}$$

then equations (21) are satisfied. Substituting (22) into (19) and (20), we get a complete minimal surface which is at the same time a simply-connected real rational algebraic surface of order $(p+2)^2$. This will be called the Enneper surface. In many ways it is the minimal surface with the "simplest" properties.

A minimal surface with the same Gauss map can be obtained by multiplying the vector $y(\zeta)$ by $e^{i\alpha}$, where α is a real constant. Two such minimal surfaces are said to be associated. For example, the helicoid and the catenoid in E^3 are associated minimal surfaces.

By (3.8) and the fact that the Gaussian curvature $K = (S/2)$, we have

$$K = -\sum_\alpha (h_{1\alpha 1}^2 + h_{1\alpha 2}^2), \tag{24}$$

so that

$$K\theta_1 \wedge \theta_2 = \sum_\alpha \theta_{1\alpha} \wedge \theta_{2\alpha} = \frac{i}{2} \sum_\alpha (\theta_{1\alpha} + i\theta_{2\alpha}) \wedge (\theta_{1\alpha} - i\theta_{2\alpha}). \tag{25}$$

The integral

$$C(M) = \int_M K\theta_1 \wedge \theta_2 \leq 0 \tag{26}$$

will be called the <u>total curvature</u> of M; it could be $-\infty$.

On Gr_p an hermitian metric can be defined by

$$d\sigma^2 = \sum_\alpha (\omega_{1\alpha} + i\omega_{2\alpha})(\omega_{1\alpha} - i\omega_{2\alpha}). \tag{27}$$

In fact, this makes Gr_p into a symmetric hermitian manifold; its Kahler form is

$$\frac{i}{2} \sum_\alpha (\omega_{1\alpha} + i\omega_{2\alpha}) \wedge (\omega_{1\alpha} - i\omega_{2\alpha}) = \sum_\alpha \omega_{1\alpha} \wedge \omega_{2\alpha} \tag{28}$$

Its integral over a holomorphic curve γ (resp. an anti-holomorphic curve) is called the <u>area</u> (resp. the negative of area) of γ. The above discussion gives the theorem:

B) The total curvature of a minimal surface in E^{2+p} is equal to the negative of the area of its image under the Gauss map.

Consider now the classical case $p = 1$, i.e., a minimal surface in E^3. Then Gr_1 is a conic in the complex projection plane. By stereographic projection from a point of the conic, the latter can be identified with the complex projective line $P_1(C)$, which in turn can be identified with the unit sphere in E^3. This explains the relation between our Gauss map and the classical one. However, even in the classical case, our definition is analytically more advantageous.

The classical Bernstein theorem was generalized by R. Osserman to the following geometrical form: Let $x: M \to E^3$ be a complete minimal surface, and let $g: M \to Gr_1 = P_1(C)$ be its Gauss map. If $\overline{g(M)} \neq Gr_1$, $x(M)$ is a plane. In this formulation the problem becomes a study of the image $g(M)$, and its "equidistribution".

In the general case we consider Gr_p to be imbedded in $P_{p+1}(C)$ (to be abbreviated to P_{p+1}) and we wish to study the relative position of $g(M)$ with respect to the hyperplanes of P_{p+1}. This is a problem on value distribution in complex function theory. A classical theorem of E. Borel can be stated as follows:

Let $g: M \to P_{p+1}$ be a holomorphic curve which does not lie in a hyperplane of P_{p+1}. Suppose M be conformally equivalent to the complex line C. To any $p+3$ hyperplanes in P_{p+1} in general position, the image $g(M)$ meets one of them.

Let P^*_{p+1} be the dual projective space of P_{p+1}, i.e., the space of all its hyperplanes. Then Borel's theorem has the consequence that the set of hyperplanes having a non-void intersection with $g(M)$ is dense in P^*_{p+1}.

In our case of a minimal surface M in E^{2+p} we can, without loss of generality, take M to be simply-connected by replacing it by its universal covering surface. The question of its type is decided by the following lemma:

C) Let M be a non-compact simply-connected complete two-dimensional Riemannian manifold. Suppose there be a function $u \geqq \epsilon = \text{constant} > 0$ satisfying

(29) $$\Delta \log u = K,$$

where K is the gaussian curvature. Then M is conformally equivalent to the gaussian plane.

By the uniformization theorem for simply connected Riemann surfaces we suppose M to be conformally the disk $|\zeta| < R \leqq \infty$, and we shall prove that $R = \infty$. Let $ds = \lambda |d\zeta|$. The hypothesis says that $\log(u\lambda)$ is harmonic. Let $\nu(\zeta)$ be its conjugate harmonic function. Then

$$w = F(\zeta) = \int_0^\zeta e^{\log(u\lambda)+i\nu} d\zeta$$

satisfies $|w'| = u\lambda \neq 0$.

In a neighborhood of $w_0 = F(0)$ we can define a branch of the inverse function $\zeta = F^{-1}(w)$. We wish to show that this branch can be extended to the whole w-plane. In fact, suppose C be a circle about w_0 beyond which the function $F^{-1}(w)$ cannot be continued analytically. There must be a point $w_1 \in C$, which is a singularity of $F^{-1}(w)$. w_1 cannot be an algebraic branch point since $F'(\zeta)$ is never zero. Thereforethe line segment γ from w_o to w_1 must correspond to an arc going to the boundary of M. Its length is

$$\int ds = \int \frac{|dw|}{u} \leq \frac{|w_1 - w_o|}{\epsilon} < \infty,$$

which contradicts the completeness of M. Therefore $F^{-1}(w)$ defines an analytic map of the whole w-plane onto $|\zeta| < R$ and $R = \infty$.

The following theorem can be considered a geometrical generalization of Bernstein's theorem to minimal surfaces in E^N:

D) (Density theorem) Let $x: M \longrightarrow E^N (N = p+2)$ be a complete minimal surface which is not a plane. Let $g: M \longrightarrow Gr_p \subset P_{p+1}$ be the Gauss map. Then the hyperplanes of P_{p+1} which meet $g(M)$ form a dense subset in the dual space P^*_{p+1}. For $p = 1$, the set $g(M)$ is dense in Gr_1.

The last statement is the assertion of the Bernstein-Osserman theorem.

To prove the theorem let w_A be homogeneous coordinates in P_{p+1}, and let L be the hyperplane with the equation $w_1 = 0$. In the set $P_{p+1} - L$ we have

$$\partial\bar{\partial} \log\left(\frac{\sum w_A \bar{w}_A}{w_1 \bar{w}_1}\right) = \partial\bar{\partial} \log\left(\sum_A w_A \bar{w}_A\right)$$

$$(30) \quad = \frac{1}{\left(\sum_A w_A \bar{w}_A\right)^2} \left\{\left(\sum_A w_A \bar{w}_A\right)\left(\sum_B dw_B \wedge d\bar{w}_B\right) - \left(\sum \bar{w}_A dw_A\right) \wedge \left(\sum w_B d\bar{w}_B\right)\right\}$$

and the expression atthe right-hand side remains invariant when the w_A's are multiplied by a common factor.

Putting

$$w^*_A = w_A / \Big(\sum_B w_B \overline{w}_B\Big)^{\frac{1}{2}}, \tag{31}$$

we have

$$\sum_A w^*_A \overline{w^*_A} = 1$$

and

$$\partial\overline{\partial} \log\left(\frac{\sum w_A \overline{w}_A}{w_1 \overline{w}_1}\right) = \sum_B dw^*_B \wedge d\overline{w}^*_B \tag{32}$$

Since

$$(e_1 + ie_2, e_1 - ie_2) = 2,$$

$\sqrt{2}w^*_A$ can be considered as the components of $e_1 + ie_2$. It follows from (13) that

$$\sum_B dw^*_B \wedge d\overline{w}^*_B = \frac{1}{2}\sum_a (\omega_{1a} + i\omega_{2a}) \wedge (\omega_{1a} - i\omega_{2a}) = -i\sum_a \omega_{1a} \wedge \omega_{2a}$$

Since the Gauss map g is anti-holomorphic, the operator $\partial\overline{\partial}$ goes into $\overline{\partial}\partial$ under g, so that if the function $\sum_A w_A \overline{w}_A / w_1 \overline{w}_1$ is considered to be on M, we have

$$\overline{\partial}\partial \log\Big(\sum_A w_A \overline{w}_A / w_1 \overline{w}_1\Big) = + iK\theta_1 \wedge \theta_2.$$

Using isothermal coordinates, we get easily

$$\Delta \log\left\{\frac{|w_1|^2}{|w_1|^2 + \ldots + |w_N|^2}\right\}^{\frac{1}{2}} = K, \tag{33}$$

provided that $g(M) \cap L = \emptyset$.

To complete the proof of the density theorem suppose the assertion untrue.* Our assumption implies that there exists a neighborhood in P^*_{p+1}

---

* By taking the universial covering surface of M there is no loss of generality in supppsing M itself to be simply connected.

of a hyperplane, which we suppose to be L, such that all the corresponding hyperplanes do not meet g(M). It follows by an elementary geometrical argument that the function

$$u = \left\{ \frac{|w_1|^2}{|w_1|^2 + \ldots + |w_N|^2} \right\}^{\frac{1}{2}} \tag{34}$$

on M is $\geqq \epsilon$ for a positive constant ϵ. By theorem C), M is conformally the gaussian plane. Since M itself is not a plane, the image g(M) is not a point. Let P_q, $0 < q \leqq p+1$, be the smallest linear space of P_{p+1}, which contains g(M). By Borel's theorem quoted above the hyperplanes of P_q which meet g(M) form a dense subset of the dual space P_q^*. Since $q > 0$, it follows that the hyperplanes in P_{p+1} which meet g(M) form a dense subset of P_{p+1}^*. But this contradicts our assumption, so that the first statement in D) is proved. The second statement is proved similarly.

§5. Minimal submanifolds on the sphere.

We consider a minimal submanifold

$$x:\ M^n \longrightarrow S^{n+p} \subset E^{n+p+1}, \tag{1}$$

where S^{n+p} is a unit sphere of an euclidean space of one higher dimension. If e_A is an orthonormal frame of tangent vectors to S^{n+p} at x, xe_A is an orthonormal frame in E^{n+p+1}, satisfying

$$(x,x) = 1,\ (x, e_A) = 0,\ (e_A, e_B) = \delta_{AB}, \tag{2}$$

where the scalar product is defined for vectors in E^{n+p+1}.
From (2) we have

$$\begin{aligned} dx &= \sum_A \omega_A \otimes e_A, \\ de_A &= \sum_B \omega_{AB} \otimes e_B - \omega_A \otimes x \end{aligned} \tag{3}$$

where

$$\omega_{AB} + \omega_{BA} = 0. \tag{4}$$

Exterior differentiation of the second equation of (3) gives

$$d\omega_{AB} - \sum_C \omega_{AC}\ \omega_{CB} = -\omega_A \wedge \omega_B. \tag{5}$$

The expression at the right-hand side of this equation gives the curvature form of the Riemannian metric on S^{n+p}. The components of the Riemann-Christoffel tensor are

$$R_{ABCD} = \delta_{AC}\delta_{BD} - \delta_{AD}\delta_{BC} \tag{6}$$

When the submanifold (1) is given, we choose, as usual, a frame field xe_A in a neighborhood of M, such that e_i are tangent vectors to M at x.

Equations (3), when restricted to this frame field, become

$$(7)\qquad dx = \sum_A \theta_A \otimes e_A,$$

$$de_A = \sum_B \theta_{AB} \otimes e_B - \theta_A \otimes x,$$

with

$$(8)\qquad \theta_\alpha = 0$$

The θ_{ij} are connection forms of the induced metric on M, so that its curvature forms are

$$d\theta_{ij} - \sum_k \theta_{ik} \wedge \theta_{kj} = -\sum_\alpha \theta_{i\alpha} \wedge \theta_{j\alpha} - \theta_i \wedge \theta_j$$

$$(9)\qquad = -\frac{1}{2} \sum_{k,\ell,\alpha} (h_{i\alpha k} h_{j\alpha \ell} - h_{i\alpha \ell} h_{j\alpha k}) \theta_k \wedge \theta_\ell - \theta_i \wedge \theta_j.$$

Its Riemann-Christoffel tensor has therefore the components

$$(10)\qquad S_{ijk\ell} = \sum_\alpha (h_{i\alpha k} h_{j\alpha \ell} - h_{i\alpha \ell} h_{j\alpha k}) + \delta_{ik}\delta_{j\ell} - \delta_{i\ell}\delta_{jk}.$$

It follows that the Ricci tensor and scalar curvature are given respectivel by

$$S_{ik} = -\sum_{\alpha, j} h_{i\alpha j} h_{k\alpha j} + (n-1)\delta_{ik},$$

$$(11)$$

$$S = -\sum_{\alpha, i, j} h^2_{i\alpha j} + n(n-1).$$

We set

$$(12)\qquad \sigma = +\sum_{\alpha, i, j} h^2_{i\alpha j} \geqq 0,$$

so that σ is the square of the norm of the second fundamental form.

Let a' be a fixed unit vector in E^{n+p+1}. We consider the height function (a, x) as a function on M. By (7) and (8), we get

$$d(x,a) = \sum (a, e_i)\theta_i$$

$$D(a, e_i) = \sum h_{i\alpha j}(a, e_\alpha)\theta_j - (x, a)\theta_i .$$

It follows that

(13) $$\Delta(x, a) = n(a, H) - n(x, a),$$

where H is the mean curvature vector. Hence we have:

A) The submanifold (1) is a minimal submanifold on the sphere, if and only if the functions (a, x) satisfy the differential equation

(14) $$\Delta(x, a) + n(x, a) = 0.$$

On S^{n+p} there are compact minimal submanifolds without boundary. The following are some examples:

Example 1. The great n-sphere, which is totally geodesic.

Example 2. Write E^{n+2} as a direct sum

(15) $$E^{n+2} = E_1^{r+1} \oplus E_2^{s+1}, \quad r+s = n,$$

so that a vector of E^{n+2} will be written uniquely as $\xi_1 + \xi_2$, with $\xi_1 \in E^{r+1}$, $\xi_2 \in E_2^{s+1}$. We define the scalar product in E^{n+2} by

(16) $$(\xi_1 + \xi_2, \eta_1 + \eta_2) = (\xi_1, \eta_1) + (\xi_2, \eta_2),$$

where the right-hand side is a sum of the scalar products in E_1^{r+1}, E_2^{s+1} respectively. Let ξ_λ be an arbitrary unit vector in E_λ, $\lambda = 1, 2$. Then $a_1\xi_1 + a_2\xi_2$, $a_\lambda > 0$, will describe a submanifold M of dimension n on S^{n+1} if

(17) $$a_1^2 + a_2^2 = 1.$$

A unit normal vector to M at $a_1\xi_1 + a_2\xi_2$ is $e_{n+1} = -a_2\xi_1 + a_1\xi_2$ (or its negative), because the latter is orthogonal to $d\xi_1$, $d\xi_2$, and to $a_1\xi_1 + a_2\xi_2$.

Hence a second fundamental form of M is

$$-(dx, de_{n+1}) = a_1 a_2 \{(d\xi_1, d\xi_1) - (d\xi_2, d\xi_2)\}.$$

On the other hand, the induced metric on M is

$$ds^2 = a_1^2(d\xi_1, d\xi_1) + a_2^2(d\xi_2, d\xi_2).$$

It follows that M is a minimal submanifold, if and only if

(18) $$\frac{r}{s} = \frac{a_1^2}{a_2^2}.$$

This gives examples where a product $S^r \times S^s$ of spheres is embedded as a minimal submanifold of S^{r+s+1}. The case $r = s = 1$ gives the Clifford surface on S^3. In the general case we will call it the <u>Clifford minimal hypersurface.</u>

<u>Example 3</u>. Let x, y, z be the coordinates in E^3, where the sphere S^2 is defined by the equation

(19) $$x^2 + y^2 + z^2 = 3.$$

It can be verified that the mapping

(20) $$u_1 = \frac{1}{\sqrt{3}} yz, \quad u_2 = \frac{1}{\sqrt{3}} zx, \quad u_3 = \frac{1}{\sqrt{3}} xy,$$
$$u_4 = \frac{1}{2\sqrt{3}} (x^2 - y^2), \quad u_5 = \frac{1}{6} (x^2 + y^2 - 2z^2)$$

defines an isometric immersion of S^2 into the unit four-sphere S^4 in E^5. Since the functions in (20) are homogeneous quadratic polynomials, the mapping is two-to-one, and we have an imbedding of the real projective plane in S^4 which is called a <u>Veronese surface.</u> As a consequence of (A) this surface is a minimal surface of S^4.

<u>Example 4.</u> Blaine Lawson has constructed closed orientable minimal surfaces of arbitrary genus in S^3 [7].

Consider the case of closed minimal hypersurfaces in S^{n+1}, i.e., the codimension $p = 1$. As usual we write

(21) $$h_{i,n+1,j} = h_{ij} = h_{ji},$$

and (7) gives

(22) $$de_{n+1} = -\sum_{i,j} \theta_{i,n+1} x e_i,$$

whe

(23) $$\theta_{i,n+1} = \sum_j h_{ij}\theta_j.$$

As in (3.11) exterior differentiation of (23) gives

(24) $$\sum Dh_{ij} \wedge \theta_j = 0,$$

where

(25) $$Dh_{ij} = dh_{ij} + \sum h_{kj}\theta_{ki} + \sum h_{ik}\theta_{kj} = \sum h_{ijk}\theta_k.$$

The covariant derivatives h_{ijk} are symmetric in any two of their indices. Hence for M minimal the contraction of h_{ijk} is zero with respect to any two indices.

Exactly the same calculation as in §3 gives

(26) $$\Delta(a, e_{n+1}) = -\sigma(a, e_{n+1}),$$

where "a" is a fixed vector in E^{n+2}.

Suppose M be oriented, so that e_{n+1} is well-defined. The mapping

(27) $$g: M \to S^{n+1}$$

which assigns to $m \in M$ the unit normal vector $e_{n+1}(m)$ is called the <u>Gauss map</u>. Then we have:

B) Let $M^n \to S^{n+1}$ be an oriented closedminimal hypersurface. If the image $g(M^n)$ of the Gauss map belongs to an open hemisphere of S^{n+1},

M^n is totally geodesic.

In fact, the hypothesis implies that there is a unit vector "a" satisfying $(a, e_{n+1}) > 0$. By (26) it follows that (a, e_{n+1}) is superharmonic. Since M is compact, this is possible only when (a, e_{n+1}) = constant. We therefore have $\sigma = 0$ and, by (12), $h_{ij} = 0$.

Remark. It would be of interest to know whether a similar result is true for arbitrary codimension p. In this case the Gauss map $g: M \longrightarrow Gr$ assigns to $m \in M$ the p-dimensional oriented subspace spanned by $e_{n+1}(m), \ldots, e_{n+p}(m)$, Gr being the Grassmann manifold. If M is compact (and oriented) and if there exists a constant decomposable p-vector A such that

$$(A, e_{n+1} \wedge \ldots \wedge e_{n+p}) > 0,$$

is M totally geodesic? Robert Reilly proved that the answer is yes, if

$$(A, e_{n+1} \wedge \ldots \wedge e_{n+p}) \geq \left(\frac{2p-2}{3p-2}\right)^{\frac{1}{2}}$$

A basic problem on closed minimal submanifolds on the sphere is whether $M^n \longrightarrow S^{n+1}$ is totally geodesic if M is diffeomorphic to the sphere S^n. The answer is affirmative for n = 2, as given by the theorem: [1], [2], [4];

C) Let $x: S^2 \longrightarrow S^3$ be a minimal surface on the unit sphere S^3. Then $x(S^2)$ is a great sphere on S^3.

To prove the theorem let

$$\alpha = \theta_1 + i\theta_2, \quad \beta = \theta_{13} + i\theta_{23}. \tag{28}$$

Then we have

$$d\alpha = -i\theta_{12} \wedge \alpha, \tag{29}$$

$$d\beta = -i\theta_{12} \wedge \beta.$$

The condition for a minimal surface can be written (cf. (4.15)):

(30) $$\beta = h\bar{\alpha} \quad , \quad h = h_{11} + ih_{12}.$$

Under a change of the frame field both α and β will be multiplied by the same complex number of absolute value one. It follows that $\bar{\beta}\alpha = \bar{h}\alpha^2$ is independent of the choice of the frame field and is a complex-valued (ordinary) quadratic differential form defined over the surface. Locally we write (cf. (4.4))

(31) $$\alpha = \lambda dz,$$

so that $\bar{\beta}\alpha = \bar{h}\lambda^2 dz^2$. Exterior differentiation of (30) and use of (29) give

(32) $$dh + 2ih\theta_{12} \equiv 0, \text{ mod } \bar{\alpha}.$$

Exterior differentiation of (31) gives

(33) $$d\lambda + i\lambda\theta_{12} \equiv 0, \text{ mod } dz.$$

Combining (32) and (33), we get

$$\frac{\partial}{\partial \bar{z}} (\lambda^2 \bar{h}) = 0.$$

Hence the coefficient $\bar{h}\lambda^2$ is holomorphic in z. When S^2 is considered as a Riemann surface, $\bar{\beta}\alpha$ is a quadratic differential. From complex function theory the latter must be zero. This proves $h = 0$ and the minimal surface is totally geodesic.

§6. Laplacian of the second fundamental form.

Consider an isometric immersion

(1) $$x: M^n \longrightarrow X^{n+p}$$

of riemannian manifolds. Let TX be the tangent bundle of X. Its induced bundle over M splits into a direct sum:

(2) $$x^*(TX) = TM \oplus (TM)^{\perp},$$

where the summands at the right-hand side are the tangent bundle and normal bundle of M respectively. The second fundamental form is a section of the bundle

(3) $$TM \otimes TM \otimes (TM)^{\perp}$$

and is symmetric in the first two factors. Relative to our usual choice of a frame field it has the components $h_{i\alpha j}$. We define its covariant derivatives $h_{i\alpha jk}$ by

(4) $$Dh_{i\alpha j} = dh_{i\alpha j} + \sum_k h_{k\alpha j}\theta_{ki} + \sum_k h_{i\alpha k}\theta_{kj} + \sum h_{i\beta j}\theta_{\beta\alpha} = \sum_k h_{i\alpha jk}\theta_k.$$

Then $h_{i\alpha jk}$ are the components of a section of the bundle

$$TM \otimes TM \otimes T(M) \otimes (TM)^{\perp}.$$

We define its covariant derivatives by:

(5) $$Dh_{i\alpha jk} = dh_{i\alpha jk} + \sum_\ell h_{\ell\alpha jk}\theta_{\ell i} + \sum_\ell h_{i\alpha\ell k}\theta_{\ell j} + \sum_\ell h_{i\alpha j\ell}\theta_{\ell k} + \sum_\beta h_{i\beta jk}\theta_{\beta\alpha} = \sum_\ell h_{i\alpha jk\ell}\theta_\ell$$

The Laplacian of $h_{i\alpha j}$ is defined to be

(6) $$\Delta h_{i\alpha j} = \sum_k h_{i\alpha jkk}.$$

We wish to establish "commutation formulas" for the differences

---

1. For the results of §§6, 7, 9, cf. [5], [11].

$$h_{i\alpha jk} - h_{i\alpha kj},$$

$$h_{i\alpha jk\ell} - h_{i\alpha j\ell k}.$$

We make the following assumptions for the rest of this section: 1) X is of constant sectional curvature C, i.e.,

$$(7) \qquad R_{ABCD} = c(\delta_{AC}\delta_{BD} - \delta_{AD}\delta_{BC})$$

2) M is a minimal submanifold, i.e.,

$$(8) \qquad \sum_i h_{i\alpha i} = 0.$$

The first commutation formula is obtained by taking the exterior derivative of the equation

$$(9) \qquad \theta_{i\alpha} = \sum h_{i\alpha j}\theta_j,$$

by which the $h_{i\alpha j}$ are defined. Under the hypothesis that X is of constant curvature this gives

$$(10) \qquad \sum Dh_{i\alpha j} \wedge \theta_j = 0,$$

from which it follows that

$$(11) \qquad h_{i\alpha jk} - h_{i\alpha kj} = 0.$$

Similarly, the second commutation formula is obtained by taking the exterior derivative of (4). It gives

$$\sum_k Dh_{i\alpha jk} \wedge \theta_k = -c\Big(\sum_k h_{k\alpha j}\theta_k \wedge \theta_i + \sum_k h_{i\alpha k}\theta_k \wedge \theta_j\Big) + \sum_{\ell, m}\Big(-\sum h_{k\alpha j}h_{k\beta\ell}h_{i\beta m}$$

$$-\sum h_{i\alpha k}h_{k\beta\ell}h_{j\beta m} - \sum h_{i\beta j}h_{k\beta\ell}h_{k\alpha m}\Big)\theta_\ell \wedge \theta_m.$$

Equating the coefficients of $\theta_\ell \wedge \theta_m$ at both sides, we get

$$h_{\alpha ijm\ell} - h_{\alpha ij\ell m} = -\sum h_{k\alpha j}h_{k\beta\ell}h_{i\beta m} - \sum h_{i\alpha k}h_{k\beta\ell}h_{j\beta m} - \sum h_{i\beta j}h_{k\beta\ell}h_{k\alpha m}$$

$$(12) \qquad +\sum h_{k\alpha j}h_{k\beta m}h_{i\beta\ell} + \sum h_{i\alpha k}h_{k\beta m}h_{j\beta\ell} + \sum h_{i\beta j}h_{k\beta m}h_{k\alpha\ell}$$

$$-c(h_{\ell\alpha j}\delta_{im} - h_{m\alpha j}\delta_{i\ell} + h_{i\alpha\ell}\delta_{jm} - h_{i\alpha m}\delta_{jl}).$$

Since M is a minimal submanifold, it follows from the definition (6) and the commutation formulas (11) and (12) that

$$\Delta h_{i\alpha j} = \sum_k h_{i\alpha kjk}$$

$$(13) \qquad = -\sum h_{k\alpha\ell}h_{k\beta\ell}h_{i\beta j} - \sum h_{i\alpha k}h_{k\beta\ell}h_{\ell\beta j}$$

$$-\sum h_{i\beta\ell}h_{k\beta\ell}h_{k\alpha j} + 2\sum h_{k\alpha\ell}h_{k\beta j}h_{i\beta\ell} + nch_{i\alpha j}.$$

If the codimension is 1, this simplifies to

$$(14) \qquad \Delta h_{ij} = -\Big(\sum_{k,\ell} h_{k\ell}^2\Big) h_{ij} + nch_{ij}.$$

We define

$$(15) \qquad < h, \Delta h > = \sum_{i,j,\alpha} h_{i\alpha j}\Delta h_{i\alpha j},$$

$$(16) \qquad < Dh, Dh > = \sum_{i,j,k,\alpha} h_{i\alpha jk}^2 \geqq 0$$

Then we can verify the formuls

$$(17) \qquad d\Big(\sum h_{i\alpha j}h_{i\alpha jk}{}^*\theta_k\Big) = \{< h, \Delta h > + < Dh, Dh >\}\theta_1\wedge\ldots\wedge\theta_n.$$

It follows that if M is compact and without boundary, we have the integral formula

$$(18) \qquad \int_M < h, \Delta h > dm = -\int_M < Dh, Dh > dm \leqq 0,$$

where dm is the element of volume of M.

§7. Inequality of Simons.

Let $M^n \longrightarrow S^{n+p}$ be an oriented minimal submanifold in the unit sphere. By (6.13) and (6.15) we have

(1) $$-\langle h, \Delta h\rangle = +\sum h_{k\alpha\ell}h_{k\beta\ell}h_{i\beta j}h_{i\alpha j} + \sum h_{i\alpha k}h_{k\beta\ell}h_{\ell\beta j}h_{i\alpha j}$$

$$+\sum h_{i\beta\ell}h_{k\beta\ell}h_{k\alpha j}h_{i\alpha j} - 2\sum h_{k\alpha\ell}h_{k\beta j}h_{i\beta\ell}h_{i\alpha j} - n\sigma.$$

We will estimate algebraically this quartic polynomial in $h_{i\alpha j}$ in terms of σ. For this purpose we introduce the quantities

(2) $$\sigma_{\alpha\beta} = \sum_{i,j} h_{i\alpha j}h_{i\beta j},$$

so that $(\sigma_{\alpha\beta})$ is a positive semi-definite symmetric matrix, and our σ defined in (5.12) is given by

(3) $$\sigma = \sum_{\alpha} \sigma_{\alpha\alpha} \geqq 0.$$

We also introduce the symmetric matrices

(4) $$H_\alpha = {}^tH_\alpha = (h_{i\alpha j}).$$

To a matrix A denote by N(A) the sum of squares of its elements. Then (1) can be written

(5) $$-\langle h, \Delta h\rangle = \sum_{\alpha,\beta} \sigma_{\alpha\beta}^2 - n\sigma + \sum_{\alpha,\beta} N(H_\alpha H_\beta - H_\beta H_\alpha).$$

We have the following algebraic lemma:

A) Let A, B be symmetric matrices. Then

(6) $$N(AB-BA) \leqq 2N(A)N(B).$$

If $A \neq 0$, $B \neq 0$, the equality sign in (6) holds if and only if there is an orthogonal matrix T such that ${}^tTAT$, ${}^tTBT$ are scalar multiples of

$$(7) \qquad \begin{pmatrix} 0 & 1 \\ 1 & 0 \end{pmatrix}, \qquad \begin{pmatrix} 1 & 0 \\ 0 & -1 \end{pmatrix}$$

(augmented by zeros). Moreover, if $A_\lambda, \lambda = 1, 2, 3$, are symmetric matrices such that

$$N(A_\lambda A_\mu - A_\mu A_\lambda) = 2N(A_\lambda)N(A_\mu), \quad \lambda, \mu = 1, 2, 3, \ \lambda \neq \mu$$

then one of them is zero.

We observe first that $N({}^tTAT) = N(A)$, where T is any orthogonal matrix. By replacing A by ${}^tTAT$ (T orthogonal), we can suppose A to be diagonal, with the diagonal elements a_i. Then we have

$$N(AB-BA) = \sum_{i \neq k} b_{ik}^2 (a_i - a_k)^2,$$

where $B = (b_{ik})$. But

$$(a_i - a_k)^2 \leqq 2(a_i^2 + a_k^2).$$

It follows that

$$N(AB-BA) \leqq 2 \sum_{i \neq k} b_{ik}^2 (a_i^2 + a_k^2) \leqq 2N(A)N(B).$$

Suppose $A \neq 0$, $B \neq 0$ and that the equality sign holds in (6). Then all the above inequalities will be equalities and we get immediately

$$b_{ii} = 0$$

For definiteness suppose $b_{12} \neq 0$. We see easily that

$$a_3 = \ldots = a_n = 0, \ a_1 + a_2 = 0.$$

Since $A \neq 0$, we have $a_1 \neq 0$, $a_2 \neq 0$. But then

$$b_{ik} = 0, \ (i, k) \neq (1, 2).$$

On the other hand, it can be verified that the equality sign holds for the matrices (7). This proves the second statement in A). The third statement also follows in an elementary way.

We now make use of A) to give an estimate of the right-hand side of (5). By definition, we have

$$N(H_\alpha) = \sigma_{\alpha\alpha} \geqq 0,$$

so that

$$- < h, \wedge h > \leqq \sum_{\alpha, \beta} \sigma_{\alpha\beta}^2 + 2 \sum_{\alpha \neq \beta} \sigma_{\alpha\alpha}\sigma_{\beta\beta} - n\sigma.$$

To simplify the right-hand side, we diagonalize $(\sigma_{\alpha\beta})$, so that $\sigma_{\alpha\beta} = 0$, $\alpha \neq \beta$. Then

$$\text{RHS} = \sum \sigma_{\alpha\alpha}^2 + 2 \sum_{\alpha \neq \beta} \sigma_{\alpha\alpha}\sigma_{\beta\beta} - n\sigma$$

$$= \left(\sum \sigma_{\alpha\alpha}\right)^2 + \sum_{\alpha \neq \beta} \sigma_{\alpha\alpha}\sigma_{\beta\beta} - n\sigma$$

$$\leqq (2 - \frac{1}{p})\sigma^2 - n\sigma.$$

From (6.18) we get the inequality of Simons:

(8) $$\int_M \{(2 - \frac{1}{p})\sigma^2 - n\sigma\} \, dm \geqq 0,$$

provided that M is compact. Formula (8) gives the theorem:

B) Let $M^n \longrightarrow S^{n+p}$ be a closed minimal submanifold in the unit sphere. Then either M is totally geodesic or $\sigma = n/q$, $q = 2 - \frac{1}{p}$, or at some $m \in M$, $\sigma(m) > n/q$.

In fact, suppose $\sigma(m) \leqq n/q$ for all $m \in M$. Then the integrand in (8) will be $\leqq 0$, and it should vanish identically, i.e., $\sigma = 0$ or n/q.

As a consequence of B) it would be of interest to study the minimal submanifolds of S^{n+p} with $\sigma = n/q$. Next to the great spheres

these can be considered to be the "simplest" minimal submanifolds.
Since σ = constant, we have

$$\sum_{i,j,\alpha} h_{i\alpha j} h_{i\alpha jk} = 0.$$

By (6.17) this implies

$$< h, \Delta h > + < Dh, Dh > = 0,$$

or

$$<Dh, Dh> = -< h, \Delta h > \leqq 0.$$

Hence $Dh = 0$ or

$$h_{i\alpha jk} = 0.$$

We have the theorem:

C) The Clifford minimal hypersurfaces and the Veronese surface in Examples 2, 3, §5 are the only minimal submanifolds on the sphere with $\sigma = \frac{n}{q}$, $q = 2 - \frac{1}{p}$.

This theorem is a local theorem. For proof, cf. the paper of Chern-do Carmo-Kobayashi. In this connection the following problems seem to be of interest:

1) Consider closed minimal submanifolds $M^n \longrightarrow S^{n+p}$ where S^{n+p} is the unit sphere, such that σ is a constant. For given n and p what are the possible values of σ? Theorem B) says that σ does not take value in the open interval $0 < t < n/q$. Is the set of values for σ discrete? What is the value next to n/q?

2) Is a closed minimal hypersurface in S^{n+1} with σ = constant uniquely determined up to an isometry in S^{n+1}? That is, let $f, g\colon M^n \longrightarrow S^{n+1}$ be closed minimal submanifolds with the same constant scalar curvature. Does there exist an isometry T is S^{n+1} such that $g = T \circ f$?

§8. The second variation.

We follow the notation of §2 and we will find a formula for the second variation

(1) $$V''(0) = \frac{\partial^2}{\partial t^2} \int_M \theta_1 \wedge \cdots \wedge \theta_n \Big|_{t=0}.$$

The analytical part consists of a computation of $d\Omega_\alpha$, with Ω_α given by (2.5). Using (1.5) we find easily that $-d\Omega_\alpha$ is equal to

(2) $$\begin{aligned} -\sum_{\beta, i} \sum_{j \neq i} \omega_\beta \wedge \omega_1 \wedge \cdots \wedge \omega_{j-1} \wedge \omega_{j\beta} \wedge \omega_{j+1} \wedge \cdots \wedge \omega_{i-1} \wedge \omega_{i\alpha} \wedge \omega_{i+1} \wedge \cdots \wedge \omega_n \\ + \sum_\beta \omega_{\beta\alpha} \wedge \Omega_\beta + \sum_\beta \tilde{R}_{\alpha\beta} \omega_\beta \wedge \omega_1 \wedge \cdots \wedge \omega_n + \text{terms quadratic in } \omega_\beta, \omega_\gamma, \end{aligned}$$

where

(3) $$\tilde{R}_{\alpha\beta} = \sum_i R_{i\alpha i\beta} .$$

Substituting into Ω_α the expressions (2.2), we can write

(4) $$\Omega_\alpha = \tilde{\Theta}_\alpha + dt \wedge \Phi_\alpha,$$

where

(5) $$\begin{aligned} \Phi_\alpha = \sum_i (-1)^i a_{i\alpha} \theta_1 \wedge \cdots \wedge \theta_{i-1} \wedge \theta_{i+1} \wedge \cdots \wedge \theta_n \\ + \sum_i \sum_{j \neq i} (-1)^j a_j \theta_1 \wedge \cdots \wedge \theta_{j-1} \wedge \theta_{j+1} \wedge \cdots \wedge \theta_{i-1} \wedge \theta_{i\alpha} \wedge \theta_{i+1} \wedge \cdots \wedge \theta_n \end{aligned}$$

By equating the terms involving dt in the equation for $d\Omega_\alpha$ we have

(6) $$\frac{\partial \tilde{\Theta}_\alpha}{\partial t} = d_M \Phi_\alpha + \sum_\beta \omega_{\beta\alpha} \wedge \Phi_\beta + \Delta_\alpha,$$

where

$$\Delta_\alpha = -\sum_\beta \tilde{R}_{\alpha\beta}{}^a{}_\beta \theta_1 \wedge \cdots \wedge \theta_n$$

(7)

$$+\sum_{\beta,i}\sum_{j\neq i} a_\beta \theta_1 \wedge \cdots \wedge \theta_{j-1} \wedge \theta_{j\beta} \wedge \theta_{j+1} \wedge \cdots \wedge \theta_{i-1} \wedge \theta_{i\alpha} \wedge \theta_{i+1} \wedge \cdots \wedge \theta_n.$$

Taking the exterior derivative of the second equation of (2.2) and using (1.5), we get

$$da_\alpha \wedge dt = \sum_i (\theta_i + a_i dt) \wedge (\theta_{i\alpha} + a_{i\alpha} dt) + \sum_\beta a_\beta dt \wedge \omega_{\beta\alpha},$$

which gives

(8)
$$-d_M a_\alpha = \sum_i (a_i \theta_{i\alpha} - a_{i\alpha}\theta_i) + \sum_\beta a_\beta \omega_{\beta\alpha}.$$

Combining (6) and (8), we have

$$\frac{\partial}{\partial t}\sum_\alpha a_\alpha \tilde{\omega}_\alpha = \sum \frac{\partial a_\alpha}{\partial t} \tilde{\Phi}_\alpha + d_M \left(\sum_\alpha a_\alpha \Phi_\alpha \right) + \sum_\alpha a_\alpha \Delta_\alpha$$

(9)

$$+ \sum_{i,\alpha} (a_i \theta_{i\alpha} - a_{i\alpha}\theta_i) \wedge \Phi_\alpha.$$

Suppose $M \times 0$ be a minimal submanifold, so that $\tilde{\omega}_\alpha \big|_{t=0} = 0$. Differentiating (2.7) with respect to t and setting $t = 0$, we get

$$V''(0) = \int_{\partial M} \frac{\partial}{\partial t} \sum_i (-1)^{i-1} a_i \theta_1 \wedge \cdots \wedge \theta_{i-1} \wedge \theta_{i+1} \wedge \cdots \wedge \theta_n + \sum_\alpha a_\alpha \Phi_\alpha$$

(10)

$$+ \int_M \sum_{i,\alpha} (a_i \theta_{i\alpha} - a_{i\alpha}\theta_i) \wedge \Phi_\alpha + \sum_\alpha a_\alpha \Delta_\alpha.$$

Suppose the variation be normal and the boundary ∂M of M be fixed, i.e.,

(11)
$$a_i(m,t) = 0, \quad m \in M, \quad t \in I,$$
$$a_\alpha(m,0) = 0, \quad m \in \partial M.$$

Then we have

$$V''(0) = \int_M \Big(\sum_{i,\alpha} a_{i\alpha}^2\Big)\, dm + \sum_\alpha a_\alpha \Delta_\alpha,$$

where $dm = \theta_1 \wedge \ldots \wedge \theta_n$ is the volume element of M. On M we have

$$\sum_\alpha a_\alpha \Delta_\alpha = \Big(-\sum_{\alpha,\beta} (\tilde{R}_{\alpha\beta} + \sigma_{\alpha\beta}) a_\alpha a_\beta\Big)\theta_1 \wedge \ldots \wedge \theta_n,$$

where

(12) $$\sigma_{\alpha\beta} = \sum_{i,j} h_{i\alpha j} h_{i\beta j}.$$

For a normal vector field a_α to M, which vanishes on the boundary ∂M we have therefore the second variation

(13) $$V''(0) = \int_M \Big\{\sum_{i,\alpha} a_{i\alpha}^2 - \sum_{\alpha,\beta} (\tilde{R}_{\alpha\beta} + \sigma_{\alpha\beta}) a_\alpha a_\beta\Big\}\, dm.$$

The $a_{i\alpha}$ have a simple meaning. For, by (8), we have

(14) $$d_M a_\alpha + \sum_\beta a_\beta \omega_{\beta\alpha} = \sum_i a_{i\alpha}\theta_i.$$

The left-hand side being the covariant differential of a_α, $a_{i\alpha}$ are the covariant derivatives of a_α. Introducing the second covariant derivatives of a_α by

(15) $$Da_{i\alpha} = \sum_k a_{i\alpha k}\theta_k$$

and the Laplacian by

(16) $$\Delta a_\alpha = \sum_i a_{i\alpha i},$$

we denote the scalar product by

(17) $$\langle a, \Delta a \rangle = \sum_{\alpha, i} a_{i\alpha i}.$$

Since

(18) $$d\Big(\sum_{i,\alpha} a_\alpha a_{i\alpha} * \theta_i\Big) = \Big(\sum_{i,\alpha} a_{i\alpha}^2 + < a, \Delta a > \Big) dm,$$

and since $a_\alpha = 0$ on the boundary ∂M, we have

(19) $$\int_M \Big(\sum_{i,\alpha} a_{i\alpha}^2 + < a, \Delta a > \Big) dm = 0$$

We introduce the operator

(20) $$La_\alpha = -\Delta a_\alpha - \sum_\beta (\tilde{R}_{\alpha\beta} + \sigma_{\alpha\beta}) a_\beta .$$

For two normal vector fields a_α, b_α to M, which vanish on the boundary ∂M, their <u>index form</u> is defined by

(21) $$I(a, b) = \int_M < La, b > dm.$$

Since

(22) $$d\Big\{\Big(\sum_i \Big(\sum_\alpha (a_{i\alpha} b_\alpha - b_{i\alpha} a_\alpha) * \theta_i\Big)\Big\} = (< \Delta a, b > - < \Delta b, a >) dm,$$

the index form is symmetric:

(23) $$I(a, b) = I(b, a).$$

By writing $I(a) = I(a, a)$, we have therefore

(24) $$V''(0) = I(a).$$

On the other hand, the normal vector fields a and b have the scalar product

(25) $$(a, b) = \int_M \Big(\sum a_\alpha b_\alpha\Big) dm.$$

From the theory of strongly elliptic operators we have the theorem:

A) The index form I(a, b) is a symmetric bilinear form on the space of C^∞ normal vector fields to M, which vanish on the boundary ∂M.

I(a, b) may be diagonalized with respect to the scalar product (25) and has distinct real eigenvalues

(26) $$\lambda_1 < \lambda_2 < \cdots < \lambda_\nu < \cdots \longrightarrow +\infty.$$

Moreover, the dimension of each eigenspace is finite

The <u>index</u> of M is the sum of the dimensions of the eigenspaces which correspond to negative eigenvalues. The <u>nullity</u> of M is the dimension of the null eigenspace.

We state the following fact without proof:

B) The totally geodesic submanifold $S^n \longrightarrow S^{n+p}$ has index p and nullity p(n + 1).

For n = 1, i.e., for a geodesic, we have $\sigma_{\alpha\beta} = 0$ and the operator L becomes

(27) $$La_\alpha = -\Delta a_\alpha - \sum_\beta \tilde{R}_{\alpha\beta} a_\beta.$$

Many theorems in global Riemannian geometry (such as the Morse index theorem, Synge's lemma, Myers' theorem, etc.) follow from the study of this operator.

§9. Minimal cones in euclidean space.

Let

(1) $$M = M^n \longrightarrow S^{n+1} \subset E^{n+2}$$

be an immersed minimal hypersurface. We denote by CM the cone over M, which consists of the points tm, $m \in M$, $0 \leqq t \leqq 1$. In general CM has a singularity at the vertex $t = 0$, which will be excluded from the following considerations. The truncated cone will be denoted by $CM_\epsilon = \{tm \mid m \in M, \epsilon \leqq t \leqq 1\}$. Then $\partial CM_\epsilon = M \cup M_\epsilon$, where $M_\epsilon = \{\epsilon m \mid m \in M\}$.

Let $e_0, \ldots, e_{n+1}$ be an orthonormal frame field over M, with e_0 describing M and e_{n+1} the unit normal vector to M at e_0. Then

(2) $$de_0 = \sum_i \theta_i \otimes e_i, \qquad de_{n+1} = -\sum_i \theta_{i,n+1} \otimes e_i,$$

with

(3) $$\theta_{i,n+1} = \sum_k h_{ik}\theta_k, \quad h_{ik} = h_{ki}.$$

The condition that M is a minimal hypersurface is expressed by

(4) $$\sum_i h_{ii} = 0.$$

A generic point on the cone CM is given by

(5) $$x = te_0, \quad 0 < t \leqq 1,$$

from which we find

(6) $$dx = dt \otimes e_0 + t\sum_i \theta_i \otimes e_i.$$

Thus e_0, e_i is an orthonormal frame in the tangent space to CM at x, and its dual coframe is

$$\tilde{\theta}_0 = dt \quad , \quad \tilde{\theta}_i = t\theta_i \tag{7}$$

The unit normal vector to CM at x is e_{n+1}, which is independent of t. In other words, the tangent hyperplane to CM remains constant along a generator, which is a geometrically obvious fact. From (2) and (6) the second fundamental form of CM is

$$-(dx, de_{n+1}) = t\sum_i \theta_i\theta_{i,n+1} = t\sum_{i,k} h_{ik}\theta_i\theta_k. \tag{8}$$

Comparing with (7), we see that CM is a minimal cone of dimension n + 1 in E^{n+2}. We wish to calculate $\tilde{\Delta} f$, where $f = f(m, t)$ is a smooth function, $m \in M$, $0 < t \leqq 1$, and $\tilde{\Delta}$ is the Laplacian on CM.

The vector fields e_0, e_i define (locally) a frame field over CM. The connection forms are the elements of an anti-symmetric matrix

$$-\begin{pmatrix} 0 & \varphi_{0i} \\ \varphi_{j0} & \varphi_{ji} \end{pmatrix},$$

defined uniquely by the conditions

$$d\tilde{\theta}_0 = \sum_i \tilde{\theta}_i \wedge \varphi_{i0},$$

$$d\tilde{\theta}_i = \tilde{\theta}_0 \wedge \varphi_{0i} + \sum_i \tilde{\theta}_j \wedge \varphi_{ji}.$$

By (7) we have

$$\begin{aligned} d\tilde{\theta}_0 &= 0, \\ d\tilde{\theta}_i &= dt \wedge \theta_i + t\sum_j \theta_j \wedge \theta_{ji} \end{aligned} \tag{9}$$

where θ_{ji} are the connection forms on M. It follows that

(10) $$\varphi_{oi} = \theta_i \ , \ \varphi_{ji} = \theta_{ji}.$$

For the function f(m, t) we put

$$df = \frac{1}{t}\sum_i f_i \tilde{\theta}_i + \frac{\partial f}{\partial t} dt.$$

The covariant differentials of the gradient field of f are

$$D\left(\frac{\partial f}{\partial t}\right) = d\left(\frac{\partial f}{\partial t}\right) - \sum_j f_j \theta_j = \frac{\partial^2 f}{\partial t^2} dt + \ldots,$$

$$D\left(\frac{f_i}{t}\right) = d\left(\frac{f_i}{t}\right) + \sum_j \frac{f_j}{t}\theta_{ji} + \frac{\partial f}{\partial t}\theta_i = \frac{1}{t}\sum_j f_{ij}\theta_j + \frac{1}{t}\frac{\partial f}{\partial t}\tilde{\theta}_i + \text{terms in } dt.$$

It follows that

(11) $$\tilde{\Delta} f = \frac{1}{t^2}\Delta f + \frac{n}{t}\frac{\partial f}{\partial t} + \frac{\partial^2 f}{\partial t^2},$$

where Δ is the Laplacian on M and Δf is computed by holding t fixed.

Our purpose is to calculate the operator in (8.20) on the normal vector field $a = fe_{n+1}$ over CM_ϵ. In the sum at the right-hand side of (8.20), we have $\tilde{R}_{\alpha\beta} = 0$ because the ambient space E^{n+2} is flat. Using (7), (8) and (8.21), we get

(12) $$I(a) \underset{def}{=} I(a, a) = \int_{CM_\epsilon} \langle \tilde{L}f, \frac{1}{t^2} f \rangle t^n dt\, dm,$$

where

(13) $$\tilde{L}f = -t^2\tilde{\Delta} f - \sigma f,$$

σ being defined as usual by

(14) $$\sigma = \sum_{i,k} h_{ik}^2 \geqq 0$$

This motivates a splitting of the operator $\tilde{L}$:

$$\tilde{L} = L_1 + L_2, \tag{15}$$

where

$$L_1:\ C^\infty(M) \longrightarrow C^\infty(M),$$
$$L_2:\ C^\infty[\epsilon,1] \longrightarrow C^\infty[\epsilon,1] \tag{16}$$

are strongly elliptic operators given respectively by

$$L_1 f = -\Delta f - \sigma f,$$
$$L_2 g = -t^2 g'' - ntg'. \tag{17}$$

The operator $\tilde{L}$ is said to have the eigenvalue μ, if the equation

$$\tilde{L}f = \mu f \tag{18}$$

has a non-trivial solution, where f: $CM_\epsilon \longrightarrow R$ is a C^∞-function which vanishes on the boundary ∂CM_ϵ. Let

$$\lambda_1 \leqq \lambda_2 \leqq \ldots \longrightarrow \infty,$$
$$\delta_1 \leqq \delta_2 \leqq \ldots \longrightarrow \infty \tag{19}$$

be respectively the eigenvalues of the operators L_1, L_2. We have the lemma:

A) Let $M \longrightarrow S^{n+1}$ be a closed minimal hypersurface in the unit sphere and let CM_ϵ be the truncated cone over M. We may choose a function f: $CM_\epsilon \longrightarrow R$ which vanishes on the boundary ∂CM_ϵ such that $I(a) < 0$, $a = fe_{n+1}$ (e_{n+1} = unit normal vector), if and only if $\lambda_1 + \delta_1 < 0$.

Let f_ν, g_τ, $\nu, \tau = 1, 2, \ldots,$ be the eigenfunctions corresponding respectively to the eigenvalues in (19). Then $f(m,t)$ has the unique expansion

$$f(m,t) = \sum_{\nu,\tau \geqq 1} c_{\nu\tau} f_\nu(m) g_\tau(t), \tag{20}$$

and we have

$$I(a) = \int_{M\times[\epsilon,1]} < \sum_{\nu,\tau=1}^{\infty} c_{\nu\tau} L_1(f_\nu) g_\tau + c_{\nu\tau} f_\nu L_2(g_\tau), t^{n-2} \sum_{\nu,\tau=1}^{\infty} c_{\nu\tau} f_\nu g_\tau > dm\, dt$$

$$= \int_{M\times[\epsilon,1]} < \sum_{\nu,\tau\geqq 1}^{\infty} c_{\nu\tau}(\lambda_\nu + \delta_\tau) f_\nu g_\tau, \; ^{n-2} \sum_{\nu,\tau\geqq 1} c_{\nu\tau} f_\nu g_\tau > dm\, dt$$

$$= \sum_{\nu,\tau\geqq 1} c^2_{\nu\tau}(\lambda_\nu + \delta_\tau) \int_M f_\nu^2 dm \int_\epsilon^1 g_\tau^2 t^{n-2} dt.$$

If $I(a) < 0$, then $\lambda_\nu + \delta_\tau < 0$ for some ν, τ, which is possible only when $\lambda_1 + \delta_1 < 0$. On the other hand, if $\lambda_1 + \delta_1 < 0$, we may simply take $f(m,t) = f_1(m) g_1(t)$.

The operator L_2 on the line t is easy to analyze, and we have the theorem, whose proof is elememtary:

B) Let $C^\infty[\epsilon, 1]$ denote the C^∞ functions on the interval $\epsilon \leqq t \leqq 1$, which vanish at the end-points. Relative to the density $t^{n-2}dt$ the operator L_2 has the eigenvalues

(21) $$\delta_1 < \delta_2 < \ldots \to \infty$$

given by

(22) $$\delta_\tau = \left(\frac{n-1}{2}\right)^2 + \left(\frac{\tau\pi}{\log\epsilon}\right)^2, \quad \tau = 1, 2,$$

Its corresponding eigenfunction is

(23) $$g_\tau = t^{-\frac{n-1}{2}} \sin\left(\frac{\tau\pi}{\log\epsilon} \log t\right).$$

In particular,

(22a) $$\delta_1 = \left(\frac{n-1}{2}\right)^2 + \left(\frac{\pi}{\log\epsilon}\right)^2.$$

For $\tau \neq \nu$, we have

(24) $$\int_{\epsilon} g_{\tau} = g_{\nu} t^{n-2} dt = 0.$$

The first eigenvalue of the operator L_1 is described by the theorem:

C) Let $M^n \to S^{n+1}$ be a closed minimal hypersurface. Then

(25)
$$\lambda_1 = 0, \text{ if } M^n \text{ is totally geodesic,}$$
$$\lambda_1 \leqq -n, \text{ otherwise.}$$

In fact, if M is totally geodesic, we have

$$L_1 = -\Delta, \quad \lambda_1 = 0.$$

The other eigenvalues are strictly positive, the eigenfunctions being the spherical harmonics.

Suppose now that M is not totally geodesic, so that σ is not identically zero. We have

(26) $$\lambda_1 \leqq \left(\int_M f^2 dm \right)^{-1} \int_M L_1(f) f \, dm,$$

for any $f \in C^{\infty}(M)$, $f \not\equiv 0$. For $\eta > 0$ we set

(27) $$f_{\eta} = (\sigma + \eta)^{(1/2)}.$$

Since M is not totally geodesic, we have

$$\lim_{\eta \to 0} \int_M f_{\eta}^2 \, dm > 0.$$

We wish to calculate Δf_{η}. Using our notation above,

$$df_{\eta} = (\sigma + \eta)^{-(1/2)} \sum_{i,j,k} h_{ij} h_{ijk} \theta_k,$$

$$D\{(\sigma + \eta)^{-(1/2)} \sum_{i,j} h_{ij} h_{ijk}\} = (\sigma + \eta)^{-(3/2)} \sum_{i,j} h_{ij} h_{ijk\ell} \sum_{\ell,q,r} h_{\ell q} h_{\ell qr} \theta_r$$
$$+ (\sigma + \eta)^{-(1/2)} \sum_{i,j,\ell} h_{ij\ell} h_{ijk} \theta_{\ell} + (\sigma + \eta)^{-(1/2)} \sum_{i,j,\ell} h_{ij} h_{ijk\ell} \theta_{\ell}.$$

It follows that

$$(28)\qquad \Delta f_\eta = -(\sigma+\eta)^{-(3/2)}\sum_k \Big(\sum_{i,j} h_{ij}h_{ijk}\Big)^2 + (\sigma+\eta)^{-(1/2)}\sum_{i,j,k} h_{ijk}^2$$

$$+ (\sigma+\eta)^{-(1/2)}\sum_{i,j} h_{ij}\Delta h_{ij}.$$

The sum of the first two terms is bounded below by

$$(\sigma+\eta)^{-(3/2)}\Big\{\sum_{i,j} h_{ij}^2 \sum_{i,j,k} h_{ijk}^2 - \sum_k \Big(\sum_{i,j} h_{ij}h_{ijk}\Big)^2\Big\} \geqq 0$$

Applying (6.14), we have

$$\Delta f_\eta \geqq (\sigma+\eta)^{-(1/2)}(n\sigma - \sigma^2),$$

and hence

$$\Delta f_\eta \cdot f_\eta \geqq n\sigma - \sigma^2.$$

It follows that

$$L_1 f_\eta \cdot f_\eta \leqq \sigma^2 - n\sigma - \sigma f_\eta^2 \leqq -n\sigma.$$

From (26) we get $\lambda_1 \leqq -n$, which is to be proved.

Combining (22a) and (25), we have, for M not totally geodesic,

$$(29)\qquad \lambda_1 + \delta_1 \leqq -n + \Big(\frac{n-1}{2}\Big)^2 + \Big(\frac{\pi}{\log \epsilon}\Big)^2.$$

This gives the theorem:

D) (Simons) Let $M^n \longrightarrow S^{n+1}$ be a closed minimal hypersurface, not totally geodesic. For $n \leqq 5$ the cone CM does not minimize area with boundary fixed.

For the Clifford hypersurface $S^3 \times S^3 \longrightarrow S^7$, Theorem C) gives $\lambda_1 \leqq -6$. But

$$L_1 = -\Delta - 6,$$

so that $\lambda_1 = -6$. We also have $\delta_1 \geq 6\frac{1}{4}$, and thus $\lambda_1 + \delta_1 > 0$. This gives the theorem:

E) Let M be the Clifford hypersurface $S^3 \times S^3 \to S^7$, and CM the minimal cone over M. Then any variation of CM holding M fixed initially increases area.

Using results of Federer-Fleming, de Giorgi, and Triscari, one gets the theorem:

F)(Bernstein-Simons) Let

$$z = z(x_1, \ldots, x_n), \text{ all } x_i,$$

be a minimal hypersurface in the euclidean space E^{n+1}. If $n \leq 7$, the function $z(x_1, \ldots, x_n)$ is linear.

BIBLIOGRAPHY

Note. This Bibliography includes only items referred to in the text. For minimal surfaces cf. [8], [9] for extensive Bibliographies.

[1] Almgren, F.J., Jr., Some interior regularity theorems for minimal surfaces and an extension of Bernstein's theorem, Ann of Math. 84 (1966), 277-292.

[2] Calabi, E., Minimal immersions of surfaces in euclidean spheres, J. of Diff. Geom. 1 (1967), 111-125.

[3] Chern, S., Minimal surfaces in an euclidean space of n dimensions, Differential and combinatorial topology (Morse Jubilee Volume), Princeton, N.J. 1965, 187-198.

[4] Chern, S., Simple proofs of two theorems on minimal surfaces, L'Ens. Math., 15 (1969), 53-61.

[5] Chern, S., do Carmo, M., and Kobayashi, S., Minimal submanifolds of a sphere with second fundamental form of constant length, Functional Analysis and Related Fields, Springer Verlag 1970, 59-75.

[6] Hsiang, W.Y., Remarks on closed minimal submanifolds in the standard Riemannian m-sphere, J. of Diff. Geom. 1(1967), 257-267.

[7] Lawson, H.B., Jr., Minimal varieties in constant curvature manifolds, Ph.D. thesis, Stanford 1968.

[8] Nitsche, J.C.C., On new results in the theory of minimal surfaces, Bull. Amer. Math. Soc. 71 (1965), 195-270.

[9] Osserman, R., Minimal surfaces (in Russian), Uspekhi Mat. Nauk 22 (1967), 55-136. An English translation of this article was published by van Nostrand Reingold Co. in 1969. A new revised edition was published by Dover in 1986.

[10] Protter, M.H. and Weinberger, H., Maximum principles in differential equations, Prentice Hall 1967.

[11] Simons, J., Minimal varieties in Riemannian manifolds, Ann. of Math. 88 (1968), 62-105.

Permissions

Springer-Verlag would like to thank the original publishers of Chern's papers for granting permissions to reprint specific papers in his collection. The following list contains the credit lines for those articles.

[123] Reprinted from *Annales Polonici Mathematici* **39,** © 1981 by Annales Societatis Mathematicae Polonae.

[126] Reprinted from *Manifolds and Lie Groups, Papers in Honor of Y. Matsushima,* © 1981 by Birkhäuser-Verlag.

[127] Reprinted from *Journal of Differential Geometry* **16,** © 1981 by Dr. Chuan Hsiung.

[128] Reprinted from *Journal of China University of Science and Technology* **11,** © 1981 by China University.

[129] Reprinted from *Bull. Amer. Math. Soc.* **6,** © 1982 by Dr. Chuan Hsiung.

[130] Reprinted from *Archiv der Math* **38,** © 1982 by Birkhäuser-Verlag.

[131] Reprinted from *American Journal of Mathematics* **105,** © 1983 by The Johns Hopkins University Press.

[135] Reprinted from *Harmonic Maps of S^2 Manifold: Proceedings of the National Academy of Sciences* **82,** (1985).

[136] Reprinted from *Asterisque,* © 1985 by Societe Mathematique de France.

[139] Reprinted by permission of Elsevier Science Publishing Co., Inc. from *Studies in Applied Mathematics,* vol. 74, pages 55–83. © 1985 by The Massachusetts Institute of Technology.

[140] Reprinted from *Aspects of Mathematics and Its Applications,* edited by Barroso, © 1986 by North Holland Physics Publishing.

[142] Reprinted from *Annals of Mathematics* **125,** © 1987 by Princeton University Press.

[144] Reprinted from *Math. Sci. Research Inst.,* © 1987 by the Mathematical Association of America.

Minimal Submanifolds in a Riemannian Manifold, reprinted with permission from the Institute for Advanced Study, © 1951.

Kansas Lecture Notes, reprinted with permission from the University of Kansas, © 1968.

The following papers were originally published by Springer-Verlag Heidelberg.

[125] Reprinted from LNM 894: *Geometry Symposium Utrecht,* edited by E. Looijenga, D. Siersma and F. Takens, © 1981.

[132] Reprinted from LNM 1007: *Geometric Dynamics,* edited by J. Palis, Jr., © 1983.

[133] Reprinted from *Differential Geometry and Complex Analysis, Volume in Memory of H. Rauch,* edited by I. Chavel and H.M. Farkas, © 1983.

[134] Reprinted from LNM 1111: *Arbeitstagung Bonn 1984,* edited by F. Hirzebruch, J. Schwermer, and S. Suter, © 1984.

[143] Reprinted from *Math. Annalen,* Volume dedicated to F. Hirzebruch 278, © 1987.

[145] Reprinted from LNM 1369: pp. 1–62, © 1989.